Protein Actions

Principles and Modeling

Ivet Bahar
Robert L Jernigan
Ken A Dill

NEW YORK AND LONDON

Garland Science
Vice President: Denise Schanck
Senior Editor: Summers Scholl
Editorial Assistant: Michael Roberts Jr.
Senior Digital Project Editor: Natasha Wolfe
Illustrator: Nigel Orme
Cover Designer: Matthew McClements, Blink Studio, Ltd.
Copyeditor: Mac Clarke
Typesetting: Nova Techset Private Limited, Bengaluru & Chennai, India
Proofreader: Technica Editorial Services
Indexer: Simon Yapp at Indexing Specialists

MIX
Paper from responsible sources
FSC
www.fsc.org
FSC® C014174

Cover Image: This image, by artist Ashley Yevick, shows the helices and beta sheets that are key building blocks of protein structure. And, it captures an aesthetic beauty of proteins – the localized symmetries of these components and the global symmetry of their overall assembly within some proteins.

ISBN 978-0-8153-4177-2

Library of Congress Cataloging-in-Publication Data

Names: Bahar, Ivet, author. | Jernigan, Robert, author. | Dill, Ken A., author.
Title: Protein actions : principles & modeling/Ivet Bahar, Robert Jernigan, Ken Dill.
Description: New York, NY : Garland Science, Taylor & Francis Group, LLC, [2017] | Includes bibliographical references.
Identifiers: LCCN 2016053076 | ISBN 9780815344766
Subjects: | MESH: Proteins–physiology | Protein Conformation | Models, Molecular | Drug Discovery
Classification: LCC QP517.P76 | NLM QU 55 | DDC 572/.633–dc23
LC record available at https://lccn.loc.gov/2016053076

Published by Garland Science, Taylor & Francis Group, LLC, an informa business, 711 Third Avenue, New York, NY 10017, USA, and 3 Park Square, Milton Park, Abingdon, OX14 4RN, UK.

Printed in the United States of America
15 14 13 12 11 10 9 8 7 6 5 4 3 2 1

Visit our web site at http://www.garlandscience.com

To our families,
Ivet dedicates this book to Izzet, Yosi, and Avi.
Bob dedicates it to Craig, Alex, Julie, Eva, and Mila.
And Ken dedicates it to Ryan, Tyler, and Jolanda.

TABLE OF CONTENTS

PREFACE

This is a textbook about the physical principles of protein molecules. It describes protein structures; their folding, binding, and aggregation equilibria and kinetics; their dynamics and mechanisms; and their role in drug discovery. We emphasize principles and models, as well as computational methods for simulating protein actions.

This book is aimed at graduate students and advanced undergraduates. We hope it will have value for a spectrum of readers from biologists, chemists, drug designers, and modelers who seek foundations and quantitative methods to physicists, engineers, material scientists, and computer scientists in other quantitative disciplines who seek an introduction to the biological and chemical properties of proteins. In some chapters, we provide appendices with advanced or embellishing material.

The 13 chapters are organized as follows. Chapter 1 introduces protein structure. Chapter 2 is a survey of some of the fascinating actions and mechanisms of proteins that will be taken up in more detail later in the book. In Chapter 3, we describe protein folding equilibria and the forces that drive folding and other actions. Chapter 4 gives the principles of binding. One of the most important principles of living systems is that the binding of a ligand can be modulated and influenced by the coupled binding of another ligand. This is the basis for molecular motors, regulation, signaling, and transduction. Chapter 5 presents some of the best-known cooperative conformational transitions: the helix–coil transformation, protein folding, and amyloid aggregation. A historical driver of protein physical chemistry was Levinthal's paradox, which puzzles over how proteins can fold so quickly to find their native states. Chapter 6 describes experiments and modeling of the mechanism and kinetics of protein folding. In Chapter 7, we consider how proteins evolve, leading to Darwinian descent with modification and even to new species.

Beginning in Chapter 8, we present the basic principles and applications of computational modeling and simulations. Chapter 8 describes the methods of bioinformatics—how to align and analyze amino acid sequences, leading to insights into structure and function. Chapter 9 introduces how protein structures and energetics are modeled in computer simulations. Chapter 10 describes molecular simulation methods such as Monte Carlo and molecular dynamics, which provide a way to explore conformational ensembles, free energies, dynamics, and mechanisms of actions. In Chapter 11, we survey some of the methods of protein-structure prediction. Chapter 12 describes the collective motions involved in protein actions, principally through the lens of simple structure-based elastic network models. Finally, Chapter 13 gives an overview of computational drug discovery, where proteins play a central role not only as targets, but also as therapeutic agents.

Our intention has been to provide the basic foundations needed for those who are—or are becoming—practitioners in the art, science, and engineering of proteins.

ACKNOWLEDGMENTS

We thank Luca Agozzino, Norma Allewell, Ahmet Bakan, Gabor Balazsi, Doug Barrick, Matthias Buck, David Case, Tom Cheatham, Alan Cheng, Mary H Cheng, Nikita Chopra, Wendy Cornell, John van Drie, Eran Eyal, Jose Luis Neira Faleiro, Michael Falvo, Angel E Garcia, Kingshuk Ghosh, Lila Gierasch, Amnon Horovitz, Ataur Katebi, Punit Kaur, Cihan Kaya, Daisuke Kihara, Jiyoung Lee, Sumudu Leelananda, Hongchun Li, Bing Liu, Tim Lohman, Justin MacCallum, Karolina Mikulska-Ruminska, Sambit Mishra, David Mobley, Daniel Otzen, Banu Ozkan, Alberto Perez, Wolfgang Peti, George Phillips, Rob Rizzo, Victoria Roberts, Geoff Rollins, Kannan Sankar, Jeremy Schmit, Madeline Shea, Guang Song, Chun Tang, Jason Wagoner, Pat Weber, Thomas Weikl, Stephan Wilkins, Lee-Wei Yang, Nese Yilmaz, She Zhang, Michael Zimmermann, and Erik Zuiderweg. We thank Hue Sun Chan and AJ Rader for considerable contributions in the early stages. We are grateful to Nigel Orme for his beautiful figures, which we regard as a centerpiece of the book. We are very appreciative of the great work, the patience, and the collegiality of the professionals at Garland Science, including Denise Schanck, Summers Scholl, Bob Rogers, Mike Morales, Michael Roberts Jr, and Natasha Wolfe. We are especially indebted to Sarina Bromberg for her engagement at all stages, her stimulating questions, and her extensive and careful assistance with production, figures, and editing. Most importantly, we thank our families for their patience, constant encouragement, and generous sharing of their time, without which this book would not have been possible.

RESOURCES FOR INSTRUCTORS

Instructor Resources are available on the Garland Science Instructor's Resource site, located at www.garlandscience.com/instructors. These resources are password-protected and available only to qualifying adopters of the text. The website provides access not only to the teaching resources for this book but also to all other Garland Science textbooks. Adopting instructors can obtain access to the site from their sales representative or by emailing science@garland.com.

Art of Protein Actions: Principles and Modeling

The images from the book are available in two convenient formats: PowerPoint® and JPEG. They have been optimized for display on a computer. Figures are searchable by figure number, by figure name, or by keywords used in the figure legend from the book.

ABOUT THE AUTHORS

Ivet Bahar works at the interface between computational and life sciences, developing models and methods rooted in fundamental principles of physical sciences and engineering. She has specialized in biomolecular systems dynamics. She is John K. Vries Chair and Distinguished Professor in the Department of Computational & Systems Biology at the University of Pittsburgh, School of Medicine. She co-founded the Joint PhD Program in Computational Biology between the University of Pittsburgh and Carnegie Mellon University. She is a member of the European Molecular Biology Organization (EMBO).

Robert L Jernigan works at the interface between structural biology, biophysics, and bioinformatics, datamining and developing models and methods, particularly to investigate mechanisms, and combining sequence and structural data. His PhD research was with Paul J Flory and his postdoctoral research with Bruno H Zimm. For most of his career he was in the intramural research program at the NIH, and he is presently Professor of Biochemistry, Biophysics, and Molecular Biology at Iowa State University.

Ken A Dill researches statistical mechanics, protein folding, and cell biophysics. He did PhD research with Bruno H Zimm and postdoctoral work with Paul J Flory. Dill is at Stony Brook University, where he is Director of the Laufer Center for Physical and Quantitative Biology, Louis & Beatrice Laufer Endowed Chair in Physical Biology, and Distinguished Professor in the Departments of Chemistry and Physics and Astronomy. He is a member of the US National Academy of Sciences and the American Academy of Arts and Sciences.

In 1676, Isaac Newton said: "*If I have seen further than others, it is by standing on the shoulders of giants.*" Some of the many giants to whom we three co-authors owe a great debt of gratitude are shown on this genealogical tree. (Genealogy provided by Academictree.org/chemistry. Image created by Ashley Yevick.)

CHAPTER 1

Proteins Are Polymers that Fold into Specific Structures

PROTEINS ARE THE MACHINES THAT PERFORM CELLULAR FUNCTIONS

Proteins are biology's workhorse molecules. In a human cell, there are about 18,000 different types of protein molecules. Think of cells as teeming factories of molecules of various types. Proteins perform many functions—as the factory workers, as the machines that produce the factory's output, and as the factory's structural framework. In contrast, nucleic acids (DNA and RNA) are the molecules that encode information, providing the instructions for making the factory products. Sugars and carbohydrates are energy sources used to run the factory. The factory walls (cell membranes) are made up primarily of lipids and polysaccharides. *E. coli* is a bacterium having about 4300 protein-coding genes and a total of about 3×10^6 individual protein molecules [1]. Figure 1.1 shows the size of a typical protein molecule compared with atoms and cells.

Proteins are targets for drug discovery and disease intervention. Discovering new drugs typically involves identifying proteins involved in a disease, and then designing drug molecules or proteins that activate or inhibit those proteins. Some drug molecules are small molecules. Increasingly, proteins are being developed within biotechnology companies, as therapeutic agents to act upon other proteins. In addition, proteins promise to play important roles in nanotechnology, as miniature engines, pumps, motors, optical transducers, and sensors, in

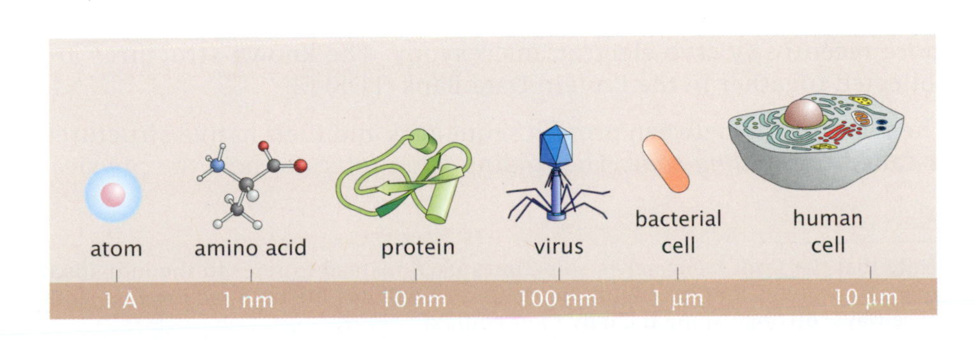

Figure 1.1 In linear dimensions, proteins are a hundredfold larger than atoms and a thousandth the size of a human cell. Note the sizes of images follow the lower size scale and are not drawn strictly to size.

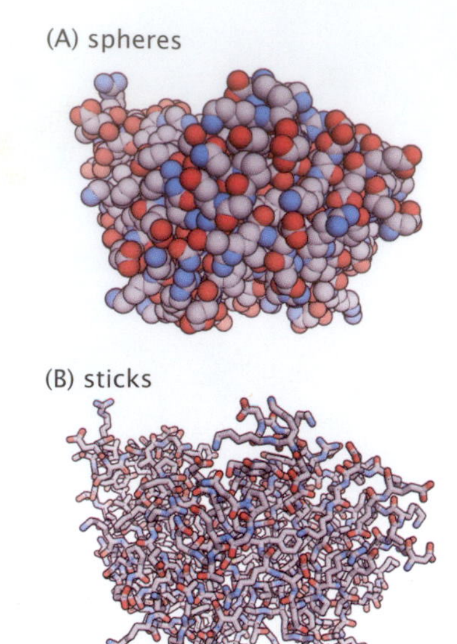

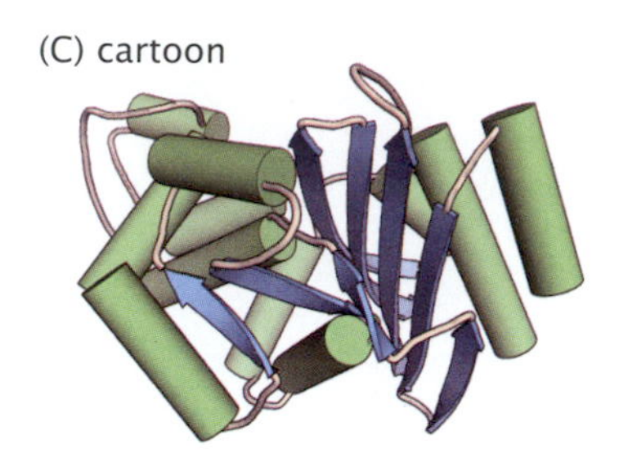

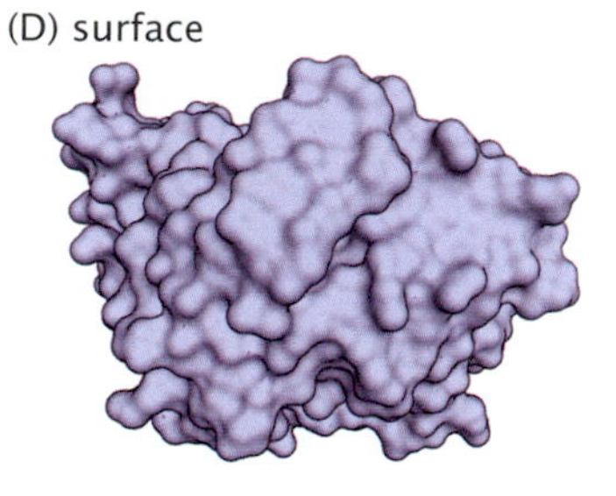

Figure 1.2 Different representations of the same protein structure illustrate different features. (A) A space-filling representation indicates each atom's position, color-coded by atom type: *gray* for carbon, *cyan* for nitrogen, *red* for oxygen, *white* for hydrogen (not shown here), and *yellow* for sulfur. (B) Sticks indicate each bond, colored according to the bonded atoms. (C) Cartoons show secondary structures. (D) Surface renderings show the shapes.

photosynthesis and other applications. Proteins are nature's miniature machines.

To understand the properties of proteins, you first need to understand their structures. Throughout this book, we use different types of images of protein structures (Figure 1.2). Sometimes you want to see chemical and atomic details. In such cases, space-filling models of the atoms or stick representations of the bonds are useful (see Figure 1.2A and B). Other times, you want to see the overall shape of the protein and not the details. In those cases, you may prefer cartoons, or surface representations (see Figure 1.2C and D).

PROTEINS HAVE SEQUENCE–STRUCTURE–FUNCTION RELATIONSHIPS

A protein is a linear polymer molecule. A polymer, like a string of beads or a pearl necklace, is composed of repeat units, called *monomers*,[1] which are covalently linked together in a linear chain-like fashion. In proteins, the repeat units are called *peptides*, so proteins are also called *polypeptides*. There are many types of polymers, natural and synthetic. Synthetic polymers—like commercial polyethylene, polystyrene, or polypropylene—are *homopolymers*: they are repeats of a single type of monomer unit strung together: –AAAAA–, for example. In this respect, proteins are different than almost all synthetic polymers: they are *heteropolymers*. In a protein chain, different types of monomers are strung together in a particular *sequence*: –ILAKW–, for example. For proteins, the units are *amino acids*, also called *residues*, because this is what is left over after the loss of water when two amino acids link together in a chain. There are 20 different types of naturally occurring amino acids. (There are additional types of amino acids, which are occasionally found in proteins, but they are rare.) Think of the set of amino acids as a set of pearls of different colors that are strung together to form a colorful pearl necklace. Or think of the amino acids as forming an *alphabet*. Stringing amino acids together in different linear sequences is a way of encoding information, like the way different sequences of letters encode words in sentences to convey meaning. Proteins have names such as lysozyme, ribonuclease, or barnase. Each protein name describes a chain molecule composed of a particular sequence of covalently bonded amino acids.

The importance of the amino acid sequence of a protein is that it encodes a particular three-dimensional (3D) structure—the protein's *native* or *folded* configuration—into which the protein balls up. In a stunning achievement, the field called *structural biology* has given us atomically detailed structures of more than 100,000 proteins, providing the relative spatial positions of their atoms. Starting with J Kendrew's study of myoglobin in 1958, protein structures have been determined by X-ray crystallography and NMR spectroscopy, and more recently by cryo-electron microscopy. The known structures are collected together in the Protein Data Bank (PDB) [2].

The relationship between protein sequences and their native structures is called the *folding code*. The folding code refers to how the physical

[1] Note that the term *monomer* is sometimes used instead to refer to the individual chains of proteins that have multiple chains. We will use the term *monomer* in both these ways, distinguishing them by their context.

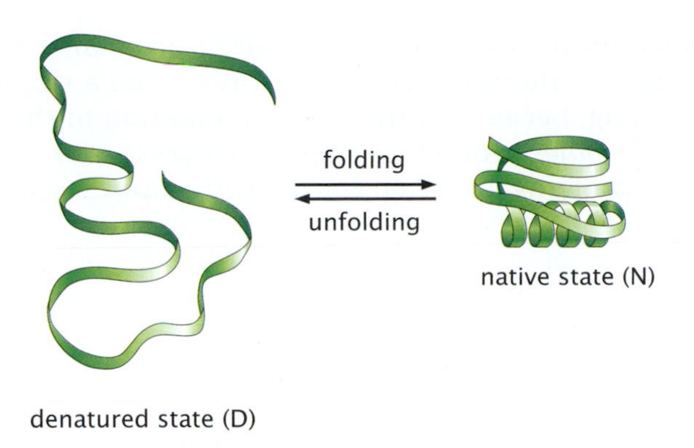

Figure 1.3 The folding process is a transition between denatured (D) and native (N) states, and requires decoding of the information carried in the protein sequence. The denatured state of a given protein is a collection of many different conformations, called an ensemble. Upon folding, each protein collapses into a relatively singular native, or folded, state regardless of its original conformation in the denatured state. The native structure is characteristic of the particular protein. It is significantly more compact than the conformations visited in the denatured state. Proteins fold or unfold depending on environmental conditions.

forces cause a protein's chain of amino acids to fold into only a single stable native structure. A polymer's ability to function because of its folding is mostly limited to proteins and some RNA molecules. A polymer's ability to function because of its folding is mostly limited to proteins and some RNA molecules. No small molecules can encode information in this way. And essentially no synthetic polymers are informational. Synthetic polymers are commercially useful for their material properties—mechanical strength, optical behavior, etc.; they are fibers, fabrics, glues, paints, tires, plastics, films. But, they do not fold into unique structures, and they do not perform the sophisticated types of functions that proteins perform in living systems. And, while DNA molecules can form some structure on the large scales of chromosomes, they have a double-helical structure that is relatively independent of sequence, so they are better suited for carrying information than performing functions. The importance of a uniquely folded structure is that it forces particular chemical groups to sit next to other chemical groups spatially in ways that are stable and functional. These spatial relationships are key to biological mechanisms.

Proteins can undergo *folding* and *unfolding* processes, from the *denatured* or *unfolded* state to the native or folded structure, and vice versa. Think of the unfolded state as a set or *ensemble* of chains that assume a huge number of different 3D arrangements, like a string that can adopt a huge number of different stringy shapes. Upon folding, all chains assume the same conformation, the native structure, but even these can change in limited ways to carry out their functions. The folding process is shown in Figure 1.3 and is described in more detail in Chapters 3, 4, and 6.

The central paradigm of protein science has long been that the amino acid sequence dictates the 3D structure, which, in turn, dictates how the protein performs its function:

SEQUENCE → STRUCTURE → FUNCTION

But, this is not the whole story. First, a protein's function can result not just from one structure, but often from transitions between structures.

Protein functionalities can result from large or small *conformational changes*. Second, "structure" does not always mean a single conformation. For one thing, because of the Brownian motion of the solvent, the native protein *wiggles* around its native structure. For another thing, parts of native proteins are often *intrinsically disordered*, whereby the chain is floppy, and does not have a single structure. Third, the biological mechanism of action of a protein can depend not only on its structure, but also on its *dynamics*. So, a more accurate statement of the protein paradigm is

SEQUENCE → STRUCTURES → MOTIONS → FUNCTION

In this chapter, we describe protein structures. In Chapter, 2, we describe protein mechanisms of action. The relationships between motions and function are described in Chapters 10 and 12. We begin with the structures and properties of a protein's building blocks, its amino acids.

AMINO ACIDS ARE THE REPEAT UNITS OF PROTEINS

Figure 1.4 shows the chemical structure of an amino acid, which is the building block unit of proteins. Every amino acid is so-called because it has an *amino* group ($-NH_2$) at one end and a *carboxylic acid* group

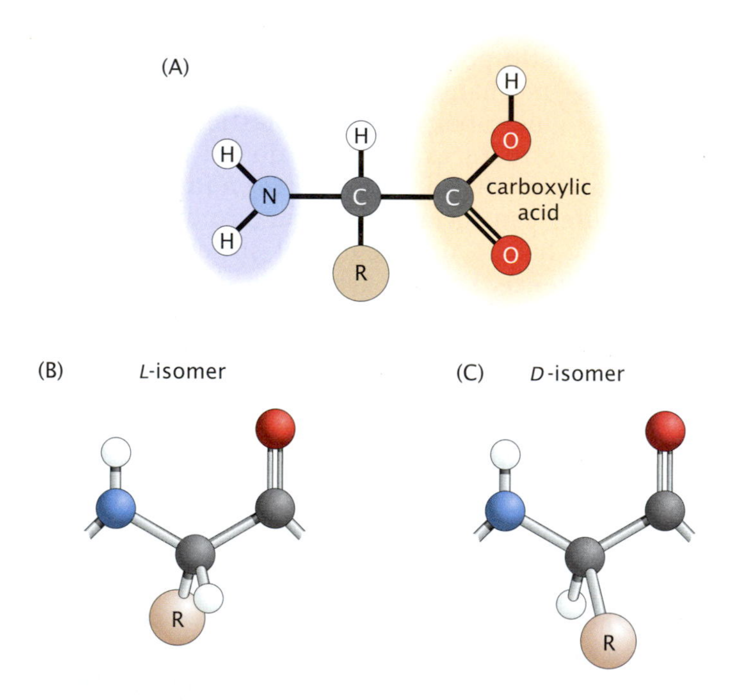

Figure 1.4 The backbone structure of amino acids determines chirality. (A) In an amino acid, the amino group ($-NH_2$ on the left) is connected to the carboxylic acid group ($-COOH$ on the right) through the C^{α}-atom, which is also bonded to the side chain (abbreviated as "R" to be general) and a hydrogen. (B) and (C) 3D view of the repeat units in proteins. Because each C^{α} is tetrahedrally bonded to four different types of atoms, amino acids are either D-isomers or L-isomers. The L-isomer is the naturally occurring form in proteins. The positions of the side chain R differ in the two isomers. In the L-isomer, it points back into the plane of the page, provided that the two flanking backbone bonds $N-C^{\alpha}$ and $C^{\alpha}-C$ lie in the plane of the page and the N-terminal end is on the left (as shown). In the D-isomer, the same arrangement of the backbone leads to an out-of-plane orientation of the R group toward the reader.

(—COOH) at the other. At physiological pH, the amino and carboxylic acid groups are both completely ionized, meaning that one end of the amino acid is a base and the other end is an acid, having a positive and negative charge, respectively. Between the amino and carboxyl ends is a C^{α} atom, also called the α-carbon. Taken together, the amino group, the α-carbon, and the carboxyl group of amino acids are collectively called the *backbone* or the *main chain* of the protein.

Figure 1.4 also shows the *side chain* of an amino acid. The side chain is labeled "R" in the figure as a placeholder to indicate that this can be any one of 20 different types of chemical groups shown in Figure 1.5. One amino acid differs from another in the side chains. The side chain is covalently bonded to the C^{α} atom. The 20 amino acid side chains differ in their chemical structures and physical properties, as described below.

Amino Acids Are Chiral Molecules

Figure 1.4B and C show a property of amino acids called *chirality* or *handedness*. An object is chiral if it does not superimpose on its mirror image. A left hand is chiral because no motion or rotation causes it to look like a right hand. Amino acids are chiral because the α-carbon is tetrahedrally bonded to four different types of groups: carbonyl C, amino N, side chain R, and hydrogen atom H. Glycine is the only natural amino acid that has no chirality, because its R group is just a hydrogen atom, with two of these protons attached to the α-carbon, and for chirality specification these are indistinguishable. There are two different stereochemical ways that the hydrogen and the R group can be situated with respect to the main chain (see Figure 1.4B and C). The two different stereochemical structures are labeled the L-isomer, also known as the left-handed (*levo*) form, and the D-isomer, which is called the right-handed (*dextro*) form. These two different structures are called *optical isomers* because they rotate the plane of plane-polarized light in opposite directions in optical spectroscopy. Biological systems contain L-amino acids almost exclusively. The evolutionary "reason" for biology's use of L rather than D is not known; it may just be a frozen historical accident. The two isomeric forms are chemically quite distinct since the only way to convert between the D- and L-isomers is to break a chemical bond. Later, we discuss how the chirality in the individual amino acids leads to handedness on a larger scale in protein structures.

The 20 Amino Acids Have Different Physical Properties

The 20 different amino acid side chains are shown in Figure 1.5. Each has a different chemical character: *charged*, *polar* (P), or *nonpolar* (also called *hydrophobic*, H). Charged and polar groups have an affinity for water (they are *hydrophilic*), while nonpolar side chains do not. At neutral pH (typical biological conditions), lysine, arginine, and histidine are *bases*; their side chains are positively charged. In contrast, aspartic and glutamic acids are *acids*; their side chains are negatively charged. Alanine, valine, leucine, isoleucine, and phenylalanine side chains are *nonpolar*; they are composed only of hydrocarbon groups (—CH—, $-CH_2-$, $-CH_3$, or aromatics, for example). Nonpolar groups are so-called because even when they are subjected to an applied electric field, they don't become very polarized (that is, they don't develop

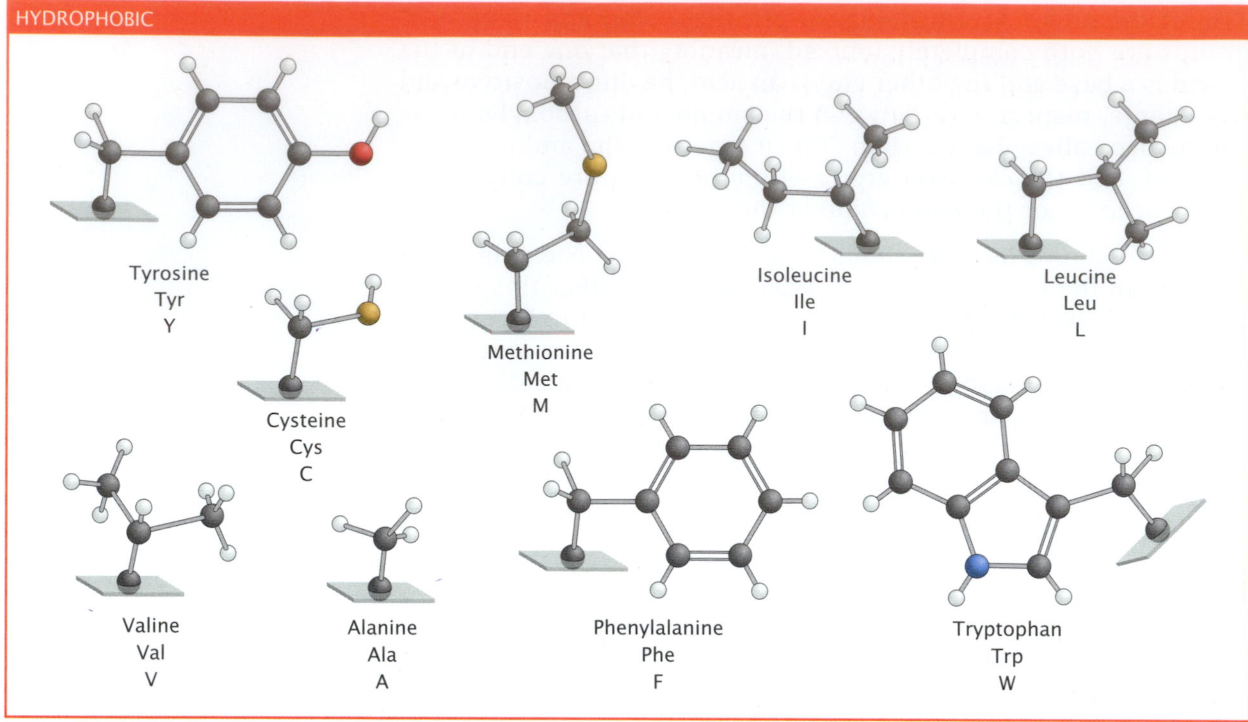

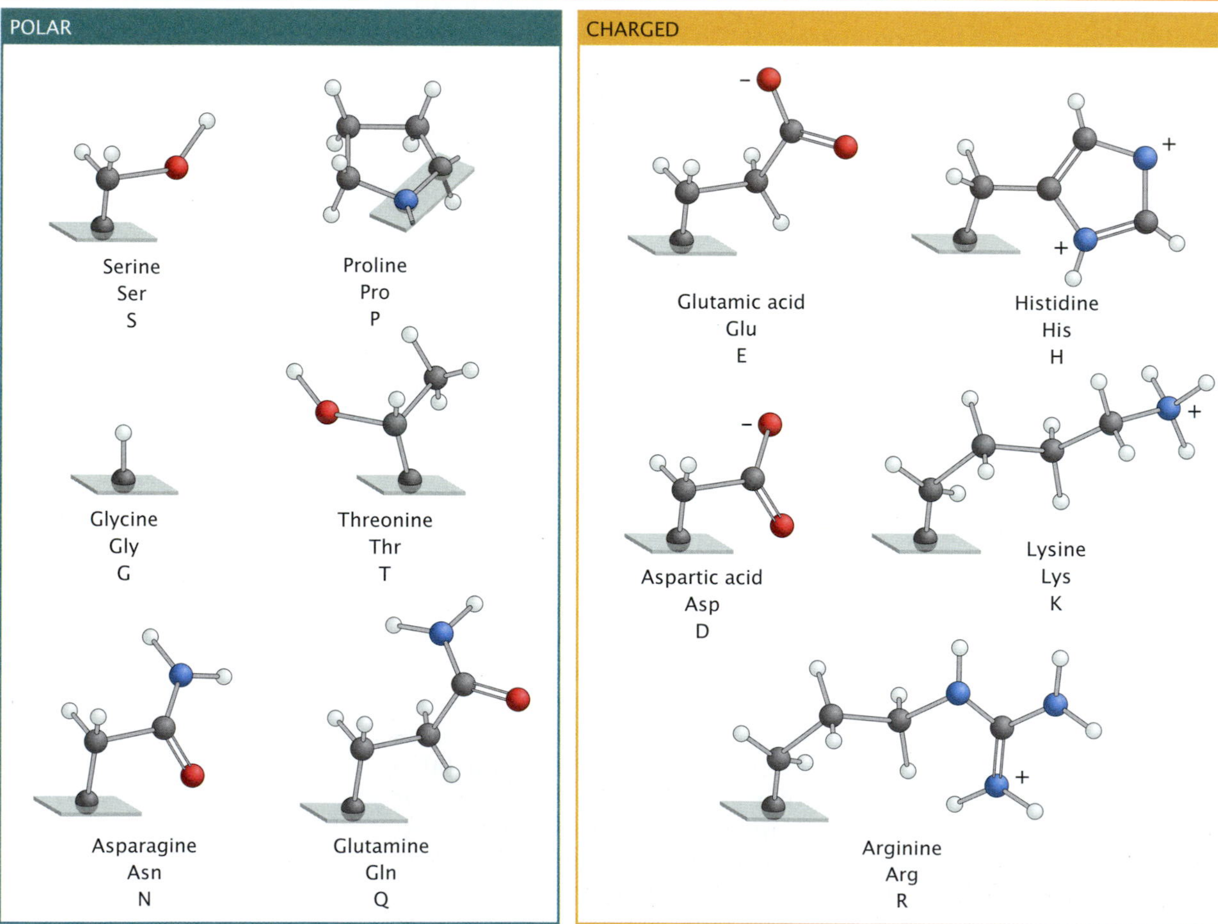

Figure 1.5 Structures of the side chains for the 20 standard amino acids, along with their three-letter and one-letter code names. α-carbons are set in the drawn planes, with the side chains reaching up from them. Amino acid side chains are classified into three broad groups based on their chemical nature: hydrophobic, polar, or charged. The charged residues become uncharged under some conditions of solution pH. Such classifications are from hydrophobicity scales, but there are many of these and some amino acids appear in different categories in different scales.

much internal charge separation or induced dipoles). Nonpolar groups are oil-like; they have an aversion to water. The side chain of glycine is just a hydrogen atom, so the polar groups in glycine's backbone dominate, giving it some limited polar character. Cysteine and methionine, despite their polar sulfur atoms, tend to be buried in the interiors of proteins with the hydrophobic residues, making them appear to be hydrophobic in character. Serine and threonine are polar due to their —OH (hydroxyl) side-chain terminal groups. Asparagine and glutamine are polar due to their amide (—(C=O)—NH_2) side-chain groups. Proline has an imino group fused to the protein backbone and histidine has an imidazole ring.

These physical classifications are neither precisely defined nor always accurate. Some amino acids have multiple personalities, with both nonpolar and polar/ionic characters, depending on their intrinsic pK_a values and the pH of the surroundings. Tyrosine has a bulky hydrophobic part, its phenyl group, but this is attached to a terminal polar —OH. Tryptophan is hydrophobic, but the N—H group on its two-ring indole group can be a hydrogen bond donor. The histidine side chain, a heterocyclic imidazole, can ionize at physiological pH, so histidine can be positively charged or uncharged depending on the pH. Lysine is charged, but it also has a chain of four methylene groups. The broad diversity of structures and biochemical mechanisms of proteins arises from the diversity of the amino acid side chains.

In Proteins, Amino Acids Are Linked Together through *Peptide Bonds*

In a protein, each amino acid is covalently linked to its neighboring amino acid along the chain by a *peptide bond* (Figure 1.6). A peptide bond is a C—N linkage between the carbonyl end of one amino acid and the amino group of the next amino acid.[2] The amino end of the whole protein is called the *N-terminus* and the carboxyl end of the protein is called the *C-terminus*. The standard convention is to number amino acids in a protein chain, $1, 2, 3, \ldots, n$, starting from the N-terminus.

Peptide Bonds Are Planar

The peptide bond, C_{i-1}—N_i, connects amino acids $i-1$ and i. This bond, together with the two backbone bonds on each side, C^{α}_{i-1}—C_{i-1} and N_i—C^{α}_i, lie rigidly within a plane, called the *peptide plane*; see Figure 1.6. The local geometry is planar because the peptide bond has double-bond character (due to the delocalization of the electrons on the carbonyl and amide units). This double-bond character results in rigidity; the peptide bond lacks torsional freedom about its own axis. So, it's possible to think of the backbone as a succession of planes that pivot relatively freely around the C^{α} carbons that join one plane to the next. Figure 1.6A shows a two-amino acid segment of a polypeptide backbone, called a *dipeptide*. Figure 1.6B shows how successive amino acids form peptide planes.

[2]The chemical process of forming a peptide bond is a *condensation* reaction, releasing a water. Thus the formation of a polypeptide containing *n residues* releases $n-1$ water molecules as described by the following polycondensation reaction: $n(\text{HNH—C}^{\alpha}\text{HR—COOH}) \rightarrow \text{H—[NH—C}^{\alpha}\text{HR—COO]}_n H + (n-1)\text{H}_2\text{O}$.

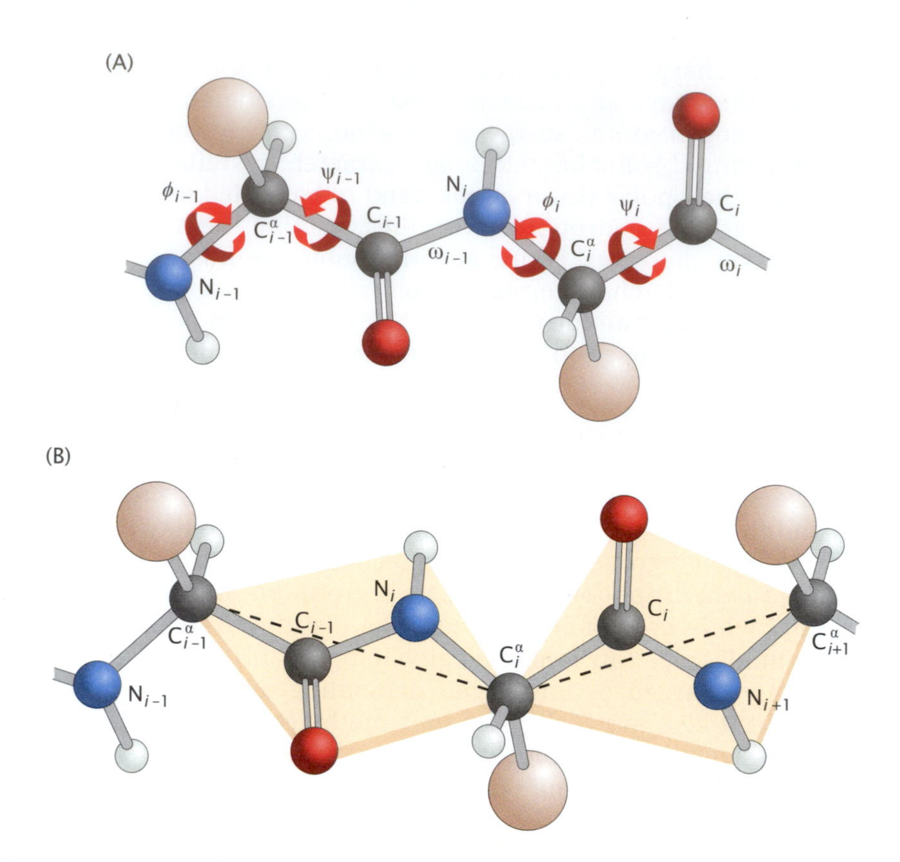

Figure 1.6 Two amino acids in sequence are connected by a peptide bond. (A) Backbone atoms are indexed by residue numbers, $i-1$ and i. Backbone torsion angles are associated with backbone bonds: ϕ_i with the N_i—C^{α}_i bond, ψ_i with the C^{α}_i—C_i bond, and ω_i with the ith C_i—N_{i+1} peptide bond. Signs of torsion angles correspond to clockwise rotations when looking from the *left* atom of the bond to the *right* atom in (A). (B) Virtual bond model representation of the protein backbone. Their lengths are fixed at 3.8 Å for the usual trans peptide bonds. *Dashed* lines are the virtual bonds connecting successive α-carbons. This representation takes advantage of the rigidity of the peptide bond in the *trans* state and the planarity of the three successive backbone bonds C^{α}_{i-1}—C_{i-1}, C_{i-1}—N_i and N_i—C^{α}_i, along with the corresponding C_{i-1}=O and N_i—H bonds. However, note that the virtual bonds do not have fixed bond angles between them.

One peptide plane can rotate relative to the next peptide plane because of the freedom of the N—C$^{\alpha}$ and C$^{\alpha}$—C bonds for torsional rotations. Rotation around the N—C$^{\alpha}$ bond defines the torsional (or dihedral) angle called ϕ, while rotation around C$^{\alpha}$—C defines an angle called ψ. The rotational angle around the peptide bond is called ω (see Figure 1.6A). The ω angle has two stable states called *trans* ($\omega = 180°$) and *cis* ($\omega = 0°$). Figure 1.6 shows the *trans* planar form, which is about 1000-fold more populated than *cis* (except for the bond preceding proline, where the ratio is only about 3-fold). A simplification that is sometimes useful is to represent a chain with *virtual bonds*, shown in Figure 1.6B with dashed lines, which are vectors that join successive α-carbons. These virtual bonds have a fixed length of 3.8 Å for *trans* peptides.

The Rotational Freedom around the Backbone Peptide Bond is Described Using Ramachandran Maps

Different amino acids have different preferred backbone torsional angles ϕ and ψ. These preferences arise from the steric collisions between close neighboring backbone and side-chain atoms [3]. A *Ramachandran map* is a plot of a ϕ angle on the x-axis and a ψ angle on the y-axis for the two bonds that flank a given α-carbon (Figure 1.7). Contours (or colors) on the Ramachandran map indicate the relative populations of the different pairs of angles. Because peptide bonds are usually in their planar *trans* conformation, you can fully specify the locations of the main-chain atoms using only these two dihedral angles, ϕ and ψ, if the bond angles and lengths are known.

Glycine has more freedom than other amino acids in its (ϕ, ψ) angles. Its side chain is only a hydrogen atom, which doesn't collide with the

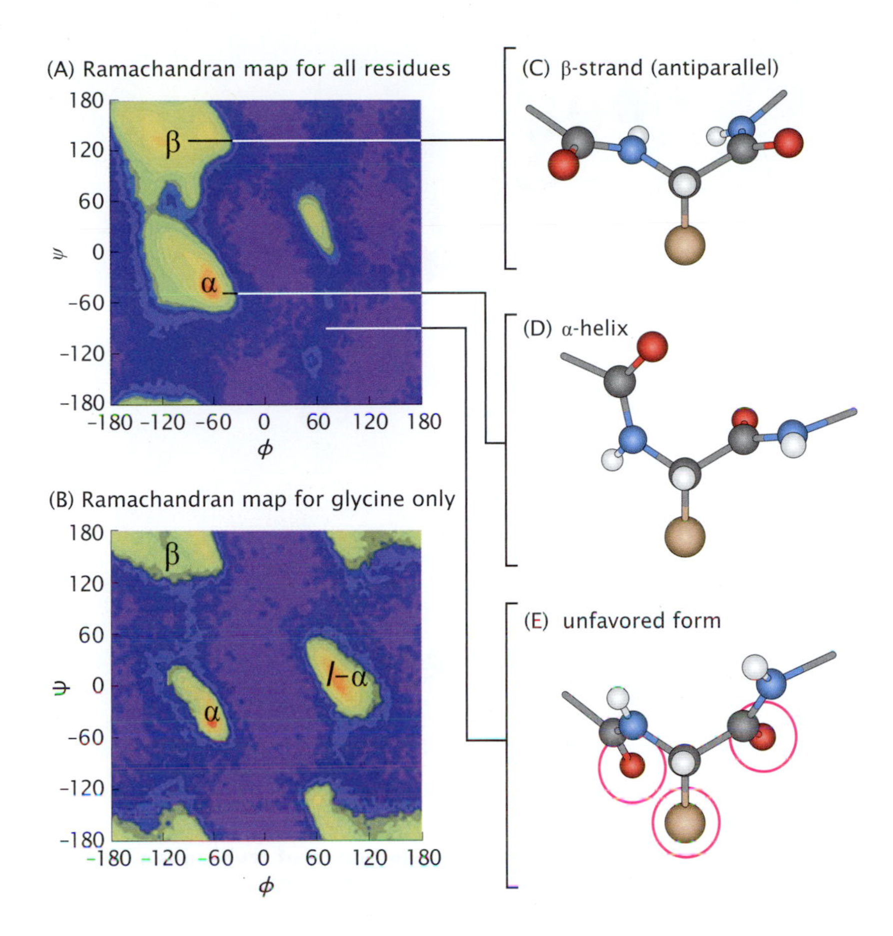

Figure 1.7 A *Ramachandran map* shows the relative populations of different (ϕ, ψ) angle pairs. The data plotted here are from a set of 593 proteins that have a structural resolution of 1.5 Å or better: (A) for all residues and (B) for glycines only. Populations are colored: *red* is the most populated, followed by *orange*, *yellow*, *green*, *blue*, and *violet*. *Violet* regions are disallowed because of steric clashes. Residues other than Gly and Pro tend to fall into one of the three labeled regions: α for right-handed α-helical conformations, β for β-stranded or *l*-α for left-handed α-helical conformations. (C) Illustrates the pair of angles and backbone geometry in the β region; (D) illustrates the local structure in the α-helical region; and (E) shows steric clashes for a left-handed conformation. (The chirality of C^α atoms is responsible for the asymmetric distribution of dihedral angles, disfavoring most left-handed ($\phi > 0$) conformations.)

backbone (see Figure 1.7). Proline has less (ϕ, ψ) freedom than other amino acids because its side-chain atoms form a covalently bonded ring with the backbone amide group, locking the backbone ϕ angle.

Side Chains Adopt Preferred Conformations

Side chains, too, have different preferred conformations. Figure 1.8 shows the rotatable bonds on the side chains of tryptophan and lysine. The convention for side chains is to name the heavy atoms by Greek letters as you move away from the C^α main chain, that is, C^β, C^γ, and so on. Similarly, the rotational angles are called χ_1, χ_2, and so on, with subscripts increasing farther away from the main chain.

Why do side chains have favored χ angles? Side chains contain short hydrocarbon chains. So, they have the same conformational preferences that hydrocarbons have. Hydrocarbon chains [for example, $(-CH_2-)_n$] in the gas phase tend to populate three *rotational isomeric states,* called *trans*, *gauche*$^+$, and *gauche*$^-$. Figure 1.9A shows

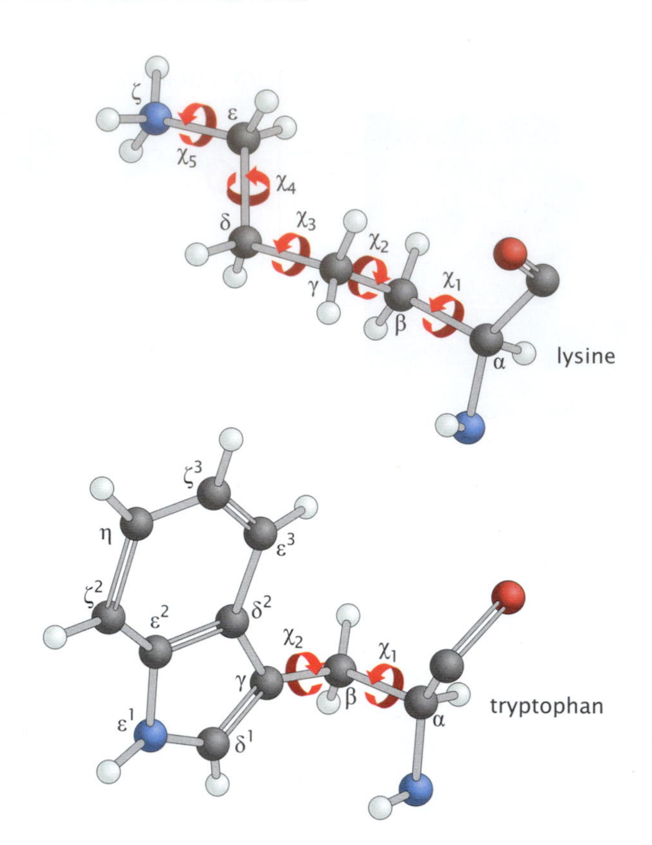

Figure 1.8 Side-chain notation and rotatable bonds of tryptophan and lysine. Side-chain carbon (*gray*) or nitrogen (*blue*) atoms are labeled $\beta, \gamma, \delta, \epsilon, \zeta, \eta$, etc., which indicate the distance (number of bonds) from the main chain C^{α} atom. The torsion angles of the rotatable side-chain bonds are labeled χ_1, χ_2, etc.

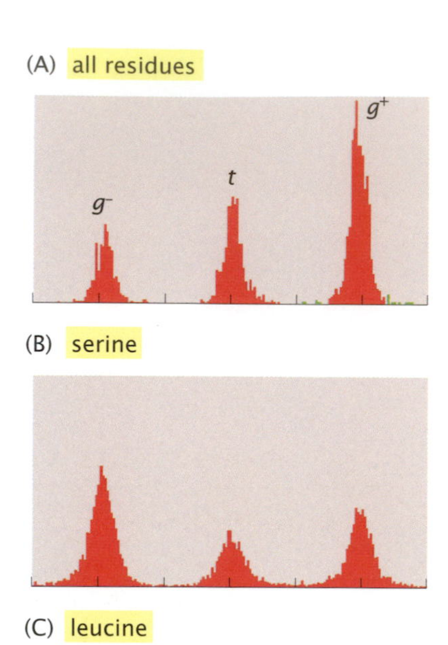

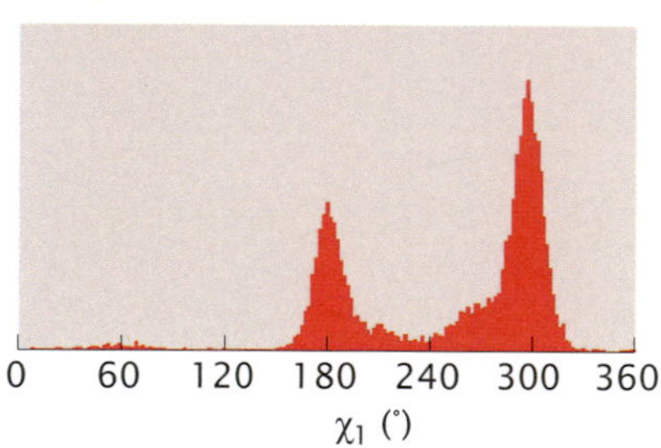

Figure 1.9 Distribution of side-chain χ_1 angles (C^{α}–C^{β} bonds) in proteins and dependence on amino acid type. (A) Frequency of occurrence of χ_1 angles for all residue types. The peaks correspond to the rotameric states, *gauche*⁻ (g^-), *trans* (t), and *gauche*⁺ (g^+). The lower probability of the g^- state is due to potential steric clashes between the backbone and side-chain atoms of the bond C^{α}–C^{β}, particularly when the backbone is α-helical. The g^- state is highly probable in serine (B) due to hydrogen bond formation propensity between the side-chain hydroxyl group and the backbone C=O group, but is completely inaccessible in leucine (C) where the branching at the C^{γ} atom gives rise to steric clash with the backbone. (A, from JM Thornton. *Protein Sci*, 10:3–11, 2001 and KS Wilson et al. *J Mol Biol*, 276:417–436, 1998; B and C, from MW MacArthur and JM Thornton. *Acta Crystallogr D*, 55:994–1004, 1999. With permission of the International Union of Crystallography.)

that the χ_1 angles in native proteins are the same three conformations. However, Figure 1.9B and C show that the relative populations of the χ_1 angles also depend, to some extent, on the identity of amino acids. While the *gauche*⁺ state is the most probable rotamer for most amino acids (see Figure 1.9A), serine prefers the *gauche*⁻ state (see Figure 1.9B). Valine and isoleucine prefer *trans* because of steric restrictions due to branching at their β-carbons. On the other hand, leucine branches at the γ-carbon, leading to steric clashes that disfavor *gauche*⁻ (see Figure 1.9C).

NATIVE PROTEINS HAVE COMPACT WELL-DEFINED 3D STRUCTURES

Proteins Come in Different Sizes and Shapes

Proteins come in different sizes, as given in terms of either their number of amino acids or their mass. Masses are in units of kilodaltons (kDa). The dalton (Da) is a universal unit of atomic mass. Defined as 1/12 the mass of a carbon atom, it is approximately the mass of a hydrogen atom. The average mass of an amino acid is 136 ± 31 Da. Proteins range in size from *peptides*, which may include up to a few tens of amino acids, to large molecules having thousands of residues. Figure 1.10 shows the distribution of protein chain lengths in yeast, where the average protein size is ~500 residues. In *E. coli*, the average protein has 360 residues. *E. coli's* average protein weighs about $136 \times 360 = 50$ kDa.

Folded proteins fall into three classes: globular, fibrous, and membrane proteins. Globular proteins are compact and roughly spherical, with axial ratios typically ranging from 1.2 to 1.4 (Figure 1.11). Globular proteins perform cellular functions, including transcription, metabolism, transport, immune responses, cell signaling, and regulation. Fibrous proteins are long and threadlike and serve roles based on their mechanical properties, in structuring cells and tissues, in collagen, silk, hair, and feathers, for example. Globular proteins are usually soluble in water, whereas fibrous proteins are not. Membrane proteins are located in membranes, an oil-like environment. Membrane proteins can transport molecules and allow the flow of ions across membranes, relay signals across membranes, or perform enzyme activities. We first focus on globular proteins.

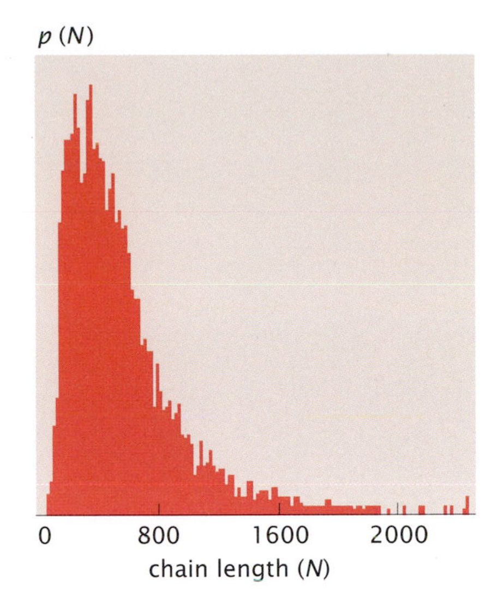

Figure 1.10 Size distribution of proteins expressed in the yeast *Saccharomyces cerevisiae* proteome (set of all proteins). The average length is 501 residues. (From J Warringer and A Blomberg. *BMC Evol Biol*, 6:61, 2006.)

Native Protein Chains Are Balled Up and Tightly Packed

There is not much free space inside a folded globular protein. Proteins are well packed. The density of a protein's atoms in its folded state is about the same as the density of atoms in liquids and solids, or about the same as the density of marbles in a jar. One way to characterize the tightness of packing is to measure the volume occupied by the individual amino acids in folded structures, and compare that volume with the total volume of the protein's constituent atoms. Even liquids

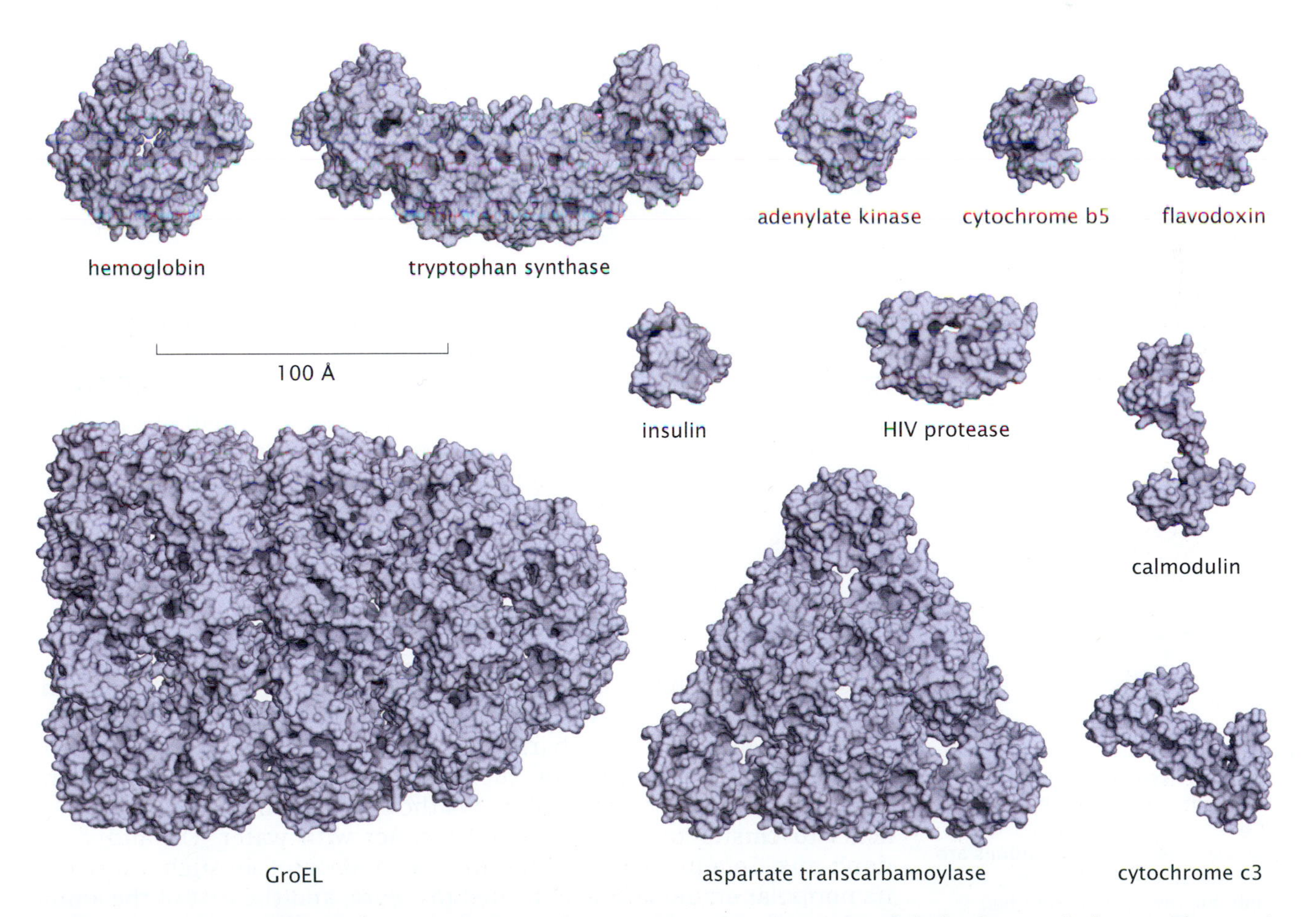

Figure 1.11 Surface representations of eleven proteins illustrate the range of their sizes and shapes. The proteins are all shown at the same magnification.

Table 1.1 Physical properties of amino acids

Residue	van der Waals volume (Å^3)[a]	Average packing density in proteins[b]	Occurrence[c] (%)
Ala	67	0.73	7.5
Arg	148	0.73	5.2
Asn	96	0.77	4.6
Asp	91	0.73	5.2
Cys *(disulfide)*	86	0.81	1.8[d]
Cys *(thiol)*		0.73	
Gln	114	0.71	4.1
Glu	109	0.70	6.3
Gly	48	0.73	7.1
His	118	0.71	2.2
Ile	124	0.73	5.5
Leu	124	0.74	9.1
Lys	135	0.79	5.8
Met	124	0.73	2.8
Phe	135	0.67	3.9
Pro	90	0.70	5.1
Ser	73	0.74	7.4
Thr	93	0.76	6.0
Trp	163	0.68	1.3
Tyr	141	0.69	3.3
Val	105	0.74	6.5

[a] Hard core volume of constituent atoms (data from FM Richards. *J Mol Biol*, 82:1–14, 1974).
[b] Calculated by dividing the van der Waals volume by the average volume of buried residues (that is, those with less than 5% of possible surface area accessible to solvent) for each residue (data from C Chothia. *Nature*, 254:304–308, 1975 and TE Creighton. *Proteins: Structures and Molecular Properties*, 2nd ed. WH Freeman, NY, 1993).
[c] Amino acid frequency from 1021 unrelated proteins (data from P McCaldon and P Argos. *Proteins*, 4:99–122, 1988).
[d] Frequency for all Cys residues.

and solids have some crevices between atoms. In the tightest packing of perfect spheres, only 74% of the volume is filled—the rest is empty cavities. Table 1.1 shows that the packing density of amino acids buried in globular proteins is also 0.74. And, typically, less than about 3% by volume inside a protein is filled by water [4]. However, while a protein core is packed *tightly*; it is not packed *uniformly*, because side chains are different, diverse, and irregular. A protein interior more closely resembles a jar of nuts and bolts than one of marbles.

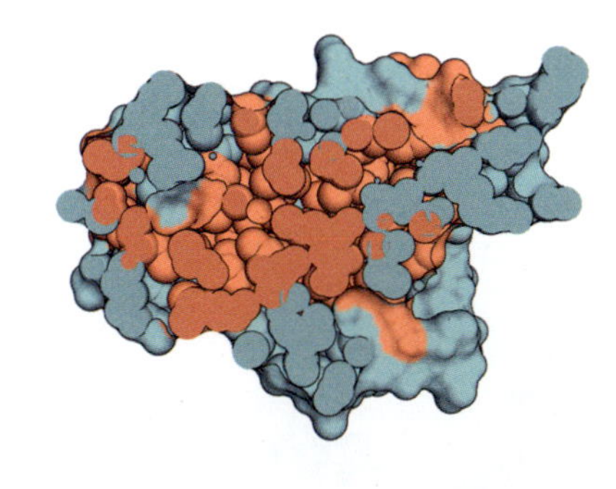

Figure 1.12 Proteins have a hydrophobic core. A cross section of interleukin-4 exposes the protein interior. Hydrophobic Trp, Phe, Tyr, Leu, Ile, Cys, Met, and Val residues are *orange*, and all others are *teal*. Hydrophobic residues tend to be buried, and polar residues lie mostly on the surface.

Proteins Have Hydrophobic Cores

In a folded protein, some amino acids are on the surface and some are buried in the protein's *core*. The surface of a water-soluble protein contains a mix of polar, charged, and nonpolar groups. As shown in Figure 1.12, the core contains mostly nonpolar amino acids, implying that a key force for protein folding is the tendency of the oil-like amino acids to cluster together to avoid contact with water. Oil and water don't mix. So a protein folds up, like an oil droplet, in such a way that its nonpolar amino acids are buried in a core, and the rest of the amino acids are on the surface. We explore the nature of the forces that drive protein folding in Chapter 3, but we introduce the types of interactions briefly here because they have a prominent role in defining structures and mechanisms.

The Amino Acids in Native Proteins Are Hydrogen-Bonded to Each Other

A folded protein chain has extensive hydrogen bonding. A hydrogen bond is a noncovalent interaction between two chemical groups that can share a hydrogen atom. The two sharing groups are called the hydrogen-bond *donor* and *acceptor*. A folded protein contains extensive networks of C═O ··· H—N backbone hydrogen bonds between different amino acids. The amide group is the hydrogen-bond donor and the carbonyl group is the acceptor. The geometry from donor to hydrogen and from hydrogen to acceptor tends to be nearly collinear.

Proteins can also form hydrogen bonds with water. Water molecules can donate hydrogens or accept hydrogens. Hydrogen bonds can also form between the side chains or between the backbone and the side-chain atoms of proteins. Hydrogen bonds are prominent features in the most common protein substructures, called α-helices and β-sheets.

Cysteines Can Form Disulfide Bonds

Cysteine is an amino acid that terminates in a *sulfhydryl* (also called *thiol*) group (—SH). Under oxidizing conditions, two different cysteine side chains can form a covalent bond with each other, called a *disulfide bond* (—S—S—). (Disulfide bonds can be broken by adding a reducing agent to a protein solution.) Disulfide bonds act as cross-links either within a protein, or between two proteins, and can impart thermal and mechanical stability.

PROTEINS HAVE HIERARCHIES OF STRUCTURE

Proteins have various forms of internal structure. The levels of structure were named primary, secondary, and tertiary by Kaj Linderstrøm-Lang [5] (Figure 1.13). The *primary structure* of a protein refers to its linear sequence of amino acids. The *secondary structure* refers to *helices* and *β-sheets*, the two main types of hydrogen-bonded regular (ordered) substructures in proteins. In an average protein, about 60% of the residues participate in α-helices or β-strands. The next higher level of organization, a protein's *tertiary structure*, refers to the 3D arrangement of its secondary structural elements and its connecting turns, loops, or coiled segments. Protein tertiary structures are described by terms such as "four-helix-bundle" or "β-barrel," reflecting the packing geometry of secondary structural elements. Tertiary structures are stabilized by *tertiary contacts* (generally hydrophobic interactions) between amino acids that are distant in the sequence but close in space.

There are also intermediate levels of structure between secondary and tertiary structures, called *structural motifs*, or *supersecondary structures*, as will be discussed later. *Quaternary structure* refers to the 3D organization of multiple chains, also called *subunits* or *monomers*, for proteins that are composed of more than one chain (here, monomer refers to one polypeptide chain, not to be confused with the amino acid

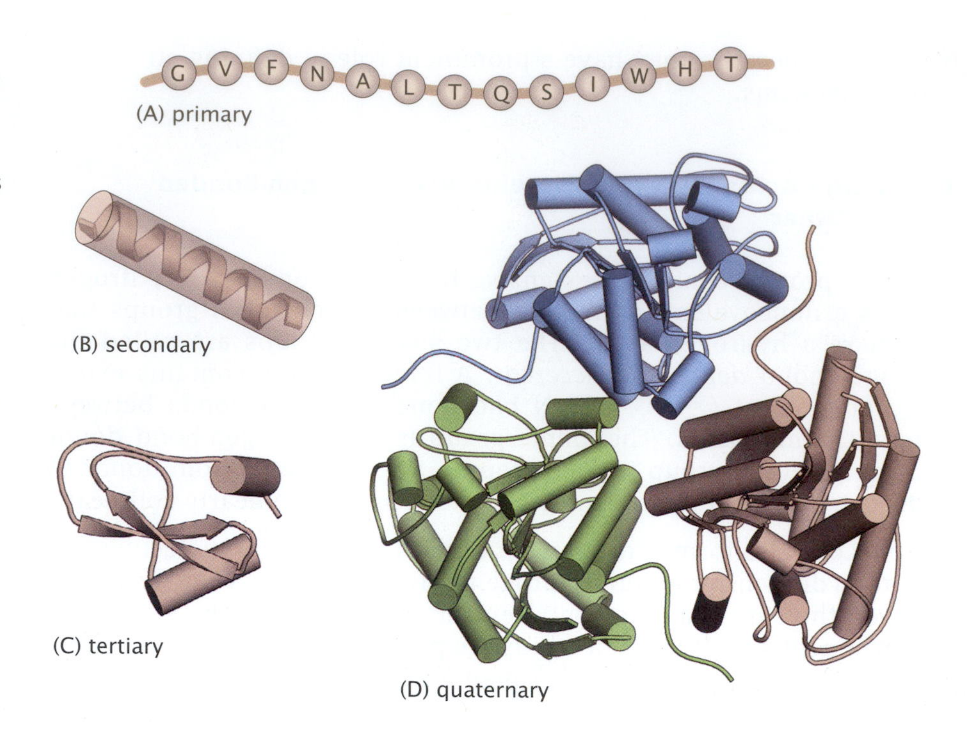

Figure 1.13 There are four hierarchical levels of structure in proteins. (A) A protein's primary structure is its amino acid sequence. (B) The secondary structures are the helices and sheets. Here, the α-helices are represented as cylinders. (C) The tertiary structure is the arrangement of all the structural elements in a single protein chain (beta-strands shown as arrows), while (D) the quaternary structure is the assembly of two or more chains within a protein. Here we show the quaternary structure of arginase (a trimeric enzyme in the urea cycle, which allows the body to dispose of ammonia).

monomers). The quaternary structures of multichain (also called *multimeric*) proteins are usually held together by noncovalent interactions between the individual *subunits*. For example, hemoglobin has a quaternary structure; it is composed of four subunits, or monomers, which are symmetrically assembled (but not covalently bonded) to form a tetrameric structure. Fibrous proteins often have quaternary structures that are stabilized by covalent bridges.

The Secondary Structures of Proteins Are Helices and Sheets

α-helices. A major component of protein structures is the *α-helix*. In an α-helix, the protein backbone spirals around its long axis (Figure 1.14). Each helical turn is composed of 3.6 amino acids, and the helix has a pitch of 5.4 Å between successive turns (or 1.5 Å rise [projected distance along the helix axis] per amino acid). There is very little empty space in the center of the helix. The broad occurrence of the α-helical structure arises from two sources. First, it is an energetically accessible conformation for these (ϕ, ψ) angles (see Figure 1.7). Second, the α-helix is stabilized by hydrogen bonds between the carbonyl oxygen of amino acid i and the amide hydrogen of amino acid $i+4$ (see Figure 1.14). α-helices can be formed by any of the amino acids (except proline) because the hydrogen bond donors and acceptors are backbone atoms.

Interestingly, the existence of the α-helix was hypothesized by Linus Pauling and his colleagues in the early 1950s, before it was discovered in nature [6]. At that time, it was not surprising that a polymer would have a helical structure. Many types of linear polymer chains tend to form various helical structures. Think about a repeated string of vectors with a fixed torsional angle between them. If that angle is 180°, the chain will be a linear rod. But for any other angle, the polymer will be a string of repetitive twist steps. That defines a helix. Typical nonbiological polymers have a single favored repeat angle (at low temperatures), so they crystallize into helical structures. About half of

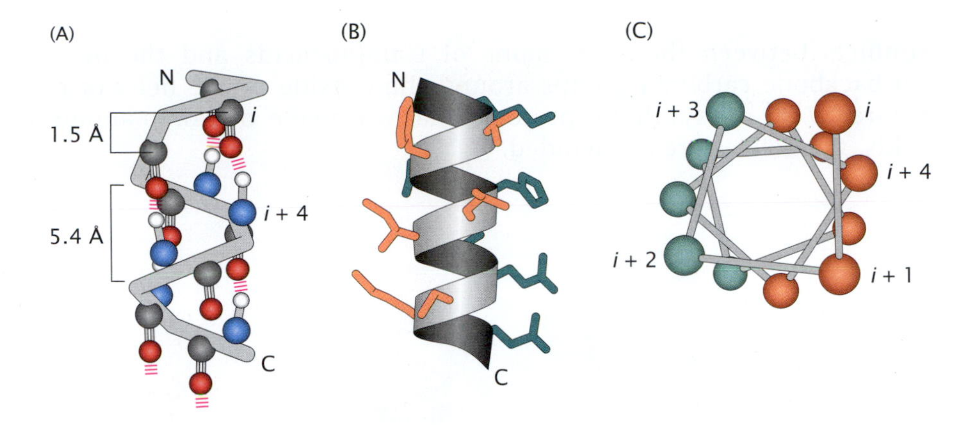

Figure 1.14 Different representations of the α-helix structure in proteins. (A) The regular main chain path is maintained by hydrogen bonds between the carbonyl O of residue *i* and the amide H of residue *i* + 4, represented here with *red* springlike connections. Successive α-carbons occupy angular positions at 100° intervals around the helical axis (to give 3.6 residues per full turn of 360°). Note that the first three N—H groups and the last three C=O groups are not hydrogen-bonded. (B) While the path of the main chain in an α-helix is fixed, the side chains (shown here as sticks) are more free to rotate. Frequently, helices are amphipathic, with hydrophobic residues (*orange*) on one side and hydrophilic residues (*teal*) on the other. (C) The amphipathic character is illustrated in a helical wheel diagram, showing the side chains as spheres viewed around the central helical axis.

the known polymer crystal structures take on one of the 22 different types of helices [7]. The surprise in Pauling's correct prediction was that protein helices would have a noninteger number of monomers per turn, 3.6 in this case. Pauling's key insight was that peptide helices would be stabilized by the hydrogen bonding between the backbone units, from the carbonyl group of one amino acid to the amide group of another.

Helices have a property called *handedness*. A helix can be either right-handed or left-handed, based on which direction it spirals (Box 1.1). Many helices in globular proteins also have a "sidedness" property. Called *amphiphilic* or *amphipathic*, those helices have a stripe of mostly hydrophobic amino acids down one side and a stripe of mostly hydrophilic amino acids down the other side. The hydrophobic stripe usually faces inward in a protein structure toward the protein core. A simple device for visualizing such patterns down the lengths of helices is a *helical wheel* diagram, shown in Figure 1.14C.

Box 1.1 Defining the *Handedness* in a Helix

Here's how a right-handed helix is defined. Align your right thumb so that it points along the helical axis that runs from the N-terminus toward the C-terminus. A helix is right-handed if it curls in the same way as your fingers curl from your palm through your fingers to your fingertips. Otherwise, it is left-handed.

In natural proteins, α-helices are right-handed (Figure 1.15). You can understand this from two basic facts: (1) side chains are on the outsides of helices (see Figure 1.14B) and (2) the naturally occurring amino acids are the L-isomers (see Figure 1.4). The L-amino acids prefer to form a right-handed helix because that minimizes steric conflicts between the side chains of L-amino acids and the helical backbone carbonyl groups around the outside of the helix (see Figure 1.7E).[3] In synthetic proteins that are made out of D-amino acids, the helices are left-handed.

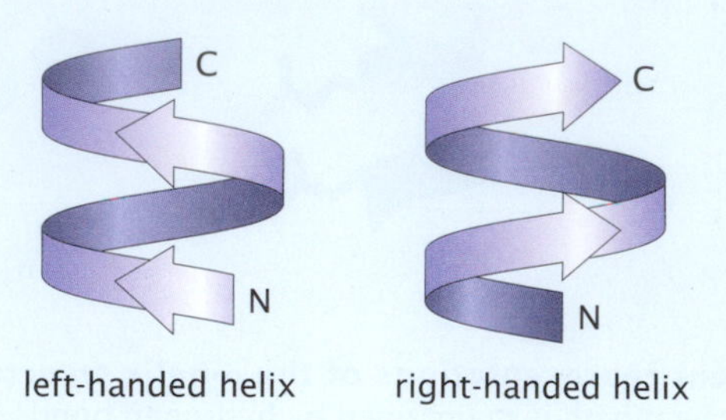

Figure 1.15 The contrast between a right- and a left-handed helix.

β-sheets. In 1951, Linus Pauling also predicted another type of secondary structure before it was found in nature [8]. *β-sheets* are train-track-like structures in which two or more segments, called *β-strands*, are aligned side-by-side (Figure 1.16). Pairs of strands are held together by amide-to-carbonyl main-chain hydrogen bonds from one strand to another. β-sheets are called *parallel* if the N → C directions of both strands are the same, or *antiparallel* if strands run alternately in opposite directions to each other. The side chains lie above or below the plane of the sheet, where they can interact with side chains from adjacent strands as well as those on neighboring sheets above or below. β-sheets are therefore intrinsically stabilized by (i) hydrogen bonds; (ii) side-chain interactions, often of nonpolar groups; (iii) favorable (ϕ, ψ) angles in the β-region of the Ramachandran map (see Figure 1.7); and (iv) by the good packing they achieve. Good packing is stabilized by so-called *van der Waals attractions.*

Less common types of secondary structure. α-helices and β-sheets are the most common—but not the only—types of secondary structure in proteins. Another type is the π-helix, in which a hydrogen bond is formed between residues $(i, i+5)$, instead of $(i, i+4)$ as in the α-helix. Another example is the 3_{10}-helix, in which hydrogen bonding is between residues $(i, i+3)$. The 3_{10}-helix has three residues per turn and ten atoms in the ring closed by the hydrogen bond. It is sometimes found in short peptides and occasionally in proteins, but its hydrogen bonds are less stable, its side-chain packing is less favorable, and its dipoles are more poorly aligned than the hydrogen bonds in the α-helix. The π-helix is rare because its rise is short, bringing side chains into close proximity. Another type of helix, called the polyproline II helix, is found in chains having high proline content, such as collagen (Table 1.2).

[3]This can be rationalized as follows: the carbonyl oxygen is larger than the amide hydrogen. Steric clashes between the side chain and the carbonyl oxygen restrict the range of possible pairs of (ϕ, ψ) values. In particular ϕ values in the range $\phi_i < 0°$ bring the R_i and $(N{-}H)_{i+1}$ groups into close proximity without causing an overlap of their atoms, while values of $\phi_i > 0°$ cause an overlap between R_i and the larger polar group $(C{=}O)_{i-1}$. The lower right quadrant of the Ramachandran map (see Figure 1.7) is almost entirely excluded due to the steric clash between R_i, $(C{=}O)_{i-1}$ and $(N{-}H)_{i+1}$.

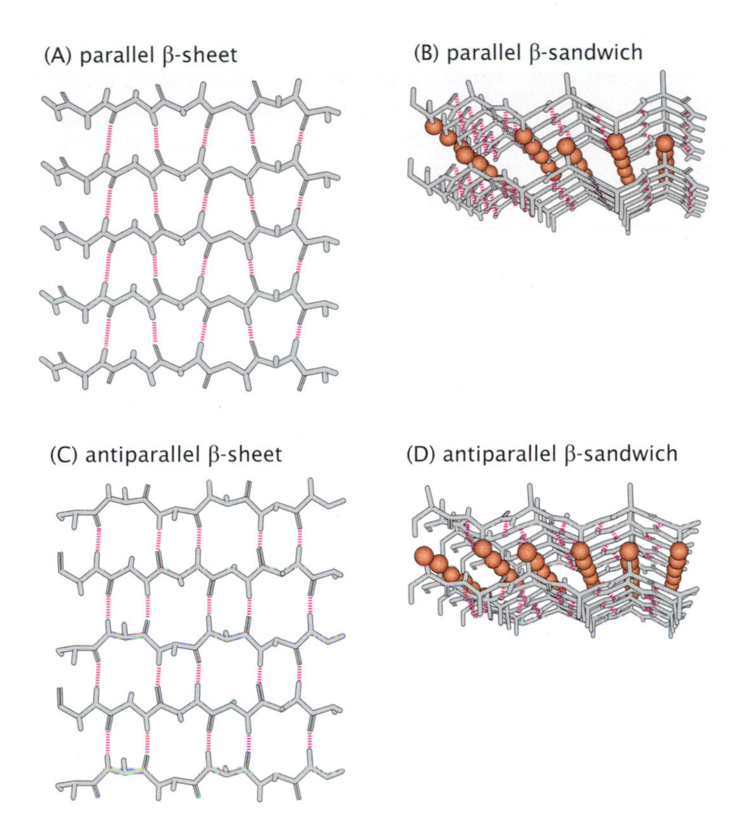

Figure 1.16 The structures of parallel β-sheets and sandwiches. The *left* column shows the hydrogen-bonding patterns in (A) parallel and (C) antiparallel β-sheets. On the *right*, pairs of sheets are tipped down so you can see the hydrophobic side-chain contacts sandwiched between the (B) parallel and (D) antiparallel sheets. (Adapted from JD Schmit, K Ghosh, and KA Dill. *Biophys J*, 100:450–458, 2011.)

Turns and loops. *Turns* and *loops* are not themselves secondary structures.[4] Turns are chain segments of three to five residues usually folded in well-defined geometries and stabilized by local hydrogen bonds. Loops are longer segments of chain that are typically less structured. Turns, also called *reverse turns*, are places where the chain reverses direction. Reverse turns tend to be composed of polar residues and glycine and proline, while loops tend not to have such preferences. Because they are polar, reverse turns tend to be located on protein surfaces. Glycine is common in turns because it contorts easily, due to the lack of side-chain hindrance. Proline is also common in turns because it has the unusual feature that its backbone is constrained in a chemical-ring structure.

Sometimes the N-terminus of a helix has an *end-capping* hydrogen bond. At the ends of helices, the backbone carbonyl and amide groups may have unsatisfied hydrogen bonds. In such cases, a side chain (for example serine) can form a hydrogen bond by folding back onto the backbone to hydrogen-bond with the unfulfilled backbone hydrogen bond.

[4]Sometimes, turns and loops are included in the definition of secondary structures. In this book, we use the term secondary structure only to refer to the regular repeating structures—helices and sheets.

Table 1.2 Geometric parameters for some regular protein conformations

Secondary structure	Residues dihedral angle (°)			# of amino acids per turn	Rise per residue (Å)
	φ	ψ	ω		
Right-handed α-helix	−57	−47	180	3.6	1.50
Left-handed α-helix	+57	+47	180	3.6	1.50
3_{10}-helix	−49	−26	180	3.0	2.00
π-helix	−57	−70	180	4.4	1.15
Polyproline II (left-handed)	−82	147	180	3.0	3.04
Parallel β-sheet	−119	113	180		
Antiparallel β-sheet[a]	−139	135	−178		
Fully extended chain[b]	180	180	180	2.0	3.6

Partly adapted from JC Kendrew, W Klyne, S Lifson, et al. *Biochemistry* 9: 3271–3479, *J Biol Chem*, 24:6489–6497, 1969, and *J Mol Biol* 52:1–17, 1969; GE Schulz and RH Schirmer. *Principles of Protein Structure*. Springer-Verlag, New York, 1979.
[a] Twisted β-sheets are abundant in proteins. There are considerable variations among the dihedral angles of twisted β-strands.
[b] Included for reference only. Fully extended chains are not commonly observed and do not form stable sheet-like chain organizations.

Supersecondary Structures, Also Called *Structural Motifs*, Are Common Combinations of Secondary Structures

Some assemblies of secondary structures appear so frequently that they are given names. As a class, they are called *supersecondary structures*. A prominent example is the *coiled coil*, in which two helices are twisted together. Another is the β-helix formed by the association of β-strands in a right-handed or left-handed helical pattern. The β-helix structure is highly stable. Notably, triple-stranded β-helical structures function as cell-puncturing devices of bacteriophages [9]. Their high stability and helical structure are essential to disrupting the host cell membrane during infection.

Supersecondary structures often have a *right-handed twist*. For example the *EF hand*, the β-hairpin, and the β–α–β motif usually exist in the right-handed form [10]. Ninety-five percent of β–α–β motifs in proteins are right-handed [11]. The chirality of supersecondary structures refers to the relative rotational orientation of these structures with respect to the chain axis, as we move along the sequence from the N- to the C-terminus. Figure 1.17 illustrates the differences between a left-handed and right-handed β-sheet and between helical motifs. The prevalence of right-handed twists for individual β-strands is explained by the intrinsic preferences of L-amino acids. Take an individual strand, with right-handed twist, and fold it into a compact structure, a loop, for example. If the loop is right-handed, it will naturally release the right-handed twist/strain of the strands. If the loop were left-handed, it would increase the strain. You can see this by first twisting a belt and then forming it either into a right-handed or left-handed loop. One way releases strain and the other increases it.

Some Substructures of Proteins Are Compact Functional *Domains*

Another type of protein substructure is called a *domain*. A domain is a piece of a polypeptide chain that can fold on its own, function on its own, evolve on its own, or can be identified by its compactness

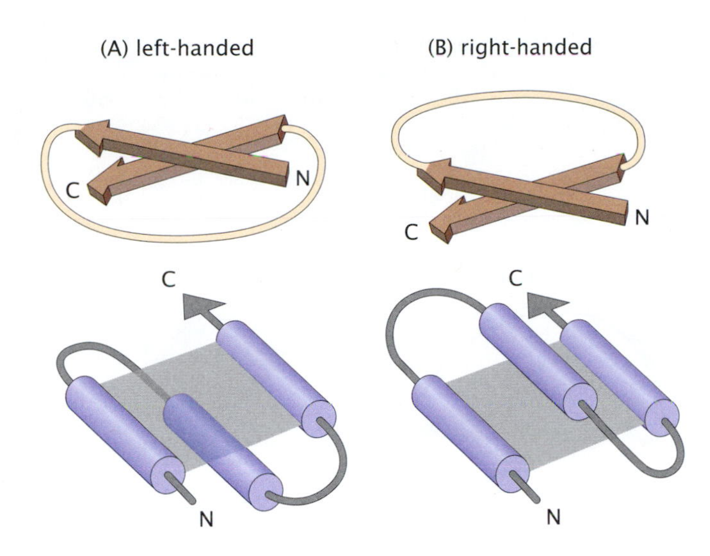

Figure 1.17 Chirality in supersecondary structures. Parallel β-strands or helical motifs may be (A) left- or (B) right-handed. The right-handed path relieves the strain of right-handed β-strands. (Adapted from JM Thornton. *Protein Sci*, 10:3–11, 2001.)

relative to neighboring parts of a chain. A domain is the smallest unit of protein function. Examples include the SH3 and ATPase domains. Often each domain has a *modular function* to perform in a protein, such as binding a ligand, spanning a cell membrane (transmembrane domains of membrane proteins), containing the active site (catalytic domains in enzymes), binding a nucleotide (DNA/RNA-binding domains in transcription factors), or providing a surface for binding other proteins (substrate-binding domains). A single protein chain may have one or more domains, often several. For example Figure 1.18 shows that pyruvate kinase has three domains. Sometimes evolution reuses domains; the same domain can appear in different proteins. Interestingly, the different domains within a protein are sometimes encoded by different regions within the genome, as a result of genetic recombination.

As of 2015, there were known structures of more than 170,000 distinct domains, ranging from 13 to more than 1000 residues.

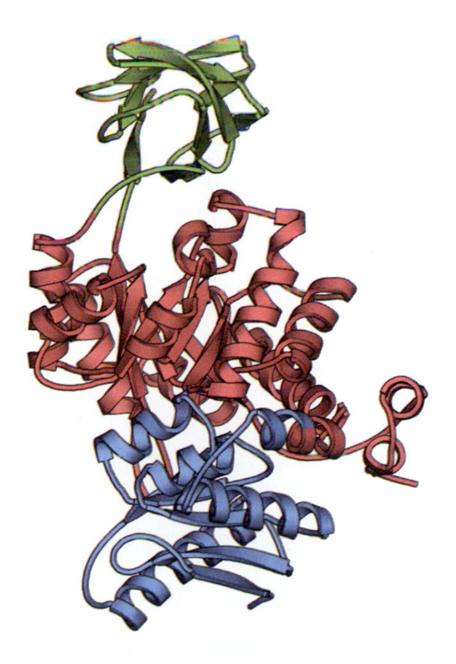

Figure 1.18 Protein domains are independently folding subunits. Pyruvate kinase consists of three domains formed by the N-terminal (*red*), PK domain (*green*), and C-terminal (*blue*) residues.

Native Protein Topologies Are Described Using *Contact Maps*

A useful way to visualize a protein's *topology*—its substructures and their positions relative to each other—is to use a *contact map* (Figure 1.19). For a chain having n residues, the contact map is an $n \times n$ matrix in which the element (i, j) is 1 (or a dot) if residues i and j are in contact, and 0 (or empty) otherwise. Contact maps are symmetrical, so usually only the upper (or lower) diagonal half of the matrix is displayed.

A contact map provides information on all inter-residue contacts in a protein's tertiary structure. In particular, secondary structures appear as simple patterns. For example, helices appear as lines of dots adjacent and parallel to the main diagonal. Parallel β-strands are also parallel, but not adjacent, to the main diagonal on the contact map. Antiparallel strands are lines of dots perpendicular to the main diagonal. Contacts are called *local* if they are near the main diagonal, and *nonlocal* if they are more distant. Helices and turns involve only local contacts, whereas sheet contacts are predominantly nonlocal. Figure 1.19 shows the contact map for the native structure of the chymotrypsin inhibitor (CI2), which has all of these elements.

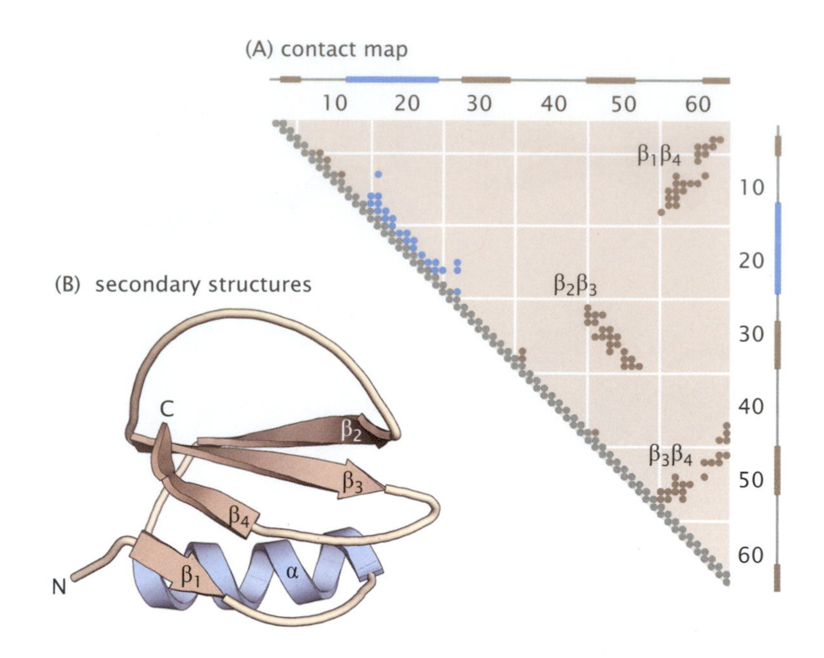

Figure 1.19 (A) Contact map for chymotrypsin inhibitor (CI2). The two axes represent the amino acid number along the sequence. A contact between two residues (represented by a dot on the map) is defined here whenever C^α or C^β atoms from different residues are within 6 Å of each other. The four large clusters of contacts indicate the main structural interactions in the protein. (B) The 3D structure and identity of β-strands in CI2. The helix is shown in *blue* and strands in *brown* in both (A) and (B). (Adapted from C Merlo, KA Dill, and TR Weikl. *Proc Natl Acad Sci USA*, 102:10171–10175, 2005.)

How Can You Classify Protein Tertiary Structures?

Suppose you discover a new protein structure and want to know if it resembles other known structures. You need a way to classify its structure, starting from its most global features and working down to its most detailed features. Proteins were first classified into four structural families in the 1970s [12]: α (mostly α-helical), β (mostly β-sheet), α/β (sequentially interspersed α-helices and β-strands), and α + β (one region of mainly α-helices joined to another region of mainly β-sheets). Now that many more protein structures are known, these initial four structural classes have become the basis for more extensive classification schemes.

Now, suppose that your protein happens to be a four-stranded β-sheet. That terminology alone is not sufficient to define the structure. There are different *topological* relationships through which the strands could come into contact with each other. Figure 1.20 shows the 12 possible topologies for four-stranded β-sheets, where the third and fourth strands (in sequence) are antiparallel. Some topologies are more common than others.

The next level of description accounts not only for the topological arrangement of secondary structural elements, but for their particular packing geometry in 3D as well. This level of description is called the *fold* of a native protein. A protein fold is a particular arrangement of secondary structures in a tertiary structure. Some common folds are shown in Figure 1.21. The *globin fold* (see Figure 1.21A) has eight α-helices. The jelly roll fold has two superimposed β-sheets (see Figure 1.21B). The *TIM barrel fold* (see Figure 1.21C) takes its name from the enzyme triosephosphate isomerase (TIM), and usually contains eight α-helices (cyan) and eight β-strands (brown). This fold is one of the most common among enzymes. Its successive β–α–β motifs are arranged into a torus. The *Rossmann fold* (see Figure 1.21D) is named after Michael Rossmann, who discovered it in the early 1970s; it binds nucleotides such as nicotinamide adenine dinucleotide [13]. Its fold is β–α–β–α–β.

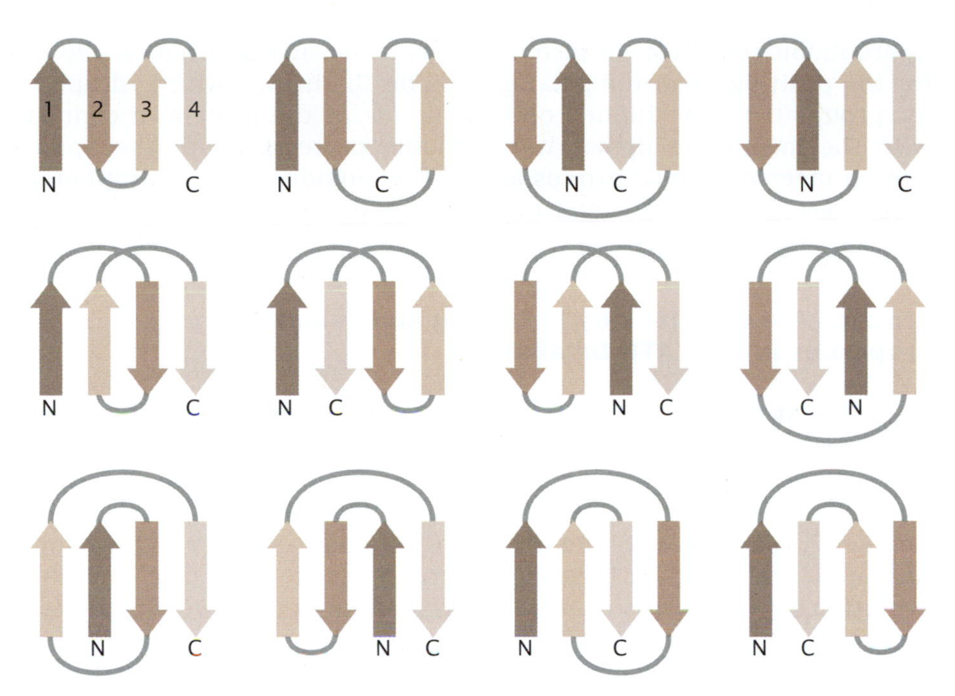

Figure 1.20 Topology diagrams for four-stranded β-sheets. Of all the possible topologies of four-stranded β-sheets, there are only 12 (shown here) that have β-strands three and four in the primary sequence antiparallel to each other and connected by three β–β hairpin loops. Not all topologies in this set are equally probable. Not all of these forms have been observed.

One of the most common protein folds is the β-*barrel*. A β-barrel consists of adjacent β-strands arranged in a cylindrical β-sheet. The cylindrical structure is favored over flat or planar β-sheets because the cylinder leaves no unsatisfied backbone hydrogen bonding groups at the edges. Often the strands are composed of alternating polar and

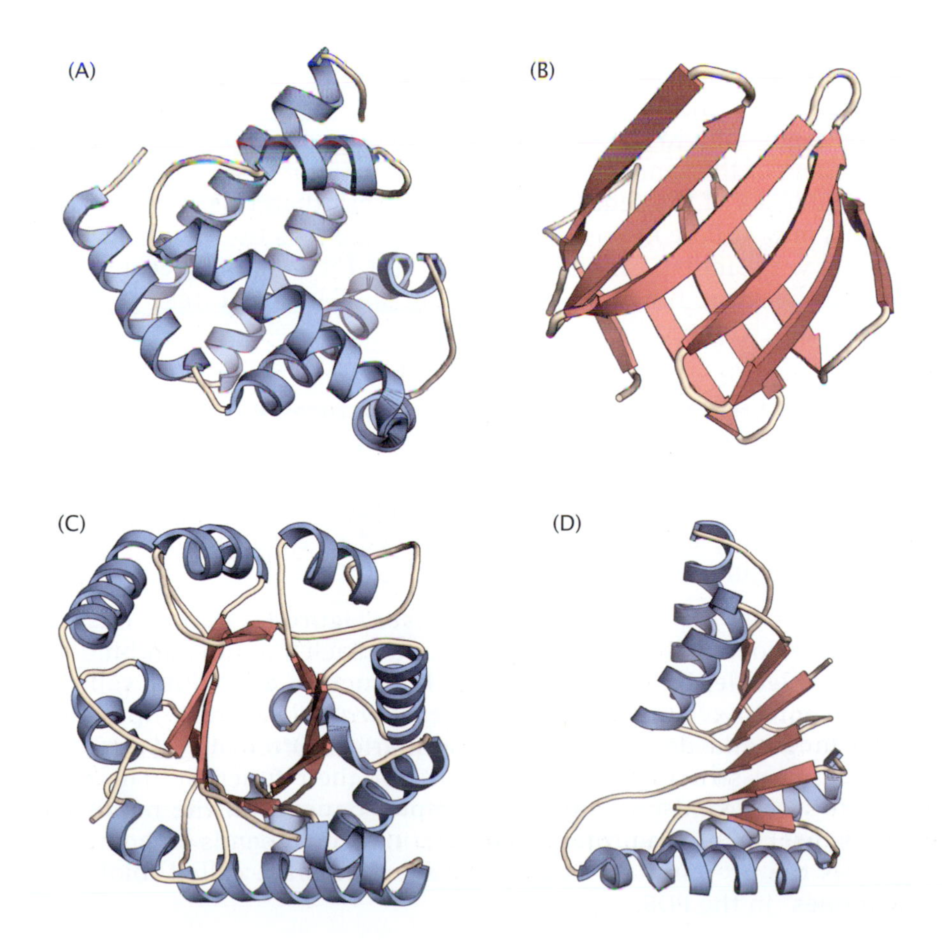

Figure 1.21 Examples of common protein folds. (A) The globin fold contains eight tightly packed α-helices packed together so they can bind an iron-containing heme group (not shown) for storing and transporting oxygen. (B) A jelly roll fold is formed by pairing two antiparallel β sheets together to form a barrel. (C) The left-handed TIM barrel is a doughnut-like shape (toroid) formed by alternating α-helices (*light blue*) and β-strands (*red*) arranged in a closed curve. (D) The Rossmann fold contains two β–α–β–α–β motifs packed to form a central six-stranded β-sheet. In this case, the β-strands (*red*) and α-helices (*blue*) are from a domain of the enzyme decarboxylase, with a flavin mononucleotide cofactor (not shown) bound onto its nucleotide-binding site.

hydrophobic residues. For membrane proteins, such alternation allows for the positioning of the polar groups on the inside, with hydrophobic groups being positioned on the outside of the protein in contact with the membrane lipids. Water-soluble proteins adopt the opposite arrangement: polar groups are on the outside, while hydrophobic groups are inside. The center of the barrel is often a binding site; for example, vitamin A (retinol) binds to the retinol-binding protein.

Proteins Are Classified by Structural and Evolutionary Properties in the CATH Database

A "fold" just describes a geometric property of a protein. You can get additional information for classifying proteins by knowing *evolutionary relationships*, that is, whether or not two proteins evolved from a single ancestor. Such relationships are discussed in more detail in Chapter 7. Here, we just note that there are databases of protein structures that are based on these considerations. Inference of a common evolutionary ancestor is usually based on how similar two amino acid sequences are to each other. A database that classifies proteins based on both their 3D native topologies and their evolutionary relationships to each other is CATH [14]. CATH assigns protein domains into subsets that belong to the same **C**lass, or that have a common **A**rchitecture or **T**opology (fold), or that belong to the same **H**omologous family (tertiary structure). Figure 1.22 gives the distribution of common CATH classes.

A protein's fold (equivalent to CATH's Topology) is purely geometric. Two proteins having the same fold need not have any particular evolutionary relationship. On the other hand, if two proteins have less than 15% sequence identity while their structures and functions suggest a common evolutionary origin, they may belong to the same *superfamily*. For example, actin, the ATPase domain of the heat-shock protein, and hexokinase belong to the same superfamily. Two proteins are regarded as members of the same *family* if they share 30% or more of their amino acid sequences. If the sequence identity is less than this, down to 15%, proteins may still be in the same family if their functions and 3D structures are very similar. Globins form a family, for example, because they perform the same function, have highly similar structures, and have high sequence identity.

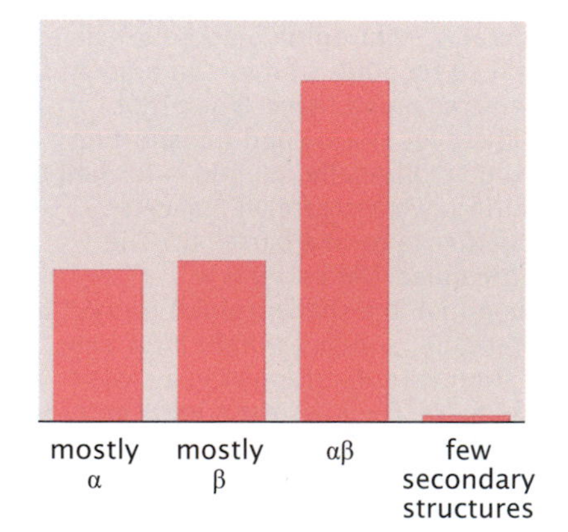

Figure 1.22 Distribution of the protein structures in the PDB in the four CATH classes.

Quaternary Structures: Higher-Order Structures Result from Noncovalent Assemblies of Multiple Chains

In contrast to the domains of tertiary structures, the subunits in a quaternary structure are not covalently bonded to each other, except for occasional disulfide bridge cross-links.

Hemoglobin has a quaternary structure (see Figure 1.11). It carries oxygen in red blood cells. It has four subunits. Each subunit can bind one oxygen molecule, but it is only the full structure of all four subunits that can bind oxygen with sufficient *cooperativity* to pick up oxygen in the lungs and deliver it to the capillaries. Often the symmetries in quaternary structures are important for their function. The cooperativities of ligand binding, for example, depend on the number of ligands that can bind a protein, and that, in turn, depends on how many subunits the protein has. Quaternary structures are called "biological assemblies" in the PDB.

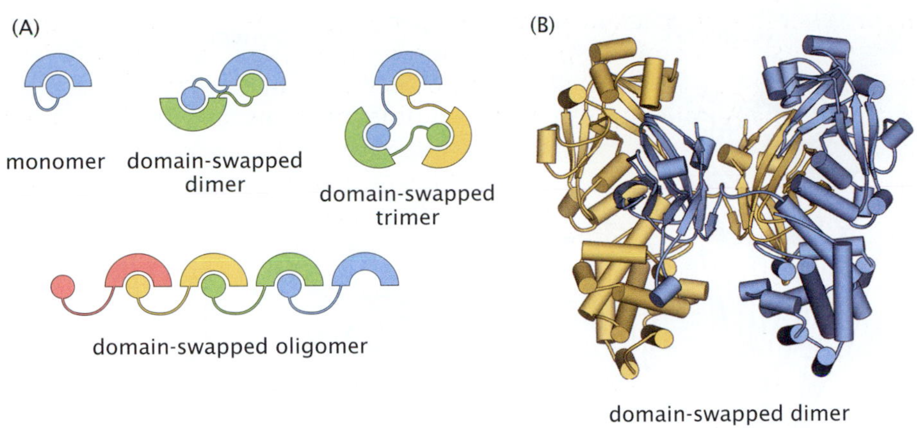

Figure 1.23 Some quaternary structures are constructed by domain swapping. (A) Protein monomers of two domains (arch and ball) are able to form stable quaternary structures ranging from dimer to trimer to oligomer. Stabilizing interactions are contributed at the interfaces formed by swapped domains from different chains. (B) The *yellow* and *blue* chains of the diphtheria toxin dimer swap β domains. (A, adapted from Gronenborn AM. *Curr Opin Struct Biol*, 19:39–49, 2009; B, adapted from MJ Bennett, S Choe, and D Eisenberg. *Proc Natl Acad Sci USA*, 91:3127–3131, 1994.)

Domain Swapping Is Another Way that Proteins Can Form Quaternary Structures

In some cases, quaternary structures are defined by the interdigitation of two or more proteins. In *domain swapping*, one domain (or secondary structure) of a monomer replaces the same domain (or secondary structure) from a different, identical monomer, giving rise to an intertwined dimer or oligomer (Figure 1.23). Sometimes, domain swapping can repeat, chaining together one protein to the next, leading to an ordered form of protein assembly. Fibronectin modules form fibrils from such chains of domain-swapped proteins.

SOME PROTEINS ARE STABLE AND FUNCTION IN THE MEMBRANE ENVIRONMENT

Membrane proteins are localized in the membranes of cells or organelles, such as mitochondria. Some membrane proteins are channels that allow the flow of ions, such as potassium. Others function as electron or proton pumps (for example, cytochrome *c* oxidase and complex IV in mitochondria, and ATPase in cell membranes), receptors (for example, G-protein-coupled receptors, GPCRs) or transporters (for example, glutamate transporters and ABC transporters) across the membrane (Figure 1.24). Some membrane proteins can function as receptors and ion channels at the same time (AMPA and NMDA). These are all *integral membrane proteins*. They are stable and functional when embedded in the lipid bilayer. They play a key role in maintaining or regulating the physiological levels of ions and substrates at the extracellular and intracellular regions, assisting with establishing appropriate concentrations or energy gradients across the membrane, and enabling signal transduction events across the membrane. *Peripheral membrane proteins* are temporarily attached (usually from the extracellular side) to cell membranes or to integral membrane proteins. Here we are referring to integral membrane proteins, unless otherwise stated. Membrane proteins are often important drug targets.

A membrane protein usually has three regions: transmembrane (TM), extracellular, and intracellular. The membrane is typically a lipid bilayer, which is a sandwich of two layers of lipid molecules. The polar

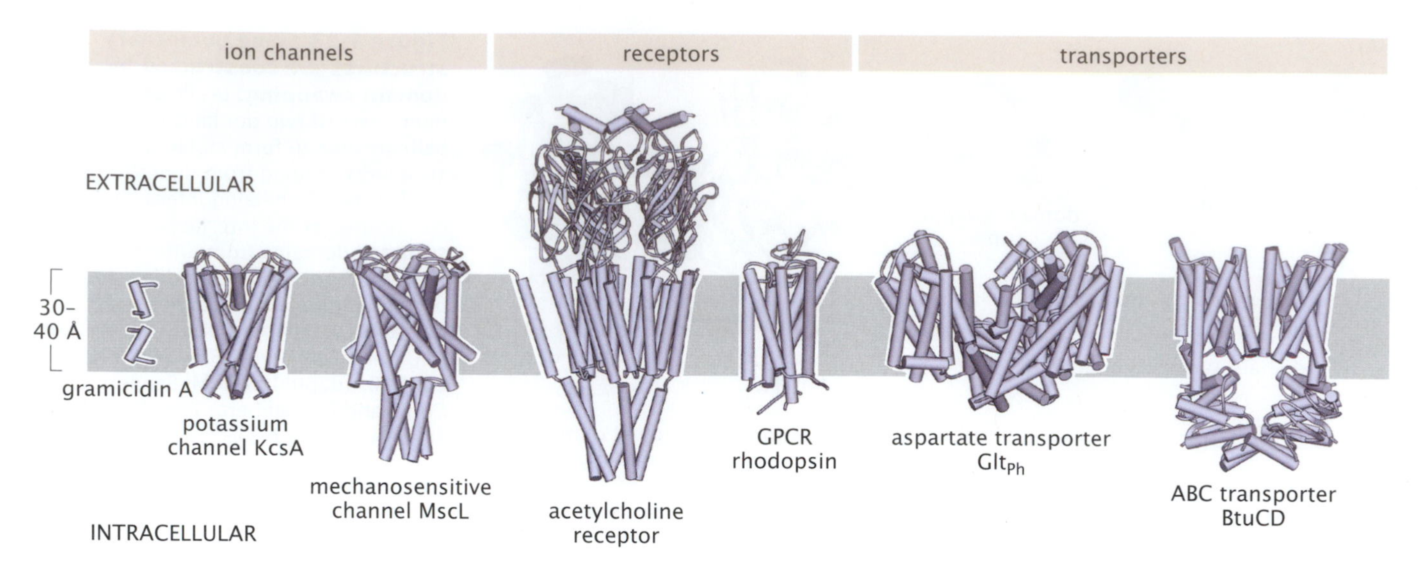

Figure 1.24 Different sizes and types of membrane proteins. The sizes range from small helical peptides in gramicidin A to multimeric proteins with large extracellular and/or intracellular domains, as in the acetylcholine receptor. (Adapted from I Bahar, TR Lezon, A Bakan, and IH Shrivastava. *Chem Rev*, 110:1463–1497, 2010.)

heads groups face outward toward the water and the hydrophobic tails face each other inside the bilayer. In contrast to globular proteins, which have hydrophobic groups buried in the core and polar/charged groups facing the surrounding water, the surfaces of membrane proteins usually contain hydrophobic residues that make favorable contacts with the surrounding lipid molecules inside the membrane.

An important class of membrane proteins is the GPCRs (G-protein-coupled receptors). GPCRs have a seven-TM-helix fold. β-barrel folds are also commonly observed, for example in porin proteins. Of the 1700 TM structures known in 2012, about 1400 are α-helical and 250 are β-barrels. Many membrane proteins are multimeric. The monomers assemble to form a central pore (for example, potassium channels, which are tetramers), or a stable scaffold that supports the cooperative transition between outward-facing and inward-facing forms (for example, glutamate transporters, which are homotrimers, and the ABC transporters, which are heterodimers).

SOME PROTEINS HAVE FIBROUS STRUCTURES

Fibrous proteins are elongated, with a single dominant type of secondary structure. For example, collagen forms a triple-helical right-handed coiled coil, so it has great mechanical strength. Fibrous protein sequences are often highly repetitive. Collagen has long stretches of repeats of the tripeptide Gly-Pro-X, where X can be any amino acid (Figure 1.25). A large fraction of the world's protein mass is fibrous. Collagen, which is the main protein of connective tissue, is the most abundant protein in vertebrates, making up 25–35% of their whole-body protein content.

The mechanical and load-bearing frameworks of organisms are constructed from fibrous proteins. Collagen forms the stress-bearing elements of skin, bone, teeth, and tendons. β-*keratin* is a two-helix coiled coil, found in fur and claws. The essential protein in silk is called *fibroin*; it forms a β-sheet. Some fibrous proteins are elastic. For

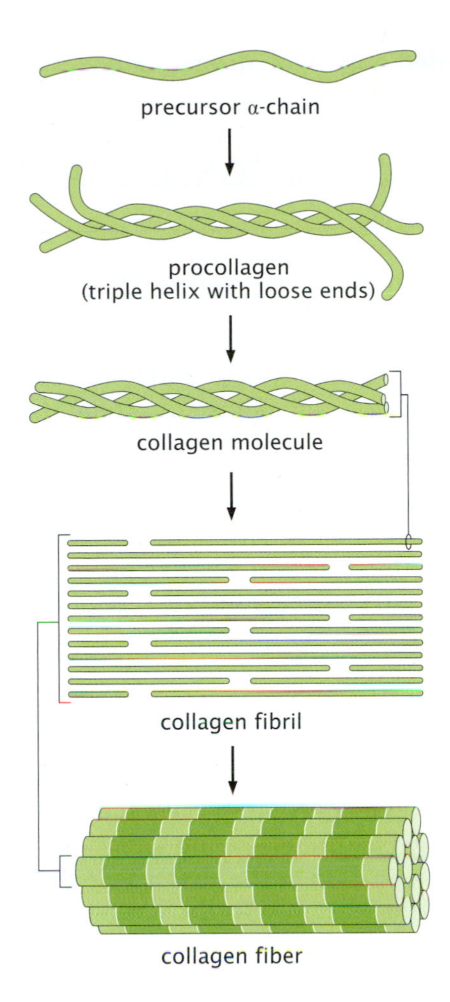

Figure 1.25 Fibrous proteins are made from α-helical coiled coils or stacked β-sheets. They are mechanically strong and can extend to macroscopic sizes. The collagen triple helix is a right-handed superhelix or coiled coil. The three helices adopt polyproline II conformations and twist around their neighbors like strands in a rope. The stability is maintained by extensive hydrogen bonding between the neighboring helices.

example, elastin, found in ligaments and the walls of the lungs and blood vessels, is primarily composed of small, nonpolar amino acids (Ala, Val, Gly), Pro, and Lys, and acts as a collection of springs that are cross-linked into an irregular assembly.

Fibrous Proteins Include Coiled Coils and β-Helices

Coiled coils occur when at least two α-helices are wound around each other in a regular twist, like the strands of a rope (see Figure 1.25). They are among the most common supersecondary structures, and can contain two, three, four, or five helices. The component helices may be aligned either antiparallel or parallel. Figure 1.25 shows a coiled coil of three helices. Coiled coils are ubiquitous, occurring in transcription factors, viral fusion peptides, and certain tRNA synthetases.

Other fibrous structures are based on β-strands. Silk fibroin has a regular amino acid sequence repeat of -(Gly-Ser-Gly-Ala-Gly)-. It has high mechanical strength because of its extensive β-structure. Silk is strong along the fiber axis because that is the direction of the chain's covalent bonding. But in the other directions, silk is more flexible because the chains are bonded only by hydrogen bonds, which are weaker than covalent bonds.

NATIVE PROTEINS ARE CONFORMATIONAL ENSEMBLES

Proteins Fluctuate around their Native Structures

So far, we have focused on single native structures of proteins. However, that is not the whole story. First, even well-defined native protein structures wiggle and shake at room temperature because they are bombarded by the Brownian motion of the surrounding solvent molecules. NMR methods can reveal the fluctuations of a protein around its native structure. Multiple chain conformations are consistent with the NMR determination of the structure of a protein, indicating the fluctuations around a native protein structure. When you refer to "the" native structure, you are referring to a representative structure from this ensemble.

A Protein Can Sample Multiple Substates under Native Conditions

Second, proteins can deviate from the "single-native-structure" ideal in another respect. They can undergo conformational changes. For example, the structure of a protein in the absence of a bound ligand is called its *apo* state, while the structure in the presence of a bound ligand is called its *holo* state. Figure 1.26 shows adenylate kinase, which undergoes changes in conformation upon ligand binding, changing from an "open" to a "closed" conformation, with ligand-binding stabilizing the closed form.

Another commonly observed structural transition, in the case of membrane proteins, is the passage from the so-called outward-facing conformation to the inward-facing conformation, known as the *alternating access model.* The outward-facing conformation is open to the extracellular environment to bind/uptake the ligand, and the inward-facing conformation releases the substrate to the intracellular region. Transitions between these two functional structures are essential to the transport of substrates by transporters (membrane proteins) across the cell membrane.

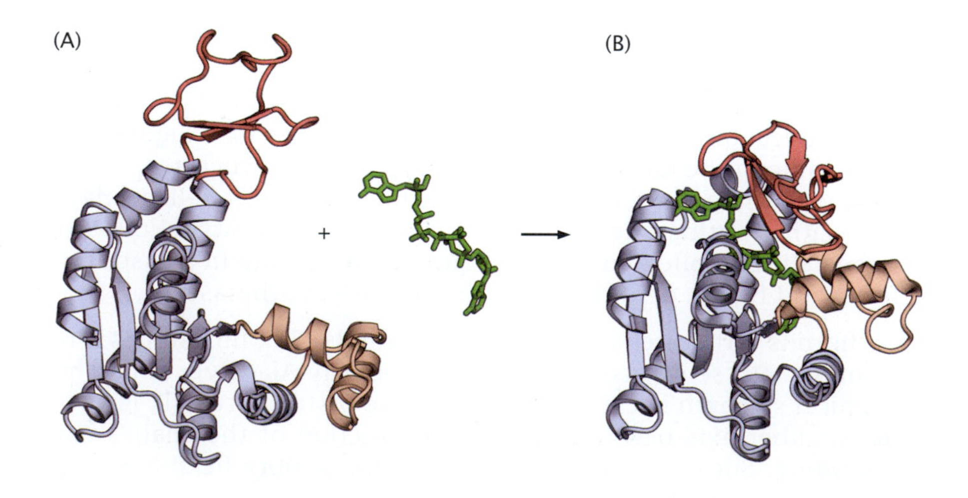

Figure 1.26 Adenylate kinase (colored by its three domains) is an example of a protein that exists in an open form (A) and a closed form (B). As in many cases, the ligand (a nucleotide, *green*) binds and the protein then closes around it. (Adapted from U Olsson and M Wolf-Watz. *Nat Commun*, 1:111, 2010. With permission from Macmillan Publishers Ltd.)

Some Proteins Are Intrinsically Disordered

A third way that proteins can deviate from the "single-native structure" paradigm is through intrinsic disorder. Some proteins have regions that are *intrinsically disordered*, meaning that those parts of the chain are not well defined in experimental structure determinations. Interestingly, intrinsic disorder can serve functional purposes. For example, a positively charged intrinsically disordered protein can bind to a negatively charged DNA molecule, causing the complex to form a unique structure upon binding, a phenomenon referred to as *folding upon binding*.

In another form of disorder, proteins can form *molten globule* states: relatively compact structures having residual native-like secondary structures but little tertiary structure. Proteins sometimes form molten globular states, for example, under acidic pH conditions.

SUMMARY

Proteins are polymer chains. They fold into compact states that are diverse in size, shape, and dynamics. They perform many different biological functions. Different proteins have different sequences of the 20 types of building-block amino acids. Some amino acids are nonpolar, some are polar, and some are positively or negatively charged. Different amino acid sequences fold into different 3D shapes. Folded proteins adopt structures on different levels: secondary structures include helices and sheets; tertiary structures are assemblies of secondary structures in well-defined folds; and quaternary structures are composed of multiple chains (or subunits). Proteins may be globular and soluble in water, or fibrous, or may be localized within membrane environments. Chapter 2 gives an overview of how a protein's biological actions are encoded in its 3D native structure and motions.

REFERENCES

[1] R Milo. What is the total number of protein molecules per cell volume? A call to rethink some published values. *Bioessays*, 35:1050–1055, 2013.

[2] HM Berman, J Westbrook, Z Feng, et al. The protein data bank. *Nucleic Acids Res*, 28:235–242, 2000.

[3] GN Ramachandran, C Ramakrishnan, and V Sasisekharan. Stereochemistry of polypeptide chain configurations. *J Mol Biol*, 7:95–99, 1963.

[4] JA Rupley and G Careri. Protein hydration and function. *Adv Protein Chem*, 41:37–172, 1991.

[5] K Linderstrom-Lang. Proteins and enzymes. *Lane Medical Lectures*, Stanford University Publications, University Series, Medical Sciences, VI:1–115, 1952.

[6] L Pauling, RB Corey, and HR Branson. The structure of proteins: Two hydrogen-bonded helical configurations of the polypeptide chain. *Proc Natl Acad Sci USA*, 37:205–211, 1951.

[7] H Tadokoro. *Structure of Crystalline Polymers*. Wiley, New York, 1979.

[8] L Pauling and RB Corey. Configurations of polypeptide chains with favored orientations around single bonds: Two new pleated sheets. *Proc Natl Acad Sci USA*, 37:729–740, 1951.

[9] S Kanamaru, PG Leiman, VA Kostyuchenko, et al. Structure of the cell-puncturing device of bacteriophage T4. *Nature*, 415:553–557, 2002.

[10] MJE Sternberg and JM Thornton. On the conformation of proteins: The handedness of the connection between parallel β-strands. *J Mol Biol*, 110:269–283, 1977.

[11] TWF Slidel and JM Thornton. Chirality in protein structure. In H Bohr and S Brunak, editors, *Protein Folds: A Distance Based Approach*, pp 253–264. CRC, Boca Raton, FL, 1995.

[12] M Levitt and C Chothia. Structural patterns in globular proteins. *Nature*, 261:552–558, 1976.

[13] ST Rao and MG Rossmann. Comparison of super-secondary structures in proteins. *J Mol Biol*, 76:241–250, 1973.

[14] F Pearl, A Todd, I Sillitoe, et al. The CATH domain structure database and related resources Gene3D and DHS provide comprehensive domain family information for genome analysis. *Nucleic Acid Res*, 33:D247–D251, 2005.

SUGGESTED READING

Branden C and Tooze J, *Introduction to Protein Structure*, 2nd ed. Garland Science, New York, 1999.

Kuriyan J and Konforti B, *The Molecules of Life: Physical and Chemical Principles*. Garland Science, New York, 2012.

Voet D and Voet JG, *Biochemistry*, 4th ed. Wiley, Hoboken, NJ, 2011.

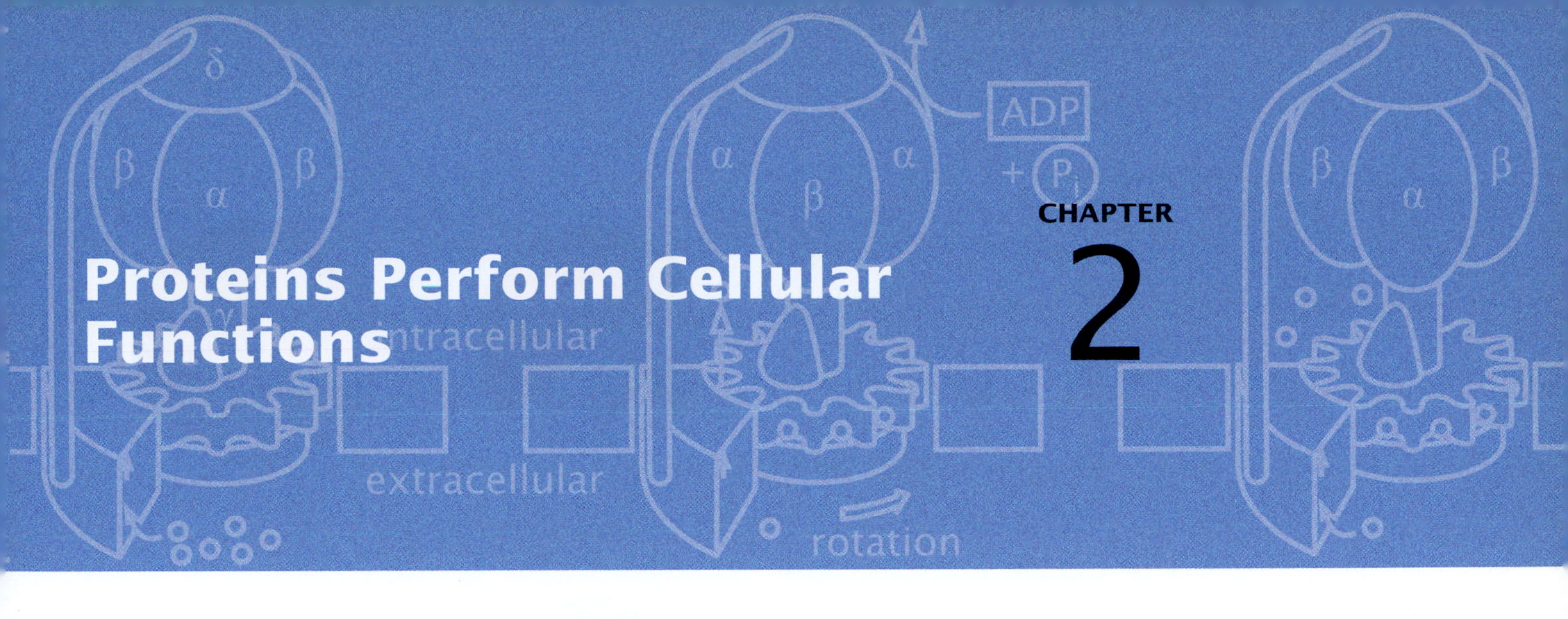

CHAPTER 2

Proteins Perform Cellular Functions

PROTEINS CARRY OUT MANY ACTIVITIES IN THE CELL

Imagine a device that performs some chemical or physical action in a repetitive, self-sustaining way. Consider the basic requirements for such a device. It needs to take in raw materials, eliminate its waste, transduce the energy it needs, and transport or create a product. An advanced version of the device might seek its own supplies and energy, so it would need mobility and sensory and regulation systems. An even more advanced version would be self-sustaining beyond its component's lifetimes. It would need to detect and repair its own broken parts, and even perhaps replicate itself, requiring memory, blueprints, and a way to translate blueprints into new component parts.

Biological cells are such devices. And most of those functions are performed by proteins. Some proteins burn fuel (metabolic enzymes), some convert light to energy (rhodopsin and the photosynthetic reaction center), some serve as batteries to run chemical processes (ATPases), some carry oxygen (globins) to locations that burn the fuel, some store and transport energy (cytochromes), some enable the channeling of ions or transport of molecules across membranes (ion channels and transporters), some form the structural infrastructure (collagen and keratin), some transduce energy to motion (actin, myosin, kinesin, and dynein), some copy the DNA and RNA blueprints (polymerases), some repair the DNA and RNA blueprints (ligases and exonucleases), some eliminate old or broken components (nucleases and proteases), some control, regulate, and signal to keep the system functioning smoothly (transcription factors, hormones, receptors, and kinases), and some (immunoglobulins and cytokines) provide defense against invaders [1].

The functionalities of proteins are listed in the Gene Ontology (GO) database (http://www.geneontology.org [2]). In GO, each protein is described with three different descriptors: its *molecular function*, its *biological process*, and its *cellular compartment* or localization within

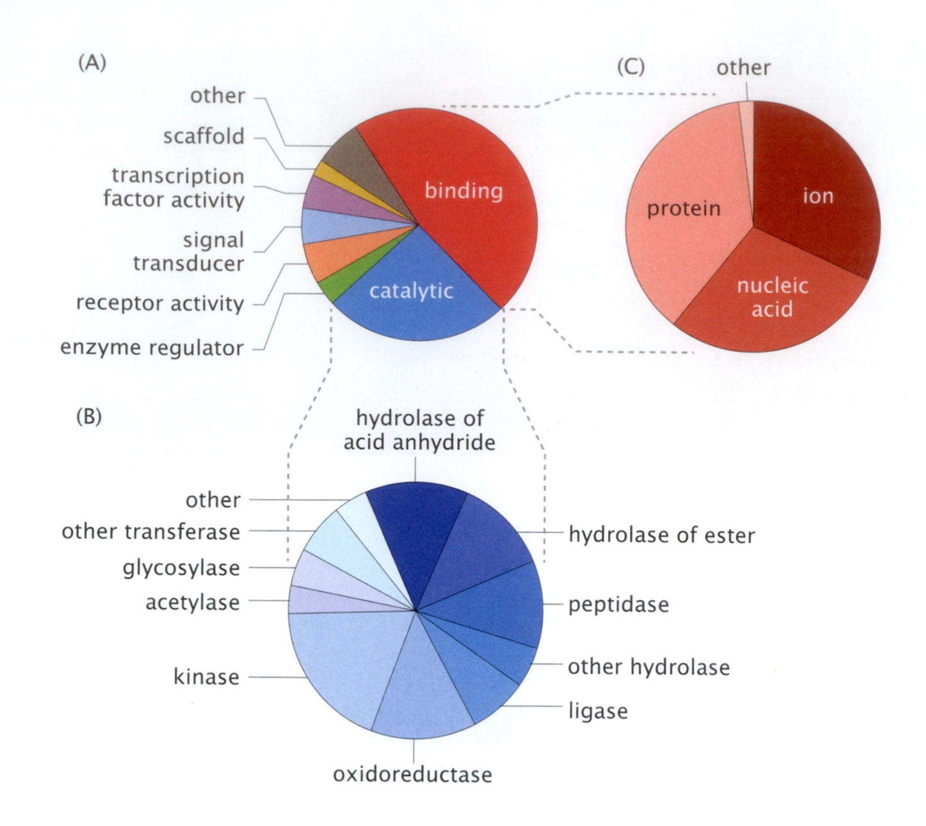

Figure 2.1 Distribution of human proteins by function: (A) by the biological processes they perform; (B) by their catalytic functions; (C) by their types of ligands. (Figures were created using Gene Ontology tools [2].)

the cell. For example, the molecular function of cytochrome *c* is oxidative reduction; the biological processes in which it is involved are apoptosis (programmed cell death) and oxidative phosphorylation, and its cellular localizations are in the mitochondrial matrix and inner membrane, although it can also be released to the cytoplasm, which is viewed as a point of no return for initiating apoptosis. Figure 2.1 lists the biological processes and molecular functions for the proteins in the human genome, as deduced from the human genome DNA nucleotide sequence of ~3 billion base pairs. The relative numbers of the different functionalities in various organisms are shown. Some proteins have multiple functions. For example, sometimes an enzyme also acts as a signaling molecule.

A PROTEIN'S FUNCTIONALITY IS ENCODED IN ITS STRUCTURE AND DYNAMICS

In the macroscopic world, you can often understand how something works—a motor, a spoon or fork, a light bulb, or a house—by looking at its structure. The same is often true at the microscopic level. Looking at the structure of a protein can give you insight about how the protein functions. You can start with the more than 100,000 three-dimensional (3D) atomic structures of proteins in the Protein Data Bank (PDB). And often you can get further insight from knowing how the protein moves and wiggles or changes into other structures. This chapter gives a survey of how protein mechanisms are encoded in structures and dynamics; later chapters give more details and describe computational modeling.

In the macroscopic world, you might build your device using a different material tailored for each purpose. You might make a chassis from plastic, an engine from metal, and a guidance system using electronics.

But consider the great advantage of having all the different component functions of your device performed by a single type of material, a material that could be *programmed* for each particular function, in the same way that you would program a computer to perform different mathematical functions. This is the power of proteins.

Different proteins perform different functions. Yet, every protein is made from the same material. A protein's 1D amino acid sequence encodes its 3D structure and its function. In order to have this sequence–structure programmability, not just any polymer would do. For example, you could also encode 1D information in a polyelectrolyte using a specific sequence of positive and negative charges, but repulsions among charges would prevent most polyelectrolytes from folding into compact functional molecules. In contrast, proteins do fold into unique compact structures. As a basis for building cellular functionalities, proteins have three key properties: they can perform different functions; they can selectively bind to each other and to other molecules (for example, DNA/RNA, metabolites, drugs, or lipids); and so protein functions can be regulated and controlled. Different protein functionalities can be hooked together, like LEGO blocks, in a modular way, to build up complex machines. While each module can fold and function by itself, the modules together can act in ways that are coupled and coordinated, like the way different parts of a machine work together. High cooperativity between different modules of complex proteins, or between different domains or subunits in multidomain or multimeric proteins, is essential for achieving biological function. High cooperativity entails efficient coupling between spatially distant structural elements, known as *allostery*, often triggered by ligand binding. Many proteins are allosteric machines regulated by such binding events. The fact that tens of thousands of different and varied actions can all be performed by a single type of molecule, namely, proteins, is a major principle underlying the success of biology. In the following sections, we give a few examples of the many roles in the lives of proteins.

PROTEINS ARE BORN

Ribosomes Manufacture Proteins

DNA carries the information for building an organism, but before proteins can be built, the information must be *transcribed* into modular units of RNA called messenger RNA (mRNA). Then, in a biological process called *translation*, proteins are synthesized on a *ribosome*, a large complex composed of RNA and proteins. The ribosome reads the modular piece of the genetic code from an mRNA and uses it as a template to covalently link one amino acid at a time into a particular polypeptide chain having an amino acid sequence that is encoded within the mRNA. One amino acid is added to a growing polypeptide chain for every three nucleotide bases (called a *triplet* or a *codon*) on the mRNA. Ribosomes are about 20 nm in diameter and composed of about one-third protein and two-thirds RNA. The ribosomes of different organisms contains 50–100 different proteins in two major subunits: a large subunit (called 50S in bacteria, where S refers to a "Svedberg unit," a unit of molecular size based on measurements in centrifugation experiments) and a small subunit (called 30S in bacteria). Figure 2.2 shows the steps

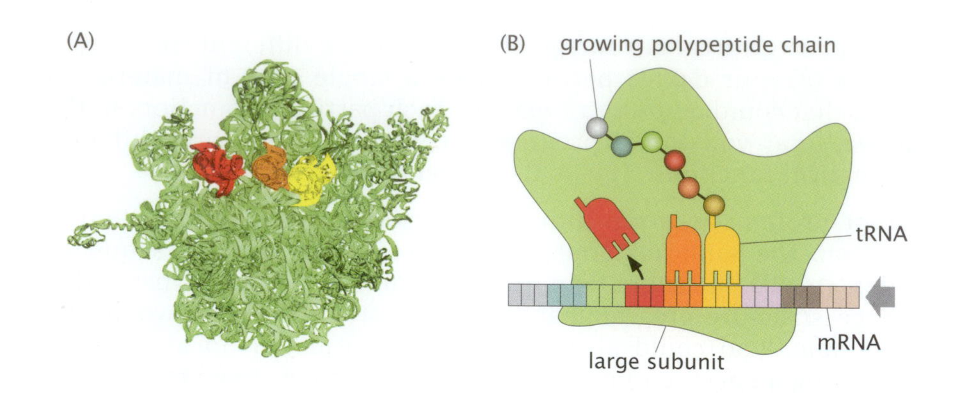

Figure 2.2 Ribosomes are factories that synthesize proteins. (A) Shown on the large subunit of a bacterial ribosome (*green*) are three tRNAs (*red*, *orange*, and *yellow*) where the tRNA molecules sequentially bind to deliver amino acids to the growing peptide chain. (From MM Yusupov, GZ Yusupova, A Baucom, K Lieberman, TN Earnest, JH Cate, and HF Noller. *Crystal structure of the ribosome at 5.5 Å resolution.* Science, 292:883–896, 2001). (B) Each bead is an amino acid (*colored spheres*) on the growing polypeptide chain, color-coded to coincide with the corresponding tRNA (various colored triplets) that delivered it. The tRNAs exit from the third binding site, left side. (Adapted from B Alberts, A Johnson, J Lewis, et al. *Molecular Biology of the Cell*, 6th ed. Garland Science, New York, 2014.)

of protein synthesis on a ribosome. Each step of adding an amino acid to the peptide involves rotation of these two subunits relative to one another. In the first step, an amino acid, which is attached to a corresponding transfer RNA (tRNA) molecule, arrives at the site of the growing polypeptide chain in the ribosome. The tRNA binds to the next upcoming codon on the mRNA, positioning its amino acid adjacent to the previous amino acid. Second, a peptide bond is formed between the new amino acid and the growing end of the polypeptide chain. Third, the empty tRNA that has just discharged its amino acid now detaches from the mRNA. Then, the structure rotates back into position for the next codon on the mRNA. The process repeats until it reaches a stop signal, which is a specific nucleic acid triplet that causes termination of the synthesis. In this way, protein chains are grown by sequential addition of one amino acid at a time.

Molecular Chaperones Are Proteins that Help Other Proteins to Fold Correctly

Chaperones are molecules that help other proteins (called *clients*) to fold correctly. Some chaperones are structured as cages that provide a sort of "private space" inside of which client proteins can fold and avoid aggregation. Chaperones were first identified as *heat-shock proteins*. When cells are subjected to various forms of stress—such as heating—proteins denature. Chaperones are able to capture denatured or misfolded proteins to help them refold while protecting them against aggregation. One of the best-understood chaperones is GroEL, from *E. coli* bacteria, and its partner co-chaperone GroES. GroEL is a complex of 14 subunits, organized in two heptameric rings, shown in Figure 2.3. A typical client protein enters into the GroEL cavity in a denatured or misfolded state. The chaperone helps the protein *to fold* by *unfolding it*. That is, inside the chaperone cavity, the client protein can be pulled apart by sticking to the interior hydrophobic walls of the chaperone. Chaperoning is an active process, in which ATP hydrolysis is required to power chaperone motions. This helps a client protein get unstuck from its current misfolded state, allowing it to try again to fold, either inside the chaperone or outside. Some proteins enter and exit chaperones many times before they fully fold successfully. This pulling-apart mechanism is generic: many different types of protein molecules in many different misfolded states can all be assisted by this single mechanism of a single type of chaperone.

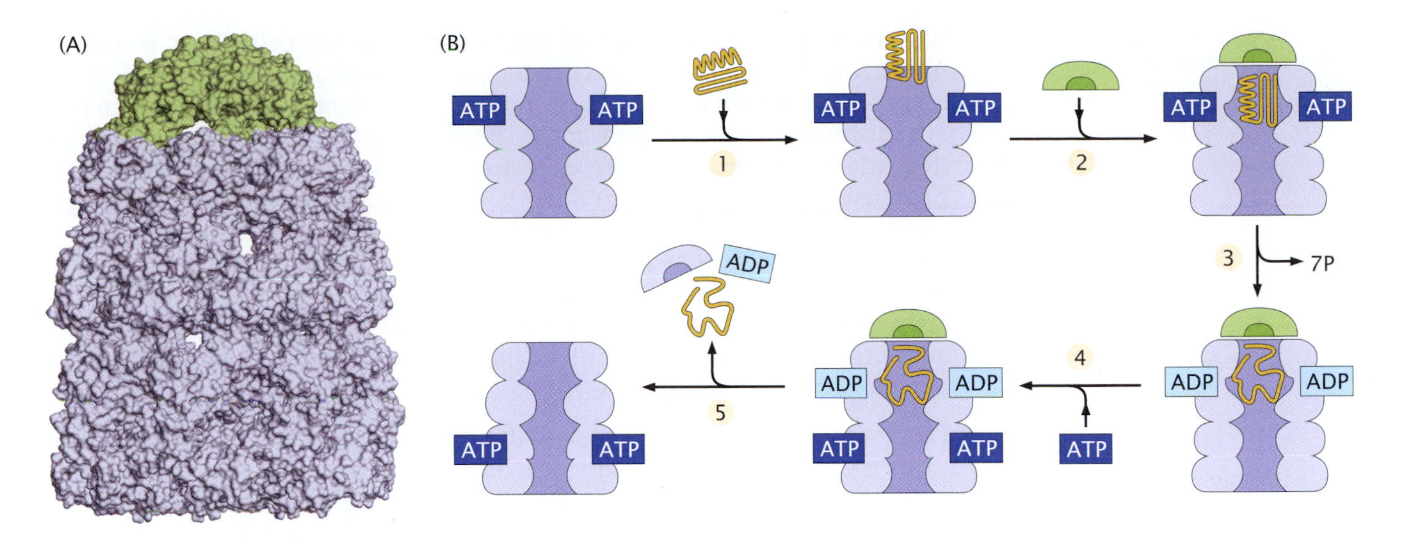

Figure 2.3 The GroEL–GroES chaperone is an ATP-regulated and ATP-powered protein machine that helps a client protein to unfold and refold and to avoid aggregation. (A) A surface view of the GroEL/GroES structure. (B) The cyclic process by which (1) the client protein is a denatured or misfolded state (shown here as misfolded) that enters into the ATP-bound ring of the chaperone; (2) the GroES cap binds; (3) the client misfolded protein can unfold inside, when ATP hydrolyzes to ADP; (4) ATP binds to the other ring that will be active in the next cycle; (5) releasing the unfolded protein from inside the cavity back out into the medium. Sometimes the client folds inside the cavity, and sometimes the client exits from the chaperone first, allowing another attempt to refold outside the chaperone. The steps are repeated if needed, with the active cavity alternating between the two rings, top and bottom (B, adapted from Z Yang, P Májek and I Bahar. *PLoS Comp Biol*, 5: e1000360, 2009).

PROTEINS WORK FOR A LIVING

Some Proteins Are Biochemical Catalysts, Called *Enzymes*

Many proteins act to catalyze biochemical reactions, speeding them up. Such proteins are called enzymes. The following are some examples.

Proteases are enzymes that break down other proteins and peptides

Proteases are enzymes that act like scissors, cutting other "client" proteins into pieces by cleaving the client's covalent backbone peptide bonds. Proteases are used for various cellular functions. First, cells break down the proteins in food to provide the raw materials for new proteins. It may seem wasteful that cells break down food proteins and then synthesize new proteins from scratch. But the proteins in food are not necessarily the right ones or present in the right amounts for what the cell needs at a given time. Second, proteases also degrade proteins that are already inside the cell, namely proteins that are *misfolded*, or oxidized, or damaged or nonfunctional, or detrimental to the cell. And third, proteases serve in the critical steps in some cascades of biochemical reactions. For example, a protease called *thrombin* cleaves a protein called *fibrinogen* to activate the formation of the fibrous protein *fibrin* that heals wounds and clots blood. Proteases play important roles in diseases. For example, activation of caspases, a type of protease, leads to *apoptosis*, also known as *programmed cell death*. In a multicellular organism, the failure of a cell to accomplish its function can trigger an apoptotic response. The cell kills itself in order to avoid harm to the whole organism. Where caspases are defective, such as in cancers, cells fail to kill themselves when they go rogue.

Why don't proteases cleave all the proteins in the cell? That would be disastrous for the cell. Two safeguards are the *regulation* and *specificity* of protease activity. Regulation refers to the fact that some proteases are normally inactive and must become activated before they can degrade other proteins. Here is one way that activation works. The protease is first synthesized with an extra peptide—called the *pro-piece*—covalently attached at its beginning. The pro-piece acts like a sheath on a knife that prevents the protease from clipping other proteins. Only when one protease molecule, call it *A*, comes along and cleaves off the *pro-piece* of another protease *B* does protease *B* become activated. In essence, one protease removes the knife's sheath from another protease to activate it.

Proteases are also specific, cleaving only at particular sequence positions of proteins. For example, *chymotrypsin* is a protease that cleaves only the peptide bond immediately following a hydrophobic amino acid. And *trypsin* cleaves only the peptide bond following a basic amino acid.

Among the best understood enzymes are the *serine proteases*. Serine proteases cleave peptide bonds using a particular spatial arrangement of three residues, called the *catalytic triad*, composed of serine, aspartatic acid, and histidine, which act in concert (Figure 2.4). The serine serves in the active site as a *nucleophile*, that is, a donor of a pair of electrons, to hydrolyze the peptide bond.

Some proteins, called *cellulases*, break down cellulose

Cellulose is a polymer of glucose monomers. Polymers that are made up of repeating sugars, such as glucose, are called *polysaccharides*. Cellulose is among the most abundant polymers on earth, making up much of the biomass of plants and trees. It gives structure to the cell walls of plants, and to wood and cotton.

The ability to break down the *glycosidic linkage* that connects one glucose monomer to the next in the cellulosic chain would have enormous commercial value for obtaining biofuels from abundant biomass. So, there have been extensive efforts to invent nonbiological catalysts that can break this linkage. However, developing such catalysts has been difficult. Nature's catalysts are *cellulase* enzymes, proteins that cleave the glycosidic bond. Cellulases are found mostly in microorganisms, such as bacteria and fungi, and in termites.

Figure 2.5 shows the mechanism by which cellulase breaks down cellulose. Cellulose chains are packed in parallel arrays, like bundles of dry spaghetti that are all stretched out and aligned together in the same direction. Cellulase converts energy into an action that pulls itself along the bundle of cellulose chains, while a small rigid "scooper" region of the enzyme digs under each cellulose molecule, peeling it away from the bundle. Coupled to the mechanical scooping process, the cellulose active site hydrolyzes the glycosidic linkage of that cellulose chain, breaking down the cellulose into its component sugars.

Some enzymes catalyze the synthesis of biomolecules

Synthesizing a new biomolecule inside a cell usually entails multiple chemical steps. Think about the efficiencies that could be gained if those different chemical steps were all close to each other spatially. In

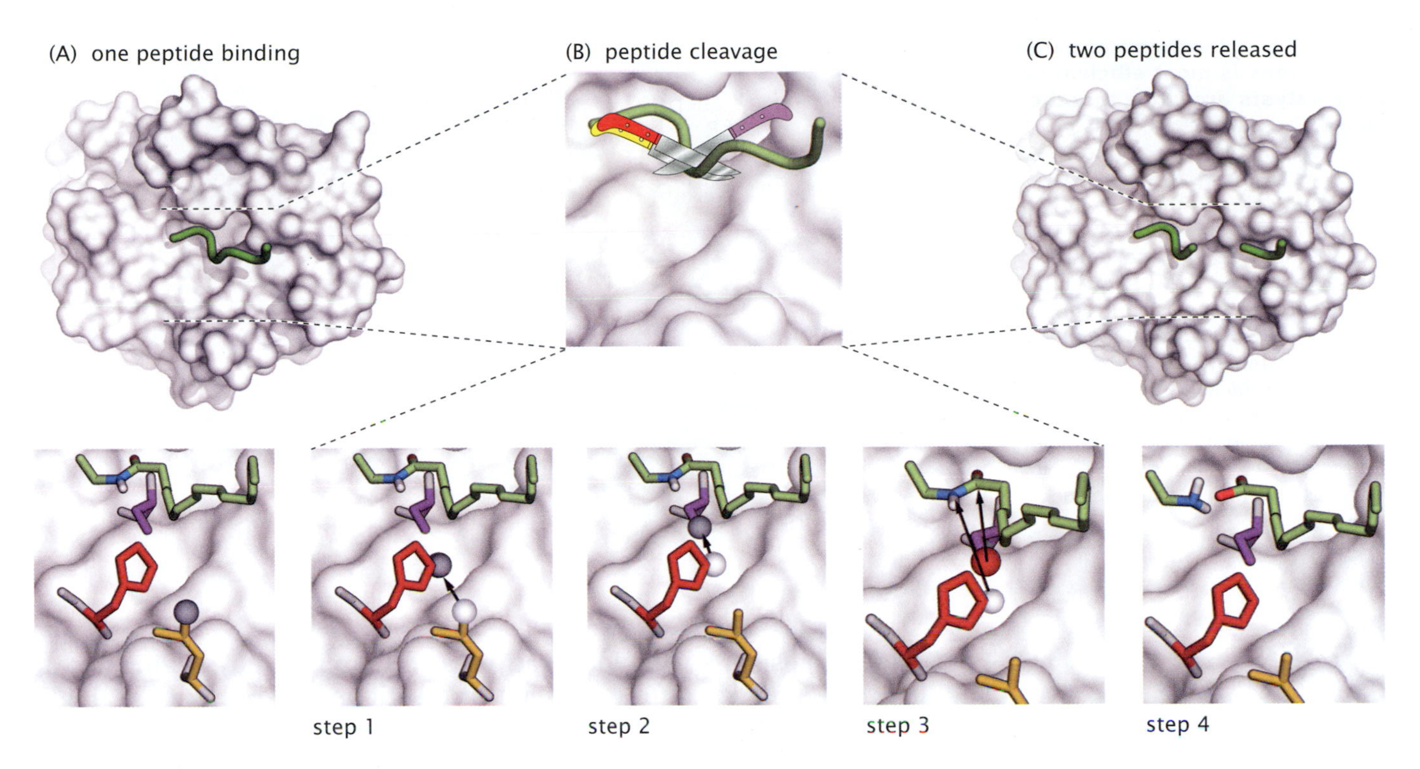

Figure 2.4 In serine proteases, the catalytic triad of amino acids cuts peptide bonds. (A) The peptide to be cleaved is shown in *green*, with the serine protease as a *white* surface; (B) Peptide cleavage, shown as a succession of four steps (bottom). The leftmost diagram shows the positioning of three catalytic residues in the active site, which provides the chemistry to cut the polypeptide (*green*, stick representation; with the nitrogen at its cleavage site shown in *blue*) by a relay passing protons from the active site amino acids to the nitrogen of the peptide bond. (The position of the proton shown as a *gray* sphere has been transferred from the position where it is shown as a *white* sphere.) This nitrogen can have only three bonds, and so it breaks the peptide bond, replacing that bond with the bond with the incoming proton. In trypsin, the biochemical cleavage is catalyzed by S195 (*purple*) and H57 (*red*), while D102 (*yellow*) supports H57 and stabilizes some intermediates. Steps 1–3 show how the three catalytic residues work together to transfer a proton to attach to the peptide bond nitrogen to cleave the bound peptide (*green*). Step 4 and panel C display the cleaved peptide. The substrate protein chain is *green*, except that the peptide bond nitrogen is *blue* and the oxygen is *red*. The *red* sphere is an oxygen, and the *white* and *gray* spheres are hydrogens. The first steps are transfer of the *gray* proton from Asp to His to Ser, and then the His proton attaches to the NH to make NH_2, and the Ser oxygen (*red*) attaches to the carbonyl carbon to yield the carboxyl group and the amino group. A water is used up to restore the amino acids His and Ser. The protease provides the platform for performing this "surgery" with high precision.

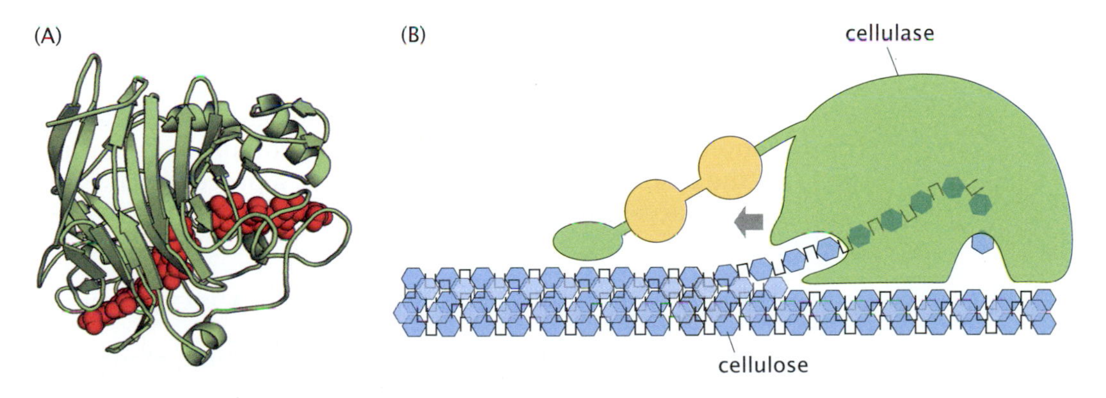

Figure 2.5 Breaking down cellulose involves coupled chemical and mechanical actions. The cellulase enzyme catalyzes the hydrolysis of the glycosidic bond that joins pairs of glucose monomers into polymeric cellulose. (A) The enzyme structure, with the cellulose chain shown as *red* spheres, and the one cut off dropping out the bottom. (B) Schema of the complete structure of the cellulase with a leading-edge carbohydrate-binding module (*green* blob) and polysaccharide (*yellow*) and peptide linkers that help to guide the cellulase. This enzyme's action is accomplished by scooping the glucose polymer up into the active site. The scooping mechanism is reminiscent of the action of a hand plane. (Adapted from http://www.nrel.gov/continuum/deliberate_science/biofuels.html.)

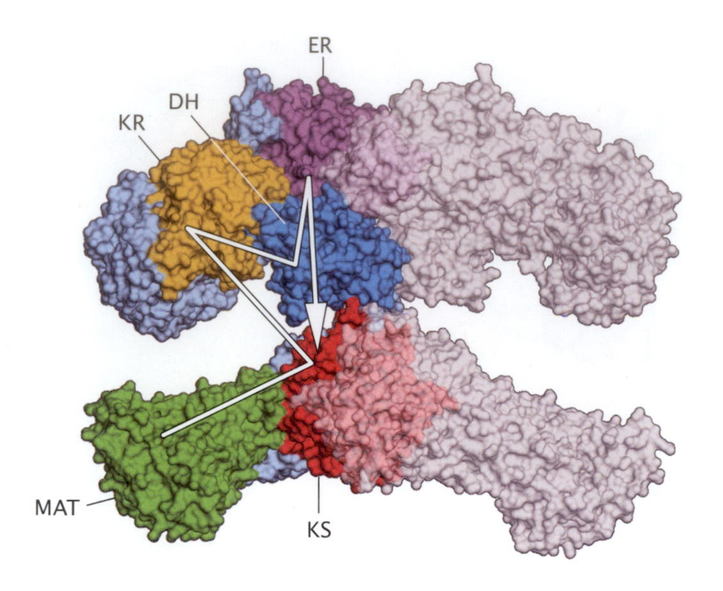

Figure 2.6 A biochemical pathway of reactions is most efficient when the catalysts are close together in space. Fatty acid synthase is a localized complex of enzymes. The *white* arrow shows five successive reactions in the synthesis of fatty acids to add two carbons to the hydrocarbon chain. All the reactions take place on the same protein, fatty acid synthase, as the substrate is passed from one domain to the next: from malonyl/acetyl transferase (MAT, *green*) to ketoacyl synthase (KS, *red*) to ketoacyl reductase (KR, *yellow*) to dehydratase (DH, *blue*) to enoyl reductase (ER, *purple*). *Gray* indicates the second identical monomer of fatty acid synthase.

some cases, such spatial localization is accomplished by having multiple active sites for different chemical reactions all located on a single enzyme molecule. Each of the active sites on the enzyme catalyzes a different step in the reaction. For example, fatty acid synthase, a dimeric enzyme that adds two carbons to lipid chains, catalyzes a sequence of reactions on its different domains (Figure 2.6). In animals, fatty acid synthase is a dimer, with all enzyme domains (each colored differently and labeled in the figure) required for fatty acid synthesis in each monomer.

Some proteins transcribe or reconfigure DNA molecules

DNA and *RNA polymerases* are enzymes that add nucleotides to growing DNA or RNA chains. RNA polymerases transcribe a DNA sequence into an RNA sequence. *Reverse transcriptase* performs the reverse operation, copying an RNA molecule to make a DNA molecule (Figure 2.7). For example, HIV reverse transcriptase (RT) has two coordinated active sites: a polymerase site that copies the viral RNA, and a second site on the RNase H domain of the protein that cuts one nucleotide at a time from the copied RNA strand. *Helicases* are energy-driven motor proteins that pull themselves along the chain of a double-stranded DNA or RNA, winding or unwinding it, thus providing access to other proteins that need to read or repair the nucleic acid chain. Figure 2.8 shows the action of *topoisomerases*, which are proteins that control the winding of the double-helical structure of DNA, a process that is needed in order to give access to proteins that make copies of DNA.

Some Proteins, Called *Motors*, Convert Energy into Motion

Some motor proteins move in a *linear* fashion, along tracks

Walking is one of the quirkiest and most recognizable ways to convert energy to motion. You lift your back left leg, swing it forward, set it down, lift your back right leg, and repeat the process. Figure 2.9 shows the protein version of this process. Kinesin is a protein that has two feet. Whenever an ATP molecule binds to kinesin, the back foot of kinesin is released from the track, driven forward, and then rebinds to

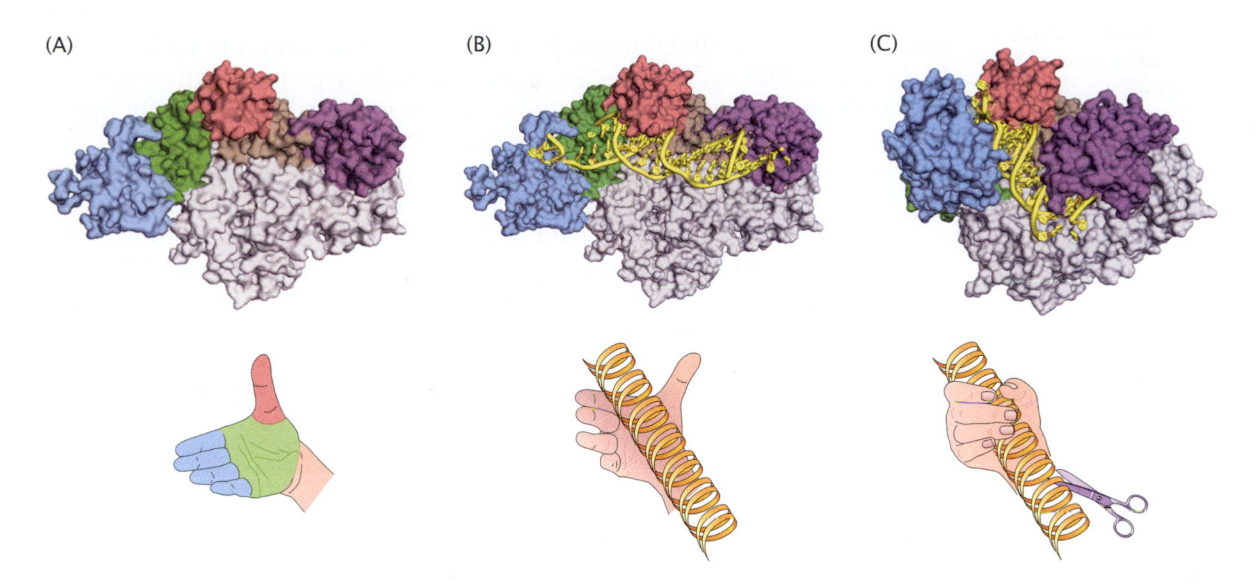

Figure 2.7 HIV reverse transcriptase (RT) copies HIV's single-stranded RNA onto single-stranded DNA, for integration into the host's genome. RT has two allosterically coupled catalytic sites: one elongates the DNA, and an RNase H breaks down the RNA. (Bottom row): The action of the DNA polymerase domain resembles a hand that pulls: fingers (*blue*), palm (*green*), thumb (*red*), connection (*light brown*). The RNase H "scissors" domain is shown in *purple*. The structural scaffold is shown in *gray*. The RT enzyme moves one base at a time along the viral RNA, to copy it onto its complement DNA, and to cut the original chain by removing one nucleotide at a time. (A, *top*) The empty enzyme. (B, *top*) The bound polynucleotide chain (*yellow*) near the polymerase site with the double helix formed from the original polynucleotide and the newly synthesized chain. (C, *top*) A closed form where the polymerase is active. Then, the chain is pulled forward by a central hinge motion in the structure.

the track, stepping down in front of the other foot. In this case, form follows function in the molecular world the same way it does in the macroscopic world of legs and feet. For historical reasons, kinesin's feet are called 'heads'.

In order for a protein to walk, it needs a track to give it traction and direction. The tracks are themselves other protein molecules. For example, the kinesin motor protein walks along a train-track-like protein quaternary complex called a *microtubule*. Microtubules are

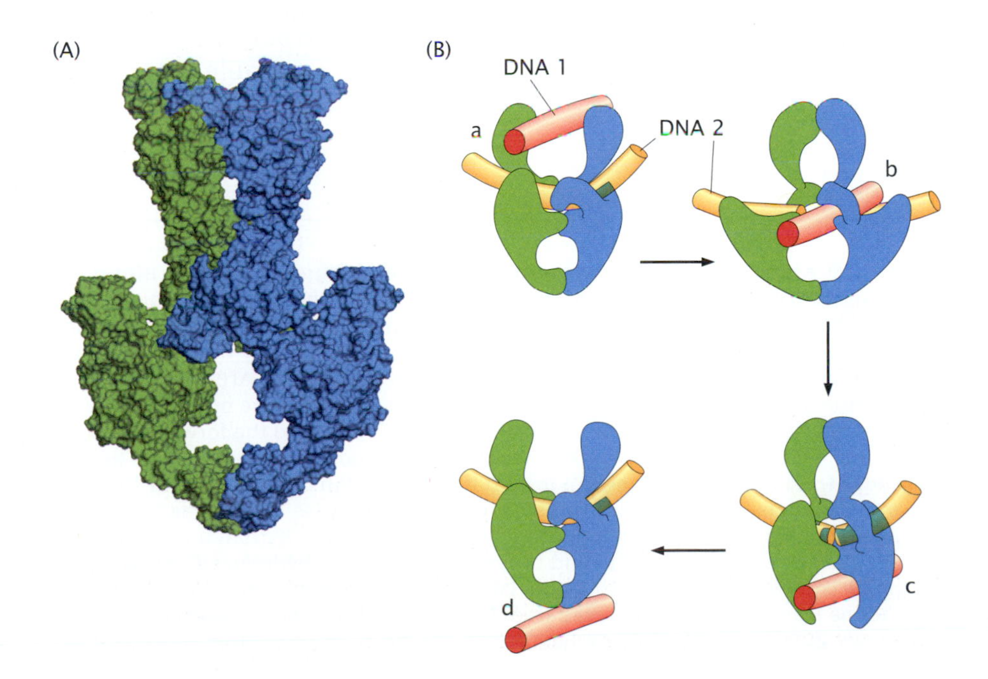

Figure 2.8 Topoisomerase II acts to "pass one DNA double helix through another double helix." (A) Structure of yeast topoisomerase II. (B) (a) DNA double helix number 1 enters at the top of the protein, while DNA helix number 2 is already bound. (b) Upon ATP hydrolysis, helix 2 is cut. (c) Helix 1 passes through the break in helix 2. (d) After passing across helix 2, DNA helix 1 is now released from the protein, and the break in helix 2 is re-ligated. The action is essential for preventing entanglements when winding or unwinding long DNA molecules.

multistranded fibrils composed of heterodimeric proteins called *tubulin* (Figure 2.10). As another example, the motor protein called *myosin* walks along tracks of *actin* proteins. The track fibrils are usually made up of multiple protein chains. To avoid traffic jams, tubulin subunits of these tracks assemble and disassemble when they are needed, based on conditions in the cell. Metaphorically, this is like roads that would appear and disappear whenever and wherever they are needed. The growth and shrinkage of microtubulin filaments is controlled by the concentration of free tubulin dimers and by the binding and hydrolysis of GTP. The fiber has a plus-end where new tubulin dimers are added, and a minus-end where dimers dissociate after hydrolysis of GTP to GDP. So, if there is plenty of free tubulin and GTP, the track will grow faster than it shrinks.

Some motor proteins convert energy to directed *rotary* motion

Other proteins are *rotary* motors. They spin, like a microscopic version of a car engine or an electric clock. Like macroscopic motors, molecular rotary motors have a stationary outside casing. The stationary part is embedded within a lipid bilayer membrane. Rotary motors also have an internal part that rotates, like the shaft of an electric motor. The rotor spins in a particular direction when the motor protein is supplied with some form of energy. Figure 2.11 shows a molecular rotary engine called F_0F_1 *ATP synthase*, more broadly called an *ATPase*. A pH gradient across the cell membrane results in a flow of protons down a proton-concentration gradient, through the motor protein, leading to turning of the rotor, which drives the synthesis of ATP from ADP and phosphate.

In the macroscopic world, some machines are reversible. For example, applying an electrical voltage drives the turning of a motor, but, conversely, if you turn the shaft of an electric motor, it creates a voltage. Some protein machines are reversible too. The F_0F_1 ATPase is reversible. When the cytoplasmic ratio of concentrations of ATP/ADP is high and the ATP hydrolysis reaction (which releases energy) is favored, the enzyme uses this chemical energy to pump protons across the membrane to create pH gradients. Or, when ATP is low, protons flow downhill, causing the protein to synthesize ATP from ADP, replenishing the cell's ATP as if the ATP supply were a rechargeable battery (see Figure 2.11).

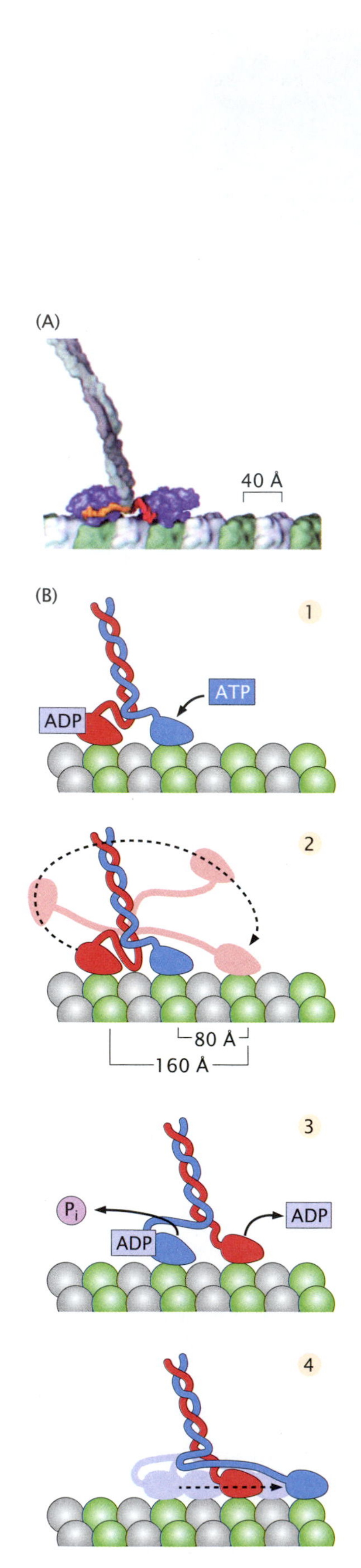

Figure 2.9 Kinesin is a motor protein that carries cargo by walking along a microtubule track. (A) The coordinated motions of the two heads of kinesin resemble walking. The head domains (*blue*) are bound to tubulin subunits as kinesin walks along a microtubulin track (*green/white*). A coiled coil (*gray*) extends away from the track to the cargo carried by the kinesin motor. (B) This motion is driven by ATP binding and hydrolysis in a cycle of four steps: (1) Both kinesin head domains are bound to tubulin. The left one also binds ADP. Motion is initiated when ATP binds to the forward head, causing the neck-linker (*blue*) to dock to tubulin. (2) The docking motion drives the other head (*red*) forward (on average) toward its next tubulin binding site. (3) This forward head then binds to the microtubule, causing a stepwise displacement of 80 Å along the filament. The binding facilitates the release of ADP from the forward head. At the same time, the ATP bound to the lagging head is hydrolyzed, releasing phosphate (P_i). (4) This is repeated by the lagging (*blue*) head moving forward after ATP binds to the leading head. (A, adapted from RD Vale and RA Milligan. *Science*, 288:88–95, 2000; illustration by Graham Johnson. With permission from AAAS.)

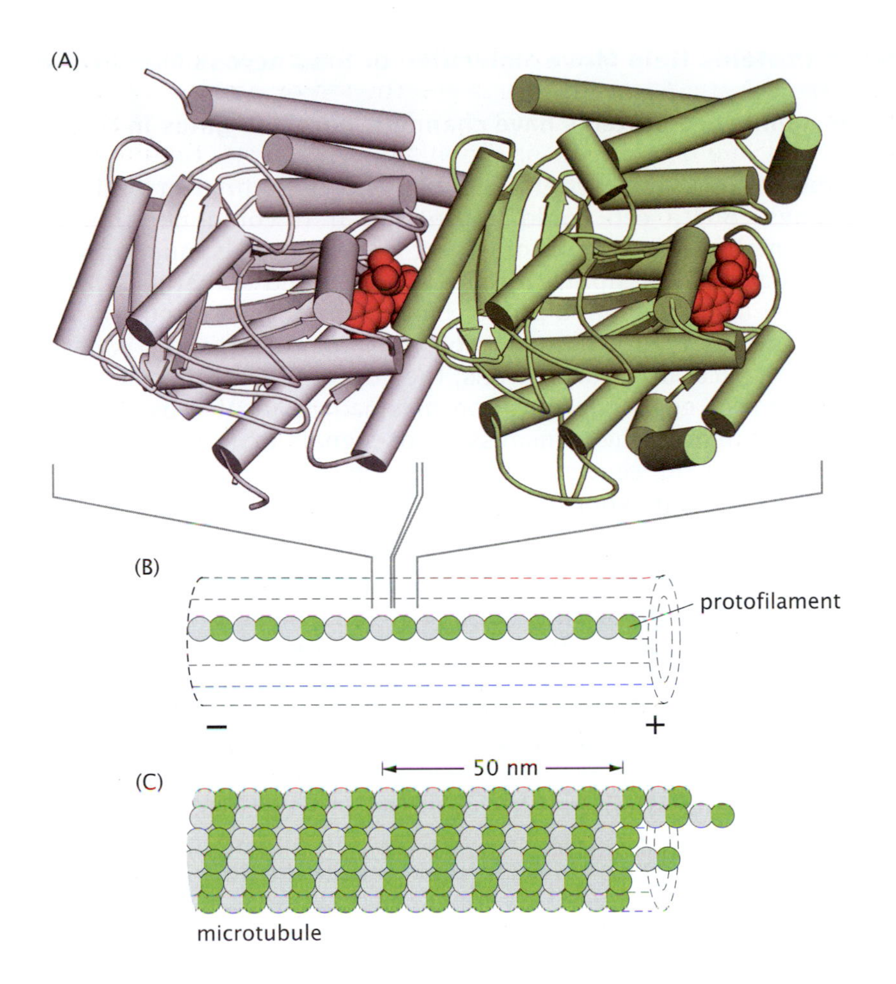

Figure 2.10 The structure of the tubulin protein and the microtubule track that it makes. (A) Tubulin heterodimers bound to GTP (*red*) assemble into (B) protofilaments from the minus to the plus end. (C) Protofilaments assemble in stiff cylindrical microtubules. (Adapted from B Alberts, A Johnson, J Lewis, et al. *Molecular Biology of the Cell*, 6th ed. Garland Science, New York, 2014.)

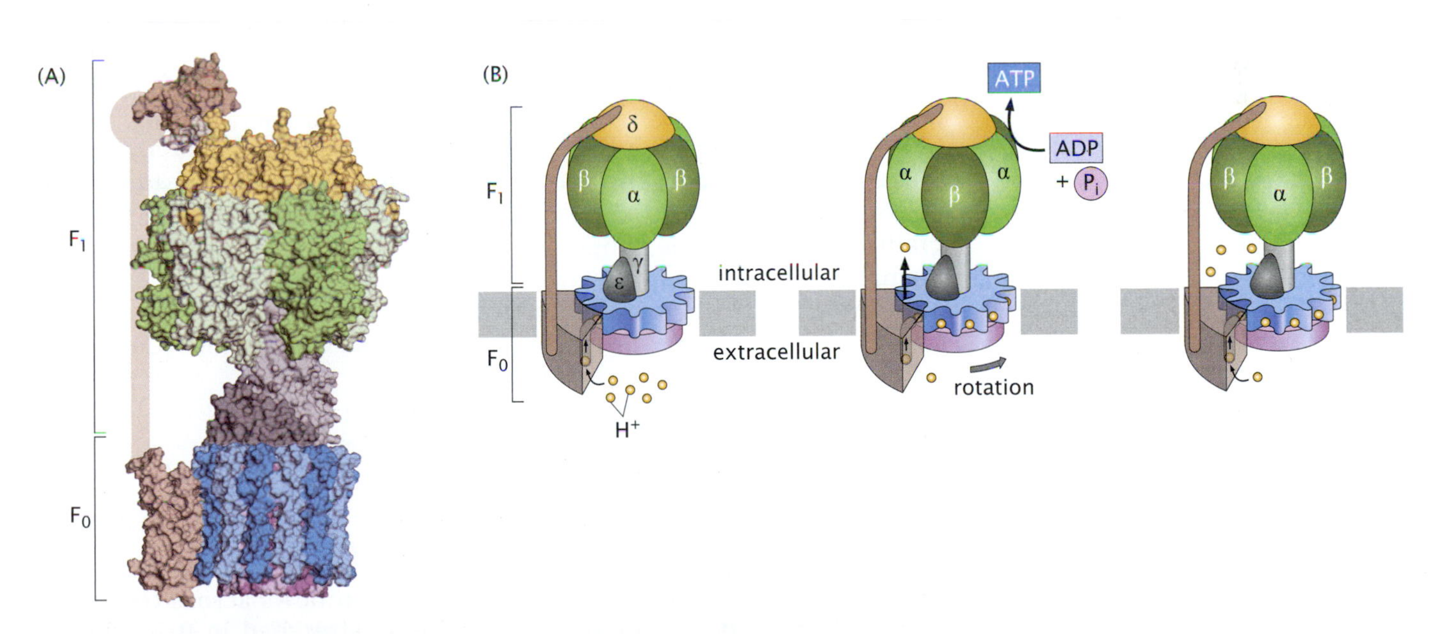

Figure 2.11 F_0F_1 ATPase is a rotary motor protein that synthesizes ATP, driven by a proton gradient. (A) Side view of the motor protein. (B) A lower pH outside the cell than inside means a higher proton concentration outside. Protons flow down their concentration gradient from outside (extracellular region) to inside (intracellular), through the motor. This proton flow causes the protein to rotate, in turn causing the protein to catalyze the conversion of ADP to ATP. Three molecules of ATP are synthesized per rotation cycle, corresponding to the three α/β dimers in the ATPase ring. (Adapted from G Oster and H Wang. *Biochim Biophysi Acta*, 1458:482–510, 2000. With permission from Elsevier.)

Some Proteins Help Move Molecules or Ions across Membranes

Some membrane proteins have channels, pores, or gates in them

Cell membranes are barriers between the cell's interior and its surroundings. But a cell couldn't function if it could not import the nutrient molecules it needs, or export its waste. Cells must import and export sugars, ions, lipids, drug molecules, and sometimes proteins. Cells must also maintain a proper balance by being selectively permeable to various ions and small molecules. Such transport is essential for the electrical firing of neurons, the contraction of muscle cells, the fertilization of eggs, the filtration of small molecules by the kidney, cellular regulation, the transmission of signals or transport of neurotransmitters in the central nervous system synapses, and the routine uptake of food and export of wastes.

Membrane proteins are proteins that reside in the lipid bilayer membrane that encases a cell. Some membrane proteins select and regulate molecular traffic into or out of cells. For some membrane proteins, it is easy to interpret the function from the structure. Those proteins have a pore through the center that acts as a conduit through the membrane. Others control transport more actively. Some membrane proteins use the energy provided by the flow of ions down their electrochemical gradient across the membrane (for example, Na^+ ions from the cell exterior to the interior) to drive movement of their cargo/substrate (for example, neurotransmitters such as glutamate and dopamine) against their concentration gradient. Those proteins couple the downhill flow of ions to uphill translocation of their substrates. These are called secondary transporters because the co-transport of cations drives the substrate transport. In contrast, primary transporters use ATP hydrolysis energy to drive the transport process.

Ion transport processes typically satisfy various criteria. They are *selective*, meaning that some ions flow in or out while others do not; *directional*, meaning that ions flow in one direction but not the other; and *regulated*, meaning that ion fluxes can be modulated by voltage differences across the membrane, pH, temperature, external forces or pressures, or binding to ligands. Ion pumps actively transport ions against their concentration gradient. Ion channels are selectively permeable to ions that diffuse passively down their electrochemical gradient. For example, potassium channels allow K^+ ion flow across membranes down their electrochemical gradient (Figure 2.12). Their selectivity filter is a spatial arrangement of amino acids that acts as a sort of chemical "sieve" to pass potassium and block other ions, even other small positively charged ions.

Mechanosensitive (MS) ion channels respond to the physical environment

Some membrane channel proteins are *mechanosensors*: they sense mechanical forces and increase or decrease ion flows across the membrane accordingly. If a bilayer membrane is stretched in its plane by the application of external forces or pressure, it can result in the opening or closing of a membrane-channel protein, leading to a change in the flow of ions across the membrane. Examples of mechanosensor membrane proteins are MscL (large-conductance mechanosensitive channel) or MscS (small-conductance mechanosensitive channel).

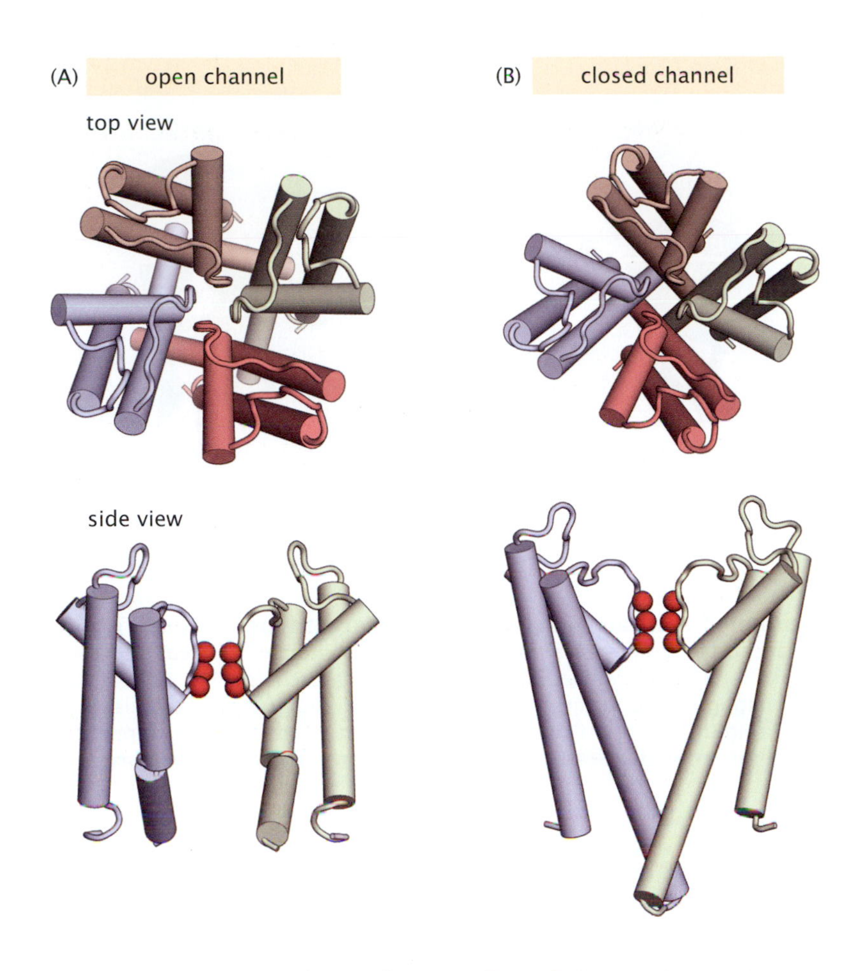

Figure 2.12 KcsA is a potassium channel whose center opens and closes to allow potassium flow into cells. KcsA is a tetramer (each subunit is shown here in a different color) that conducts only potassium ions from the outside to the inside of a cell. The transmembrane helices of KcsA are shown in the open (A) and closed (B) states. The top row shows the open and closed states of the tetramer as seen from the top looking down through the pore. The bottom row shows side views of just two monomers, indicating the *red* carbonyls that act as the "selectivity filter" to allow K^+ ions to pass through and block other ions. (For more details, see [3].)

Mechanosensitive ion channels are found in various tissues and organisms. They play a role in touch, hearing, balance, and pain. Mechanosensor proteins also serve as osmotic pressure "release valves" when the salt concentrations inside cells get too high, bloating the cell, stretching the membrane, and causing the mechanosensor channels to open up to ion flux and reduce the osmotic pressure.

ATP binding cassette (ABC) proteins transport small molecules and ions through membranes

There are thousands of different types of molecules inside and outside the cell, from small to large, from charged to nonpolar. Molecular transport into and out of cells must be selective—the cell needs to retain certain molecules and expel others. Imagine the challenge of designing such a highly specific filter. It must be choosy. And, not all the "cargo" molecules that should be transported can be anticipated in advance (such as toxins that the cell has not seen before).

It would be impossible to have a different type of receptor for every type of cargo. ABC transporters are membrane proteins that import different types of good molecules to the interiors of cells, without opening the floodgates for bad molecules. These proteins import a wide range of molecules, such as sugars, vitamins, or metals. ABC transporters also export hydrophobic molecules, such as lipids, toxins, or drugs. Some pharmaceuticals are designed to enter into pathogenic bacteria and kill them. But the ABC transporters of those bacteria can expel the drugs, limiting their effectiveness.

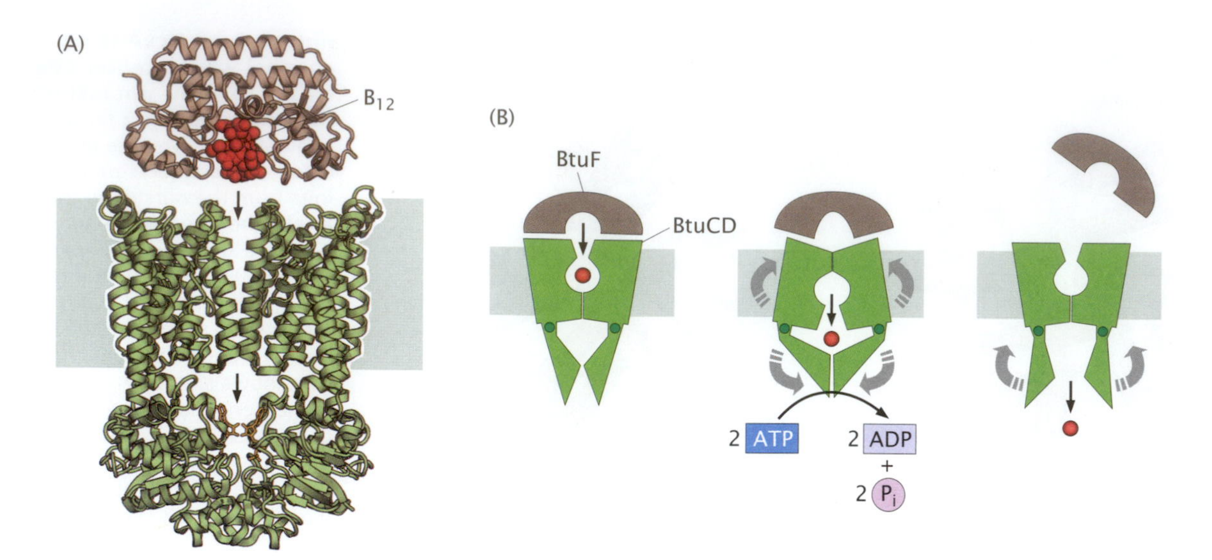

Figure 2.13 ABC transporters are membrane proteins that can transport many different types of small molecules across cell membranes. (A) Side view of an ABC transporter (*green*). This transporter imports vitamin B_{12} (*red*) into the cell. (B) This transporter is versatile. It can bind to various different docking-module proteins (*brown*). Each docking module itself recognizes and binds to particular small molecules to import or export them. Once docked, the cargo (small molecule) is inserted into the transporter by an active ATP-driven process. Using this energy, a conformational change opens the pore to release the ligand inside the cell, then the transporter releases the carrier module. (Adapted from KP Locher. *Curr Opin Struct Biol*, 14:526–531, 2004. With permission from Elsevier.)

Figure 2.13 shows how the vitamin B_{12} transporter of *E. coli* works. The transport mechanism has two components. First, the BtuF protein is an independent "mobile vehicle" that captures and binds vitamin B_{12} outside the cell. BtuF hauls its cargo to the cell membrane, where it attaches to the docking-module membrane protein BtuCD. After docking, BtuCD shuttles the vitamin B_{12} across the membrane and releases it inside the cell. The process is powered by a nucleotide-binding domain that provides the energy source.

Some protein complexes, called *translocons*, haul entire proteins across membranes

Sometimes, cells must import or export large molecules, such as whole proteins, across membranes. This job is performed either by protein complexes that are called translocases (in the mitochondrial membrane) or translocons (in the endoplasmic reticulum (ER) membrane). These proteins can haul nascent polypeptides across membranes while the peptides are being synthesized on the ribosome (*co-translational translocation*). Or, translocons can haul whole proteins after synthesis (*post-translational translocation*). If you were designing such a channel, you would immediately face two challenges. First, the native proteins to be transported come in various sizes and shapes. It would be most efficient if the many different types of cargo proteins—all having different sizes and shapes—could be hauled across the membrane by a single type of translocase. Second, transporting a large molecule, such as a protein, must be done without allowing small molecules to leak in or out at the same time.

The mechanism of co-translational translocation is as follows. The ribosome inside the cell binds to the translocon in the ER membrane, leading to export of the growing peptide chain through the membrane to inside the ER. The translocon thus captures the protein molecule

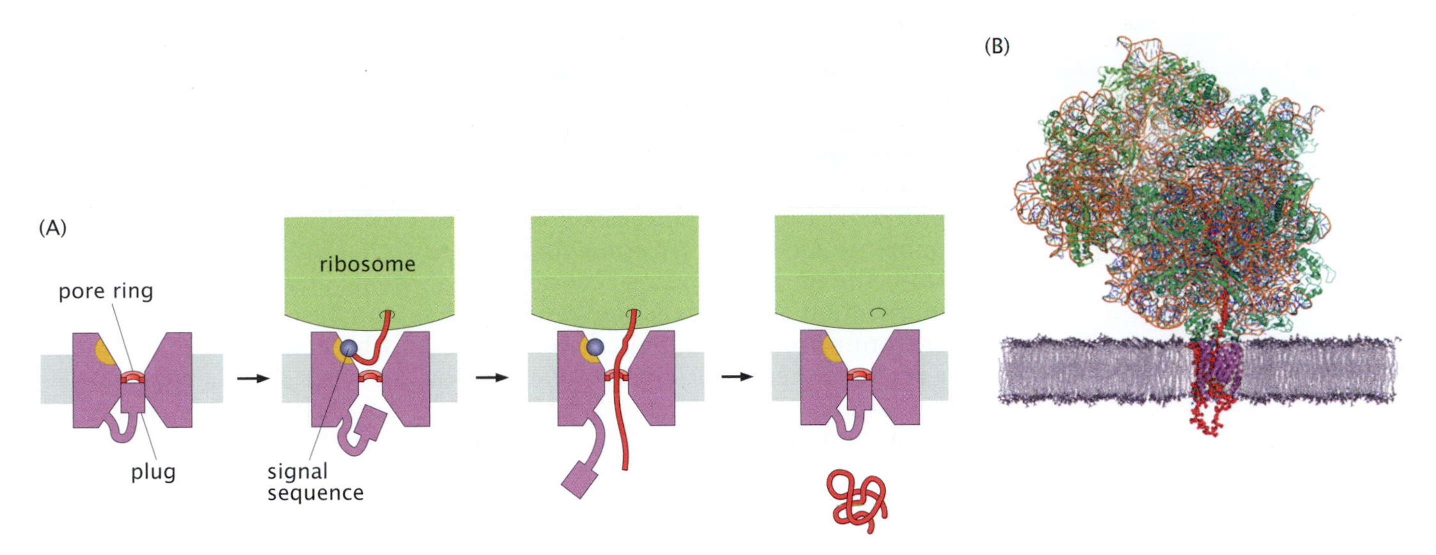

Figure 2.14 Translocation channels are protein complexes that can transport large molecules, such as proteins, across membranes. (A) From left to right: The closed (plugged) form of the core component (sec61 in humans) of the translocon. The protein that is being synthesized on the ribosome enters the sec61 complex. The plug is removed, allowing the protein through. Once the protein has crossed the membrane, sec61 becomes re-plugged to prevent other molecules from passing through. (B) Molecular modeling and cryo-EM data [4] show the ribosome (ribosomal RNA in *orange* and ribosomal proteins in *green*) docked onto the sec61 (*purple*) in a lipid bilayer with a newly synthesized protein chain (*red*) passing through. (B, adapted from J Frauenfeld, J Gumbart, EO Sluis, et al. *Nat Struct Mol Biol*, 18:614–21, 2011. With permission from Macmillan Publishers Ltd.)

as it is being synthesized before it has folded up. The translocon has a narrow conduit through which the unfolded protein chain snakes across the membrane before it is folded. This "spaghetti slurping" process explains how a single type of translocon can translocate a wide variety of proteins, no matter the sizes or shapes of their native structures. And, in order to prevent unwanted leakage of smaller molecules, the translocon pore is stoppered by a removable molecular plug, like the plug on a bathtub drain. Figure 2.14 shows the translocation process, which is coupled to protein synthesis: (1) a protein is synthesized on a ribosome; (2) the growing protein chain exits a tunnel into the translocon; (3) the protein is exported across the membrane by the translocon.

Some Proteins Have Signaling and Regulatory Actions

What have been described so far are mostly chemical and mechanical actions of proteins. However, proteins also serve in other important actions, namely, to regulate, control, receive, transduce, and send signals that can coordinate the biochemical, biophysical, and genetic actions of cells and organisms.

Cellular biochemistry is coordinated by regulatory networks of proteins

A cell's internal biochemical and genetic activities must be coordinated, regulated, and controlled. When a cell has too much or too little of some molecule, the cell must increase or decrease the rates of synthesis or degradation of this molecule by positive or negative feedback mechanisms. And, a cell must alter its behaviors in response to stresses, to changes in environment, changes in amounts or types of food, or to attacks by pathogens, for example. The balancing of biological activities is called *homeostasis*.

Cells have interconnected *networks* of proteins that link together biochemical pathways. There are three general classes of pathways: *metabolic pathways* for converting food into energy; *gene-regulation pathways* for producing proteins and for controlling growth and development; and *signal transduction pathways* that help trigger responses from the immune and central nervous systems and maintain the communication across cells and within cells. Many biochemical pathways are collections of chemical reactions in which the product of one enzymatic reaction is the substrate for the next. Biochemical *cycles* are chains of reactions in which the first and last steps coincide. An example is the Krebs cycle (also called the citric acid cycle), a cyclic metabolic pathway that converts carbohydrates, fats, and proteins into the cell's energy, with carbon dioxide and water as the waste products. In these networks, the action of one protein can affect the actions of other proteins. In signaling *cascades*, a messenger molecule binds to a receptor, causing the activation of other proteins, which can cause the activation of still additional proteins, leading to amplification of the signal.

Signal transduction proteins control the flow of information

Signalling entails passing information from one molecule to another. Imagine a biological action that requires protein *A* bound to protein *B*. Here's one mechanism. First, phosphorylate a tyrosine on protein *B*. Second, attach an SH2 domain to protein *A*. An *SH2 domain* is a small modular conserved peptide of about 100 residues. SH2 acts as a sort of "flashlight" for the two proteins to find each other, because SH2 provides an affinity of its protein *A* for phosphorylated protein *B*.

SH2 can attach to many different proteins. For example, it is found in more than 100 different human proteins. Here is how it is used in signaling. A membrane protein receptor "senses" a small-molecule signal in the extracellular compartment, causing the phosphorylation of a tyrosine in the intracellular compartment. This tyrosine phosphorylation then activates a cascade of protein–protein interactions inside the cell whereby SH2-domain-containing proteins are recruited to tyrosine-phosphorylated sites. Such sequences of events can lead to altered patterns of gene expression or other cellular responses.

G-protein-coupled receptors help transmit information from outside to inside the cell

G-protein-coupled receptors (GPCRs) are membrane proteins. They have a seven-transmembrane (TM)-helix structure. GPCRs are *signaling* molecules: a small molecule from outside the cell binds to a GPCR, causing a conformational change through the membrane and sending a signal deeper into the cell. This signaling process often entails two *messengers*. One messenger is the small signaling molecule from outside the cell that activates the GPCR protein, causing a conformational change in the GPCR that propagates all the way through the membrane to the α-subunit of the G protein bound at the intracellular surface of the GPCR. This prompts the exchange of its GDP for GTP, which drives the dissociation of the three subunits of the G protein. The dissociated α-subunit may initiate several signaling pathways, each involving

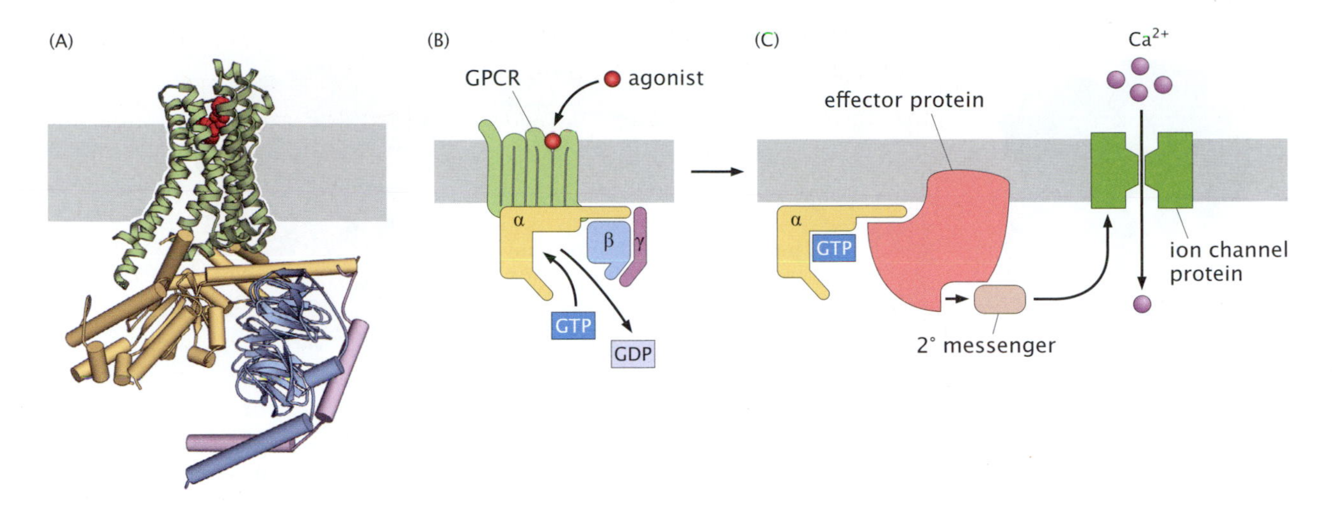

Figure 2.15 A GPCR binding event leads to calcium influx into the cell. (A) Molecular structure of a GPCR (*green*) in the presence of an agonist (*red*), complexed with the G protein heterotrimer (α, β, and γ subunits, colored *yellow*, *blue*, and *purple*, respectively). (B) Cartoon of a signaling cascade: A hormone (agonist, *red*) binds to the extracellular region of the GPCR (*green*) and triggers a conformational change in the GPCR. The change propagates to a G protein on the cytosolic side. The GPCR helps the G protein α-domain (*yellow*) exchange GDP for GTP, allowing its inhibitory $\beta\gamma$-subunit complex (*blue* and *purple*) to dissociate. (C) Gα can then bind and activate a cascade of signaling events that can open the calcium channels to promote the influx of Ca^{2+} into the cell, for example, or the influx of Ca^{2+} from the ER (not shown). (A, adapted from SG Rasmussen, BT DeVree, Y Zou, et al. *Nature*, 477:549–555, 2011. With permission from Macmillan Publishers Ltd.)

different types of second messengers (such as cyclic AMP or inositol triphosphate) (Figure 2.15). GPCRs are among the most important targets for drug discovery because of their roles in maladies including heart disease, asthma, color blindness, and metabolic and growth diseases.

Regulation and control are often performed in biology by *coupled binding*. Suppose a ligand *A* binds to a protein *P*, causing some biological action. Now, if another ligand *B* can also bind to the protein *P*, changing the action of the bound complex *PA*, you have the basis for regulation and transduction in biological systems. Examples are given below.

A protein can be regulated by binding a small molecule

A widely studied example of biological regulation occurs in *hemoglobin*, the protein inside red blood cells that binds and transports oxygen molecules (O_2) throughout the bloodstream. Hemoglobin (Hb) has four subunits, each of which has a heme group that binds an oxygen molecule. Hemes are examples of additions to proteins called *prosthetic groups*. Prosthetic groups are often small molecules that attach to proteins and assist in the protein's mechanism of action in some way. Hemoglobin binds four different oxygen molecules cooperatively. That is, the second O_2 molecule binds to Hb more easily than the first O_2 does, for example (Figure 2.16). Moreover, the binding affinities between O_2 and Hb are regulated by the pH of the surrounding solution. Because of differences in pH in different parts of the body, hemoglobin takes up oxygen where the oxygen concentration is high, and releases oxygen where the oxygen concentration is lower. This is one way that oxygen transport is regulated and controlled.

Calmodulin (CaM) is a small protein that regulates the actions of other proteins. CaM binds to a larger target protein (CaM-dependent kinase II), modulating its activities (for example, during the synaptic

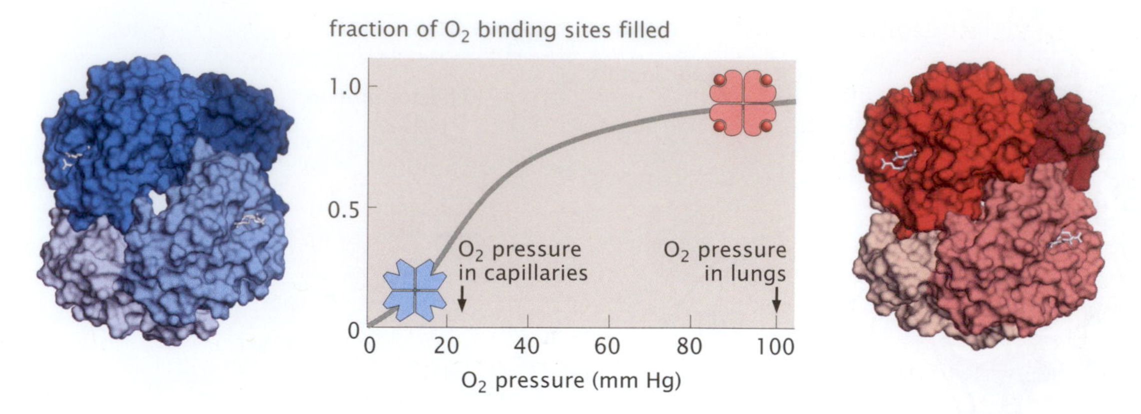

Figure 2.16 Hemoglobin is a protein that binds oxygen in the lungs, and releases the oxygen in peripheral tissues. (*Blue*) Hemoglobin has no oxygens bound at low O_2 concentrations. (*Red*) Hemoglobin has four oxygens bound at high O_2 concentrations. The binding is *cooperative*, meaning that binding one oxygen molecule increases the affinity for binding to the next oxygen molecule. This cooperativity (sigmoidal curve shape) means that hemoglobin acts like a sensitive switch: a small change in blood pH or oxygen concentration triggers a switch from fully unbound to fully bound, and vice versa. Cooperativity results from an allosteric change in the conformation of the protein, which is encoded in the amino acid sequence. (Plot adapted from J Darnell, H Lodish, and D Baltimore. *Molecular Cell Biology*, 2nd ed. WH Freeman, San Francisco, 1990. With permission from WH Freeman.)

transmission events that lead to memories in the brain). When calcium binds to CaM, it changes the target-protein interaction, changing the biological action of the target protein. Figure 2.17 shows how calcium binding changes the conformation of CaM. The binding of four Ca^{2+} to CaM's two domains causes the formation of a central long 21-residue helix that exposes the otherwise-buried methionine-rich hydrophobic patches. The exposure of these hydrophobic regions affects how CaM binds to CaM-binding proteins. An example of a CaM-binding protein is myosin-light-chain kinase, an enzyme that phosphorylates and activates myosin.

Regulation can be achieved by chemical modifications

Biological actions can also be modulated by *covalent chemical modification* of a protein. Sometimes, proteins undergo *post-translational* modifications—chemical alterations that happen after the protein is

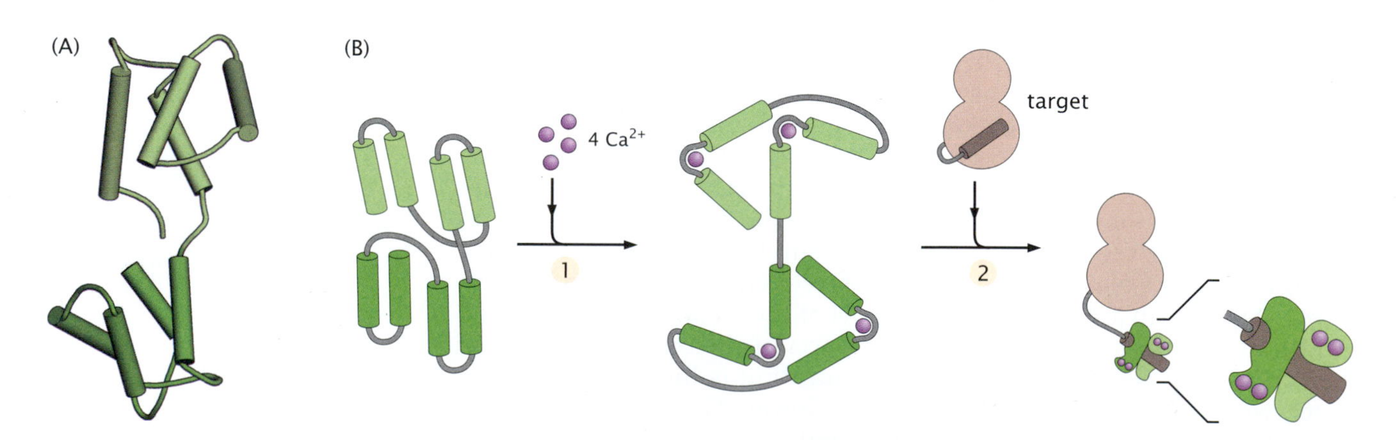

Figure 2.17 The binding of calcium ions induces a large conformational change in calmodulin (CaM). (A) CaM has two domains. Each is a four-helix bundle. (B) When CaM is saturated with Ca^{2+}, the pairs of helices in each domain open up into EF-hands with Ca^{2+} bound (step 1). The hands are now competent to bind to a helix that inhibits target CaM-dependent protein kinase II (step 2), causing the kinase to pop open into an active state.

synthesized on the ribosome. The most common post-translational modification is phosphorylation. A protein can be *phosphorylated* or *dephosphorylated* by the action of another protein. Phosphorylation is the covalent addition of a phosphate group to a protein, through a reaction with adenosine triphosphate (ATP):

$$\text{protein} + \text{ATP} \rightarrow \text{protein-PO}_3^{-2} + \text{ADP}$$

Phosphorylation reactions are catalyzed by proteins called *kinases*. A kinase helps add a phosphate group to a protein. In eukaryotic proteins, kinases add phosphate groups to serine, threonine, or tyrosine residues. Adding a phosphate group, which is negatively charged, to a polar amino acid in a protein changes the chemical character of the phosphorylation site from hydrophilic to charged, and can sometimes result in a conformational change of the protein.

Dephosphorylation is the removal of a phosphate ion by hydrolysis:

$$\text{protein-PO}_3^{-2} + H_2O \rightarrow \text{protein-H} + \text{HPO}_4^{-2}$$

It is catalyzed by proteins called *phosphatases*.

Proteins can also be covalently modified by *acetylation*. In this case, an acetyl group, $-COCH_3$, is added to a nitrogen, either at the protein's N-terminus, forming $CH_3(CO)$—NH–protein, or to the end of a lysine side chain. Acetylation often occurs on the lysine residues of *histones*. Histones are cell nucleus proteins that bind tightly to DNA to condense and protect it. The tightly wound complexes between histones and DNA are called *nucleosomes* (Figure 2.18). Histone acetylation is important because it participates in regulating the levels of genes and proteins that are expressed in cells. *Epigenetics* is the term given to alterations in gene expression that follow from environmental factors, such as covalent modifications, that are not directly encoded within the genome itself.

Myristoylation is the covalent attachment of a myristoyl fatty acid group to the α-amino group of an N-terminal glycine residue, yielding a $CH_3(CH_2)_{12}$—(CO)—NH–protein. *Prenylation* is the addition of a

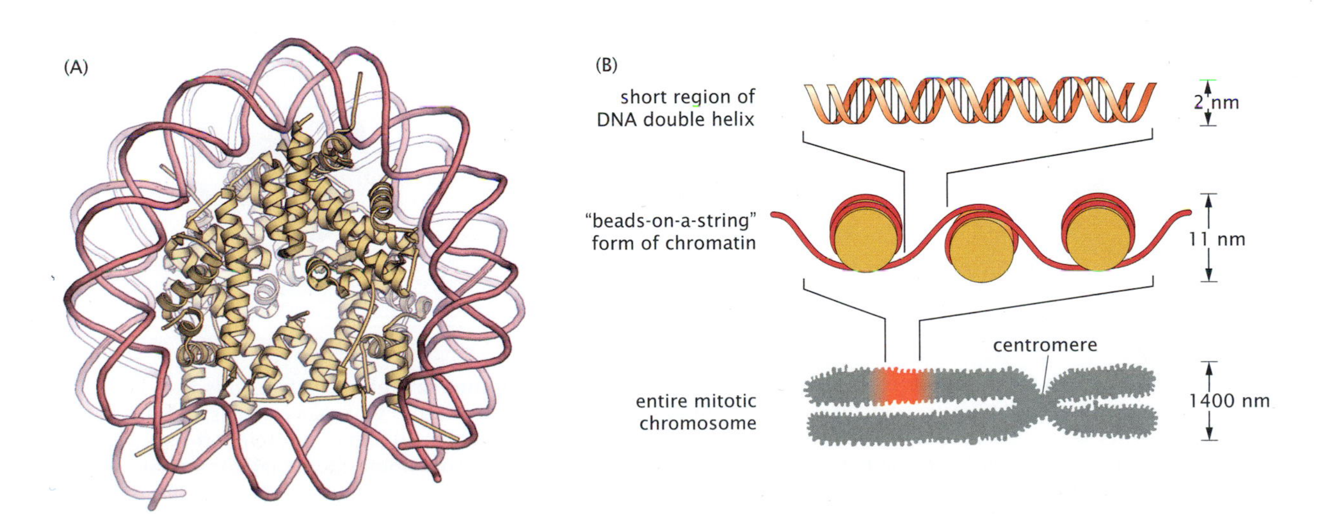

Figure 2.18 Chromosomes have hierarchical structures of DNA and proteins. (A) Histone proteins (*yellow*) assemble and are wrapped with DNA (*red*) to constitute nucleosomes. (B) The different structural scales of DNA. On the the largest scale, nucleosomes are condensed to form chromosomes.

hydrophobic group to a protein. *Palmitoylation* is the attachment of a palmitoyl group to cysteine residues via thioester linkages, to form the side group $CH_3(CH_2)_{14}$—(CO)—S —CH_2—). *Glycosylation* is the covalent attachment of sugars to proteins. In all these cases, attachment or removal of the covalent groups can affect the conformations of the proteins. This serves to switch on or off various activities, regulating the actions of those proteins, or targeting them to specific locations in the cell.

Some Proteins Are Structural and Protective Materials

Some proteins regulate DNA packaging and processing

Some proteins, such as histones, interact with DNA or RNA. Groups of eight histone proteins, which are positively charged, come together to form a nucleosome core particle upon which negatively charged DNA is wrapped and condensed. Histones help cells pack a large amount of DNA into a small space, and protect the DNA from degradation.

Viral capsid proteins transport and deliver viral molecules

Viruses are particles in which RNA or DNA molecules are encapsulated within a protein container called the *viral capsid*. A viral capsid is a highly symmetrical assembly of hundreds of copies of a capsid protein. The proteins forming these capsids can also play a role in packaging and/or disassembling the genetic material of the virus. In the case of some bacteriophages such as HK97 flu virus (named for an epidemic in Hong Kong in 1997), the capsid undergoes a conformational transition from a procapsid into a mature capsid. The nucleic acid chain is squeezed into the viral head, expanding the capsid volume, and changing its shape (Figure 2.19).

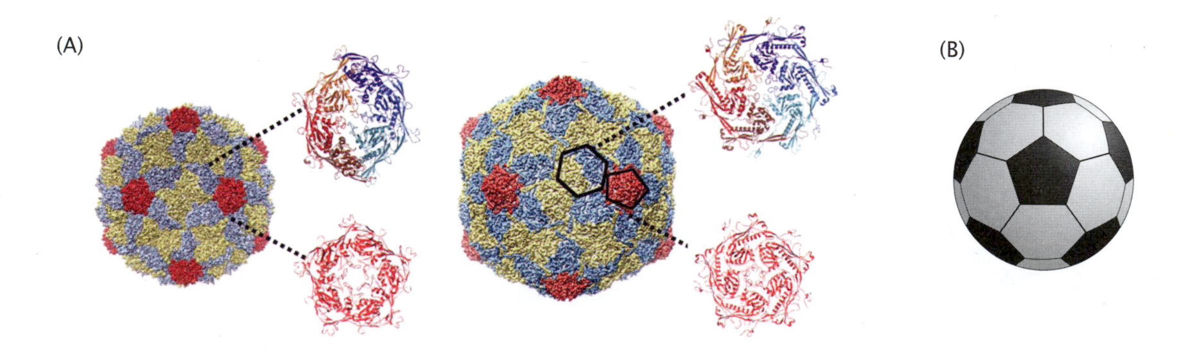

Figure 2.19 The capsid of the HK97 virus is composed of 420 copies of the same protein. The capsid is the shell that encapsulates the genetic material of the virus. Capsids can spontaneously arise from self-assembly of many copies of a single protein, due to the symmetry of the protein–protein interactions. (A) The two states of the capsid and its proteins: the compact state without genetic material inside (left) and the expanded state of the fully active virus containing the genetic material (right). Both structures include centers that have six or five of the same subunits interacting, shown in the blowups. Note the unique expansions of these groups between the empty and filled capsid. This is another case of an important protein conformational transition, and a remarkable case of the use of different protein interactions to stabilize two different capsid forms. (B) A soccer ball, illustrating how hexagons and pentagons can self-assemble into an icosahedron of pentamers (*black*) and hexamers (*white*). (A, adapted from I Gertsman, L Gan, M Guttman, et al. *Nature*, 458:646–650, 2009. With permission from Macmillan Publishers Ltd.)

PROTEINS ARE HEALTHY OR SICK OR DIE

Proteins Can Misfold, Sometimes in Association with Disease

In much of this chapter, protein actions are described in terms of protein structures and motions. Many native proteins are highly structured. So, our interpretations have been "mechanistic": you observe a structural feature and infer how that might affect its physical or chemical actions, in the same way that you might deduce how a macroscopic machine works. But not all protein actions are so mechanistic. Instead, some biological actions arise from disordered structures of proteins. And sometimes that disorder itself plays a functional role. Intrinsically disordered proteins (IDPs) constitute a large class of proteins. One important example is transcription factors—small proteins that assist in the process of transcribing DNA to RNA. Transcription factors bind to DNA, near coding regions, and assist in attracting other proteins that are involved in transcription or regulation, such as those needed for DNA repair or to initiate transcription (for example, polymerases and DNA repair enzymes). The binding of a transcription factor to DNA often involves a disordered chain segment, which becomes ordered upon binding. Intrinsic disorder can be of value in imparting flexibility so that a given peptide sequence can recognize more than one particular DNA sequence, or in providing conformational control and switching when the protein is in the presence of regulatory molecules, for example.

On its own, a protein may have a well-defined native structure. But in a concentrated medium of other proteins, a protein may misfold or associate with other proteins. Many maladies, including cystic fibrosis and neurodegenerative diseases such as Alzheimer's, Parkinson's, and Huntington's, are associated with protein misfolding. In misfolding diseases, proteins associate with each other into large aggregates or insoluble fibrils, collectively called *amyloids*. Creutzfeldt–Jakob and Mad Cow disease are *prion* diseases. A prion (PRoteinaceous Infectious particle) is a particle of misfolded protein that infects the host with a particular disease. Although the disease mechanism is not known, it is thought that some proteins misfold, somehow nucleating the misfolding of other proteins, often into large-scale β-sheets. Amyloid structures can be quite stable. They accumulate in brain or neural tissue and can be associated with cell death.

Proteasomes Digest Proteins

The *proteasome* is a multiprotein complex that serves as a "garbage disposal" for degrading "bad" proteins, that is, those that are misfolded or oxidized or damaged in some way. Figure 2.20 shows that a proteasome has a central core particle where proteolysis occurs, and two peripheral, regulatory particles. To prevent indiscriminate digestion of proteins, a proteasome acts only to degrade proteins that have been properly tagged for degradation. The tag is itself a small protein, of 76 residues, called *ubiquitin*, because it is ubiquitous. The covalent attachment of ubiquitin to a protein is called *ubiquitination*. Only proteins that have one or more ubiquitins attached to them will be degraded by proteasomes. The 2004 Nobel Prize in Chemistry was awarded to A Hershko, I Rose, and A Ciechanover for discovering this process.

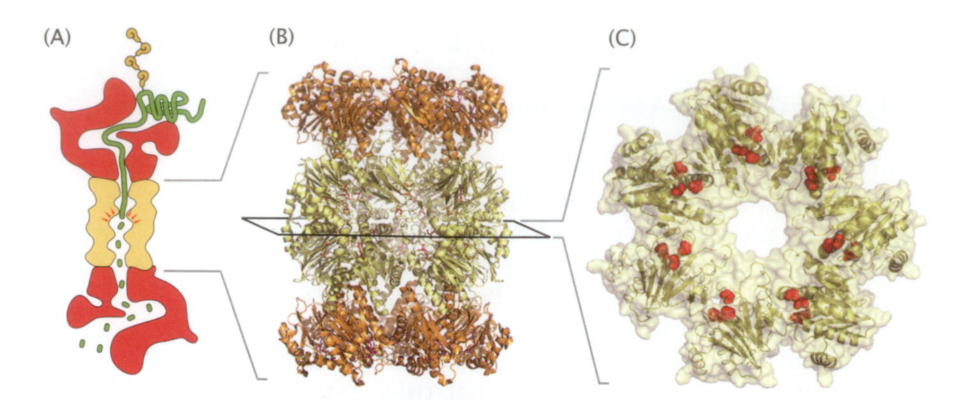

Figure 2.20 The proteasome is a multiprotein complex that degrades proteins targeted by the addition of ubiquitin. (A) Schematic showing the protein being degraded (*green*), its targetting tag protein (ubiquitin, *yellow–brown*), which directs the client to the proteasome's unfolding proteins (*red*), which feed it to the core proteasome (*light yellow*) for degradation. The complete proteasome structure consists of a central core particle cylinder where proteolysis occurs and two regulatory particle lids. (B) The core particle chamber for degrading proteins inside a proteasome is formed by stacking heptameric (sevenfold) donuts made of α (*orange*) and β (*yellow*) subunits. Two identical regulator particles (*red* in A, not shown in B) bind at each end of this chamber in order to recognize and initiate unfolding for proteins that are ubiquitin-tagged for degradation. (C) A cross-sectional view of the molecular surface of the chamber interior with the seven protease active sites (*red* spheres) indicated.

SUMMARY

Proteins perform many biochemical and biophysical functions in cells. Their mechanisms of action can often be inferred from their molecular structures and dynamics. Protein motors convert chemical energy into forces within muscles and into directed motion to transport materials within cells. Small molecules, protein molecules, and chemical signals flow across membranes, mediated by membrane proteins. Some membrane proteins contain gates that control the passage of small molecules, ions, and metabolites into or out of cells; others have channels that allow for the selective passage of ions. Electrical signals flow in nerves, brain, and muscle because of the flows of ions and the binding of neurotransmitters. Enzymes are proteins that catalyze biochemical reactions. Enzymes break down other proteins; they break down cellulose into sugars; they unwind, transcribe, process, and package DNA; and they add covalent modifications to proteins and DNA, as one basis for regulation. Some proteins function as molecular machines in large multimolecular complexes. For example, ribosomes synthesize proteins; fatty acid synthase synthesizes fatty acids; chaperones assist in folding and disaggregating proteins; and proteasomes chemically degrade proteins that are tagged for destruction. Actin filaments and microtubule tracks act as conveyor belts for intracellular transport processes.

REFERENCES

[1] GA Petsko and D Ringe. *Protein Structure and Function.* New Science Press, London, 2004.

[2] M Ashburner, CA Ball, JA Blake, et al. Gene Ontology: tool for the unification of biology. The Gene Ontology Consortium. *Nat Genet*, 25:25–29, 2000.

[3] S Bernèche and B Roux. Energetics of ion conduction through the K^+ channel. *Nature*, 414:73–77, 2001.

[4] JF Ménétret, J Schaletzky, WM Clemons Jr, et al. Ribosome binding of a single copy of the SecY complex: Implications for protein translocation. *Mol Cell*, 28:1083–1092, 2007.

SUGGESTED READING

Alberts B, Johnson A, Lewis J, Morgan D, Raff M, Roberts K, and Walter P, *Molecular Biology of the Cell*, 6th ed. Garland Science, New York, 2015.

Dobson CM, The structural basis of protein folding and its links with human disease. *Philos Trans R Soc Lond B Biol Sci*, 356:133–145, 2001.

Frank J (ed.), *Molecular Machines in Biology: Workshop of the Cell*. Cambridge University Press, New York, 2011.

Kuriyan J, Konforti B, and Wemmer D, *The Molecules of Life*. Garland Science, New York, 2013.

Roux B (ed.), *Molecular Machines*. World Scientific, New York, 2011.

Williamson M, *How Proteins Work*. Garland Science, New York, 2012.

Proteins Have Stable Equilibrium Conformations

CHAPTER 3

NATIVE AND DENATURED STATES ARE STABLE STATES OF PROTEINS

An important feature of proteins is their *stable equilibrium* conformations. A protein has a given native structure under given native conditions, and it has a given denatured structure under given denaturing conditions. That structure does not depend on the kinetic process by which the protein reached that state. So, a protein that was folded on a ribosome in a cell is in the same state as a protein that was refolded from an unfolded structure in a test tube, as long as it is under the same conditions. That means that protein properties can usually be expressed by equilibrium thermodynamics, without the need for pathway information. There are some exceptions, however. The structures of protein crystals or aggregates sometimes do depend on the initial conditions or how fast they were formed. Here, we explore some stable states of proteins and the forces that stabilize them.

Protein stability is important for several reasons. First, it matters to the cell. A cell maintains *proteostasis* (the folding homeostasis of the full set of all its proteins, its proteome). Cells have developed sophisticated mechanisms for maintaining and regulating protein stability and conformational equilibria in the face of protein degradation or of misfolding and aggregation. Mechanisms include the use of chaperones (proteins that help other proteins fold), proteases, the proteasome, and control of protein synthesis and degradation rates. The stabilities of proteins are factors in amyloid and other diseases, aging, and cancer. Second, protein stability gives insights into the physical forces of protein folding. Third, when developing biotechnological therapeutics (protein drugs), it is essential to formulate solution conditions that manage protein folding, degradation, aggregation, solubility, crystallization, and fibrillization.

Under biological conditions, a protein is typically *folded*, or *native*.[1] Under harsher conditions—in acids or bases, at high temperatures, or in chemical denaturants—a protein can be *unfolded* or *denatured*. Changes in protein structure are either *reversible* or *irreversible*. Reversibility means that if you perturb a protein away from its initial state—even through a large perturbation—and you then re-establish the initial conditions, the protein will return to its initial state. The advantage of finding conditions for reversibility is that you can interpret the results using the powerful tools of equilibrium thermodynamics.

Some transformations of proteins are irreversible. High temperatures can covalently degrade proteins, for example, by hydrolyzing the amide side-chain groups of glutamine and asparagine, removing the NH_2 group and converting the remaining side chain into glutamic acid or aspartic acid. Protein backbones can be covalently degraded by proteases. Covalent degradation is irreversible because lowering the temperature or removing the protease after covalent bonds have been broken does not return the protein to its original structure.

At high concentrations, proteins may crystallize or aggregate, either reversibly or irreversibly. Irreversible aggregation among proteins may arise either from covalent bonding between proteins, for example through disulfide bond formation, or when chains become so highly entangled that they cannot disengage from each other on the experimental timescale. In this chapter, we focus on reversible processes.

Anfinsen's Hypothesis: Native States Are Thermodynamically Stable

Until the 1960s, a key question was whether proteins could fold and unfold reversibly. Previously, protein science had been hindered by an inability to purify proteins. Experimentalists had often inadvertently studied irreproducible processes such as aggregation. It had not been clear that protein structures were thermodynamically stable states of matter. Experiments on bovine pancreatic ribonuclease A by Christian B Anfinsen in the early 1960s gave the first proof that folding was reversible and that native states were thermodynamically stable [1]. Anfinsen broke the native disulfide bonds, denatured the proteins, then reestablished native conditions, and found that the protein refolded correctly. At that time, disulfide bonding was the method of choice because disulfides were uniquely trappable and identifiable. His work, for which he was awarded the 1972 Nobel Prize in Chemistry, showed that the native structure of a protein could be fully encoded within its amino acid sequence (that is, it is thermodynamically stable), and thus successful folding does not require a special processing history or kinetic sequence of events. On the other hand, protein fold-

---

[1] Throughout this book, we use the terms *native* and *folded* interchangeably, and we use *unfolded* and *denatured* interchangeably. We use the terms *conformation* and *configuration* interchangeably, consistent with standard usage in statistical mechanics. We use the term "state" in a sense that is macroscopic, not microscopic. A state corresponds to a signal that you can see in some experiment. Fluorescence or circular dichroism can distinguish the native state (N) from the denatured (D) state, for example. Therefore, by our definition, a "state" encompasses an *ensemble* of microscopic chain conformations—often a huge number of such *microstates*.

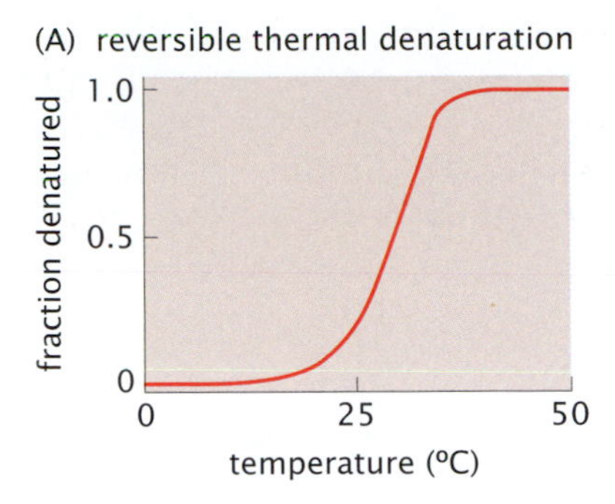

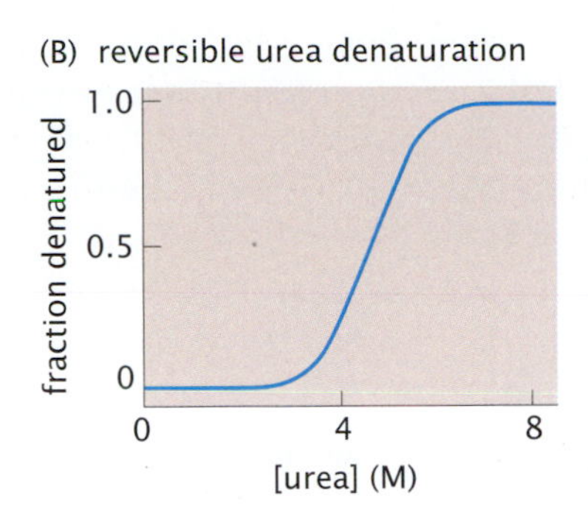

Figure 3.1 Proteins are denatured by increasing temperature or denaturant concentration. (A) The thermal denaturation of ribonuclease A, as determined by spectroscopic and viscosity measurements. (B) The urea-induced denaturation of ribonuclease T1, as determined by spectroscopic methods. (A, from A Ginsburg and WR Carroll. *Biochemistry*, 4:2159–2174, 1965; B, from JA Thomson, BA Shirley, GR Grimsley, and CN Pace. *J Biol Chem*, 264:11614–11620, 1989. © 1989 The American Society for Biochemistry and Molecular Biology.)

ing in the cell can be assisted by other molecules. For example, some proteins are known to undergo *co-translational folding*; their folding happens as they are being released by the ribosome. And, the folding of many proteins is often facilitated *in vivo* by the class of proteins called *chaperones*, which help other proteins fold inside cells and prevent their misfolding and aggregation. Nevertheless, many small single-domain proteins are routinely unfolded and refold reversibly in test tubes, without biological helper molecules. Hence, native structures of proteins are stable equilibrium states of matter.

The Basic Experiment of Protein Stability Is Equilibrium Denaturation

The basic experiment that measures protein stability is equilibrium denaturation. You make up a series of different protein solutions, $1, 2, \ldots, M$, having different amounts of denaturing agent $x_1, x_2, \ldots, x_M$. By "denaturing agent," we mean either temperature or a chemical, such as guanidinium hydrochloride (GuHCl), urea, alcohol, or acids or bases, for example. Then, for each particular solution, having denaturing agent in amount x, you measure (typically by some form of spectroscopy) the proportion $f_N(x)$ of native protein molecules and the proportion $f_D(x) = 1 - f_N$ of denatured protein molecules (assuming you observe only the two states, a common situation). Increasing the denaturing agent increases the population of D relative to N. In the absence of denaturing agent, the protein is fully native (except for fluctuations). In high concentrations of denaturing agent, the protein is denatured. Figure 3.1A shows denaturation by high temperatures; Figure 3.1B shows denaturation by high concentrations of urea. Such plots, called *melting* or *denaturation* curves, are typically sigmoidal in shape, as a function of denaturing agent x. At the midpoint, a small change in denaturant concentration shifts the distribution of protein conformations. In this sense, protein denaturation is a miniaturized version of a phase transition in a larger system; for example, a small change in temperature, at the right temperature, causes water to boil or freeze.

To get insight from a denaturation profile, you need to convert it into a quantity called the *free energy*,[2] G. In order to compute the *folding free*

[2]There are two types of free energy: the Gibbs free energy G and the Helmholtz free energy F. F differs from G in how pressure–volume effects are treated. Pressure–volume changes, which can be large for gases or gas-liquid systems, are typically small for protein solutions, so the enthalpy H and internal energy U are interchangeable, $H \approx U$, and the Gibbs free energy $G(T, P, N) = H - TS \approx F(T, V, N) = U - TS$ is

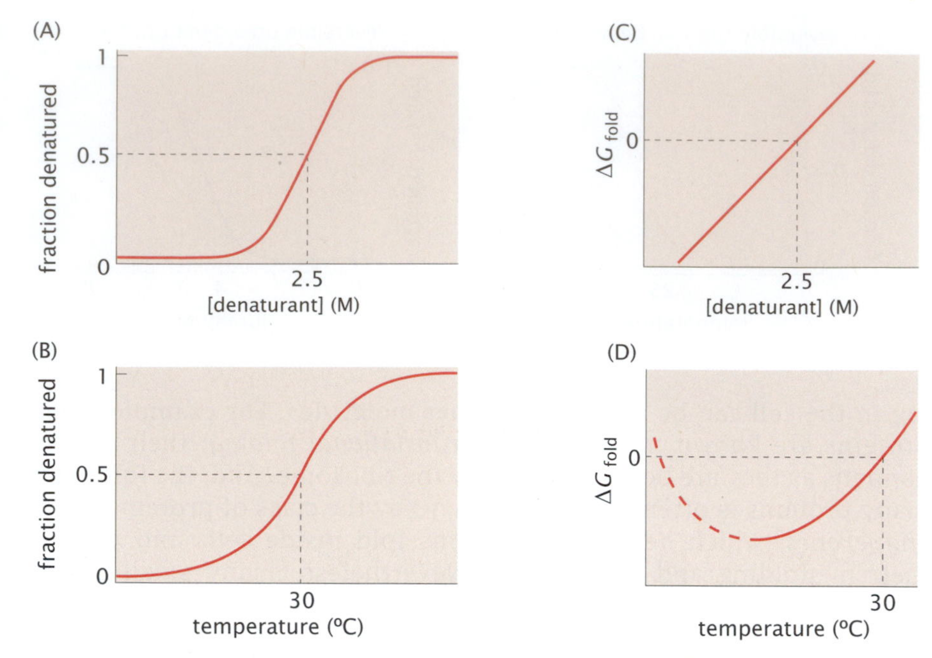

Figure 3.2 Free energies of folding can be determined from denaturation curves using Equations 3.1 and 3.2 (A and B). The free-energy dependence on denaturants such as urea or GuHCl is linear (C), so ΔG_{fold} in the absence of denaturant can be extrapolated. In contrast, the temperature dependence (D) is curved.

energy, $\Delta G_{\text{fold}} = G_{\text{N}} - G_{\text{D}}$ for the folding of a protein from its denatured state (D) to its native state (N), define a folding *equilibrium constant* K as

$$K = \frac{f_{\text{N}}}{f_{\text{D}}}, \tag{3.1}$$

where f_{N} and f_{D} are the fractions of native and denatured states. You can express the free energy of folding in terms of K as

$$\Delta G_{\text{fold}} = -RT \ln K, \tag{3.2}$$

where T is the absolute temperature (in Kelvin) and R is the gas constant. For unfolding, you have $\Delta G_{\text{unfold}} = -\Delta G_{\text{fold}}$.[3] Next, we describe how to use ΔG_{fold} to get insights into the driving forces of folding.

Figure 3.2A illustrates chemical denaturation, and Figure 3.2C shows $\Delta G_{\text{fold}}(c)$, the dependence of the free energy of folding on the concentration c of GuHCl denaturant. $\Delta G_{\text{fold}}(c)$ is typically a linear function of c. The slope of $\Delta G_{\text{fold}}(c)$ versus c is called the *m-value*.

Another way to denature proteins is by heating them (see Figure 3.2B). $\Delta G_{\text{fold}}(T)$ is usually a curved function of temperature T (see Figure 3.2D). You can measure the properties of proteins as a function of temperature in a *calorimeter*. A differential scanning calorimeter applies a series of different temperatures to a protein solution at equilibrium. A calorimeter measures the heat taken up or given off by the protein solution at each temperature. The heat absorption in protein unfolding rises to a maximum with increasing temperature, then decreases (Figure 3.3). The temperature, T_m, at which the heat

---

interchangeable with the Helmholtz free energy. Experimental studies usually report the free energy G and enthalpy H.

[3] $R = 8.314\,\text{J mol}^{-1}\,\text{K}^{-1}$) is the gas constant. $R = kN_{\text{avo}}$, where k is Boltzmann's constant, $1.38 \times 10^{-23}\,\text{J K}^{-1}$ per molecule, and $N_{\text{avo}} = 6.022 \times 10^{23}$ molecules mol^{-1}. At room temperature ($T = 300$ K $= 27°$C), the product RT is approximately $0.6\,\text{kcal mol}^{-1}$ using $R = 1.987 \times 10^{-3}\,\text{kcal mol}^{-1}\,\text{K}^{-1}$). (Note the conversion 1 cal = 4.184 J.)

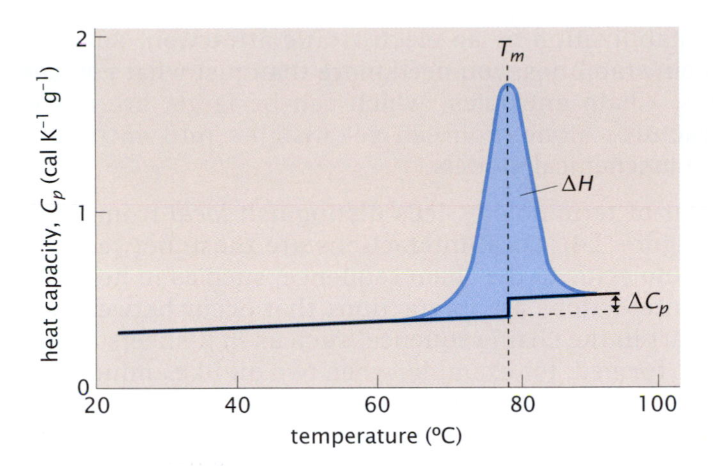

Figure 3.3 Calorimetry measures the heat capacity of protein unfolding with temperature. Increasing the temperature of a protein solution leads to a peak in the heat capacity, indicating that energy is absorbed as the protein unfolds. T_m is the temperature of the peak of the transition. ΔC_p is the heat capacity of unfolding, measured at constant pressure. It is the difference between the pre-transition and post-transition baselines. ΔH, the calorimetric enthalpy of unfolding, is given by the area under the peak, extrapolating to T_m from the baselines. (Adapted from PL Privalov and NN Khechinachvili. *J Mol Biol.* 86(3):665–84, 1974. With permission from Elsevier.)

absorption is maximal—called the *melting temperature* or *denaturation temperature*—is the midpoint of the denaturation transition of the protein. You want to know the *excess* heat capacity of the unfolding transition: the heat capacity change from unfolding minus the heat capacity change of the pure solvent. As a practical matter, you can obtain ΔC_p at the denaturation point by either subtracting the sloping baseline or integrating to get the enthalpy and finding the slope, $\Delta C_p = dH/dT_m$.

Calorimetry experiments show that protein denaturation resembles a *melting process*. At its melting point, a material's energy and entropy both increase sharply with temperature, as bonds break. Similarly, at the midpoint temperature of thermal denaturation, a protein's energy and entropy increase. The increased energy indicates that some intrachain interactions are broken, and the increased entropy indicates that the system gains conformational freedom. To explore this more deeply, we will now develop models. To do this, we need the language of thermodynamics and statistical mechanics, which we review later. But first, let's see how to convert from experimental denaturation curves to free energies. From the temperature dependence of $\Delta G_{\text{fold}}(T)$, you can get two component quantities: the change in the enthalpy upon folding, $\Delta H_{\text{fold}}(T)$, and the change in the entropy upon folding, $\Delta S_{\text{fold}}(T)$, through the thermodynamic relationship [2]

$$\Delta G_{\text{fold}} = \Delta H_{\text{fold}} - T\Delta S_{\text{fold}}. \tag{3.3}$$

The curves $\Delta G_{\text{fold}}(c)$ and $\Delta G_{\text{fold}}(T)$ give insights into the forces that drive protein conformational changes.

Stabilities and Structures Give Insights about the Driving Forces of Folding

To understand the forces that stabilize native proteins, first look at native structures. Native structures are compact. And their cores are mostly hydrophobic. This indicates the importance of hydrophobic interactions in a folded protein. Look at secondary structures, α-helices and β-sheets. A key feature is their hydrogen bonding. Also, proteins tend to be well packed, indicating the importance of van der Waals interactions. Some native proteins have *salt bridges*, where a positively charged atom is situated near a negatively charged atom, indicating

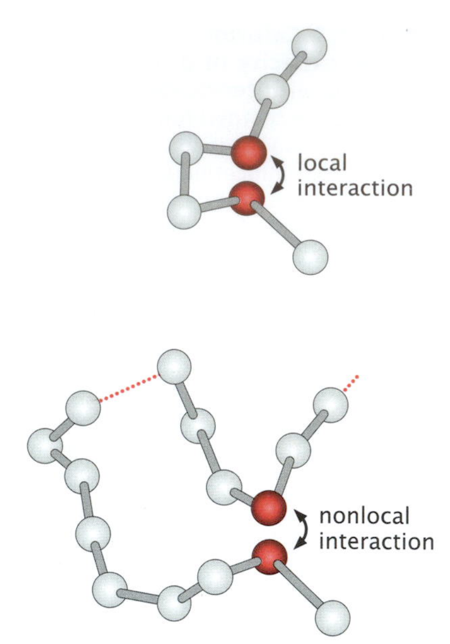

Figure 3.4 Interactions are defined as local or nonlocal, based on their separation in the chain sequence.

possible stabilization by an electrostatic attraction. But to fully interpret protein stabilities, you need more than just what you see in native structures. Chain entropies, which can be large, are not observable from structures alone. You can get insights into entropies by using statistical mechanical models.

As a matter of terminology, let's distinguish *local* from *nonlocal* interactions (Figure 3.4). Local interactions are those between close neighboring amino acids in the chain sequence, such as in helices and turns. Nonlocal interactions are interactions that occur between amino acids farther apart in the chain sequence, such as in β-sheets. Nonlocal interactions are formed, for example, when two oil-like amino acids displace water to come into contact with each other. Protein stability comes from both types of interactions. Local versus nonlocal is terminology that refers to the *chain separation* between the contacting monomers, to be distinguished from *short-ranged versus long-ranged*, which refers to the dependence on the distance through space between the interacting monomers. Coulombic interactions are long-ranged through space (the energy depends on distance r as $1/r$) and van der Waals attractions are short-ranged through space (depending on $1/r^6$), for example, irrespective of the chain separation. Here is a brief summary of the types of noncovalent intrachain interactions in proteins.

Hydrophobic interactions are important for folding

The *hydrophobic effect* refers to the tendency of oil and water to separate. About half of the 20 types of amino acid side chains have oil-like, or nonpolar, character. In its native structure, a protein's nonpolar amino acids tend to be buried within its core, implying that folding is driven, at least partly, by the tendency of a protein's nonpolar amino acids to hide from water. Here are two indications of the importance of hydrophobic interactions in protein folding: (1) proteins are unfolded by solvents, such as GuHCl, urea, alcohols, and surfactants, which weaken the hydrophobic driving force for folding, and (2) there is a large positive heat capacity of unfolding, $\Delta C_{p,\text{unfold}} > 0$, for typical small proteins. A large positive heat capacity is a signature of the hydrophobic effect in simple systems. A characteristic fingerprint of hydrophobic interactions is that the transfer of nonpolar molecules such as benzene, toluene, or alkanes from their pure liquid state to water increases the heat capacity, $\Delta C_p = C_{p,\text{with solute}} - C_{p,\text{without solute}}$ (Table 3.1).

Table 3.1 The transfer of nonpolar molecules into water at 23°C is unfavorable (the free energy is positive), dominated by the entropy, and associated with a positive change in heat capacity

Substance	Surface area (Å^2)	ΔG^{hyd} (kcal mol^{-1})	ΔH^{hyd} (kcal mol^{-1})	ΔS^{hyd} (cal K^{-1} mol^{-1})	ΔC_p^{hyd} (cal K^{-1} mol^{-1})
Benzene	240	4.64	0.497	−13.88	53.8
Toluene	275	5.45	0.413	−16.9	62.9
Ethylbenzene	291	6.26	0.483	−19.4	76.0
Cyclohexane	273	6.74	−0.024	−22.7	86.0
Pentane	272	6.86	−0.478	−24.47	95.6
Hexane	282	7.77	0.00	−26.08	105.0

(Adapted from RL Baldwin. *Proc Natl Acad Sci USA*, 83:8069, 1986; PL Privalov and SJ Gill. *Pure Appl Chem*, 61:1097, 1989.)

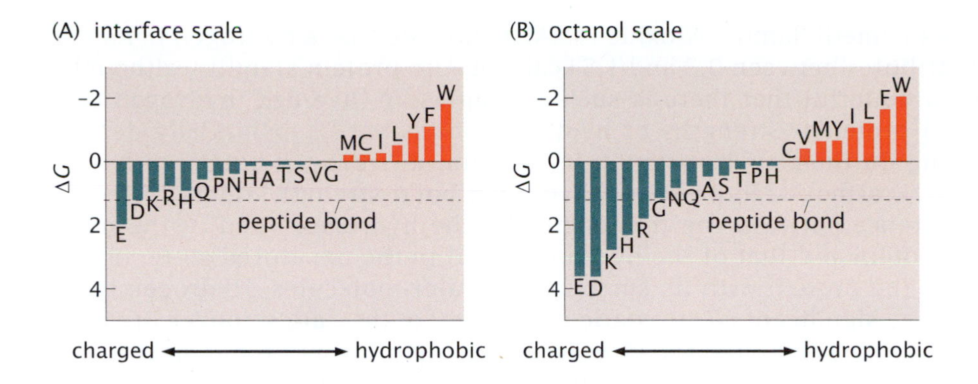

Figure 3.5 Two hydrophobicity scales for amino acids. A 5-mer peptide has one of the 20 different amino acids in the middle position. The peptide is partitioned from water into (A) a lipid bilayer interface or (B) a bulk octanol medium. The *orange* bars show the (hydrophobic) residues that favor the oil-like environment relative to water, and the *green* bars show the (charged and polar) residues that favor being in water. The horizontal *dashed* line indicates the contribution of the peptide bond to all amino acids. Histidine (H) is shown twice, for its charged and uncharged forms (the charged form is on the left). (A, data from WC Wimley and SH White. *Nature Struct Biol*, 3:842, 1996; B, data from WC Wimley, TP Creamer, and SH White. *Biochemistry*, 35:5109–5124, 1996.)

The hydrophobic effect is a consequence of water–water hydrogen bonding. Roughly speaking, hydrogen bonding between two water molecules is strong, so water molecules tend to configure around solutes in ways that maximize their hydrogen bonding with other water molecules. Said differently, two nonpolar molecules that are put into water will associate with each other to minimize their exposure to water, which maximizes water–water hydrogen bonding. *Hydrophobicity scales* express approximately the relative tendencies of different types of molecules to partition from water into oil-like environments. Two such scales for amino acids are shown in Figure 3.5. There are many different hydrophobicity scales for amino acids, depending on the type of oil and the conditions of measurement. In general, molecules that register as hydrophobic on one scale tend to register as hydrophobic on the other scales, but there are variations among scales. Figure 3.6 illustrates one application of hydrophobicity scales; it shows that strings of hydrophobic amino acids in a protein sequence can identify the membrane-spanning parts of membrane proteins.

Hydrogen bonds stabilize native protein structures

A *hydrogen bond* occurs when a hydrogen-bond donating group, such as an amide, —N—H, shares its hydrogen with an accepting group, such as a carbonyl oxygen, O═C—, in this case leading to —N—H···O═C—. Hydrogen bonding is extensive in native protein structures, mainly among backbone carbonyls and amides, particularly in α-helices and β-sheets. Mutational studies, as well as experiments using *osmolytes* such

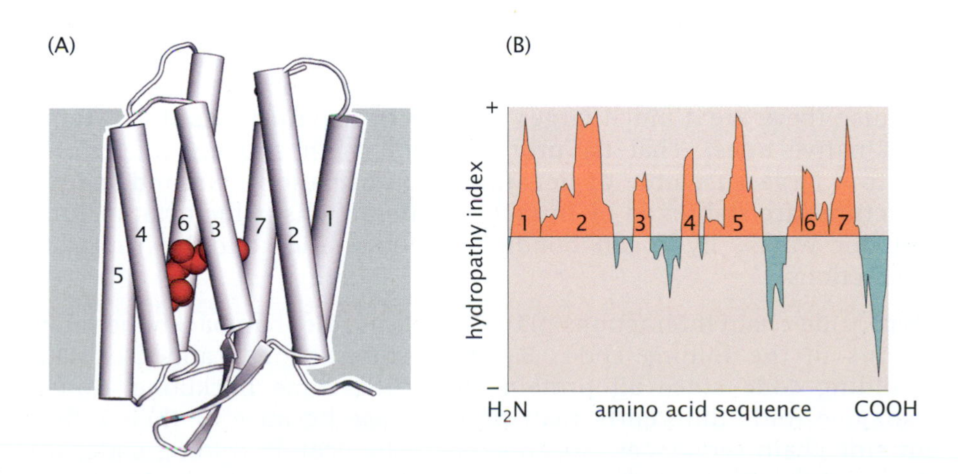

Figure 3.6 A hydropathy plot identifies the hydrophobic stretches of sequence that define membrane-spanning regions of a membrane protein. A hydropathy plot uses a hydrophobicity scale to show the hydrophobicity as a function of sequence position. *Orange* indicates hydrophobic regions, while *green* indicates polar or charged regions. (A) Bacteriorhodopsin has seven membrane-spanning α-helices corresponding to (B) seven *orange* peaks in the hydropathy plot. (Adapted from D Eisenberg. *Annu Rev Biochem*, 53:595–623, 1984. Modified with permission from the Annual Review.)

as trimethylamine *N*-oxide (TMAO), indicate that a hydrogen bond contributes between 0.3 and 1.5 kcal mol^{-1} to protein stability, although it is doubtful that there is such a thing as an "average" hydrogen bond in proteins. Strengths of hydrogen bonds vary substantially depending on their environments. In vacuum, a hydrogen bond can be 5 to 10 kcal mol^{-1}. In proteins, "hydrogen bond strength" refers to the difference between the free energy of the hydrogen bond in the native protein and that of the hydrogen bond in the denatured state, usually of the protein with its surrounding water molecules. Hydrogen bonds have significant electrostatic character. So, they are stronger in oil-like environments (like the interior of a protein), where the dielectric constant is low (estimated to be 2–12), than in waterlike environments, where the dielectric constant is high (around 80).

van der Waals interactions contribute to tight packing in proteins

Amino acids pack tightly within native protein cores. Tight packing results mainly from van der Waals interactions. These are (1) *short-ranged attractions* that draw atoms together and (2) even *shorter-ranged repulsions* that prevent two atoms from occupying the same space. van der Waals attractions and repulsions are largely responsible for the close steric fitting together of amino acids in protein cores. Mutational studies show that the tendency to fill space efficiently is about as strong as the hydrophobic interaction. For example, adding a methyl group to an empty cavity stabilizes a protein by about 0.9 kcal mol^{-1}.

Close packing is not difficult to achieve. Shake up nuts and bolts in a jar, and they'll pack well too. Interestingly, the fraction of space filled inside a protein is around 74%, which is about the same as for the maximum-density packing of identical spheres. How can a protein interior be so densely packed? It turns out that by mixing together small and large and irregular objects, you can sometimes pack space even more tightly than 74% by filling the nooks and crannies effectively. Think about how marbles can fill the small spaces between well-packed bowling balls, for example. Similarly, amino acids can fill space well too, because of their different sizes and shapes. In the text that follows, we describe how these various types of forces contribute to protein stability.

To understand how these forces contribute to protein folding and unfolding, we start from a few basic premises. First, proteins are chain molecules. Chain molecules can adopt many different conformations because different backbone torsional states typically have similar energies, separated by small barriers. Proteins denature, at least in part, because there are a much larger number of denatured conformations than native ones. That is, proteins gain chain entropy by unfolding to a large ensemble of denatured conformations. Second, native structures are stabilized by intrachain interactions, which can include van der Waals, hydrogen bonding, electrostatic, and hydrophobic interactions.

Third, side-chain interactions play a different role than backbone interactions in the folding code. Backbone interactions cannot explain a folding code since all proteins have the same backbone atoms: lysozyme folds differently than ribonuclease because of their different side-chain sequences. So, to explain the protein folding code, we focus on the hydrophobic interactions that explain the hydrophobic cores of proteins. Charge interactions are also encoded in side chains, but there are relatively few charge–charge interactions in most proteins, and they tend to be located on a protein's surface, essentially in water, where they interact weakly. Charge interactions are described in more detail in Appendix 3A and in Chapter 9. Fourth, the sharpness of protein folding transitions is a nearly universal feature of globular proteins.

Before discussing the physics of folding, we first give a brief review of statistical mechanics, mainly the *Boltzmann distribution law* [2].

STATISTICAL MECHANICS IS THE LANGUAGE FOR DESCRIBING PROTEIN STABILITIES

Statistical mechanics says that you can obtain the free energy in terms of a microscopic description of the system, called the *partition function* Q, through the relation

$$G = -RT \ln Q. \tag{3.4}$$

The partition function is the sum of the relative *statistical weights* over all $j = 1, 2, 3, \ldots$ microscopic states accessible to the system:

$$Q = \sum_{j=1}^{s} \omega(\epsilon_j) e^{-\beta\epsilon_j}, \tag{3.5}$$

where ϵ_j is the energy of state j, $\beta = (RT)^{-1}$, and $\omega(\epsilon_j)$ is the *density of states*, that is, the number of distinct microscopic configurations that have a given energy ϵ_j. In this way, $\omega(\epsilon_j)$ is the count of the number of states having that particular energy, and is called the *degeneracy* of that state. The quantity $e^{-\beta\epsilon_j}$ is called the *Boltzmann factor* for state j.

To make a statistical mechanical model, you must first know the microscopic states of the system, their energy levels ϵ_j, and their densities $\omega_j = \omega(\epsilon_j)$. Then you can compute the probability p_j of any state j using the Boltzmann distribution law:

$$p_j = \frac{\omega(\epsilon_j) e^{-\beta\epsilon_j}}{Q}. \tag{3.6}$$

The power of statistical mechanics is that it provides a way to express physically observable properties. Experiments usually reflect the average properties of large numbers of molecules. So, to compare with experiments, we want to compute *ensemble averages* of quantities such as the energy or the fraction of protein molecules folded. For a property A with a value of A_j in state j, the ensemble average is given by

$$\langle A \rangle = \sum_j A_j p_j, \tag{3.7}$$

so that the value for each state is weighted by its Boltzmann probability p_j.

In this chapter, we combine the Boltzmann distribution law with some simple models to explore the equilibrium states of proteins. Why are proteins compact and globular, with hydrophobic cores? What dictates a protein's native secondary structures? What is the folding code? That is, what intermolecular interactions can explain how a tertiary structure is encoded in an amino acid sequence? Why do proteins denature at high temperatures, in denaturants, or in acidic or basic solutions?

(A) square lattice, $z = 3$

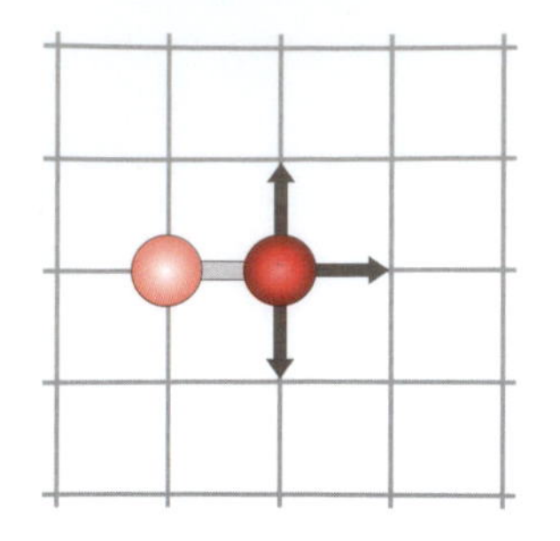

(B) cubic lattice, $z = 5$

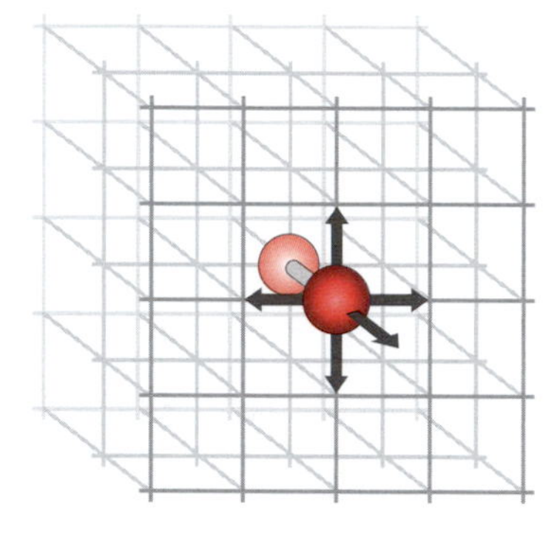

Figure 3.7 The basic unit step used in lattice models of proteins. Chain conformations are restricted by chain self-avoidance and lattice geometry. (A) On the 2D square lattice, after the first step is taken, the maximum number of possible directions for the next step is $z = 3$ (shown as arrows), because the chain cannot double back on itself. (B) On the 3D cubic lattice, $z = 5$.

Why do some proteins also denature at low temperatures (a process called *cold denaturation*)? We begin with the *HP model*, arguably the simplest model of protein folding.

Why Do Proteins Fold and Unfold? The HP Model

Under native conditions, proteins fold because many of the chain's amino acids are more strongly attracted to each other than they are to water. Under denaturing conditions, proteins unfold: (1) to increase the chain's conformational entropy and (2) because denaturing solvents effectively weaken the interactions between amino acids that hold the native protein together. As a simple model, let's represent a protein molecule as a string of monomer beads on a lattice (Figure 3.7). Each monomer is a single bead centered on one lattice site such that no lattice site can have more than one monomer. The beads are linked together by rigid covalent bonds. Consider a short chain having just six beads. The chain can adopt different possible configurations on a 2D square lattice. The *conformational space* is the set of all the viable configurations of the chain. This is often referred to as the set of *self-avoiding walks* because the chain can only follow bond directions that match the lattice edges and self-reversals are not permitted.

In this *HP model*, the 20 different types of amino acids are approximated using a simple *binary code* of just two monomer types: hydrophobic (H) or polar (P). A contact is defined when two monomers are adjacent to each other in space but not adjacent in the sequence. Whenever two H monomers form a noncovalent contact (that is, an *HH contact*), there is a favorable interaction energy $\epsilon_0 < 0$; all other types of contacts are assumed to have an energy of zero [3].

As an example, consider the 6-mer sequence in which monomers 1, 4, and 6 are hydrophobic (H) and the rest are polar (P). Each of the 36 lattice configurations shown in Figure 3.8 is called a microscopic state, or a *microstate*. Microstates are the finest-grained description a model provides of its accessible states. The microstates live on different rungs of an *energy ladder*, depending on how many hydrophobic contacts they make. You can collect microstates together to define *macrostates* in whatever ways are convenient or useful or experimentally measurable. For example, in Figure 3.8, the three different energy levels provide a natural definition of three macrostates: the native state N is the collection of all microstates having two hydrophobic contacts (there is only one such microstate, for this HP sequence); the intermediate state I is the collection of all microstates having one HH contact (there are seven such microstates for this HP sequence); and the denatured state D is the collection of all the other 28 microstates (those states having no HH contacts). There is an inherent symmetry to many of these possible walks, but because we are concerned only with the contacts between pairs of beads, we are free to redefine our origin and consider only unique walks. Thus, in the enumeration of unique walks, one proceeds by considering only conformations that make a first step down and only make a step to the right after some previous step to the left. The single folded conformation has a lower energy than the unfolded conformations, so the single microstate occupies the lowest rung of the energy ladder. The lowest-energy level is called the *ground state*. The next higher rung is called the first *excited state*, and the highest rung for this model is called the second excited state. The

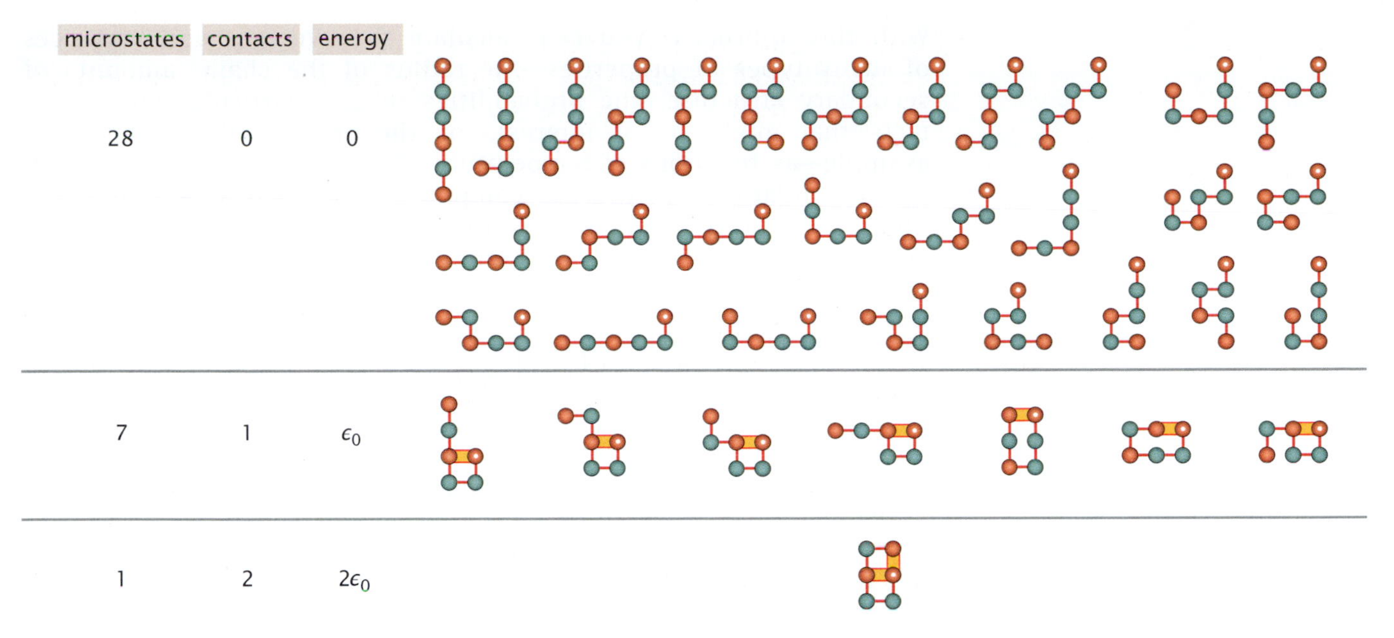

Figure 3.8 Energy ladder of a six-bead model sequence: HPPHPH. This six-bead binary sequence model has 1 low-energy conformation with two HH contacts, 7 conformations with one HH contact, and 28 conformations with no HH contacts. Hydrophobic residues are *orange* circles, polar residues are *green*, and HH contacts are indicated by *orange* bands. A *white* dot indicates the first bead of the chain. Note the geometrically clustered *orange* hydrophobic cluster of three in the lowest-energy form. ϵ_0 is a negative number, and so this state with $2\epsilon_0$ energy is the most energetically favorable state. (From KA Dill and S Bromberg. *Molecular Driving Forces: Statistical Thermodynamics in Biology, Chemistry, Physics and Nanotechnology*. 2nd ed. Garland Science, New York, 2011.)

degeneracies, that is, the numbers of microstates in each macrostate, are 1, 7, and 28, respectively.

According to Equation 3.5 the partition function Q, summing over all the weights for the states of this model HP sequence, is

$$Q = 28 + 7e^{-\epsilon_0/RT} + e^{-2\epsilon_0/RT} = 28 + 7x + x^2, \tag{3.8}$$

where we have simplified the notation by using $x = e^{-\epsilon_0/RT}$. According to Equation 3.6 the populations of the folded, intermediate, and unfolded states are

$$\begin{aligned} p_{\mathrm{N}} &= x^2/Q, \\ p_{\mathrm{I}} &= 7x/Q, \\ p_{\mathrm{D}} &= 28/Q. \end{aligned} \tag{3.9}$$

Here is how you apply statistical mechanics. First, you specify the contact energy ϵ_0 and the temperature T. Then you compute the populations, which are given by Equation 3.9 for this model. Finally, you test your model predictions against experiments. Using Equation 3.7 for a property A that has the value A_{D} in the denatured state, A_{I} in the intermediate state, and A_{N} in the native state, the ensemble average is computed from the populations as

$$\langle A \rangle = \sum_j A_j p_j = A_{\mathrm{D}} p_{\mathrm{D}} + A_{\mathrm{I}} p_{\mathrm{I}} + A_{\mathrm{N}} p_{\mathrm{N}}. \tag{3.10}$$

For example, the average energy is

$$\langle \epsilon \rangle = 0 p_{\mathrm{D}} + \epsilon_0 p_{\mathrm{I}} + 2\epsilon_0 p_{\mathrm{N}}. \tag{3.11}$$

With this approach, you can calculate the averages and variances of many types of properties—the radius of the chain, amounts of secondary structure, the probabilities of any particular chain contacts that might be of interest, or the end-to-end distance, for example—as functions of temperature. For instance, using a thermodynamic relationship, you can compute the heat capacity as $d\langle\epsilon\rangle/dT$. Another property of interest is the free-energy difference between any of the three states. For example, the free energy of folding, $\Delta G_{\text{fold}} = G_{\text{N}} - G_{\text{D}}$, is

$$\Delta G_{\text{fold}} = -RT \ln\left(\frac{p_{\text{N}}}{p_{\text{D}}}\right) = 2\epsilon_0 + RT \ln 28. \tag{3.12}$$

The model predicts a *folding transition* between the protein's ensemble of open conformations and its single lowest-energy native state. The transition happens where $p_{\text{N}} = p_{\text{D}}$, that is, where $\Delta G_{\text{fold}} = 0$. So, you can compute the midpoint temperature of the folding transition for this model using Equation 3.12:

$$T_m = \frac{-2\epsilon_0}{R \ln 28}. \tag{3.13}$$

Figure 3.9 shows the predicted populations of N, I, and D as functions of temperature. At low temperatures, the folded state is the most populated (most stable). At high temperatures, the model protein is unfolded. Temperature controls the balance. At low temperatures, the chain folds because the HH "sticking energy," which drives the system into the folded state, contributes more to ΔG_{fold} than the chain entropy does. At high temperatures, the chain entropy, which drives the system into the unfolded state, contributes more to ΔG_{fold} than the HH sticking energy does. The denaturation curve is sigmoidal because thermal energies at low temperatures are not sufficient to break bonds, thermal energies at intermediate temperatures break HH bonds, and at high temperatures, no further bond breakage happens because the bonds are already all broken. Consistent with experiments, this behavior leads to a peak in the heat capacity, which is, by definition, just the point at which the incremental energy absorption versus temperature is maximal.

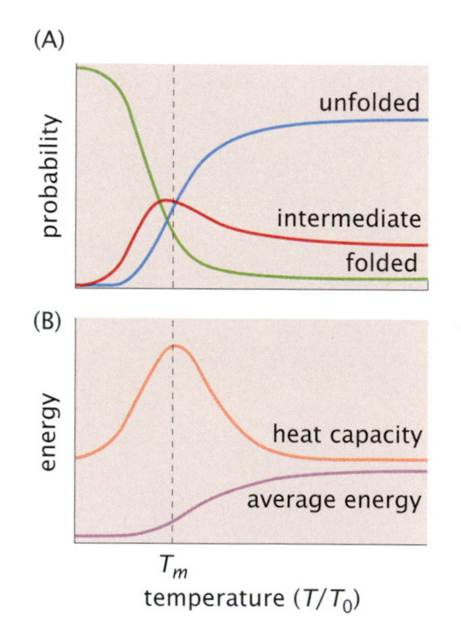

Figure 3.9 Heating denatures the model sequence HPPHPH. (A) Relative populations p_{N} of folded (2 HH contacts), p_{I} of intermediate (1 HH) and p_{D} of unfolded states (0 HH contacts) versus temperature using Equations 3.8 and 3.9. (B) Average energy $\langle\epsilon\rangle = 2p_{\text{N}}\epsilon_0 + p_{\text{I}}\epsilon_0 + 0p_{\text{D}}$ (*purple*). Heat capacity = $d\langle\epsilon\rangle/dT$ (*orange*). A fixed HH sticking energy ϵ_0 defines a constant $T_0 = \epsilon_0/R$, which has units of temperature, so T/T_0 on the x-axis is a dimensionless temperature scale.

In this model, there are three states: N, I, and D. This model sets the stage to discuss protein-folding *cooperativity* in Chapter 4. But, alternatively, you are free to collect together and label states in whatever ways are convenient for your problem of interest. For example, if your experiment measures only native and non-native states, then you could define the population of non-native molecules as $p_{\text{non-native}} = p_{\text{I}} + p_{\text{D}}$.

Proteins Have a Folding Code

This toy lattice model shows the nature of the *protein folding code*, that is, how different amino acid sequences can encode the folding of chain molecules into different specific native structures. In the HP model, a protein folds to a compact state with a hydrophobic core because such structures maximize the number of HH pairings among the amino acids. The sequence HPPHPH encodes one particular folded conformation at low temperatures: its single native structure is the conformation having the maximal number of HH contacts. For a different HP sequence, the maximization of HH contacts will lead to a different native structure. (Some sequences, however, do not encode a unique fold; for example, the all-P sequence PPPPPP does not fold at all.) This model shows that a simple *binary code* (that is, sequence of hydrophobic and polar monomers) is sufficient to cause a chain molecule to adopt a single compact conformation from among a large space of alternative possible conformations.

Evidence that the protein folding code is dominated by the binary HP patterning comes from the experiments of Kamtekar et al. [4]. In those experiments, Hecht and coworkers randomized the sequence of amino acids in a four-helix-bundle protein, subject only to the constraint that interior residues must be H and exterior residues must be P. They found that these HP sequences all folded stably into the four-helical-bundle structure, indicating the importance of the HP sequence pattern in specifying the native fold.

This simple binary-code lattice model captures the basic ingredients of folding: namely, that a chain has many non-native conformations and only one native conformation, and that conformations cannot violate *excluded volume* (that is, two amino acids cannot occupy the same space) (see Figures 3.7 and 3.8.) This and other types of models have shown that for many (but not all!) properties of proteins, atomic details are not needed. In fact, details are sometimes more distracting than helpful. For example, where entropies are important, as in folding, it is usually more crucial that a model be able to count conformations correctly over the whole conformational space than capture high-resolution atomic detail. Moreover, lattice models are also useful in "theoretical experiments" in which principles are explored by turning interactions on and off; you can explore entropies in such control experiments since all possible states can be enumerated. Of course, for other properties, other details can be important.

However, if folding is dominated by the HP code, how do we explain the prevalence of secondary structures, like helices and sheets? To explain secondary structures, we need two additional factors. First, hydrogen bonding, a signature of secondary structures, must play a role. But, helices and β-hairpins, when covalently snipped out of proteins and put by themselves into water, usually don't form very stable secondary structures. So, hydrogen bonding alone is not sufficient to explain the stabilities of short helices and sheets in native proteins. Tertiary forces must help to stabilize these forms. Next we describe how chain compactness stabilizes secondary structures, and how compactness reduces the number of conformations.

The Collapse of the Chain Helps Stabilize Secondary Structures

When a polymer chain is confined within a small volume, it will preferentially populate secondary structures, such as α-helices and β-sheets. This is a simple geometric consequence of confining any 1D rope-like object in a tight space at high density. The only systematic and regular way to pack a chain or a rope into a small space is to use helix-like or sheet-like conformations (**Figure 3.10**). Think of the lines of people in an airport security line. The lines run back and forth in regular patterns, resembling a squared-up β-sheet in 2D; there are no other regular arrangements that can pack a line of people into a tight space. Look at Figure 3.8; you see that only 4 of the 36 possible configurations would fit into a 2×3 compact lattice, and they have more secondary structures than the others. Figure 3.10 shows results for 41 different

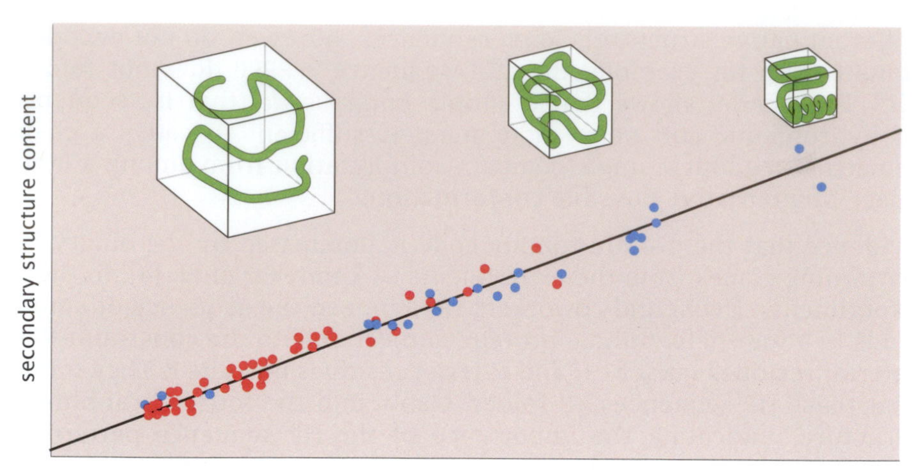

Figure 3.10 Increasing the chain compactness increases the secondary structure content. Polymer chains can fill tight spaces only if they form regular structures such as helices and sheets. The number of conformational microstates diminishes in going from the largest to smallest confining space. Plotted here is the relative compactness of the unfolded state versus secondary structure. Compactness is measured by the inverse of the Stokes radius and secondary structure by circular dichroism. *Red* circles represent proteins that were studied in fully unfolded states; *blue* circles represent proteins studied in their stable partially unfolded states. (Adapted from VN Uversky and AL Fink. *FEBS Lett*, 515:79–83, 2002.)

proteins, each under a condition that causes it to have some intermediate compactness (the inverse of the average volume occupied, that is, of the radius cubed), neither fully unfolded nor fully folded. It shows that the amount of secondary structure a protein adopts is proportional to the chain's compactness. So, secondary structures in proteins are partially stabilized by cooperation with the hydrophobic interactions that stabilize the compact structures.

Protein Folding Energy Landscapes Are Funnel-Shaped

A protein folding *energy landscape* illustrates the balance between a folding protein's interactions and its chain entropy. An energy landscape is a mathematical expression of an energy or free energy as a function of the various *degrees of freedom*, that is, of the conformational options that are available to the molecule. Even for the simple models discussed previously, an energy landscape is complex and of high dimensionality. There are two main types of cartoons that help to visualize such mathematical functions. One way to visualize energy landscapes is to show the *density of states* versus the energy $n\epsilon_0$, where $n = 0$ defines the denatured state, $n = 1$ the intermediate state, and $n = 2$ the native state (see Figure 3.8). For this HP sequence, we have 28 microstates for $n = 0$, 7 microstates for $n = 1$, and 1 microstate for $n = 2$ (Figure 3.11A). Turn this plot on its side to see the funnel shape

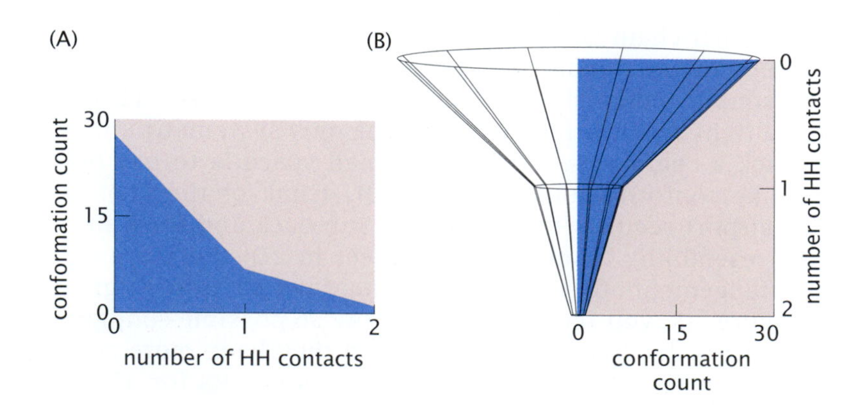

Figure 3.11 An energy landscape funnel describes the density of states. (A) The density of states for our HPPHPH 6-mer. (B) Now, rotate the axes of (A) to see the energy landscape. It has a funnel shape. There are many conformations having high energy (few HH contacts), so these states have high conformational entropy. There are few conformations having low energy (more HH contacts), so these are low-entropy states. The native (N) state has the maximum number of HH contacts in this model.

of n versus the density of states. Figure 3.11B shows that the number of states is large at the top (there are many open conformations having high energy), smaller in the middle (representing the fewer different states of intermediate energy), and fewest at the bottom (representing the small number of states that have the lowest possible energy). While the exact shape of this curve will depend on the monomer sequence and the type of model used to represent proteins, the basic physical principle of the funnel shape of the energy landscape applies to all proteins in any model—namely that there are always more conformations of high energy than of low energy.

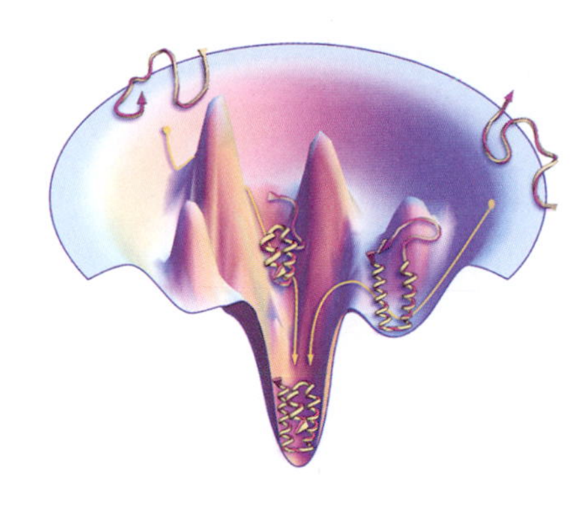

Figure 3.12 Another type of energy landscape also shows the funnel shape for protein folding. This type of cartoon energy landscape represents the free energy (y-axis) versus the chain degrees of freedom (x-axis). Like Figure 3.11, it shows the many unfolded states at the top and the few native-like states at the bottom. But this landscape also shows schematically the hills and valleys in energy of the different chain conformations. (From KA Dill and JL MacCallum. *Science*, 338:1042–1046, 2012. Reprinted with permission from AAAS.)

There is also another type of diagram for visualizing an energy landscape. In this case, we plot the free energy of the chain as a function of the many different degrees of freedom (bond angles, bond lengths, etc.). Figure 3.11 shows a two-dimensional version of this type of high-dimensional energy landscape. Such pictures sometimes show landscapes with bumps and wiggles: in such cases, the independent variable that is being represented is not the single density-of-states value (as in Figure 3.11), but rather one of the large number of possible actual conformational coordinates (as in Figure 3.12). Such pictures provide a way of thinking about trajectories from some particular chain conformation to another, indicating, metaphorically, how the skiers reach the bottom (native state) of a mountainside that has trees, bumps, gulleys, and other obstacles.

SIMPLE PROTEINS DENATURE WITH TWO-STATE THERMODYNAMICS

Now, let's explore a different simple model—called the *two-state model*—for describing experiments on the folding stabilities of small globular single-domain proteins. To explain the term "two-state," imagine that you could reach into a protein solution and pull out protein molecules, one at a time. For materials such as proteins that undergo two-state transitions, at the denaturation midpoint (where the fractions are equal, $f_N = f_D = \frac{1}{2}$ and therefore $\Delta G = 0$), you would find that half of the protein molecules are fully folded and half are fully unfolded. The alternative is that you might find *intermediate states*; that is, some individual molecules would be *partially folded*. Small proteins usually have two stable states and no partially folded equilibrium intermediates. Processes are also called *cooperative* if they undergo two-state transitions.

You can determine the cooperativity from calorimetry (see Figure 3.3). You use two measurements. First, calorimetry experiments give a quantity called the *calorimetric enthalpy*, ΔH_{cal}. This direct measurement requires no model-based interpretation. Second, you measure two other quantities from a calorimetry experiment: T_m is the melting midpoint temperature and $\Delta C_{p,\max}$ is the peak of the heat capacity (after subtracting the baseline value). Now, combining the latter two quantities with a *model assumption that the system in your calorimeter has a two-state transition* gives a prediction of the *van't Hoff enthalpy*, $\Delta H_{\text{vH}} = 2T_m \Delta C_{p,\max}$. Therefore, if these two enthalpies are equal, $\Delta H_{\text{vH}} = \Delta H_{\text{cal}}$, this means that the transition is two-state, involving no intermediates. Such data provide one form of evidence that typical small single-domain globular proteins undergo two-state folding transitions. Another indication that a transition is two-state is when two different experimental quantities, such as circular dichroism measurements of secondary structure and ultraviolet spectroscopy measurements of the burial of tryptophan groups, have denaturation curves that superimpose on each other. If those curves do not superimpose, it indicates the presence of intermediates. We explore the physical origins of protein-folding cooperativity in Chapter 4.

There is extensive experimental data on the thermal stabilities for reversible folding and unfolding of small proteins: the free energies $\Delta G_{\text{fold}} = G_{\text{D}} - G_{\text{N}}$, enthalpies ΔH, entropies ΔS, and heat capacities, ΔC_p. These thermal quantities mainly depend just on the number N of amino acids in the chain. So far, extensive studies have shown no other strong dependence of the thermal properties of folding stability. Stability does not appear to depend on the amount of secondary structure or types of tertiary structure, or numbers of hydrophobic amino acids or hydrogen bonds, or counts of salt-bridging ion pairs, for example. This is remarkable because other important properties of proteins—such as their native structures and biochemical mechanisms—can depend strongly on such details. This simplicity allows us to readily capture the observed dependencies of folding stability on temperature, denaturants, pH, salts, and the effects of protein confinement within tight spaces. In the next section, we assume that proteins fold with two-state cooperativity. In Chapter 4, we explore the physical basis of two-state cooperativity.

We begin with the chain entropy. As illustrated in Figure 3.13, for a chain having N amino acids, the number of conformers in the denatured state will be approximately $Q = z^N$, where z is the number of different conformations per amino acid. So, we approximate the chain entropy in the denatured state as

$$S_D = R \ln Q = NR \ln z. \tag{3.14}$$

For a 2D square lattice, you would use $z = 3$ (see Figure 3.7). For a 3D simple cubic lattice, you would use $z = 5$. These are the numbers of lattice step directions a bond can take from a given lattice site to a neighboring site without landing on top of the preceding bond at that site (but not accounting for other longer-range overlaps).

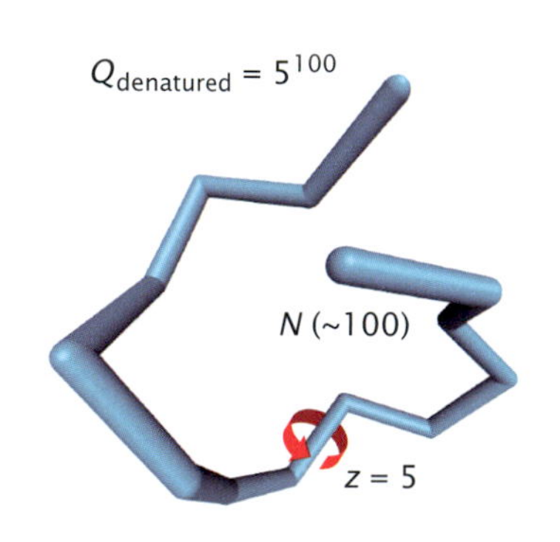

Figure 3.13 To compute the chain entropy, count the conformations and take the logarithm. z represents the number of rotational conformers around each peptide bond and N is the total number of amino acids. So, in a simple approximation, the folding from the denatured state ($Q_{\text{D}} = z^N$) into the native state ($Q_{\text{N}} = 1$) has a chain entropy decrease of $\Delta S_{\text{fold}} = S_{\text{N}} - S_{\text{D}} = -NR \ln z$.

Next, let's represent the energetics of folding using a *transfer model*: in the folding process, an amino acid is transferred from being solvated in water to being buried in the protein core. Then the folding free energy will be approximately

$$\Delta G_{\text{fold}}(N) = N(g + RT \ln z), \tag{3.15}$$

where $g < 0$ is the free energy of transferring an amino acid from water into a hydrophobic-core environment (resembling ϵ_0 in our previous lattice model (see Figure 3.8). The main points embodied in Equation 3.15 are that (i) the total contact free energy and the entropy of folding for the whole protein should scale linearly with the number of residues, (ii) the native structure is stabilized by the residue–residue attractions captured in g (note that g is negative), and (iii) the chain entropy opposes collapse.

Protein Stability Depends Linearly on Denaturant Concentration

Chemical agents such as guanidine hydrocholoride and urea tend to denature native proteins. Other chemical agents, such as glycerol, sugars, some salts, TMAO, and sarcosine, tend to have the opposite effect; they stabilize native proteins. Added denaturants cause a water solution to be a more favorable environment for amino acids, helping to unfold proteins. Added stabilizer compounds cause a water solution to be a less favorable environment for amino acids, driving proteins more strongly toward their native structures. We can use Equation 3.15 to describe how chemical agents affect protein stability.

In the transfer model, adding denaturants or stabilizers linearly weakens or strengthens the interactions experienced by each residue (with their surroundings):

$$g(c) = g_0 + m_1 c. \tag{3.16}$$

In the absence of denaturant ($c = 0$), the free energy of dissociation of a residue from its neighbors in the protein is g_0. The reason for the linear dependence on denaturant concentration, $m_1 c$, is that increasing the amount of denaturant in the solution increases proportionally to the amount of denaturant in the first solvation shell around each nonpolar solute molecule (or solvent-exposed residue in our case), decreasing proportionally to the solute–solute attraction. So, you can think of denaturant molecules as being distributed randomly in every available space in solution, including at sites in the first shells around nonpolar solutes (called a *mean-field* approximation). Substituting Equation 3.16 into Equation 3.15 predicts a linear dependence of the folding free energy on denaturant concentration,

$$\Delta G_{\text{fold}}(c) = N(RT \ln z + g_0 + m_1 c). \tag{3.17}$$

Equation 3.17 predicts that the slope, $m = Nm_1$, called the *m-value*, of $\Delta G_{\text{fold}}(c)$ versus c is a product of the increased surface area of monomers exposed upon denaturation multiplied by the free energy per unit surface area. Figure 3.14 shows that the experimentally observed m-values increase linearly with chain length N, as predicted.

For denaturing agents, $m_1 > 0$; that is, increasing the concentration of the chemical in solution decreases the protein's stability. On the other hand, for chemical agents that stabilize proteins, $m_1 < 0$ in this case, increasing the chemical concentration stabilizes the folded protein. Stabilization can arise from different physical causes. Some stabilizers act to increase the polarity of an aqueous solution. Others stabilize by excluded volume: such agents occupy space in an otherwise inert way,

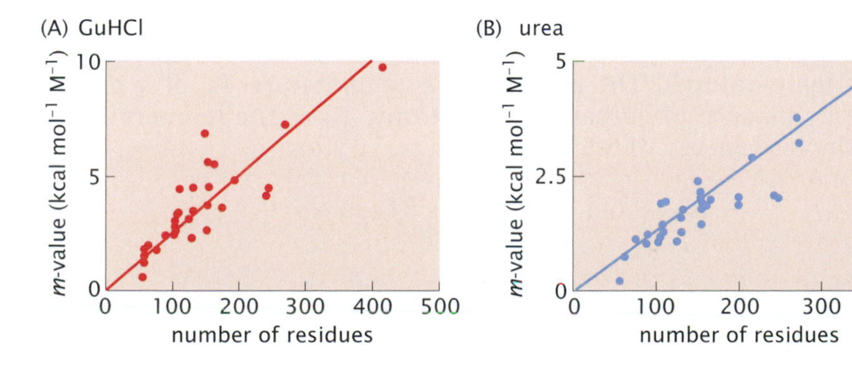

Figure 3.14 The denaturant *m*-value depends linearly on chain length. Denaturation by GuHCl (A) or urea (B) shows that the m-values depend linearly on the chain length N as indicated by Equation 3.17 [5]. (Data from JK Myers, CN Pace, and JM Scholtz. *Protein Sci*, 4:2138–2148, 1995.)

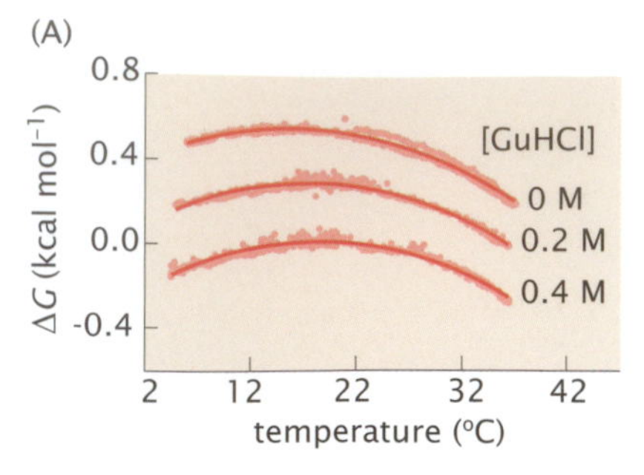

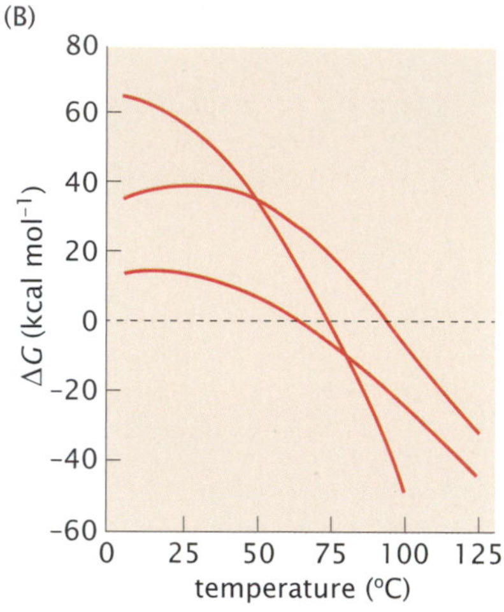

Figure 3.15 Protein folding free energies depend on temperature. (A) Denaturation curves of yeast ferricytochrome *c*. In this experiment, denaturation is also assisted by added guanidine hydrocholoride, so that the downward curvature on the left is observable. This illustrates *cold denaturation*, a temperature *below which* the protein will denature. (B) Unfolding free energies of three different proteins, indicating their unfolding at high temperatures. (A, from CJ Nelson, MJ LaConte, and BE Bowler. *J Am Chem Soc*, 123:7453–7454, 2001. With permission from American Chemical Society; B, from GI Makhatadze and PL Privalov. *Adv Protein Chem*, 47:307–425, 1995. With permission from Elsevier.)

restricting the conformations that the denatured protein could adopt, therefore effectively stabilizing the native state. For example, polysaccharide chains are sometimes covalently linked onto biotechnologically important proteins, such as erythropoetin (EPO), to give stable solution formulations. Next, let's consider how protein stabilities depend on temperature.

Protein Stability Is a Nonlinear Function of Temperature

Figure 3.2 and Figure 3.15 show that protein stability is not a linear function of temperature. Extensive experiments on the transfer of amino acids from water into oil-like media show that the residue-residue interaction free energy, which is g in our simple transfer model, depends on temperature.

When two nonpolar groups dissociate in water, there is a large positive change in the heat capacity per amino acid, Δc_p, that is approximately independent of temperature. To capture experimental data on model compound transfer data requires three parameters: Δc_p; T_h, the temperature at which the enthalpy of transfer is zero; and T_s, the temperature at which the entropy of transfer is zero. Model-compound transfer experiments can be modeled as

$$\Delta g(T) = \Delta h - T\Delta s = \Delta c_p(T - T_h) - T\Delta c_p \ln\left(\frac{T}{T_s}\right). \tag{3.18}$$

Now, by substituting Equation 3.18 into Equation 3.15 and using $g = g_0 + \Delta g(T)$, our folding model gives

$$\Delta G_{\text{fold}}(T) = N\left[RT\ln z + g_0 + \Delta c_p(T - T_h) - T\Delta c_p \ln\left(\frac{T}{T_s}\right)\right]. \tag{3.19}$$

Equation 3.19 predicts that $\Delta G_{\text{fold}}(T)$ is a curved function of temperature. It predicts two denaturation temperatures, that is, points at which $\Delta G_{\text{fold}} = 0$. One denaturation midpoint occurs at a high temperature: above that point, the conformational entropy dominates the free energy, and the chain unfolds, as shown also in the HP lattice model from before. Proteins can also be denatured, in principle, at low temperatures. Cold denaturation is a peculiarity of hydrophobic interactions in water. At low temperatures, a protein would unfold because the HH interactions become weaker as temperature is lowered (but this is often obscured by the freezing point of water).

This transfer model predicts that that the entropy, enthalpy, heat capacity, and free energy of protein folding should depend linearly on the chain length N, consistent with the experiments shown in Figure 3.16 [6].

Proteins having stronger intramolecular attractions have higher denaturation temperatures. The denaturation temperature T_m of a protein reflects its balance of enthalpy and entropy. T_m is the temperature at which the free energy of folding is zero,

$$\Delta G_{\text{fold}}(T_m) = N[g(T_m) + RT_m \ln z] = 0, \tag{3.20}$$

so

$$T_m = \frac{-g(T_m)}{R\ln z}. \tag{3.21}$$

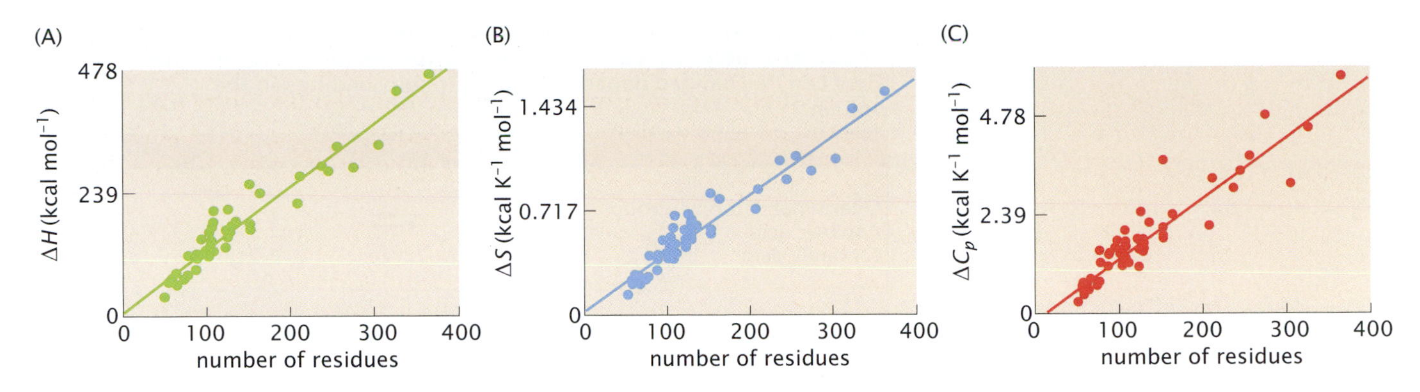

Figure 3.16 Thermal properties of proteins depend linearly on chain length. (A) ΔH, enthalpy of unfolding at 373.5 K, (B) ΔS, entropy of unfolding at 385 K, and (C) ΔC_p, temperature-independent heat capacity of unfolding, determined for 59 proteins, are plotted as functions of chain length N. The lines are linear regressions. (Adapted from AD Robertson and KP Murphy. *Chem Rev*, 97:1251–1267, 1997.)

Recalling that $g < 0$ (residue–residue attractions are favorable in the absence of denaturant), you see that stronger attractions lead to a higher T_m. As an example, if you substitute the value $z = 7.54$ for an average (ideal) protein [5], and a typical melting temperature, say, $T_m = 353\,\text{K}$, you find that $g(T_m) = -1.43\,\text{kcal mol}^{-1}$.

Folding is driven by a small difference between a large chain entropy that opposes folding and a large residue–residue contact free energy that favors folding. At T_m, the net folding free energy is zero, $\Delta G_m = 0$. At other temperatures (such as room temperature or physiological temperatures), a protein's stability is small: ΔG is usually around 5–20 kcal mol^{-1}. This small stability may be important biologically because, if proteins were too stable, they might be unable to respond to changes in the environment, or to undergo conformational changes that are essential for functioning and recycling of amino acids.

PROTEINS TEND TO UNFOLD IN ACIDIC OR BASIC SOLUTIONS

Proteins can denature if they are in acidic or basic solutions. This is because such solutions cause proteins to have a net charge. Proteins become more positively charged in acid solutions. Proteins become more negatively charged in basic solutions. The consequence is that if a protein has a net charge on it (either positive or negative), there will be net charge repulsions among pairs of charged residues. Those charge repulsions tend to unfold proteins because unfolding relieves the high charge density in the folded state.

The charges on proteins come from the acidic amino acids (glutamic acid and aspartic acid) or the basic amino acids (arginine, lysine, and histidine). For example, the protonation of an acidic side chain can be expressed as the following equilibrium:

$$HA \xrightarrow{K_a} \text{H}^+ + A^-, \tag{3.22}$$

where A represents an acidic group, such as an aspartic or glutamic acid side chain, H^+ is a dissociated proton and HA is the protonated uncharged form. If the pH of the solution around an acidic side chain is higher than about 4.1, these acidic side chains will have a net negative charge because they give up their protons to the surrounding

Table 3.2 Typical pK_a values of ionizable groups in proteins. These are intrinsic values these groups would have in water. They can be changed depending on their surroundings in the protein

Group	Acid		Base	pK_a
C-terminal carboxyl group: Aspartic acid Glutamic acid	$-C(=O)-O-H$	⇌	$-C(=O)-O^-$	3.1 4.1 4.1
Histidine*	imidazolium (N^+–H, N–H)	⇌	imidazole (N, N–H)	6.0
N-terminal amino group	$-NH_3^+$	⇌	$-NH_2$	8.0
Cysteine	$-S-H$	⇌	$-S^-$	8.1
Tyrosine	phenyl–O–H	⇌	phenyl–O^-	10.9
Lysine	$-C-NH_3^+$	⇌	$-C-NH_2$	10.8
Arginine	$-N-C(=NH_2^+)-NH_2$	⇌	$-N=C(-NH_2)-NH-H$	12.5

* Histidine is the only amino acid that has a pK_a in the common range of experiments. So, in some locations in protein structures it exists in charged form and in others it is uncharged.

solution. Or, if the pH of the solution is lower than about 10, the basic amino acids will have a net positive charge because they take up protons from the solution. Histidine can also have a net charge if the pH is below about 6 (Table 3.2). So, if the pH of a solution is sufficiently high or low, a given protein can have a net charge, due to net charges on whatever numbers of these types of side chains it has. If the protein has a sufficiently high net charge, those charges will repel each other, causing an expansion of the chain, denaturing the protein. This is how acids or bases denature proteins. A simple model is given in Appendix 3A.

Two important consequences follow from these electrostatic contributions to protein folding stability. First, Figure 3.17 shows that a protein's melting (denaturation) temperature, a quantity that reflects

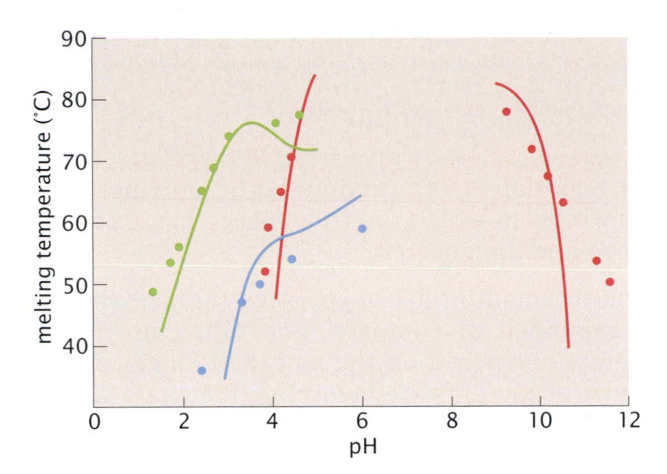

Figure 3.17 Protein melting temperature depends on pH. When proteins are put into solutions of low pH (acids) or high pH (bases), their denaturation temperature is reduced. This is because the protein develops net charge, and because charges repel, driving unfolding. The *red* points and line represent acid and base destabilization of myoglobin. The *green* points and line represent lysozyme, and the *blue* ones represent RNase A. The lines on this plot are calculated by combining the hydrophobic free energy and chain entropy in Equation 3.19 with the electrostatics in Equation 3.A.10. (From K Ghosh and KA Dill. *Proc Natl Acad Sci USA*, 106:10649–10654, 2009.)

the protein's folding stability, will be reduced in acidic or basic solutions.

Second, if a protein has a net charge (that is, the protein is not at its *isoelectric point*), adding salt to the solution will typically stabilize the folded state. (Adding salt increases the charge shielding or κ in Equation 3.A.10.) The salt molecules swarm around the protein, shielding charges, thus weakening the electrostatic repulsions among the fixed charges on the protein.

Proteins that have a large net charge and low hydrophobic content usually lack ordered structure (Figure 3.18) [7].

Now, we go beyond the simple idea that all z^N denatured conformations are equivalent to each other. Understanding some properties of proteins requires that we take into account that some denatured-state conformations can have different radii and different free energies than others. The denatured state is an ensemble of conformations.

A DENATURED STATE IS A DISTRIBUTION OF CONFORMATIONS

In this section, we make three points about the denatured states of proteins. First, a denatured state is a broad distribution of microstates. Second, the denatured state distribution can shift its average size under different denaturing conditions. Third, denatured chains that are located in confined or restricted spaces have different properties than those that are unrestricted. Let's first consider the distribution of the different end-to-end distances r of the denatured protein. Modeled as a random-flight polymer chain, the free energy of a denatured conformation is [2]

$$\frac{G_D}{RT} = -N \ln z + \frac{3}{2}\left(\frac{r}{r_0}\right)^2 + \frac{aN^2 v}{r^3}, \tag{3.23}$$

where $r_0 = Nb^2$ is the average end-to-end distance of a random-flight or freely jointed chain of N links, each of length b [2]. The second term in Equation 3.23 expresses the *elastic energy*, the basis for the retractive force in rubber bands, which arises because the end-to-end distance of

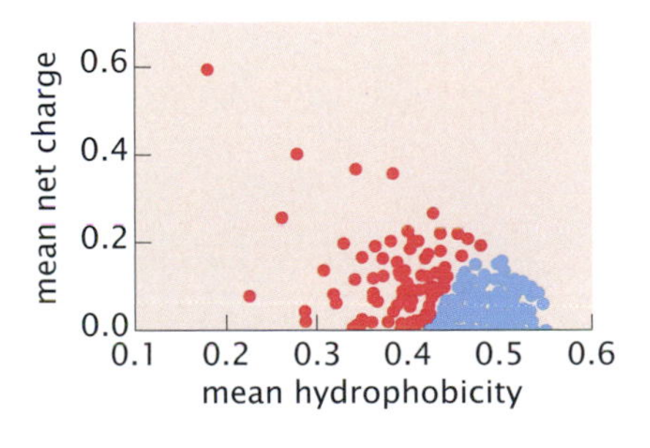

Figure 3.18 Natively unfolded proteins (*red*) tend to have little structure because they are more charged and less hydrophobic than folded proteins (*blue*). (From VN Uversky, JR Gillespie, and AL Fink. *Proteins: Struct., Funct., Genet.*, 41:415–427, 2000.)

a freely jointed polymer chain follows a Gaussian distribution about r_0:

$$G_{\text{elastic}} = -TS_{\text{elastic}} = -RT\ln(e^{-3r^2/(2r_0^2)}) = \frac{3}{2}RT\left(\frac{r}{r_0}\right)^2. \tag{3.24}$$

According to Equation 3.24, random-flight polymer chains act like Hooke's-law springs, in which the free energy has a square-law dependence on end-to-end distance r.

The third term in Equation 3.23 expresses the fact that chains prefer to be either expanded or compact, depending on nonlocal solvent-mediated interactions based on the so-called Flory approximation [8]. If intramolecular attractions are favorable, the chain will collapse. But, if interactions with solvent are more favorable, the chain will populate open expanded conformations. v is the volume of a chain residue. Nv/r^3 is the volume fraction of space that is occupied by the residues, or the probability of neighboring a given residue, which, multiplied by N gives the expected number of residue–residue contacts per residue. a is an interaction energy between a pair of residues ($a = 1 - g_0(T)$ in the model given previously). When $a < 0$ (poor-solvent conditions), residues are attracted to each other strongly, and the third term favors chain collapse. When $a > 0$ (good-solvent conditions), residues prefer to interact with the solvent, and the third term favors chain expansion.

The Radius of the Denatured Chain Grows as $N^{0.6}$

Different proteins will have different denatured-state radii r_D, depending on N, the number of amino acids in the chain. To find the approximate radius of the denatured state, r_D, use Equation 3.23 to compute the radius that minimizes the free energy,

$$\frac{d}{dr}\left(\frac{G_D}{RT}\right)\bigg|_{r_D} = 0, \tag{3.25}$$

which leads to

$$\frac{r_D}{r_0^2} - \frac{aN^2v}{r_D^4} = 0. \tag{3.26}$$

Rearranging Equation 3.26 and substituting the definitions of r_0 and a gives

$$r_D^5 = aN^2vr_0^2 = (1-g)b^2vN^3. \tag{3.27}$$

Equation 3.27 predicts that $r_D \sim N^{3/5}$. This prediction is in excellent agreement with experiments. Figure 3.19 shows the results of small-angle neutron scattering experiments on 28 different proteins, which indicate that $r_D \sim N^{0.598}$ [9].

What is the importance of this 3/5-power dependence? Classical polymer theory predicts three different regimes for the radius of a polymer molecule versus chain length: (1) an exponent of 1/3 ($R \sim N^{1/3}$) implies solvent conditions (called a Flory *poor solvent*) in which a chain is collapsed; (2) an exponent of 1/2 implies solvent conditions (called a Flory *theta solvent*) in which the chain conformations follow an ideal random flight, where the intrachain attractions exactly balance the excluded-volume repulsion. (3) an exponent of 3/5, as we have here, implies solvent conditions (called a Flory *good solvent*), in which the chain is expanded, and the excluded-volume repulsions are greater

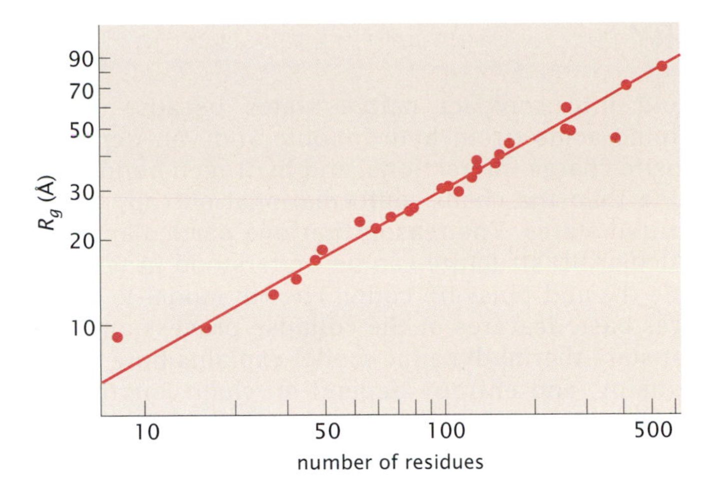

Figure 3.19 Average radii of denatured proteins increase with chain length. The radii of gyration R_g (radius measured from the center of mass) for chemically denatured, cross-link-free, and prosthetic-group-free proteins and peptides were determined by small angle X-ray scattering [9]. The R_g, or average end-to-end distances for denatured proteins, scales with N, the number of amino acids in the chains, as $r_D \sim N^{0.598}$. This agrees with the relation predicted by Flory for polymers in a good solvent, $R_g \sim N^{3/5}$ (From JE Kohn, IS Millett, J Jacob, et al. *Proc. Natl. Acad. Sci USA*, 101:12491–12496, 2004. Copyright (2004) National Academy of Sciences, USA.)

than the intrachain attractions. Therefore, proteins in native conditions resemble collapsed polymers in poor solvents, while proteins in strong denaturants act like expanded polymers in good solvents. Box 3.1 summarizes these scaling relationships.

Box 3.1 Geometric Scaling Relationships

	Native	Intermediate	Denatured
Radius	$R \sim N^{1/3}$	$R \sim N^{1/2}$	$R \sim R^{3/5}$
Surface area	$S \sim N^{2/3}$	$S \sim N^{1/3}$	$S \sim N^{6/15}$
Volume	$V \sim N^{1}$	$V \sim N^{3/2}$	$V \sim N^{9/5}$

Individual proteins can vary. They are not perfect spheres [10].

We are now in a position to explain how proteins behave in situations where they are in crowded conditions or in confined spaces. A protein is said to be *confined* when it is contained within a fixed space, for example inside a chaperone or ribosome cavity. A protein is *crowded* when it is in a solution with other inert molecules (such as polysaccharides or sometimes other proteins) that restrict the volume available to it. Proteins are crowded in normal cell environments, where nearly 25% of the volume is occupied by other protein molecules.

Confinement or Crowding Can Increase a Protein's Folding Stability

Putting a protein in a tight space can often increase its folding stability. Consider a protein inside a space that is *inert*, that is, a space that imposes only steric constraints, and does not otherwise interact energetically with the protein. Steric confinement limits the possible conformations that the denatured state of the protein could have. Some of the denatured conformations that would otherwise have been highly expanded when the protein is free of constraints will not be allowed within the confinement volume. This means that confinement will reduce the conformational entropy of the denatured chain, increasing the free energy of the denatured chain. That is, confining the chain causes the difference, $\Delta\Delta G_{\text{fold}} = (\Delta G_{\text{fold}})_{\text{confined}} - (\Delta G_{\text{fold}})_{\text{free}} < 0$. So, confining the chain effectively makes the folded state more stable (Figure 3.20). Therefore, confinement also increases a protein's denaturation temperature (Figure 3.21).

unconfined

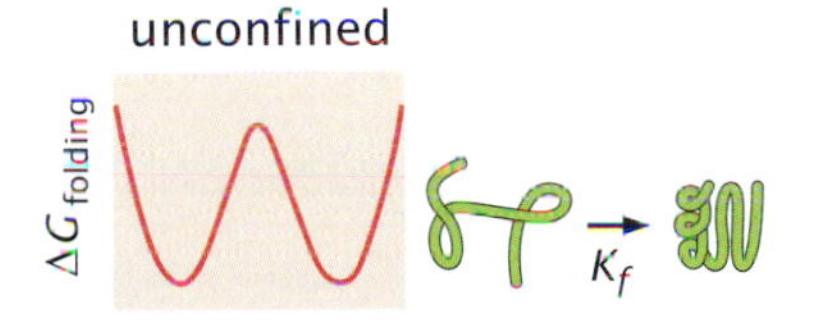

confined in pores

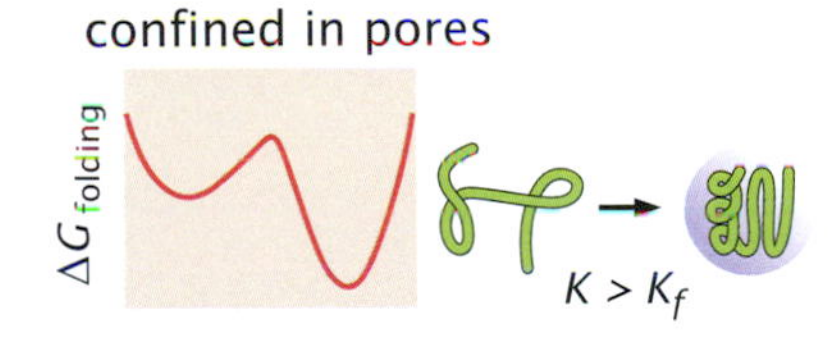

confined at surfaces

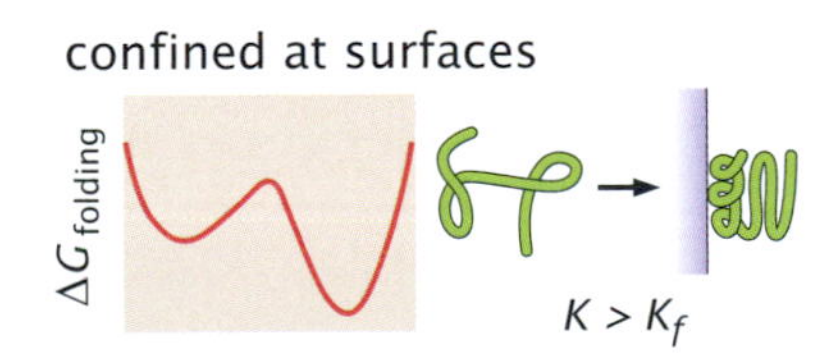

Figure 3.20 A protein in a tight space has greater folding stability, because its denatured state has fewer conformations and thus less conformational entropy. (Adapted from H-X Zhou and KA Dill. *Biochemistry*, 40:11289–11293, 2001.)

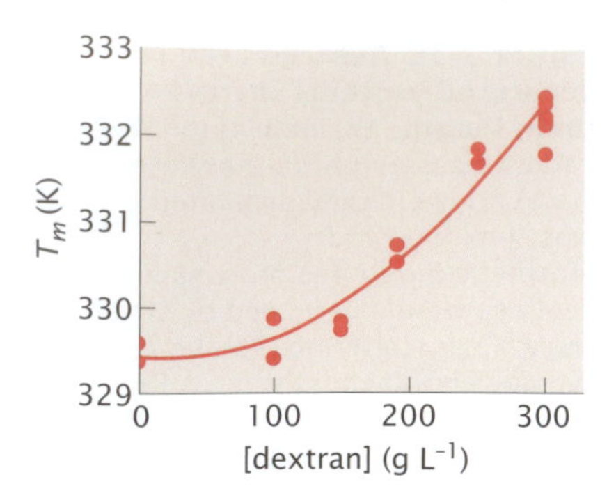

Figure 3.21 Crowding stabilizes lysozyme. The thermal denaturation temperature T_m of hen egg-white lysozyme, determined by circular dichroism, increases with the concentration of dextran, an inert crowding agent. The line indicates a square-law dependence of T_m on the volume excluded by the dextran, consistent with a decrease in the entropy of the unfolded state. (From K Sasahara, P McPhie, and AP Minton. *J Mol Biol*, 326:1227–1237, 2003. With permission from Elsevier.)

SUMMARY

Proteins fold into compact native states because the attractions between amino acids (from hydrophobic and van der Waals interactions, opposite-charge interactions, and hydrogen bonding) are greater in magnitude than the chain conformational entropy that favors the open denatured states. The reason that one particular sequence folds to one particular 3D structure can be understood in terms of a binary hydrophobic (H) and polar (P) code. The HP model is a lattice model that captures basic features of the collapse process and the sequence code. A two-state thermodynamic model explains how the folding free energy, enthalpy, and entropy depend on chain length, temperature, and denaturants. The denatured state has both higher energies (fewer contacts) and higher entropies. Ensembles of denatured conformations are not all the same. Changing solution conditions can expand or contract the denatured state. The scaling powers depend on the state of a protein: $N^{1/3}$ for the native state and $N^{3/5}$ for the denatured state. Confining a protein molecule in an inert container can stabilize it. When a protein is in an acidic or basic solution, its titratable side chains can be charged, and if the net positive or negative charge on the protein is sufficiently high, the protein will denature because of like-charge repulsions.

APPENDIX 3A: A SIMPLE ELECTROSTATIC MODEL OF DENATURATION BY ACIDS AND BASES

How do acids and bases destabilize folded proteins? First, to model the effects of solution *pH* on the side-chain protonation equilibria, let q_a be the net charge on each of the N_a acidic groups. Then the fractional charge per acidic group, f_a, is

$$f_a = \frac{q_a}{N_a} = \frac{[A^-]}{[A^-]+[HA]} = \frac{\alpha}{1+\alpha}, \tag{3.A.1}$$

where $\alpha = [A^-]/[HA]$. The acid dissociation constant K_a is

$$K_a = \frac{[A^-][H^+]}{[HA]} = \alpha[H^+]. \tag{3.A.2}$$

Now, we switch to the notation $\mathrm{p} = -\log$. That is, $\mathrm{pH} = -\log[\mathrm{H}^+]$ and $\mathrm{p}K_a = -\log K_a$. Side-chain p$K_a$ values are given in Table 3.2. Taking the logarithm (base 10), Equation 3.A.2 becomes

$$\alpha = 10^{\log\alpha} = 10^{\mathrm{pH}-\mathrm{p}K_a}. \tag{3.A.3}$$

Substituting Equation 3.A.3 into Equation 3.A.1 gives

$$f_a = \frac{q_a}{N_a} = \frac{10^{\mathrm{pH}-\mathrm{p}K_a}}{1+10^{\mathrm{pH}-\mathrm{p}K_a}}. \tag{3.A.4}$$

Equation 3.A.4 gives the *titration curve* for a given type of acidic group. Examples are shown in Figure 3.A.1.

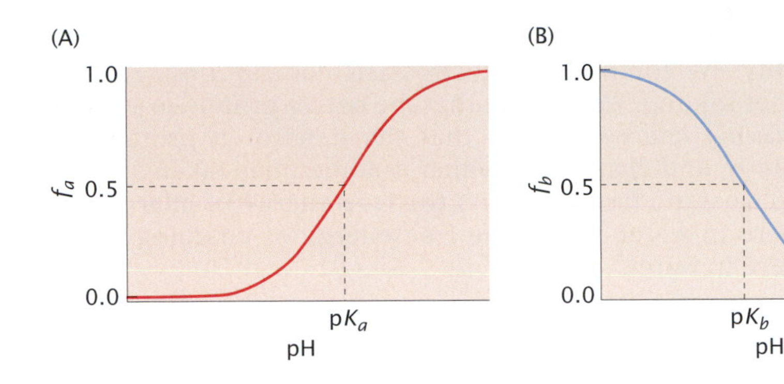

Figure 3.A.1 Titration curves calculated for (A) acids and (B) bases by Equations 3.A.4 and 3.A.6. When the pH of the solution equals the pK_a or pK_b, half of the ionizable groups carry a charge.

When the pH of the solution is lower than the pK_a of an acidic group, you have $K_a < [H^+]$. In this case, the acidic group will be mostly protonated (uncharged). On the other hand, when $pH > pK_a$, the acid will be mostly deprotonated (charged).

The same type of analysis applies to a basic group. The deprotonation equilibrium for a base is

$$BH^+ \xrightarrow{K_b} B + H^+, \tag{3.A.5}$$

where the fractional charge per basic group, f_b, is

$$f_b = \frac{[BH^+]}{[BH^+] + [B]} = \frac{q_b}{N_b} = \frac{10^{pK_b - pH}}{1 + 10^{pK_b - pH}}, \tag{3.A.6}$$

leading to a titration curve of the form shown in Figure 3.A.1B.

Basic groups become deprotonated (that is, they take the form B, rather than BH^+) when $pH > pK_b$. Basic groups that are deprotonated have no net charge; basic groups that are protonated have a net positive charge.

Now let's apply this reasoning to protein denaturation. Suppose a protein has N_a acidic and N_b basic groups. To keep the math simple, suppose all the acids have identical pK_a values and all the bases have identical pK_b values. The protein will then have q_b positive charges and q_a negative charges, where

$$q_a = N_a \left(\frac{10^{pH - pK_a}}{1 + 10^{pH - pK_a}} \right) \tag{3.A.7}$$

and

$$q_b = N_b \left(\frac{10^{pK_b - pH}}{1 + 10^{pK_b - pH}} \right). \tag{3.A.8}$$

The net charge on the protein will be

$$q_{net} = q_b - q_a. \tag{3.A.9}$$

The particular solution condition that causes a protein to have a net charge of $q_{net} = 0$ is called the *isoelectric point* of the protein. To put a protein at its isoelectric point, you adjust the pH of the solution. Different proteins have different isoelectric points because of their different collections of charged side chains.

Here is an approximate model for how a protein's electrostatic charges affect its stability. We compute an electrostatic contribution, ΔG_{es}, to the free energy of folding. We assume that the native protein is a sphere of radius r_{N} having charge Q_{N} and that the denatured protein is a sphere of radius r_{D} and charge Q_{D} (which is sometimes taken as equal to Q_{N}). The electrostatic free energy is the free energy of charging up the native sphere in water minus the free energy of charging up the denatured sphere in water.

The total electrostatic free-energy difference between native and denatured states is[4]

$$\frac{\Delta G_{\text{es}}(T, pH, c_s)}{RT} = \frac{q_{\text{N}}^2 l_b}{2r_{\text{N}}(1+\kappa r_{\text{N}})} - \frac{q_{\text{D}}^2 l_b}{2r_{\text{D}}(1+\kappa r_{\text{D}})}, \tag{3.A.10}$$

where q_{N} is the total charge on the native protein, q_{D} is the total charge on the denatured protein, and l_b is a constant (for fixed temperature) called the *Bjerrum length.* κ is the inverse of the *Debye length.* The Debye length describes the distance through space over which interactions between charges diminish due to an intervening salt solution. The Debye length gets shorter with increasing salt concentration c_s according to the following relation:

$$\kappa^2 = 2c_s l_b. \tag{3.A.11}$$

The justification for using the dielectric constant of water for the native protein is that charged side chains are mainly on the protein's surface, and are largely solvated in both native and denatured states. The value of r_{N} can be determined by knowing the radius of gyration in the native state. The radius of gyration for the denatured state, r_{D}, can be obtained from experimental measurements, such as those shown in Figure 3.19.

[4]The electrostatic free energy for charging a sphere of radius r in a solvent having dielectric constant ϵ from zero charge to a charge of Q is

$$\Delta G_{\text{es}} = \frac{1}{2}\int \sigma\psi \, ds,$$

where $\sigma = q$ is the final surface charge. The integral is taken over the surface of the sphere, and ψ is the electrostatic surface potential [2]:.

$$\psi(r') = \frac{Cq}{\epsilon r(1+\kappa r)} e^{-\kappa(r'-r)},$$

where r' is the distance from the center of a sphere of radius r, $Ce^2\mathcal{N}$ is an electrostatic constant equal to $1.386 \times 10^{-4}\,\text{J m mol}^{-1}$. Integrating gives the free energy of charging the sphere [2]:

$$\frac{\Delta G_{\text{es}}(r)}{RT} = \frac{q_{\text{net}}^2 l_b}{2r(1+\kappa r)},$$

where l_b is the Bjerrum length,

$$l_b = \frac{1.386 \times 10^{-4}\,\text{J m mol}^{-1}}{\epsilon RT}.$$

R is the gas constant, T is the absolute temperature, and ϵ is the dielectric constant of water. The Bjerrum length is the distance over which the electrostatic energy between two unit charges in a given solution diminishes to RT. At room temperature, $l_b = 7.13\,\text{Å}$, using the dielectric permittivity of water, $\epsilon = 78$, at room temperature.

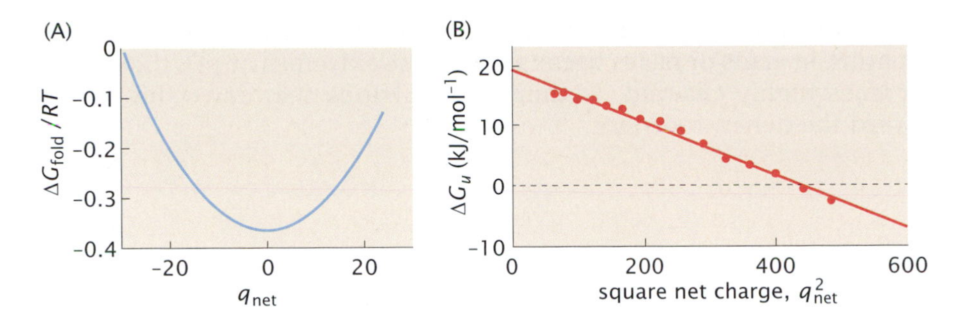

Figure 3.A.2 Protein stability depends on the square of the net charge. (A) The free-energy difference between folded and unfolded states is a function of the protein's net charge. The net charge on metmyoglobin is calculated from the pK_a values of its ionizable groups (B) The free energy of unfolding the ferricytochrome *c* molten globule depends on the net charge. Free energies of thermal unfolding of the acid-induced molten globule state were determined by scanning calorimetry for various charge states of horse ferricytochrome *c*. The line is fit to the linearized square of the net charge. (pK_a values in (A) are from E Breslow and FRN Gurd. *J Biol Chem*, 237:371–381, 1962. (B) From Y Hagihara, Y Tan, and Y Goto. *J Mol Biol*, 237:336–348, 1994. With permission from Elsevier.)

The pK_a of an acidic or basic side chain, as given in Table 3.2, is a property of its protonation equilibrium measured in water. But sometimes, in folded structures, protonatable side chains are buried in native hydrophobic cores. In those cases, the pK_a of a side chain can be different than its value in water. Those altered values of pK_a should be used in the given model, if they are known.

Equation 3.A.10 shows how to interpret charge effects on protein stability. It says that if you systematically vary the net charge, $q_{net} = q_b - q_a$ (Equation 3.A.9) by mutating a protein's acidic or basic amino acids, the protein's stability will decrease in proportion to the square of the net charge, q^2_{net}, on the protein. Figure 3.A.2A shows this prediction that the more net charge on the protein, the less stable the folded state will be. This is why extremes of pH, which lead to extremes of net charge on the protein, can unfold proteins (Figure 3.A.3). Experimental confirmation of this prediction through random acetylation experiments [11] is shown in Figure 3.A.2B.

In summary, at low pH (high concentrations of H^+ ions), both the acidic and basic side chains are protonated—the acidic side chains are mostly in the form of –COOH and the basic side chains are in the form of $-NH_3^+$, so the protein has a net positive charge. At high pH (low concentrations of H^+ ions), both the acidic and basic groups are deprotonated—mostly in the forms of $-COO^-$ and $-NH_2$, giving a net negative charge. In the middle range of pH, both acidic and basic side chains are charged, in the forms of $-COO^-$ and $-NH_3^+$. So, if the numbers of acidic groups is about the same as the number of basic groups, the net charge on

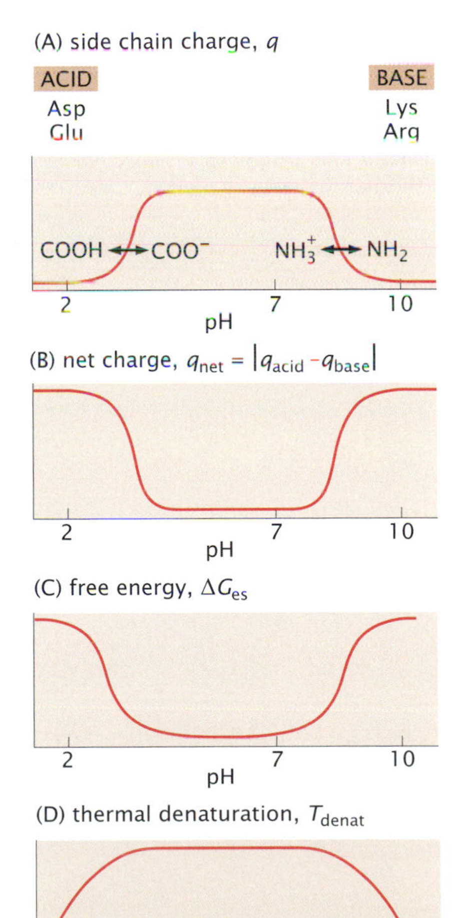

Figure 3.A.3 The charge and stability of a protein depend on solution pH. (A) As the solution is titrated from acidic to basic pH, most of the acidic –COOH side-chain groups become deprotonated. At high pH, basic $-NH_3^+$ side-chain groups give up a proton. (B) At very low pH, the protein has both –COOH and $-NH_3^+$ groups, so it is positively charged. At very high pH, the protein has both $-COO^-$ and $-NH_2$ groups, so it is negatively charged (C) The electrostatic free energy follows a pattern similar to the net charge. (D) These charge effects destabilize the protein for thermal denaturation at both extremes of pH.

the protein will be small in the mid-range of pH. Hence, proteins will denature in acids or bases because, at those extremes of pH, the protein becomes highly charged, leading to repulsions that drive the protein toward the denatured state.

REFERENCES

[1] CB Anfinsen, E Haber, M Sela, and FH White Jr. The kinetics of formation of native ribonuclease during oxidation of the reduced polypeptide chain. *Proc Natl Acad Sci USA*, 47:1309–1314, 1961.

[2] KA Dill and S Bromberg. *Molecular Driving Forces: Statistical Thermodynamics in Biology, Chemistry, Physics and Nanoscience*, 2nd ed. Garland Science, New York, 2011.

[3] KA Dill, S Bromberg, K Yue, et al. Principles of protein folding—a perspective from simple exact models. *Protein Sci*, 4:561–602, 1995.

[4] S Kamtekar, JM Schiffer, H Xiong, et al. Protein design by binary patterning of polar and nonpolar amino acids. *Science*, 262:1680–1685, 1993.

[5] K Ghosh and KA Dill. Computing protein stabilities from their chain lengths. *Proc Natl Acad Sci USA*, 106:10649–10654, 2009.

[6] L Sawle and K Ghosh. How do thermophilic proteins and proteomes withstand high temperature? *Biophys J*, 101:217–227, 2011.

[7] VN Uversky, JR Gillespie, and AL Fink. Why are "natively unfolded" proteins unstructured under physiologic conditions? *Proteins*, 41:415–427, 2000.

[8] PJ Flory. *Principles of Polymer Chemistry*. Cornell University Press, Ithaca, NY, 1953.

[9] JE Kohn, IS Millett, J Jacob, et al. Random-coil behavior and the dimensions of chemically unfolded proteins. *Proc Natl Acad Sci USA*, 101:12491–12496, 2004.

[10] D Flatow, SP Leelananda, A Skliros, et al. Volumes and surface areas: Geometries and scaling relationships between coarse-grained and atomic structures. *Curr Pharm Design*, 20:1208–1222, 2014.

[11] Y Hagihara, Y Tan, and Y Goto. Comparison of the conformational stability of the molten globule and native states of horse cytochrome *c*. Effects of acetylation, heat, urea and guanidine-hydrochloride. *J Mol Biol*, 237:336–348, 1994.

SUGGESTED READING

Dill KA and Bromberg S, *Molecular Driving Forces: Statistical Thermodynamics in Biology, Chemistry, Physics, and Nanoscience*, 2nd ed. Garland Science, New York, 2011.

Flory PJ, *Principles of Polymer Chemistry*, Cornell University Press, Ithaca, NY, 1953.

CHAPTER 4

Protein Binding Leads to Biological Actions

INTRODUCTION

What are the principles that underly protein actions? The mechanisms of proteins in motors, pumps, transporters, and signal and energy transducers, and in processes of regulation, catalysis, and self-assembly can seem extraordinarily clever and diverse. Yet, such complex actions are often simply the consequences of elementary underlying binding or unbinding events. Here, we describe binding processes, including coupled binding and allostery, which serve as a language for biological mechanisms, actions, and machines. This chapter draws from [1], where additional details can be found.

Proteins often act like miniature machines. Machines *transduce* one type of action into another; they act in *repetitive cycles*; and they are *reusable*. Your car engine converts (transduces) the explosive conversion of liquid to gas into pressure, leading to mechanical work and motion, in repetitive cycles. Electromagnetic motors transduce electrical currents into magnetic forces, also leading to work and motion in repetitive cycles.

Many proteins are transducers too. But proteins are neither heat engines nor electromagnetic devices. Protein transduction arises from *chemical potential* differences, that is, from out-of-equilibrium concentration imbalances, for example of ATP relative to ADP, or from concentration gradients, typically of ions across membranes. Proteins mediate coupled actions often through a binding event at one site, resulting in a conformational or *allosteric* change, affecting another binding event at a different site on the protein. For example, myosin is a molecular motor protein that moves along a track-like protein called actin. When ATP binds to myosin, myosin undergoes an allosteric change, dissociating from actin, allowing myosin to move relative to the track, on average in a single direction, resulting in muscle contraction.

Or, consider signaling. A small-molecule ligand outside a cell binds to a signaling protein that is in the cell membrane. It triggers a protein conformational change that leads to a change in binding of the protein to a second ligand inside the cell. In this way, chemical information is transmitted across the cell membrane. As another example, a transcription factor (TF) is a protein that binds DNA, and regulates the transcription of genetic information to RNA molecules that results in making specific proteins. Yet a third component—an activator or inhibitor protein (or a cofactor)—can regulate the amount of protein produced by binding to the TF or the DNA. And drug discovery is often a business of finding a third-party small molecule (the drug) that inhibits or enhances the binding of a protein to its normal substrate/metabolite. In this chapter, we express binding processes in terms of simple equilibria and rates; later (Chapters 10, 12, and 13), we describe the modeling of binding at a more structural level.

BINDING CAN BE MODELED USING BINDING POLYNOMIALS

We begin with *binding polynomials*, the most basic description of binding and coupled binding. In experiments, you can measure *binding curves*—the numbers of ligand molecules that bind to a protein as a function of the concentration of ligand molecules in solution. To extract useful information about the *binding affinities*, you make a model of the binding processes involved.

Binding Polynomials Are Used to Compute Binding Curves

Consider the general process of *multiple binding equilibria* of a ligand X with $i = 1, 2, 3, \ldots, t$ different binding sites on a protein P:

$$\begin{aligned} P &+ X &\overset{K_1}{\rightleftharpoons}\; & PX, \\ P &+ 2X &\overset{K_2}{\rightleftharpoons}\; & PX_2, \\ & & \vdots & \\ P &+ iX &\overset{K_i}{\rightleftharpoons}\; & PX_i, \\ & & \vdots & \\ P &+ tX &\overset{K_t}{\rightleftharpoons}\; & PX_t. \end{aligned}$$

Each equilibrium constant K_i is defined by

$$K_i = \frac{[PX_i]}{[P]x^i},$$

where $x = [X]$ is the concentration of free ligand, $[P]$ is the concentration of free P molecules, and $[PX_i]$ is the concentration of those P molecules that have i ligands bound. There can be many different ligation states of P molecules in solution at the same time. Some P molecules contain no ligand; their concentration is $[P]$.

What are the fractions of all the P molecules in solution that have i ligands bound? To compute fractions, you need a denominator. So, first, we sum up the concentrations of all of the species of P, which we denote by $Q[P]$:

$$\begin{aligned} Q[P] &= [P] + [PX_1] + [PX_2] + \ldots + [PX_t] \\ &= [P](1 + K_1x + K_2x^2 + \ldots + K_tx^t) \end{aligned}$$

$$\Rightarrow Q = 1 + K_1x + K_2x^2 + \ldots + K_tx^t$$
$$= \sum_{i=0}^{t} K_ix^i, \tag{4.1}$$

where $K_0 = 1$. This sum is called the *binding polynomial Q*. We define Q in this way (factoring out $[P]$) for convenience, to be dimensionless. The fraction (or population) P of molecules that are in the i-liganded state is $p(i) = [PX_i]/(Q[P]) = K_ix^i/Q$, for $i = 1, 2, \ldots, t$.

A key quantity for comparing models with experiments is the average number $\nu(x) = \langle i \rangle$ of ligands bound per P molecule, as a function of the ligand concentration x. Using the definition of average, $\langle i \rangle = \sum_0^t ip(i)$, gives

$$\nu(x) = \langle i \rangle = \frac{K_1x + 2K_2x^2 + 3K_3x^3 + \ldots + tK_tx^t}{1 + K_1x + K_2x^2 + K_3x^3 + \ldots + K_tx^t}$$
$$= Q^{-1}\sum_{i=0}^{t} iK_ix^i$$
$$= \frac{x}{Q}\frac{dQ}{dx} = \frac{d\ln Q}{d\ln x}. \tag{4.2}$$

The binding polynomial Q is the sum over all possible ligation states. Q resembles a partition function. Whereas a partition function is a sum over all possible energy levels, the binding polynomial Q is the sum over all possible ligation states. Equation 4.2 shows that the derivative of Q with respect to x gives an average number of ligands bound per P molecule.

Other properties can also be computed from the binding polynomial. For example, you can find the *fluctuations* in the number of ligands bound from the second moment $\langle i^2 \rangle$. The variance is $\sigma^2 = \langle i^2 \rangle - \langle i \rangle^2$, where you get $\langle i \rangle$ from the expressions above, and you take the second derivative of the binding polynomial to get

$$\langle i^2 \rangle = \frac{\sum_{i=0}^{t} i^2K_ix^i}{\sum_{i=0}^{t} K_ix^i} = \frac{x}{Q}\frac{d}{dx}\left(x\frac{dQ}{dx}\right). \tag{4.3}$$

One Ligand Can Bind to, and Saturate, One Site

Let's consider a simple process of one ligand molecule binding to one site on a protein. If a ligand of type X is put into a solution in equilibrium with a protein P, then the two can form a "bound complex" PX:

$$X + P \overset{K}{\rightleftharpoons} PX. \tag{4.4}$$

Summing all the species of P gives

$$[P]Q = [P] + [PX] = [P](1 + Kx) \Rightarrow Q = 1 + Kx, \tag{4.5}$$

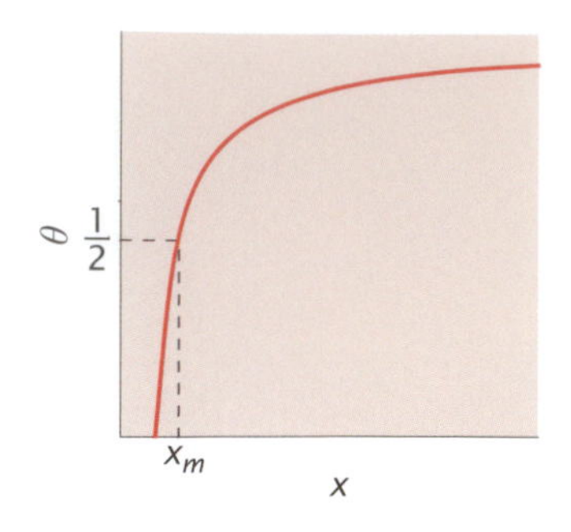

Figure 4.1 Estimating binding affinity from a binding curve. The concentration x_m of ligand x at which half the binding sites are saturated, $\theta = \frac{1}{2}$, gives the binding affinity, $K = 1/x_m$.

where we have simplified the notation by using $x = [X]$. So, the fraction of P binding sites that are filled is

$$\theta = \frac{d \ln Q}{d \ln x} = \frac{x}{Q}\frac{dQ}{dx} = \frac{Kx}{1+Kx}, \tag{4.6}$$

where we have used $dQ/dx = K$. Equation 4.6, known as the Langmuir equation (originally introduced for the adsorption of small molecules on a surface), is general, and applies whether the concentration x refers to mole fractions, molarities, molalities, or any other units. The unit you use for x determines the units of the binding constant, also called the affinity, K.

Two prominent features of binding are captured by Equation 4.6. First, at low concentrations, the amount of bound complex in solution, $\theta = Kx$, increases in proportion to the ligand concentration x in solution. Second, at high ligand concentrations, the binding *saturates*: all the binding sites on P become occupied. You can estimate the binding affinity K from a binding curve. Find the midpoint value $x = x_m$ for which half the P sites are bound, $\theta = \frac{1}{2}$. Then, the affinity is $K = 1/x_m$ (Figure 4.1).

Binding Polynomials Can Be Constructed Using the Rules of Probability

If you want to model a binding process, you first construct its binding polynomial. Often, you can do this simply by inspection, skipping the step of writing out the various equilibria. To do this, you use the addition and multiplication rules of probability [1]. According to the addition rule, if two states are mutually exclusive (bound and unbound, for example), then you can sum their statistical weights. Use 1 as the statistical weight for the unbound state (since we take all other states to be relative to it) and Kx as the statistical weight for the bound state, and add them to get $Q = 1 + Kx$ for single-site binding. Similarly, according to the multiplication rule, if two ligands bind independently, then the contribution to the binding polynomial is the product of the two statistical weights. For example, for two independent sites, $Q = (1 + K_a x)(1 + K_b x)$. Box 4.1 shows how to model n independent sites.

Box 4.1 Multiple Independent Binding Sites

Suppose a protein has n independent binding sites, all having the same affinity K. Using the addition rule, the statistical weight for each site is $1 + Kx$. Because there are n independent sites, the binding polynomial is the product

$$Q = (1 + Kx)^n. \tag{4.7}$$

Taking the derivative according to Equation 4.2 gives ν, the *occupancy*, the average number of ligands bound per P molecule:

$$\nu(x) = \frac{x}{Q}\left(\frac{dQ}{dx}\right) = \frac{nxK(1+Kx)^{n-1}}{(1+Kx)^n} = \frac{nKx}{1+Kx}. \tag{4.8}$$

So far, we have only described ligand binding *equilibria*. But often binding equilibria are also the basis for *kinetics*. The rates of some reactions can depend on how much ligand is bound to a site. For example, in Michaelis–Menten kinetics, described below, a reaction reaches its maximum speed when the concentration of substrate is high enough to saturate its binding to its protein site.

Michaelis–Menten Kinetics Arises from an Underlying Binding Step

Often a rate process, such as enzyme catalysis or the transport of molecules through protein channels, can be well modeled as involving two steps: first an equilibrium step of binding the ligand to the protein, then a kinetic step. Consider the conversion of a substrate X into a product Z, catalyzed by enzyme P through a binding event, followed by a conversion event:

$$P + X \overset{K}{\rightleftharpoons} PX \xrightarrow{k_{\mathrm{cat}}} P + Z. \tag{4.9}$$

This mechanism is called *Michaelis–Menten kinetics*. Here PX is the enzyme with substrate bound, and Z is the product. Figure 4.2 shows the time course of change of concentrations $[P]$, $[X]$, and $[PX]$ according to this mechanism.

Assume that the rate constant k_{cat} is small enough for the binding of substrate to enzyme (the first reaction) to be at equilibrium. The equilibrium constant is

$$K = \frac{[PX]}{[P]x}, \tag{4.10}$$

where $x = [X]$ is the substrate concentration, $[PX]$ is the concentration of the enzyme–substrate complex PX, and $[P]$ is the concentration of free enzyme. The rate of the enzyme-catalyzed reaction is in principle the rate of production of Z, dZ/dt, also called the velocity v:

$$v = \frac{dZ}{dt} = k_{\mathrm{cat}}[PX] = k_{\mathrm{cat}}K[P]x. \tag{4.11}$$

The total number of enzyme molecules will be conserved during the reaction, in either bound or unbound form, so

$$[P_T] = [P] + [PX] = [P](1 + Kx). \tag{4.12}$$

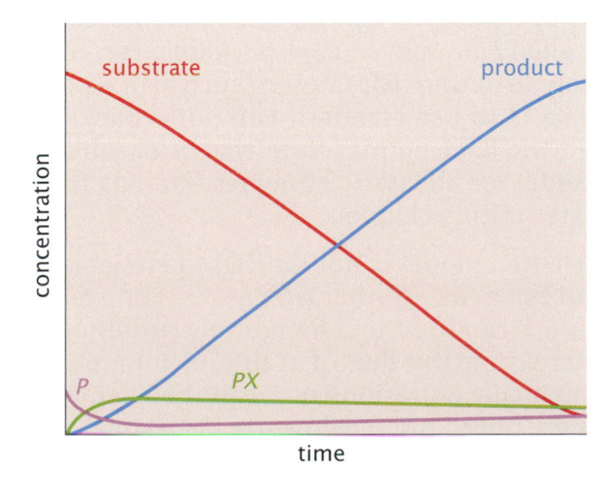

Figure 4.2 Time course of a Michaelis–Menten process, showing the depletion of substrate over time, the increase of product, and the small concentration $[PX]$ throughout. (Adapted from R Garrett, CM Grisham, and M Sabat, *Biochemistry*, 3rd ed. Brooks/Cole, Cengage Learning, Belmont, CA, 2007. With permission of Cengage Learning SO.)

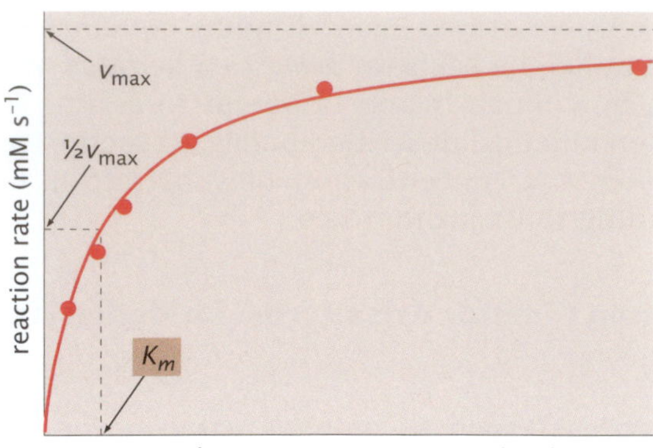

Figure 4.3 Michaelis–Menten kinetics: the reaction rate increases, then saturates, wth substrate concentration. The rate of product production increases with increasing substrate concentration, until saturating at a maximum speed. K_m is the concentration of the substrate when the reaction velocity equals one-half of the maximum velocity for the reaction.

Now, you can express v in terms of the measurable constant quantity P_T, which allows you to eliminate the hard-to-measure transient concentration $[PX]$. You can get the rate per enzyme molecule by dividing Equation 4.11 by Equation 4.12:

$$\frac{v}{[P_T]} = \frac{k_{cat}K[P]x}{[P](1+Kx)} = \frac{k_{cat}Kx}{1+Kx}. \tag{4.13}$$

Product formation is fastest when the enzyme is fully saturated by substrate: $Kx/(1+Kx) = 1$. Then the maximum rate is $v_{max} = k_{cat}[P_T]$. Expressed in terms of v_{max}, the rate of the reaction is

$$\frac{v}{v_{max}} = \frac{Kx}{1+Kx} = \frac{x/K_d}{1+x/K_d}, \tag{4.14}$$

where the last equality is expressed in terms of the *dissociation constant* $K_d = 1/K$ (examples are given in Appendix 4A).

Equation 4.14 indicates how there is a maximum rate (*kinetic saturation*) of an enzyme-catalyzed reaction because of saturation in an underlying binding step (Figure 4.3). The substrate must bind the enzyme before it can catalyze the reaction. If the concentration of substrate in solution is sufficiently high, it fills the available enzyme sites, so the reaction becomes limited by the concentration of enzyme. Sometimes, Equation 4.14 is expressed alternatively as

$$\frac{1}{v} = \left(\frac{K_d}{v_{max}}\right)\frac{1}{x} + \frac{1}{v_{max}}, \tag{4.15}$$

which allows you to plot your data as a straight line, of $1/v$ against $1/x$, where the slope will give K_d/v_{max}, and the intercept will be $1/v_{max}$. Such plots are called *Lineweaver–Burk* or *double-reciprocal* plots. These days, a better way to extract parameters such as K_d and v_{max} from your experimental data is to use standard computer packages that account for the errors in the data points. Note that Michaelis–Menten kinetics is not a good model for allosteric enzymes; models for those enzymes are described later in this chapter.

The Michaelis–Menten model is not limited to enzymatic catalysis. It is also often useful for treating other processes. For example, sometimes a ligand molecule X can bind to a membrane protein channel, opening the channel, increasing the flux of traffic of ions or small molecules through it. At high concentrations of X, all the channels are bound, so all the channels are open, and the flowrate is maximal. Another example is given in Box 4.2.

Box 4.2 The Michaelis–Menten Mechanism Describes Kinesin Walking Speeds

A linear motor protein called kinesin walks along fixed track proteins called microtubules (see Figure 2.10). Kinesin walks faster as the ATP concentration increases, up to a point beyond which further ATP no longer causes additional increases in speed. This relationship is well modeled by Michaelis–Menten kinetics as shown in Figure 4.4.

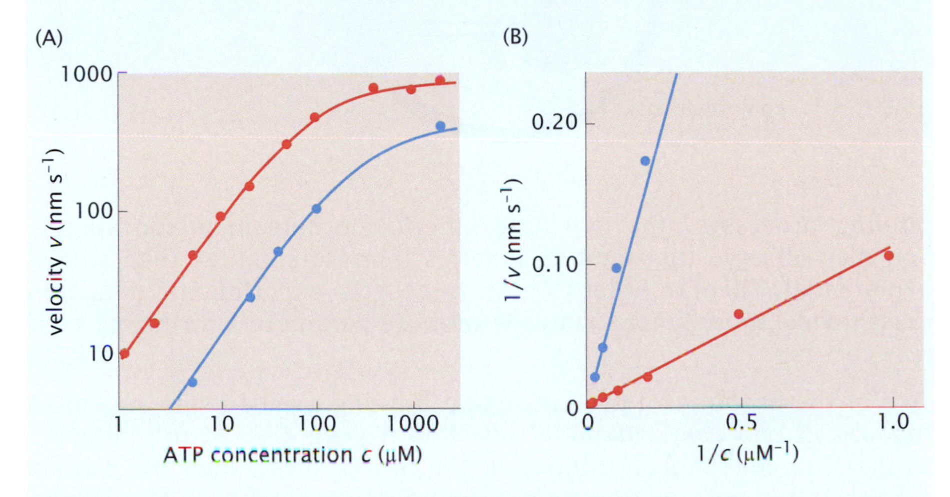

Figure 4.4 The Michaelis–Menten mechanism applies to the speed of the motor protein kinesin. Single-molecule molecular force clamp experiments measured walking rates of kinesin at loads of 1.05 pN (*red*) and 5.6 pN (*blue*). (A) The speed increases with ATP concentration, then saturates. (B) Replotting the data gives straight lines on a double-reciprocal plot. (A, adapted from K Visscher, MJ Schnitzer, and SM Block. *Nature*, 400:184–189, 1999. With permission of Macmillan Publishers Ltd.)

Ligand Binding Rates May Follow Two-State Kinetics, or Be More Complex

We have described how the rate of some process results from a binding event, where the binding step itself is fast enough to equilibrate before a final slower kinetic step. But suppose instead you are interested in the binding kinetics itself of some process. Drug discoverers sometimes want to know the off-rate of a drug that binds to a protein. The off-rate is a measure of the residence time the drug spends bound to the protein, and this is often a critical measure of the biological effectiveness of a drug. How can you model the binding and unbinding kinetics? Consider one ligand binding to one site, $P + X \rightarrow PX$. In some cases, the kinetics of binding can be simply modeled by two-state kinetics:

$$P + X \underset{k_{\text{off}}}{\overset{k_{\text{on}}}{\rightleftharpoons}} PX, \tag{4.16}$$

where the two rate parameters k_{on} and k_{off} represent the on- and off-rates, respectively. In Chapter 6, we describe how models such as this can be expressed as differential equations that can be solved to give the time dependence of the reaction. For now, we just note that this two-state model leads to single-exponential kinetics. So, if you observe single-exponential kinetics of binding and unbinding, then this two-state model will be sufficient to model it, and you can obtain the on- and off-rate coefficients by fitting to the data. However, for other

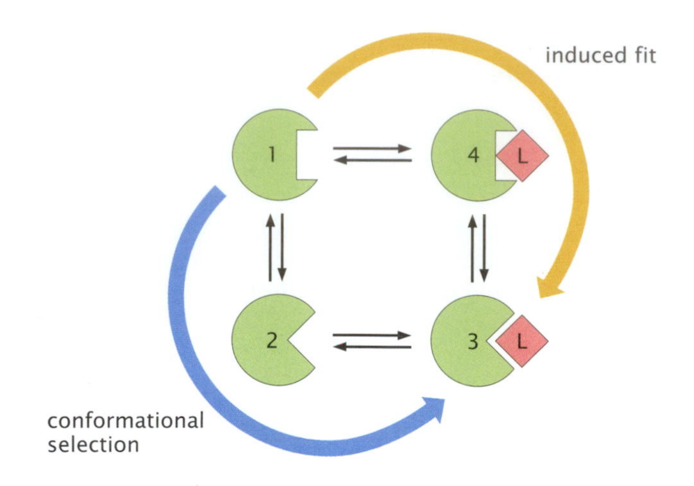

Figure 4.5 Induced fit vs. conformational selection. In the induced-fit mechanism, the ligand binds first, then the protein conformation adjusts. In conformational selection, the protein populates an ensemble of conformations, and the ligand binds to a selected one (adapted from AD Vogt and E Di Cera. *Biochemistry*, 51:5894–5902, 2012. With permission of American Chemical Society.).

binding processes, the two-state kinetic mechanism is too simple. You may observe that binding kinetics is more complex than single-exponential. Then, in order to infer rates from experimental data, you must invoke more kinetic states with more parameters for fitting your rate curves.

Two of the best-known protein–ligand binding mechanisms are called *induced fit* and *conformational selection* (Figure 4.5). In the induced-fit mechanism, the ligand binds to the protein first, then the protein readjusts its conformation to accommodate the ligand. In the conformational selection mechanism, the protein samples an ensemble of conformations and the ligand selects the conformation that permits the most favorable interaction upon binding. These two mechanisms are shown as two different kinetic routes from the unbound to the bound state in Figure 4.5.

For example, the conformational selection mechanism has been modeled as [2]

$$P_1 \underset{k_2}{\overset{k_1}{\rightleftharpoons}} P_2 \underset{k_u}{\overset{xk_b}{\rightleftharpoons}} P_2S. \tag{4.17}$$

This mechanism requires four parameters: k_1, k_2, k_u, and k_b. The quantity x indicates that forward rate of the second reaction also depends on the concentration of the ligand X. Alternatively, a variant of this two-step reaction can express the induced-fit mechanism. Or a ligand can bind through a combination or through other mechanisms. Studying the structural basis for binding mechanisms is an active area of research, described in Chapter 12.

IN ALLOSTERY, BINDING IS COUPLED TO CONFORMATIONAL CHANGE

An important principle of protein mechanisms is that one binding event can drive a protein to conformational change, in turn modulating a different binding event. In an *allosteric process* or *allosteric regulation*, an effector ligand that binds to a site on the protein may cause actions elsewhere in the protein, increasing or decreasing binding events elsewhere in the protein (called *positive* or *negative cooperativity*, respectively). ("Allosteric" comes from the Greek *allos* for "other" and *stereos* for "structural" or "shape-based.") Binding of a ligand to the allosteric site triggers a cooperative conformational change, which in turn affects the ability of a distal site to bind or unbind another ligand [3]. We begin with the simplest example of coupled binding.

Binding Can Be Cooperative between Two Binding Sites

Suppose that one ligand X can bind to P with equilibrium constant K_1, and a second copy of the ligand X can bind PX with equilibrium constant K_2:

$$P + X \overset{K_1}{\rightleftharpoons} PX_1 \qquad \text{and} \qquad PX_1 + X \overset{K_2}{\rightleftharpoons} PX_2.$$

When the reactions are written in this form, the equilibrium constants are

$$K_1 = \frac{[PX_1]}{[P]x} \qquad \text{and} \qquad K_2 = \frac{[PX_2]}{[PX_1]x} = \frac{[PX_2]}{K_1[P]x^2},$$

Now the binding polynomial is

$$Q = 1 + K_1 x + K_1 K_2 x^2, \tag{4.18}$$

and the binding curve describing the ligation state as a function of ligand concentration is

$$\nu(x) = \frac{d \ln Q}{d \ln x} = \frac{K_1 x + 2K_1 K_2 x^2}{1 + K_1 x + K_1 K_2 x^2}. \tag{4.19}$$

This equation describes the occupancy, or the expected number of ligands per P molecule in general. The equilibrium constants K_1 and K_2 determine $\nu(x)$. In some situations, you might observe $K_2 = K_1$. In that case, the binding of one ligand is *independent* of the binding of the other. In other situations, you might observe $K_2 > K_1$ (called *positive cooperativity*): the first ligand binding facilitates or activates the second. In another class of situations, you might observe $K_2 < K_1$ (*negative cooperativity*): the first ligand binding hinders or inhibits the second. Positive and negative cooperativities are the basis for inhibitors and activators and regulatory mechanisms of proteins. The Hill model describes highly cooperative binding of multiple ligands.

The Hill Model Describes a Type of Cooperative Binding

If ligand binding is highly cooperative, that is, if P binds either zero molecules or exactly n ligand molecules at a time, then the Hill model is useful:

$$P + nX \overset{K}{\rightleftharpoons} PX_n, \qquad \text{where} \qquad K = \frac{[PX_n]}{[P]x^n}. \tag{4.20}$$

In this case, n, which is called the Hill coefficient, is a measure of the cooperativity. The sum over the states of P is given by

$$[P]Q = [P] + [PX_n] = [P](1 + Kx^n), \tag{4.21}$$

so the binding polynomial is $Q = 1 + Kx^n$. Then $dQ/dx = nKx^{n-1}$, and the average number of ligands bound per P molecule is

$$\nu(x) = \frac{d \ln Q}{d \ln x} = \left(\frac{x}{Q}\right)\left(\frac{dQ}{dx}\right) = \frac{nKx^n}{1 + Kx^n}. \tag{4.22}$$

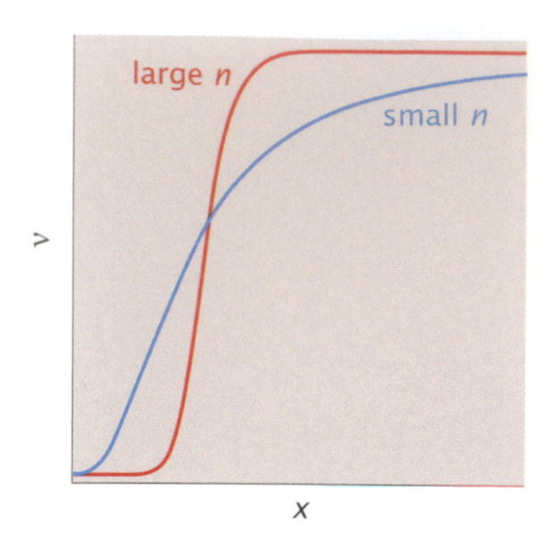

Figure 4.6 The Hill model of binding cooperativity. This plot shows a sigmoidal increase of ν, the fraction of sites filled by ligand, versus the bulk ligand concentration x. A larger Hill coefficient n gives a steeper curve. (From KA Dill and S Bromberg. *Molecular Driving Forces: Statistical Thermodynamics in Biology, Chemistry, Physics, and Nanoscience*, 2nd ed. Garland Science, New York, 2011.)

For a Hill process, a plot of ν versus x will be sigmoidal (Figure 4.6). The steepness of the transition increases with n.

Here is the general strategy for modeling binding. To fit your experimental data, start with the simplest model (that is, a model having the fewest parameters) that is consistent with your data and with any known underlying molecular structural symmetries. Then, using the binding polynomial for that model, compute the number of bound ligand molecules as a function of the bulk ligand concentration, adjusting the model parameters to best fit the data. If that model does not fit the data within acceptable error, you try a different model.

By themselves, binding polynomials do not give information about molecular structures. However, combining binding-polynomial modeling with known symmetries can offer insights about structures in binding. Below, we show this approach for hemoglobin, an important classic example.

Oxygen Binding to Hemoglobin Is a Cooperative Process

Hemoglobin is a tetrameric protein that binds and transports oxygens around the body. Each of the four subunits of hemoglobin can bind one oxygen molecule. Suppose you assume four identical independent binding sites on hemoglobin, each with affinity K, giving a binding polynomial

$$Q = (1 + Kx)^4 = 1 + 4Kx + 6(Kx)^2 + 4(Kx)^3 + (Kx)^4,$$

where x represents the oxygen concentration (pressure). This is just the Langmuir model for four independent identical sites with $n = 4$, that is, $\nu = 4Kx/(1 + Kx)$. The binomial coefficients in this expression reflect the numbers of ways you can bind a given number of ligands to the four sites: $4!/(4!0!) = 1$ is the number of arrangements in which the tetramer has either no ligands or four ligands bound. There are $4!/3!1! = 4$ different ways to distribute one or three ligands on the four sites, and $4!/2!2! = 6$ ways to distribute two ligands. You would fit your binding curve by using a single adjustable parameter K. But early experiments showed that oxygen binding to hemoglobin is cooperative, and therefore does not fit the independent-site model. For example, the fourth oxygen binds with 500-fold higher affinity than the first.

Better models were developed by Monod, Wyman, and Changeaux (MWC) [4] and by Koshland, Némethy, and Filmer (KNF) [5]. They provided the earliest insights into the molecular mechanism of allostery. Both models use a structure-based description of binding, and both assume that the four subunits of the tetrahedral hemoglobin each have two conformational states: unliganded and liganded. For hemoglobin, these states are called T (tense) and R (relaxed), corresponding to the oxygen-free and oxygen-bound states of the individual subunits. First, we consider the MWC model.

The MWC Model Describes Allostery

In the MWC model, the T (tense) and R (relaxed) conformational states coexist, following the equilibrium

$$R \overset{L}{\rightleftharpoons} T, \qquad \text{where} \qquad L = \frac{[T]}{[R]}.$$

The ligand X has different affinities for the two protein conformations (Figure 4.7):

$$R + X \overset{K_R}{\rightleftharpoons} RX \quad \text{and} \quad T + X \overset{K_T}{\rightleftharpoons} TX.$$

In the absence of ligand, T is more stable than R. That is, $L > 1$. Suppose the ligand binds R much more tightly than T, or $K_R \gg K_T$. In that case, adding ligand shifts the system from the T state toward the R state.

To illustrate the principle, consider the simplest case, the binding of one ligand. There are four possible states of the system: R, T, RX, and TX, where R and T have no ligand and RX and TX each have one ligand bound. The binding polynomial Q is the sum over all ligation states of both T and R species:

$$[P]Q = [R] + [RX] + [T] + [TX], \tag{4.23}$$

where $[P] = [R] + [T]$, $[R]$ is the concentration of unbound R conformers, $[T]$ is the concentration of unbound T conformers, and $[RX]$ and $[TX]$ are the concentrations of ligand-bound R and T molecules, respectively. In terms of the equilibrium constants, $[T] = L[R]$, $[RX] = K_R[R]x$ and $[TX] = K_T[T]x = K_T L[R]x$, where x denotes the concentration $[X]$ of ligand. Equation 4.23 becomes

$$Q = \left(\frac{1}{1+L}\right)\{(1 + K_R x) + L(1 + K_T x)\}. \tag{4.24}$$

The fraction f_R of proteins in the R state (ligand-bound or unbound) is

$$f_R = \frac{[R] + [RX]}{[R] + [RX] + [T] + [TX]} = \frac{1 + K_R x}{1 + K_R x + L(1 + K_T x)}, \tag{4.25}$$

and the fraction f_T in the T state is

$$f_T = \frac{[T] + [TX]}{[R] + [RX] + [T] + [TX]} = \frac{L(1 + K_T x)}{1 + K_R x + L(1 + K_T x)}. \tag{4.26}$$

Dividing Equation 4.26 by Equation 4.25 gives

$$\frac{f_T}{f_R} = \frac{L(1 + K_T x)}{(1 + K_R x)}.$$

Figure 4.8 shows that increasing the oxygen concentration shifts the equilibrium from T to R, provided that $K_T/K_R \ll 1$. The two states have equal occupancy at ligand concentration $x = (L - 1)/(K_R - LK_T)$, and the R state becomes more populated at higher x.

So far, we have considered only one binding site on hemoglobin. In a single subunit, hemoglobin can shift from one state (T) to (R) with increasing concentration of the ligand that stabilizes the R state. However, most allosteric proteins are multimeric, (like hemoglobin), leading to a cooperativity between the $T \rightarrow R$ transitions of the subunits. The MWC model assumes that the molecule undergoes an *all-or-none transition*, where all subunits are either in the R state or in the T state (Figure 4.9A). The MWC model differs from the KNF model in this respect. The KNF model assumes a *sequential transition* between T and R states, one subunit at a time (see Figure 4.9B), such that subunits in the T and R states may coexist in the same protein. Experiments show that in the case of hemoglobin, the tetrameric protein has indeed two accessible states, all-R or all-T for the four subunits, consistent with

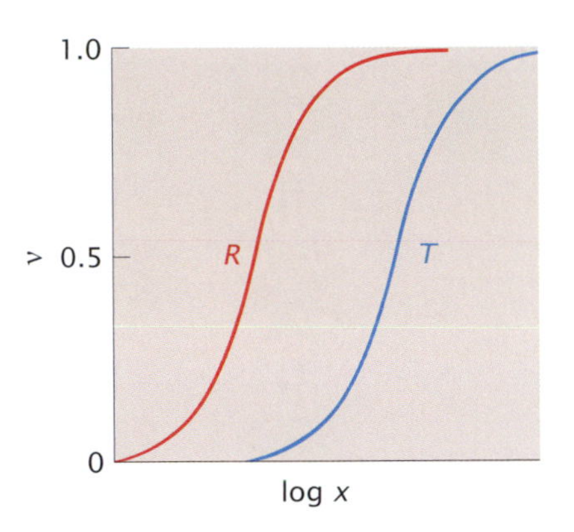

Figure 4.7 In hemoglobin, the ligand (oxygen) has greater affinity for the *R* conformation than for *T*. (Adapted from J Wyman and SJ Gill, *Binding and Linkage.* University Science Books, Mill Valley, CA, 1990.)

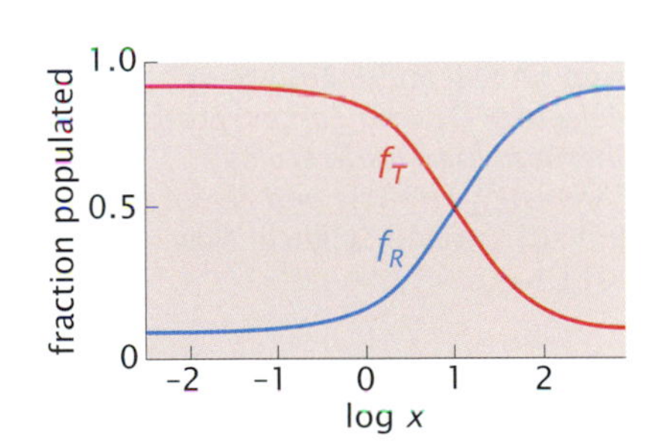

Figure 4.8 Increasing ligand concentration x causes a conformational change from T to R (Equations 4.25 and 4.26). The fraction f_T of molecules in the T state falls off as the fraction of molecules in the R state increases. (From KA Dill and S Bromberg. *Molecular Driving Forces: Statistical Thermodynamics in Biology, Chemistry, Physics, and Nanoscience*, 2nd ed. Garland Science, New York, 2011.)

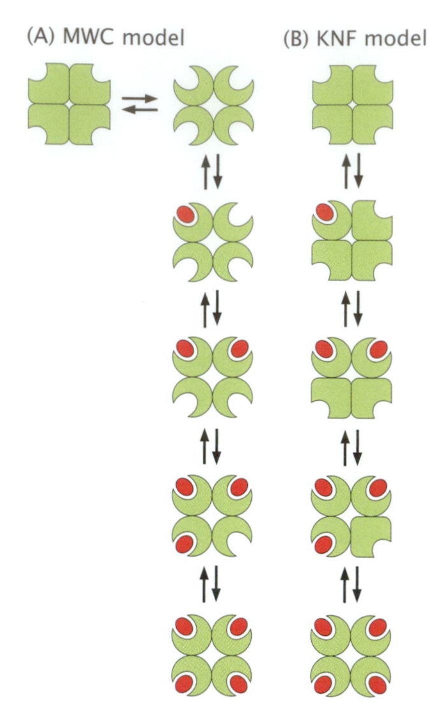

Figure 4.9 Two different models of allostery: MWC and KNF. The T state of each hemoglobin subunit is square; the R state is round. (A) MWC model. All four subunits are in state T or all four are in state R. The ligand selects and stabilizes the conformation in which all subunits are in the R state. In that conformer, the affinities of all the subunits for additional ligands subunits are increased, appearing as if the subunits switched in a *concerted* way. (B) KNF model. Binding a ligand induces only one T monomer to shift to the R state. To switch the next subunit requires a second ligand. Hence, the subunits in this model switch in a *sequential* way. (Adapted from KA Dill and S Bromberg. *Molecular Driving Forces: Statistical Thermodynamics in Biology, Chemistry, Physics, and Nanoscience*, 2nd ed. Garland Science, New York, 2011.)

the MWC model. While simultaneous conformational conversion of all subunits does not necessarily imply simultaneous binding (rather, it implies a conversion into a state that has high affinity for binding subsequent ligands), for simplicity, we assume cooperative binding, too, between all ligands. Thus, the MWC binding polynomial for hemoglobin becomes

$$Q = (1 + L)^{-1}\{(1 + K_R x^4) + L(1 + K_T x^4)\}. \tag{4.27}$$

In that case, the average number ν of ligands bound is

$$\nu(x) = \frac{d \ln Q}{d \ln x} = \{Q(L + 1)\}^{-1}(4K_R + 4LK_T)x^4. \tag{4.28}$$

The MWC model can explain *chemical amplification*: small changes in oxygen concentrations in solution can lead to large changes in the amount of oxygen bound to hemoglobin, for example. This amplification comes from a *concerted action*, according to the MWC model: The four subunits all fluctuate between T and R. The first ligand bound stabilizes the all-R state, favoring the binding of additional ligands to the protein. The average number of ligands bound increases as the fourth power of ligand concentration. As a result, the experimentally observed states are dominated by T or RX_4, with a negligible population of intermediate-ligation states. This all-or-none transition can be facilitated by structural symmetry: all subunits of the protein tend to move in concert when the structural change in one propagates, as a domino effect, to all others, due to tight packing and overall coupled architecture.

Another example of MWC chemical amplification is given in Box 4.3, whereby the action of binding a neurotransmitter leads to opening an ion-channel protein receptor in a nerve cell, allowing current flow through the channel.

Box 4.3 Some Neurotransmitters Are Ligands that Control the Gating of Ion Channels: An MWC Mechanism

The MWC model describes how a binding event couples to a conformational switching event. Acetylcholine (ACh) is a neurotransmitter (a small molecule). ACh binds to a protein called the nicotinic ACh Receptor (nAChR) in neuronal membranes. nAChR also functions as a ligand (ACh)-gated ion channel. The binding of two ACh to AChR cooperatively stabilizes the open form of the pentameric protein channel, allowing ions to flow across the neuronal membrane. The model here, given by Phillips et al. [6], expresses how the closed state of AChR is favored when no ACh ligand is present, and that ACh preferentially binds to the AChR open state. So, according to the MWC model of this system, the binding polynomial is

$$Q = f_c(1 + K_c x)^2 + f_o(1 + K_o x)^2, \tag{4.29}$$

where f_c is the relative population of closed receptors, and f_o is that of open receptors in the absence of ligand, K_c is the binding affinity of ACh to closed AChR and K_o is the affinity for the open state. So, the population P_o of open AChR increases with ACh concentration as

$$P_o = \frac{f_o(1 + K_o x)^2}{f_c(1 + K_c x)^2 + f_o(1 + K_o x)^2}. \tag{4.30}$$

In short, increasing the ACh concentration "pulls" the receptor toward the open state as in Figure 4.10.

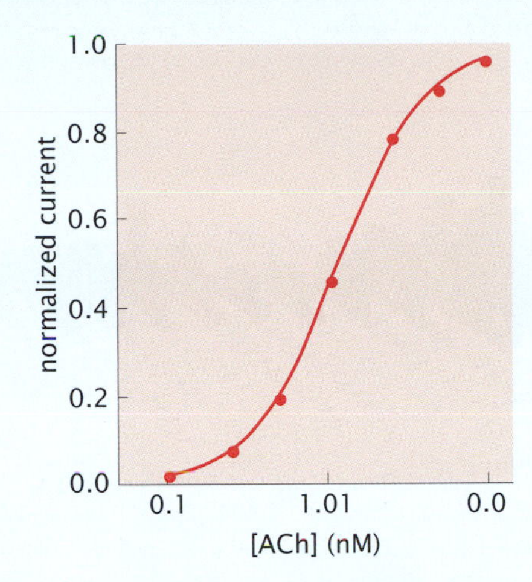

Figure 4.10 Probability that the channel is open as a function of acetylcholine concentration. (Adapted from W Zhong, JP Gallivan, Y Zhang, et al. *Proc Natl Acad Sci USA*, 95:12088–12093, 1999. Copyright (1998) National Academy of Sciences, USA.)

Another MWC-like cooperative process is observed when chaperone machines bind to ATP (Box 4.4).

Box 4.4 The GroEL Chaperone Is a Cooperative MWC-Like "Protein Folding Machine"

Chaperones are protein machines that act with MWC-like cooperativity. The bacterial chaperonin GroEL is a molecular machine composed of 14 identical protein subunits, organized in two heptameric rings. GroEL is like a barrel that captures client proteins that are unfolded or misfolded, binds them to unfold them, assist in their correct folding, and then releases them. The process entails an ATP-regulated allosteric cycle (see Figure 2.3 in Chapter 2). GroEL can bind up to 14 ATP molecules. The high-affinity and low-affinity conformers of the individual subunits for binding ATP (the ligand) are called the R and T states (as in hemoglobin). Binding of ATP molecules triggers a cooperative conformational change in all seven subunits in a ring toward the R state, opening the machine to entry by the misfolded client protein. The client protein binds to the inside walls of the barrel, unfolding the client. Then, ATP is hydrolyzed to ADP, switching the chaperone machine to the T state, releasing the client protein from the barrel walls, allowing the client protein to attempt to refold.

A mass-spectrometry study shows all the individual binding states as a function of ATP concentration [7]. Figure 4.11 shows the highly cooperative binding transition from R to T as a function of ATP concentration. The states with zero or 14 bound ATP molecules show the highest populations (although intermediate occupancies of ATP-binding sites are possible, albeit at low probabilities).

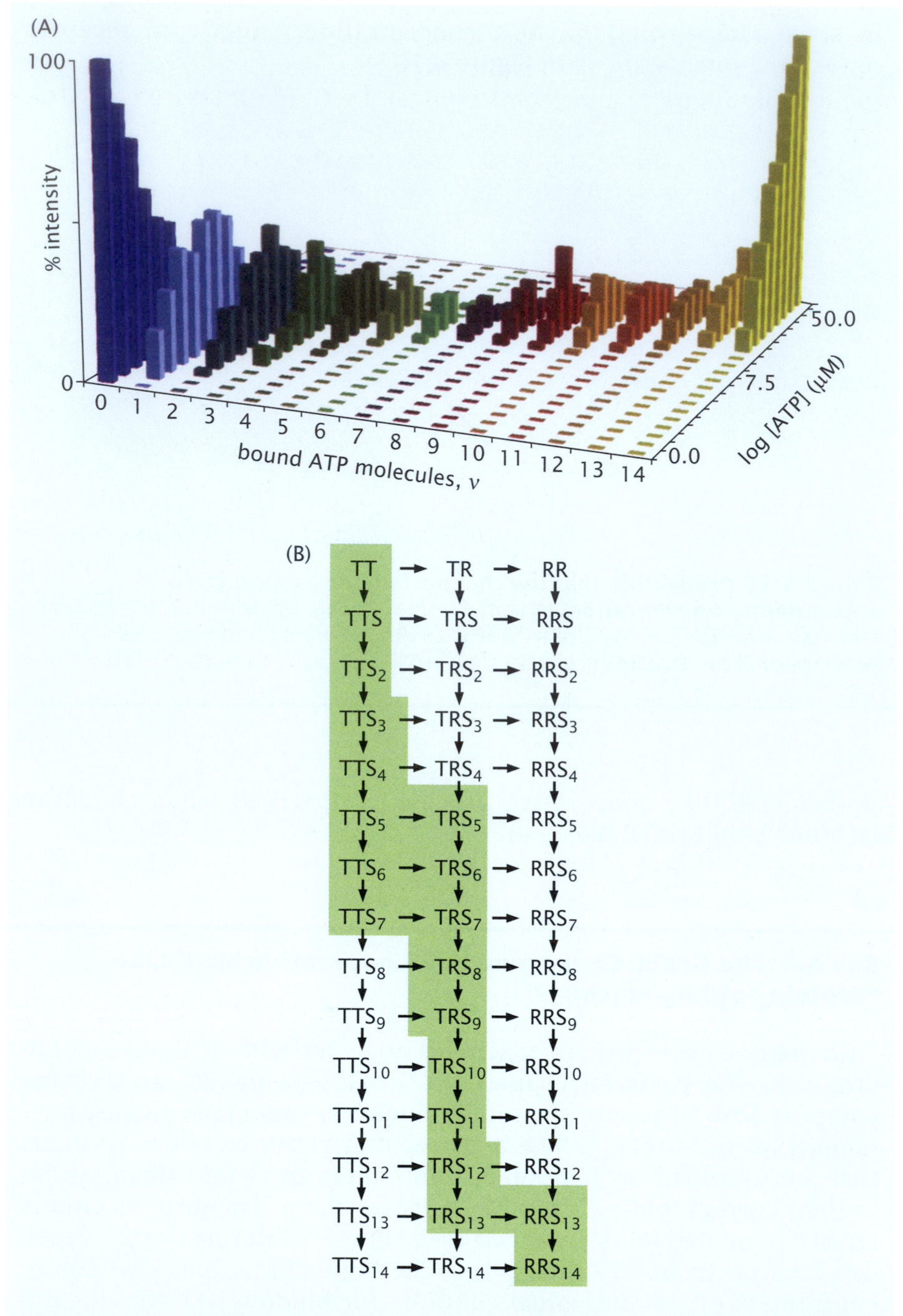

Figure 4.11 Dependence of ligation state ν of GroEL on ATP concentration *x*. (A) The plot shows a gradual decrease in the population of free GroEL ($\nu = 0$) with increasing ATP levels, and a gradual increase in the population of a fully bound GroEL ($\nu = 14$) at high ATP levels. Intermediate ligation states exhibit a concomitant increase and then decrease in their populations with increasing ATP. (B) The schema of states and their transitions, with the most populated states indicated in *green* shading. (Adapted from A Dyachenko, R Gruber, L Shimon, et al. *Proc Natl Acad Sci USA*, 110:7235–7239, 2013. Copyright (1998) National Academy of Sciences, USA.)

Binding Affinities Have Their Molecular Basis in Free Energies

In this chapter, a key quantity is a protein's affinity for its binding partner, which we have expressed as quantities such as K. You determine such binding affinities by fitting experimental *binding curves*, which give the amounts of ligand bound as a function of the amount of ligand in solution. Now, a binding affinity K also has a basis in molecular structures and energies. K results from displacing the solvent in the protein's binding site and from the ligand's stable interatomic attractions with the atoms of the protein's binding site. To relate a binding affinity K to molecular structures and energetics, you can express it in terms of the *free energy* ΔG of its corresponding binding process as

$$K = e^{-\Delta G/RT}. \tag{4.31}$$

By expressing binding in terms of the free energy, we now have a language for understanding how binding events can couple to other kinds of protein processes—mechanical forces and work, electrical potentials, light and radiation, or chemistry, for example. This is because free energies are sums of energetic and entropic factors, and proteins are sensitive to them. Figure 4.12 is a cartoon of how mechanical forces lead to changes in the products of force $\times$ displacement, pressure $\times$ area, and charge $\times$ electrostatic potential, for example. The energies and entropies of these processes, and the methods for computing them from first-principles physical models, are described elsewhere (see Chapters 9 and 10 of [1]).

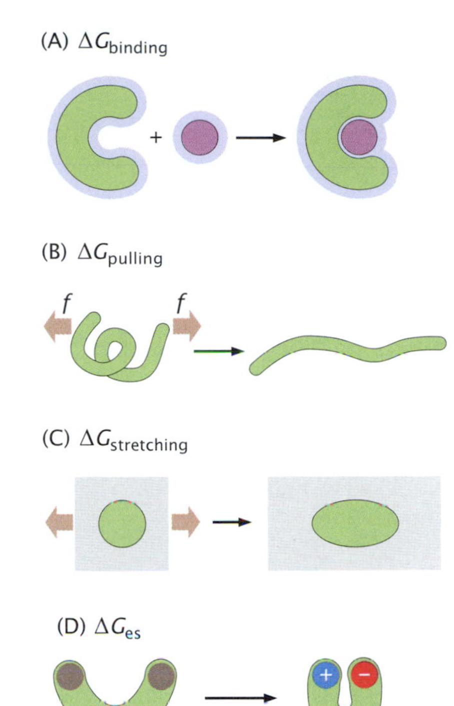

Figure 4.12 Binding processes can couple to other types of forces. Different types of free energies can couple to each other, leading to transduction that modulates one action by another. (A) The binding process. Mechanical forces lead to changes in the products of (B) force $\times$ displacement, (C) pressure $\times$ area, and (D) charge $\times$ electrostatic potential.

INHIBITORS AND ACTIVATORS CAN MODULATE OTHER BINDING ACTIONS

So far, our considerations of multiple ligation have just involved a given type of molecule. In the case of hemoglobin, we considered only the binding of one to four oxygens. Now, we consider situations involving different chemical types of ligands. Hemoglobin binds oxygen, but it can also bind carbon monoxide. Binding to carbon monoxide, for example, inhibits hemoglobin's binding to oxygen, causing toxicity and asphyxiation. In the following, we consider binding processes that involve different types of ligands (Figure 4.13).

Some Effector Molecules Are Competitive

Suppose a ligand of type X binds with affinity K to a site on molecule P. A different type of ligand molecule, Y, can bind to the same site on P with affinity R:

$$P + X \xrightarrow{K} PX,$$
$$P + Y \xrightarrow{R} PY.$$

Assume the solution concentrations of the two ligands are $x = [X]$ and $y = [Y]$. The binding polynomial will be

$$Q = 1 + Kx + Ry, \tag{4.32}$$

(A) competitive binding: *P* binds *X* or *Y*

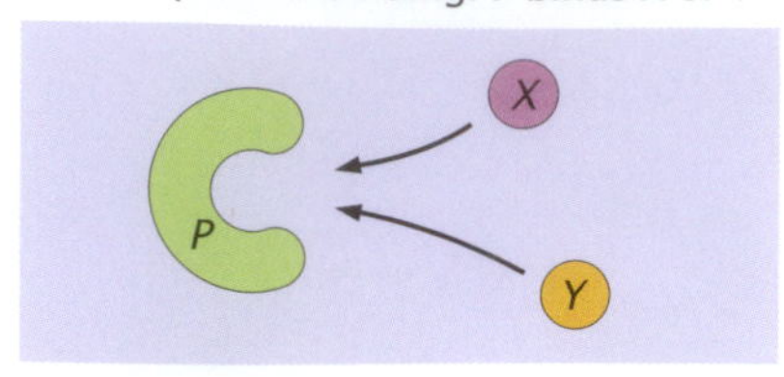

(B) prebinding: *P* binds *X* first, then *Y*

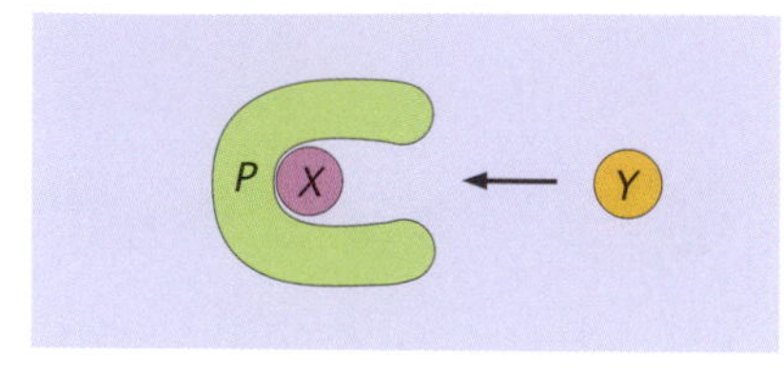

(C) noncompetitive binding: *P* binds *X* or *Y* or both

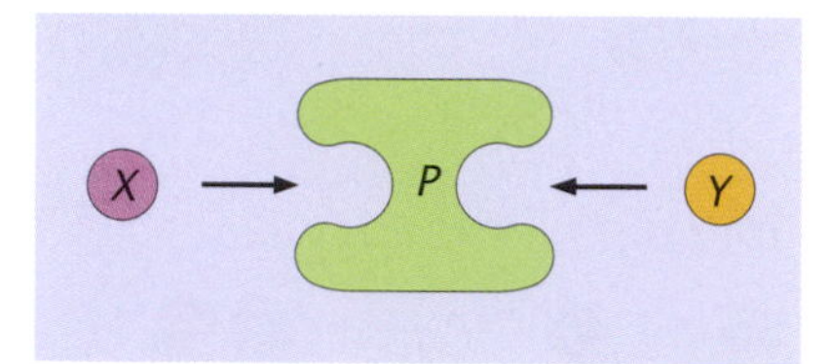

Figure 4.13 Types of inhibition. (A) Competitive inhibitor *Y* competes for the site where *X* binds to *P*. (B) Uncompetitive inhibitor *Y* does not bind unless a ligand *X* also binds. (C) Noncompetitive inhibitor *Y* has no effect on the binding of *X* to *P*, for example, because it binds at an independent site. (Adapted from KA Dill and S Bromberg. *Molecular Driving Forces: Statistical Thermodynamics in Biology, Chemistry, Physics, and Nanoscience*, 2nd ed. Garland Science, New York, 2011.)

because the *P* binding site can either be free (with a statistical weight of 1), OR it can bind *X* (with weight Kx) OR it can bind *Y* (with weight Ry). The average fraction of *P* sites that are filled by *X* is $\nu_X = (\partial \ln Q / \partial \ln x)$:

$$\nu_X = \frac{Kx}{Q} = \frac{Kx}{1 + Kx + Ry}. \tag{4.33}$$

Similarly, the fraction of *P* sites filled by *Y* is $\nu_Y = (\partial \ln Q / \partial \ln y) = Ry/Q$. *Y* is called a *competitive inhibitor*, because *Y* competes with *X* for the binding site. Figure 4.14 shows this inhibition: the amount of *X* bound is reduced as *Y* increases in concentration or affinity (ν_X decreases as Ry increases). Equation 4.33 also applies to impurities that can "poison" a surface or catalyst by displacing its ligand.

Binding polynomials tell you populations of all the possible states, bound and unbound. But they don't tell you which binding states are active. Consider two different scenarios. First, in the previous model, suppose the biological action of *P* happens only when *P* binds to *X* but does not bind to *Y*. Then, biological action will correlate with ν_X. And *Y* will inhibit that action. But another possibility is that the biological action happens when *P* binds to either ligand *X* or *Y*. In that case, the biological action will correlate instead as

$$\text{action} \propto \frac{Kx + Ry}{1 + Kx + Ry}. \tag{4.34}$$

In this second mechanism, *Y* is now an *activator*, not an inhibitor. An activator is a molecule that promotes the binding of another; see the example of recruitment in Box 4.5. In order to compute the properties of molecular machines, you need to know which terms in a binding polynomial are responsible for the action of interest. The binding polynomial itself does not tell you this; you need additional knowledge of the mechanism of action.

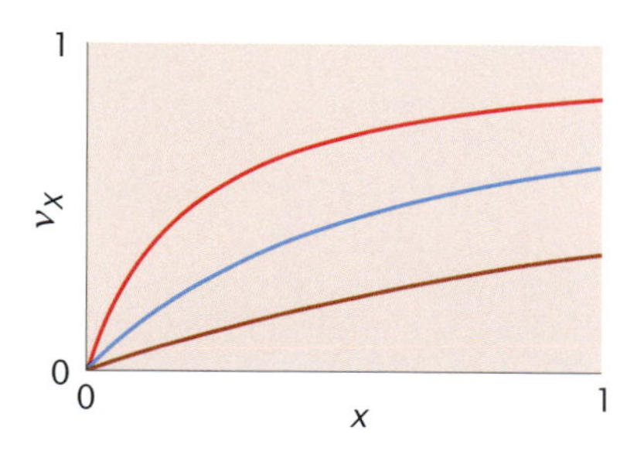

Figure 4.14 Binding isotherms for a ligand *X* in the presence of a competitive inhibitor *Y*, where the curves are for three levels of *Y*: (*brown*) highest *Y*, (*red*) lowest *Y*, and (*blue*) intermediate *Y*. (From KA Dill and S Bromberg. *Molecular Driving Forces: Statistical Thermodynamics in Biology, Chemistry, Physics, and Nanoscience*, 2nd ed. Garland Science, New York, 2011.)

Box 4.5 An Example of Activated Binding: Recruitment of RNA Polymerase to DNA

Some biological processes are expressed in terms of *recruitment*—the binding of a molecule *X* to *P* 'recruits' the binding of another molecule *Y*. This is simply an expression of activated binding. RNA polymerase (RNAP) is a large protein complex that binds to DNA molecules and transcribes the DNA code to synthesize a proper sequence in messenger RNA. RNAP by itself binds to DNA at a *promotor site* with affinity K_{RNAP}. Another protein, called an activator protein, binds to DNA with affinity K_{act} in the absence of other molecules. But, when both the activator protein and the RNAP complex bind to DNA, they do so with positive cooperativity, that is, with a greater overall affinity than for the two proteins individually. This cooperative binding is called *recruitment* because the net consequence is that the presence of the activator protein is positively correlated with the presence of RNAP (Figure 4.15). This is a biological example of positive cooperativity in binding.

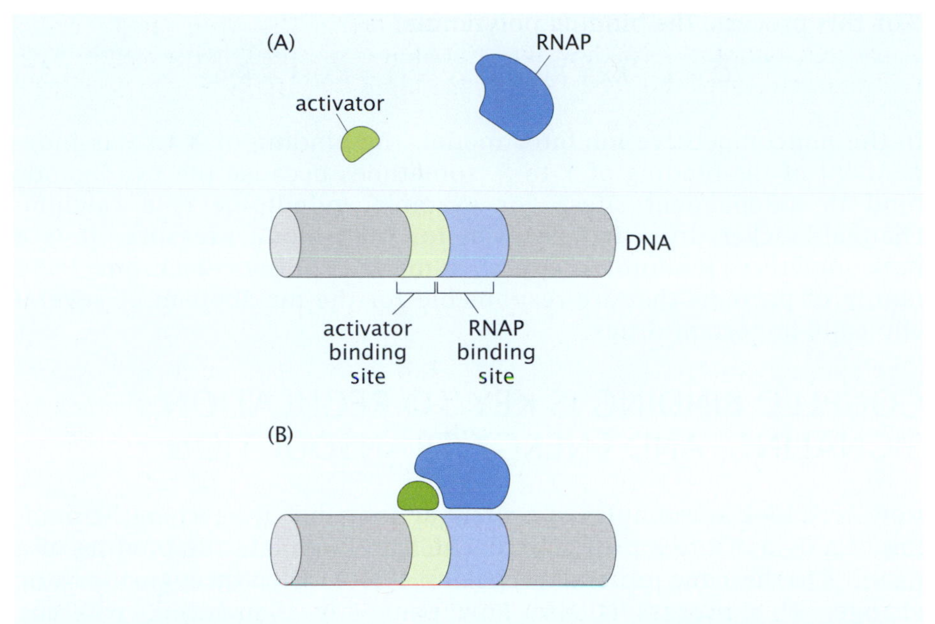

Figure 4.15 ***Recruitment*** **is an example of activated binding.** (A) DNA has binding sites for both RNAP and activator protein. (B) Recruitment is a term that expresses that the two binding events have positive cooperativity, so when one binds, it enhances the affinity of DNA for the other.

Here's another type of coupled ligand binding.

Prebinding: Ligand *Y* Binds Only If *X* Also Binds

Suppose that molecule Y can bind to P only if X is already bound to P:

$$P + X \xrightarrow{K} PX,$$
$$PX + Y \xrightarrow{R} PXY.$$

Now the binding polynomial is

$$Q = 1 + Kx + KRxy, \tag{4.35}$$

because the binding site can be empty (statistical weight of 1) OR occupied by X (statistical weight Kx) OR bind both (X AND Y) (statistical weight $KRxy$). Ligand Y cannot bind alone. The average number of X molecules bound per P is

$$\nu_X = \frac{Kx + KRxy}{1 + Kx + KRxy}. \tag{4.36}$$

Some Effectors Are Noncompetitive

A noncompetitive inhibitor can be described by the process

$$P + X \xrightarrow{K} PX,$$
$$P + Y \xrightarrow{R} PY,$$
$$P + X + Y \xrightarrow{KR} PXY.$$

For this process, the binding polynomial is

$$Q = 1 + Kx + Ry + KRxy = (1 + Kx)(1 + Ry). \qquad (4.37)$$

In the noncompetitive inhibitor model, the binding of X to P is independent of the binding of Y to P, sometimes because the two ligands bind to independent sites. For example, nifedipine is a calcium-channel-blocker drug that is given for high blood pressure; it is a noncompetitive inhibitor of cyp2c9, a member of the cytochrome P450 family of proteins that are responsible for the metabolism of several clinically important drugs.

COUPLED BINDING IS KEY TO REGULATION, SIGNALING, AND ENERGY TRANSDUCTION

Now, let's look at examples of protein actions that arise when the binding of a ligand A to a protein molecule can modulate the binding of a ligand B to the same protein in repeated cycles, often through allosteric change. This process is also how energy is transduced, whereby ATP or concentration gradients cause some other energetically uphill action.

Biochemical Engines Harness Energetically Downhill Processes to Drive Uphill Processes

Now let's explore how a molecular machine can create motion or force or haul cargo by transducing energy. Batteries power energy-utilizing (uphill) processes by coupling them to downhill chemical reactions. Similarly, biological systems power uphill processes, such as the "recharging of biology's batteries," producing ATP from ADP by harnessing the downhill flow of ions or protons from regions of high to low concentration. Another example is the transport of substrates by transporters (membrane proteins) against their concentration gradients, enabled by the opposite co-transport of ions down their electrochemical (electrical potential and chemical concentration) gradient, a common mechanism of neurotransmitter transport in chemical synapses. How is this energy transduced? Biological *engines* transduce energy through coupled binding in mechanistic cycles. While the core of a macroscopic engine might be a chamber for burning fuel, the core of a biochemical engine can be just an individual protein or DNA molecule that has coupled binding affinities. Often, the transduction action itself that results from coupled binding is a consequence of conformational transitions or motions.

Figure 4.16 shows a protein engine called *F_0F_1 ATPase*. This protein does the uphill energetic work of phosphorylating ADP to create ATP inside mitochondria, by coupling that process to the energetically downhill process of the flow of protons from outside to inside the mitochondrion. Protons flow from a region of high concentration (H) to low concentration (L). The F_0F_1 ATPase motor catalyzes this coupled process through the following steps: ADP binds to the empty protein, then a proton from H binds also to the protein, this catalyzes the conversion of ADP to ATP, then the ATP is released, then the proton is released into the medium L, and the protein returns to the empty state again.

Let's express these states in terms of a binding polynomial. Let y be the concentration of protons Y in the high-concentration region. And

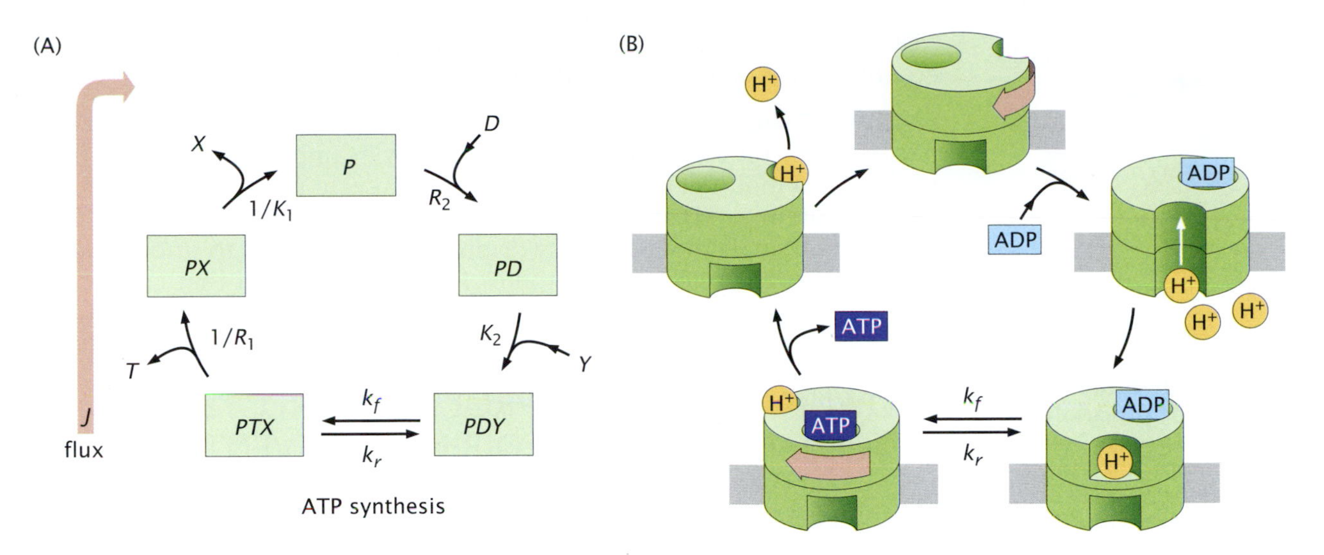

Figure 4.16 F_0F_1 ATPase is a cyclic machine that couples ATP with proton pumping. The protein *P* converts ADP to ATP by coupling to the downhill flow of protons from high concentration *y* to low concentration *x*, in cycles. (A) shows the direction the motor spins and defines the quantities in the binding polynomial. (B) shows a cartoon of the structural actions. See Figure 2.11 in Chapter 2 for the structures. (B, adapted from KA Dill and S Bromberg. *Molecular Driving Forces: Statistical Thermodynamics in Biology, Chemistry, Physics, and Nanoscience*, 2nd ed. Garland Science, New York, 2011.)

let x be the concentration of protons X in the low-concentration region. Let t be the concentration of ATP (which we also call T). And, let d be the concentration of ADP (called D). The binding polynomial is the sum of all the states of the binding partners in association with the ATPase protein P (see Figure 4.16),

$$[P]Q = [P] + [PX] + [PTX] + [PDY] + [PD]$$

$$\Rightarrow Q = 1 + K_1 x + K_1 R_1 tx + K_2 R_2 yd + R_2 d, \qquad (4.38)$$

where the K's and R's are equilibrium constants illustrated in Figure 4.16. $[PX] = [P]K_1 x$ is the concentration of protein bound only to X, $[PTX] = [P]K_1 R_1 tx$ is the concentration of protein bound to both X and ATP, $[PD] = [P]R_2 d$ is the concentration of protein bound to ADP, and $[PDY] = [P]K_2 R_2 yd$ is the concentration of protein bound to ADP and Y.

The ATPase will spin clockwise if the proton gradient is large enough. To see this, define the flux $J/[P]$ as the number of conversions of T to D per second per engine (that is, per P molecule). To illustrate the idea, assume that $D \rightarrow T$ is the rate-limiting step and that all other steps happen much faster. Then at steady state, the flux will be determined by the relative rates k_f for ATP synthesis and k_r for hydrolysis on the bottom step of Figure 4.16:

$$\begin{aligned} \frac{J}{[P]} &= \frac{k_f[PDY]}{[P]Q} - \frac{k_r[PTX]}{[P]Q} \\ &= \frac{k_f K_2 R_2 yd}{Q} - \frac{k_r K_1 R_1 tx}{Q} \\ &= \frac{k_f K_2 R_2 yd - k_r K_1 R_1 tx}{1 + K_1 x + K_1 R_1 tx + K_2 R_2 yd + R_2 d}. \end{aligned} \qquad (4.39)$$

This model shows how coupled binding leads to a transduction of energy. If $y/x > k_r K_1 R_1 t/(k_f K_2 R_2 d)$, then a downhill gradient (of protons in this case) drives the "recharging of the cell's battery," converting

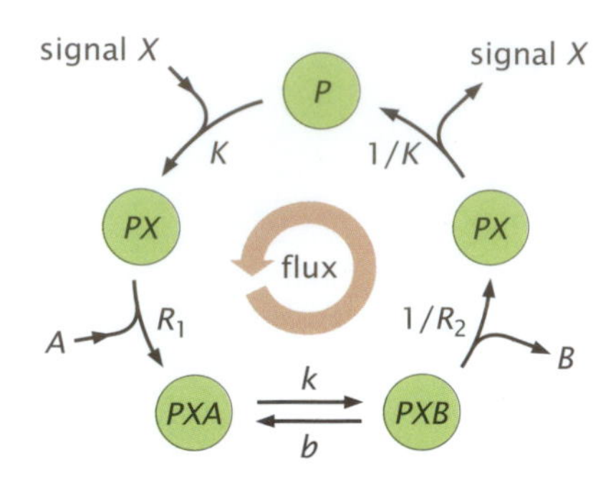

Figure 4.17 Signaling can be modeled as a cyclic process of coupled binding actions. The concentration of signal molecule X binding to protein P drives the conversion of molecule A to B, for example by phosphorylating it. (Adapted from KA Dill and S Bromberg. *Molecular Driving Forces: Statistical Thermodynamics in Biology, Chemistry, Physics, and Nanoscience*, 2nd ed. Garland Science, New York, 2011.)

ADP to ATP. However, this engine stops spinning or producing ATP at equilibrium: $k_r K_1 R_1 tx = k_f K_2 R_2 yd$.

Figure 4.17 shows a thermodynamic cycle for transducing chemical signals (concentrations of small molecules, for example), converting them to chemical modifications of proteins. This cycle resembles that of the motor protein in Figure 4.16. Signaling entails binding and unbinding of a signal small molecule X to a protein P, coupled to another process such as the phosphorylation or methylation event that converts a different protein from form A to B.

BROWNIAN RATCHETS PRODUCE DIRECTED MOTION FROM COUPLED BINDING EVENTS

An interesting puzzle is how biological systems produce *directed motion*. Some proteins are rotary motors that spin in predominantly one direction. Some proteins are linear motors that walk along other track-like proteins primarily in a single direction. These directed actions are needed in order to haul molecular cargo inside cells to their destinations, or to pump molecules into or out of cells in a directed way, or to cause muscles to push or to pull. Cells are so small that molecules inside them will generally move in random directions, due to Brownian motion. How do biomolecules achieve directed motion? Directed motion can be achieved by coupling binding and unbinding events to energetically downhill ATP-to-ADP binding and conversion events. Such processes are sometimes called *Brownian ratchets*.

Figure 4.18 shows how Brownian ratchets give directed motion [1]. Consider a particle bound to a track molecule with a sawtooth-shaped energy function. The leftmost image (A) shows the particle bound in a well at a point of minimum energy. (B) shows the release of the particle from its low-energy bound state, which requires the input from some external energy source, such as the conversion of ATP to ADP. (C) shows how the unbound particle now diffuses randomly left and right. The Gaussian distribution in (C) indicates the lateral distribution

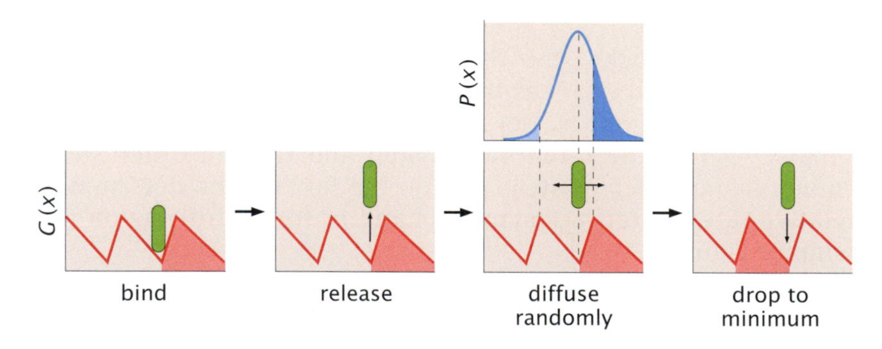

Figure 4.18 A Brownian ratchet creates directed motion by repeated cycles of binding and unbinding. A particle binds to a track with an asymmetric interaction, shown here as a "sawtooth" potential. When it is freed from the track, the particle diffuses randomly horizontally. But, when it re-binds to the track, the particle falls more often into the energy well to the right than the well to the left because of the asymmetry of the potential. The shading under the Gaussian diffusion curve $P(x)$ shows that the probability of falling into next well to the right is higher than falling into the well to the left of the current well. (Adapted from KA Dill and S Bromberg. *Molecular Driving Forces: Statistical Thermodynamics in Biology, Chemistry, Physics, and Nanoscience*, 2nd ed. Garland Science, New York, 2011.)

of the particle after some average diffusion time, with no preference for either direction. The gray shading in (C) shows how rectification happens: because the sawtooth energy function is asymmetric, there is a higher probability that the particle will fall into the next well on the right than that it will fall into the well on the left. (D) shows that the particle usually falls into the well to the right of where it started. Hence, repeated cycles of binding and unbinding of the particle to the track, coupled to energy utilization on each cycle, leads to the particle's average motion to the right. Box 4.6 shows an example of helicases, which are motor proteins that slide along DNA unidirectionally, like a train along tracks, to separate the two strands of DNA by unwinding them. Helicase action enables the transcription of RNA from DNA and other processes such as DNA repair and recombination.

Box 4.6 Helicases Are Directional Motors

Helicases are proteins that move along DNA double helices (which act like a track). This action unwinds the double-stranded DNA into its single strands. Here, we just consider how the energy of ATP is used to propel the helicase in a single direction along the DNA. Figure 4.19 shows a Brownian-ratchet-like model in which helicase binding to DNA is coupled to the hydrolysis of ATP to ADP, to create this directed motion.

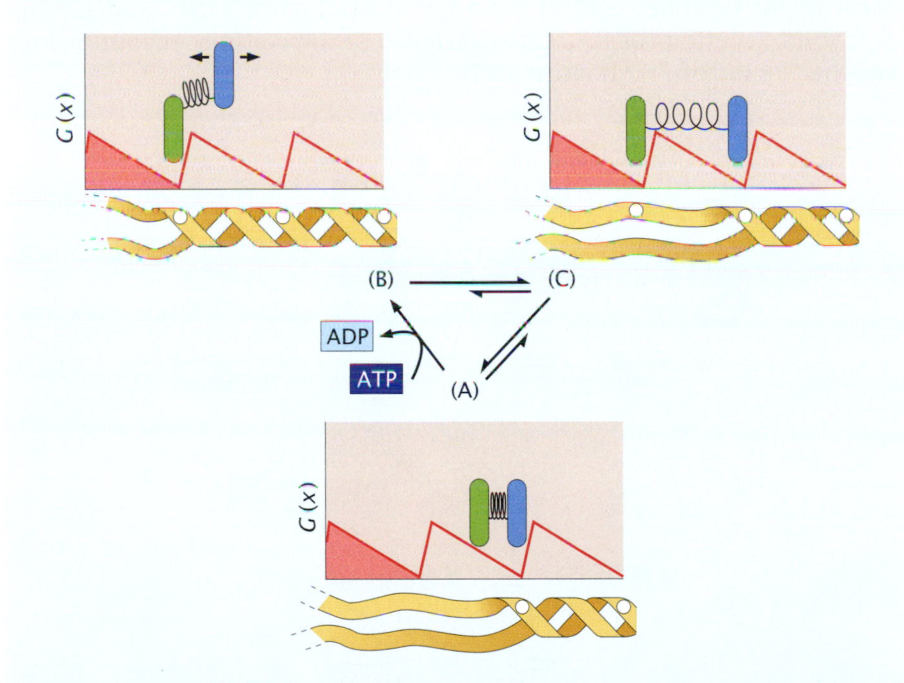

Figure 4.19 The helicase motor cycle. Helicase has two subunits. (A) Helicase is bound to the DNA. (B) One subunit (*blue*) unbinds, driven by converting ATP to ADP, and moves one step to the right, during which the DNA is unwound. (C) The other subunit (*green*) is now pulled to the right by the first one. The whole helicase protein has now taken one step to the right, and the process begins again. The white dots are markers to indicate the front and following sites of the helicase bound to the DNA. (Adapted from EJ Tomko, CJ Fischer, A Niedziela- Majka, and TM Lohman. *Mol Cell*, 26:335–347, 2007. With permission from Elsevier.)

Another example of Brownian ratcheting is called *kinetic proofreading*. This is a molecular process of quality control. It ensures accuracy, for example when RNA polymerase transcribes DNA into RNA, or when the ribosome translates mRNA molecules to synthesize proteins. In such polymerization processes, new monomers are added to a growing chain, using a template that directs the process to put the right monomers into the right sequence [8]. In kinetic proofreading, errors are reduced by coupling this polymerization to the binding and conversion of GTP to GDP, as shown in Box 4.7.

Box 4.7 *Kinetic Proofreading* Reduces Copying Errors When Proteins Are Synthesized

During protein synthesis, amino acids are added, one at a time, to a growing protein chain. A tRNA carrying a particular amino acid diffuses in the cell and binds to the mRNA on the ribosome that is guiding the protein synthesis. The 20 different tRNA molecules attempt to bind to the same nucleic acid mRNA codon (N). However, it is crucial that only the correct tRNA (C) should ultimately bind its cognate mRNA. In cells, this selectivity is highly accurate. Incorrect anticodons (I) bind to mRNA chains at only a frequency of about 1 mistake for every 10,000 units added. Yet, these anticodons and codons have remarkably similar structures and energetics. How is this matching carried out so accurately? The answer appears to involve two actions: (1) the correct binding affinities alone account for a factor of about 100 of the selectivity (Figure 4.20), and (2) an activated kinetic process, called kinetic proofreading, accounts for another factor of 100 (Figure 4.21).

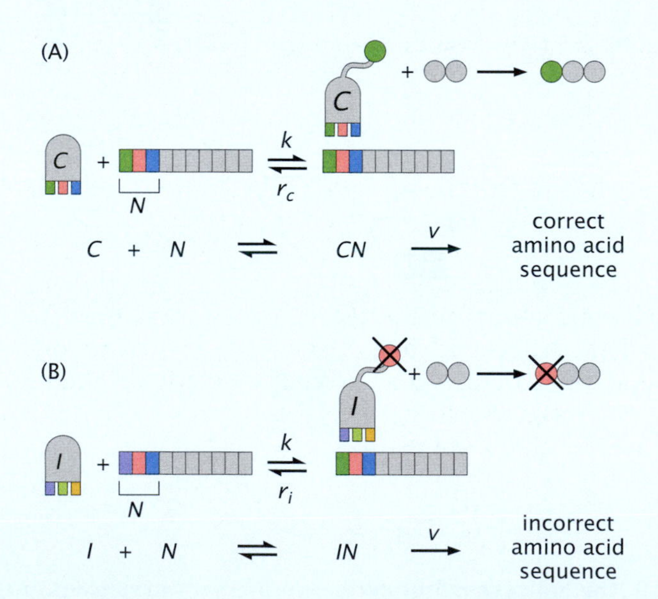

Figure 4.20 An mRNA codon (N) binds either to (A) its correct (or cognate) tRNA (C) or (B) an incorrect tRNA (I). After the mRNA binds the tRNA (C or I), the protein chain elongates at velocity v. Experiments show that C binds 100 times more tightly than I. But this is not sufficient for high-accuracy copying. Kinetic proofreading improves the accuracy.

First, consider the simplest selectivity mechanism: a difference in binding affinities. Figure 4.20 shows a fast equilibrium binding step, followed by a protein chain elongation step that happens at rate v. So, we have $[CN] = K_C[C][N]$ and $[IN] = K_I[I][N]$. And the binding affinities K_C and K_I for correct and incorrect anticodons, respectively, are

$$K_C = \frac{k}{r_C}, \quad (4.40)$$

$$K_I = \frac{k}{r_I}, \quad (4.41)$$

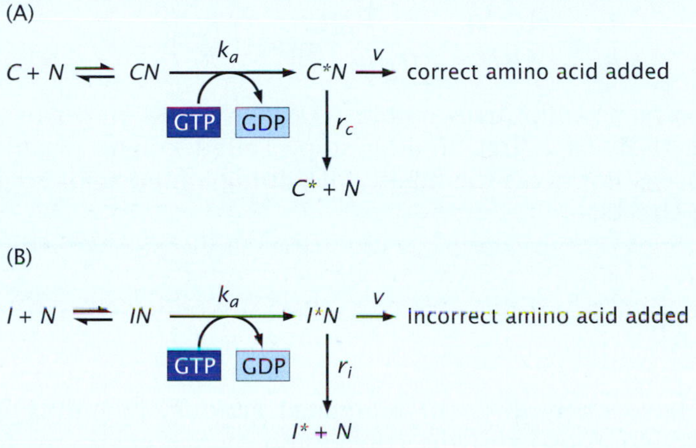

(A)

$C + N \rightleftharpoons CN \xrightarrow{k_a} C^*N \xrightarrow{v}$ correct amino acid added

GTP GDP

$C^*N \xrightarrow{r_c} C^* + N$

(B)

$I + N \rightleftharpoons IN \xrightarrow{k_a} I^*N \xrightarrow{v}$ Incorrect amino acid added

GTP GDP

$I^*N \xrightarrow{r_i} I^* + N$

Figure 4.21 Kinetic proofreading achieves high-accuracy copying by coupling amino acid addition to GTP binding. (A) for correct match between codon and anti-codon; (B) for incorrect match between codon and anti-codon.

in terms of the off-rates r_C or r_i for the correct or incorrect tRNAs to fall off the mRNA. These reactions have about the same forward rates k, because tRNAs are carried by a protein (EF-Tu in bacteria) by diffusion to their mRNA targets on the ribosome. This diffusion rate does not depend much on the tRNA. In contrast, the affinities of C and I are different because their off-rates, r_C and r_i, are different. Off-rates depend on structural and energetic details of binding. C binds the mRNA more tightly, so it comes off more slowly. The selectivity is given by the ratio of the incorporation rates:

$$\frac{R_C}{R_I} = \frac{v[CN]}{v[IN]} = \frac{K_C[C][N]}{K_I[I][N]} = \frac{K_C}{K_I} = \frac{r_I}{r_C} \approx 100. \quad (4.42)$$

We have assumed that $[C] \approx [I]$ because the 20 different tRNAs will all be available in roughly equal concentrations. The value of 100 comes from experiments. Equation 4.42 says that correct binders have 100-fold slower off-rates than incorrect binders.

Now, the cellular process achieves greater selectivity through an added biochemical step (see Figure 4.21). In the cell, the binding of tRNA to mRNA entails an additional step of chemical modification by GTP activation, which is irreversible (hence the term "kinetic" in kinetic proofreading). So, now the tRNA has a second opportunity to fall off the mRNA; namely when it is activated. The rate of falling off an mRNA is not dependent on whether the tRNA is activated or not, so the fall-off rates of the activated complexes are also approximately r_C and r_I. The activation rate k_a is determined by the GTP hydrolysis, so it doesn't depend on whether the substrate is C or I. Under steady-state conditions, the inflow will equal the outflow of activated complex, so

$$k_a[CN] = r_C[C^*N], \tag{4.43}$$

$$k_a[IN] = r_I[I^*N]. \tag{4.44}$$

Now, combining Equations 4.43 and 4.40 gives the selectivity of this activated drop-off mechanism as

$$\frac{R_C}{R_I} = \frac{v[C^*N]}{v[I^*N]} = \frac{(r_C/k_a)[CN]}{(r_I/k_a)[IN]} \approx \left(\frac{K_C}{K_I}\right)^2 = 10^4. \tag{4.45}$$

This shows how the irreversible GTP-hydrolysis step can amplify the selectivity of a first binding step. Called kinetic proofreading, this process improves the fidelity of adding amino acids to growing peptide chains.

SUMMARY

Proteins have many different biological actions, including catalysis, the transduction of light, the conversion of ATP energy to force and motion, gating, regulation, signaling, oxygen transport, the winding and unwinding of DNA, the transcription of DNA into RNA, the synthesis of other proteins, and more. Many of these actions are a simple consequence of the basic principle that when an effector ligand binds to a protein, it triggers a conformational change, modulating a different binding event in the protein, often distant from the site of effector binding. And many of these actions entail cooperative binding. Binding properties and kinetics can often be described by simple models using binding polynomials. Examples of cooperativity include the Hill, the MWC, and KNF models.

APPENDIX 4A: TYPICAL DISSOCIATION CONSTANTS FOR PROTEINS

Binding affinities are commonly expressed as a dissociation constant K_d, in molar units (M). The value of K_d corresponds to the concentration of ligand where the binding site on a protein is half occupied, that is, the concentration of ligand at which $[PX] = [P]$. The smaller a dissociation constant, the more tightly bound is the ligand, since this equal condition of bound and unbound forms can be achieved with lower concentrations of ligand. Thus a ligand with a nanomolar (nM) dissociation constant binds more tightly to a particular protein than a ligand with a micromolar (μM) dissociation constant. Table 4.A.1 shows the broad ranges of K_d for different binding processes, as well as the range of free-energy changes for these processes. Likewise, the two *Michaelis–Menten* parameters, K_m and v_{max}, vary over broad ranges. The ratio k_{cat}/K_m is usually considered to be a measure of enzyme efficiency, and may vary from $\sim 10^1$ to $\sim 10^{10}\,M^{-1}\,s^{-1}$.

Table 4.A.1 Binding values by interaction type

Interaction type	K_d (M)	$\Delta G_{binding}$ (kcal mol^{-1})*
Enzyme–ATP	1×10^{-3} to 1×10^{-6}	−4 to −8
Signaling protein–target	1×10^{-6}	−8
Transcription factor–specific sequence DNA	1×10^{-9}	−12
Protein–inhibitor (drug)	1×10^{-9} to 1×10^{-12}	−12 to −17
Avidin–biotin (strongest known noncovalent interaction)	1×10^{-15}	−21

(From J Kuriyan, B Konforti, and D Wemmer. *The Molecules of Life: Physical and Chemical Principles.* Garland Science, New York, 2012.)

*$\Delta G_{binding} = -RT \ln K_a = RT \ln K_d$, where $RT = 0.6$ kcal mol^{-1} at $T \sim 300$K, and K_a is the ligand association constant.

REFERENCES

[1] KA Dill and S Bromberg. *Molecular Driving Forces: Statistical Thermodynamics in Biology, Chemistry, Physics, and Nanoscience*, 2nd ed. Garland Science, New York, 2011.

[2] AD Vogt and E Di Cera. Conformational selection or induced fit? A critical appraisal of the kinetic mechanism. *Biochemistry*, 51:5894–5902, 2012.

[3] I Bahar, C Chennubhotla, and D Tobi. Intrinsic dynamics of enzymes in the unbound state and relation to allosteric regulation. *Curr Opin Struct Biol*, 17:633–640, 2007.

[4] J Monod, J Wyman, and JP Changeaux. On the nature of allosteric transitions: A plausible model. *J Mol Biol*, 12:88–118, 1965.

[5] DE Koshland Jr, G Némethy, and D Filmer. Comparison of experimental binding data and theoretical models in proteins containing subunits. *Biochemistry*, 5:365–385, 1966.

[6] R Phillips, J Kondev, J Thiriot, and HG Garcia. *Physical Biology of the Cell*, 2nd ed. Garland Science, New York, 2013.

[7] A Dyachenko, R Gruber, L Shimon, et al. Allosteric mechanisms can be distinguished using structural mass spectrometry. *Proc Natl Acad Sci USA*, 110:7235–7239, 2013.

[8] SC Blanchard, RL Gonzalez, HD Kim, et al. tRNA selection and kinetic proofreading in translation. *Nat Struct Mol Biol*, 11:1008–1014, 2004.

SUGGESTED READING

Dill KA and S Bromberg, *Molecular Driving Forces: Statistical Thermodynamics in Biology, Chemistry, Physics, and Nanoscience*, 2nd ed. Garland Science, New York, 2011.

CHAPTER

5

Folding and Aggregation Are Cooperative Transitions

aggregation nuclei

native-like aggregates

PROTEINS CAN UNDERGO SHARP TRANSITIONS IN THEIR STRUCTURES OR PROPERTIES

Proteins can undergo cooperative changes. In a cooperative change, a small perturbation leads to a large consequence. One example was already described in Chapter 4: the binding of a ligand drives allostery or large conformational changes in a protein. Here, we consider two other examples. One is the folding process. In test tube experiments, a protein can be driven from a denatured state to its native structure sometimes by a very small change in temperature or denaturant concentration. Another example is protein aggregation, where a small increase in a protein's concentration, under the right conditions, can drive individual proteins into large-scale multimolecular association in the form of aggregates, precipitates, crystals, or amyloid fibrils.

What is cooperativity? Consider three different types of change: **(1) A gradual change**, such as when the density of water changes only a little bit when you heat it by a few degrees around room temperature. **(2) A cooperative change**, as when the density of water undergoes a large and sharp change at its boiling point, from liquid density to gas density. Cooperative transitions are also called *two-state* or *all-or-none*. At its boiling point, you will find clusters of water in one of two states: either liquid or steam. **(3) A large continuous "noncooperative" change**, such as near a *critical point*. When water is heated at its critical point, some clusters of water are dense like a liquid, other clusters have low density like steam, and still other clusters have densities that are intermediate between those of water and steam. A noncooperative transition—also called a *higher-order transition* in macroscopic systems—is not two-state. How would you know if a transition has two-state cooperativity or not? Next, we address this question in three steps, for different types of transitions. What are the molecular structures of clusters of molecules? What experiments can tell the difference? And how can you recognize the distinction using energy landscapes of models?

What molecular structures define a cooperative transition? Consider a series of beakers, each containing protein, in which there are

Figure 5.1 Is protein folding cooperative? You can tell by checking for intermediates. (A) From left to right, increasing denaturing agent leads from native to denatured proteins. In a two-state cooperative transition, the middle beaker is a mix of native plus denatured proteins. There are no intermediate structures. (B) In a noncooperative transition, the middle beaker has structures that are intermediate between native and denatured. (C) The denaturation profile (fraction of native protein versus x, denaturant concentration in this case) cannot distinguish between (A) and (B), because it is an average over all the molecules in each beaker.

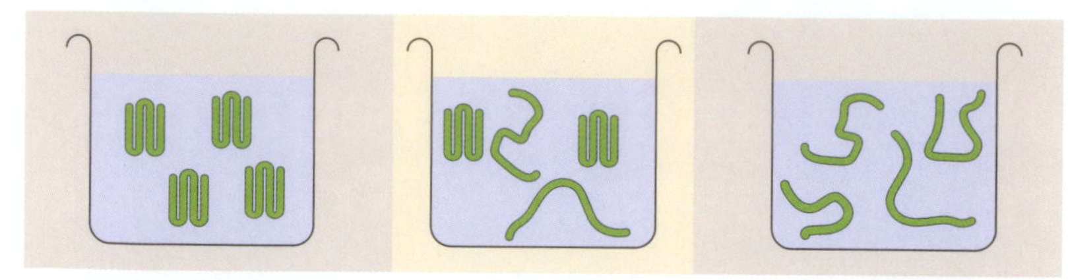

(B) not cooperative

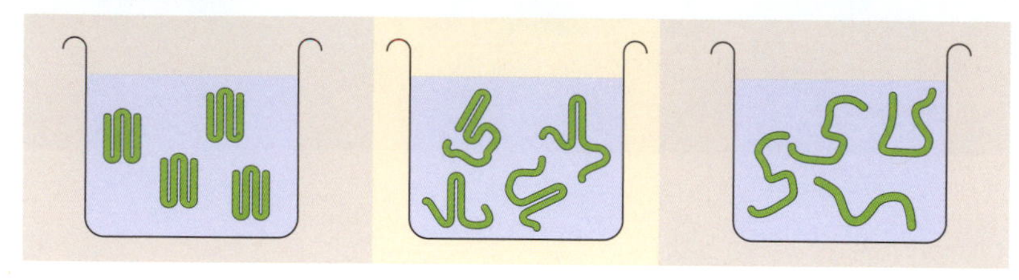

(C) sigmoidal melting curve

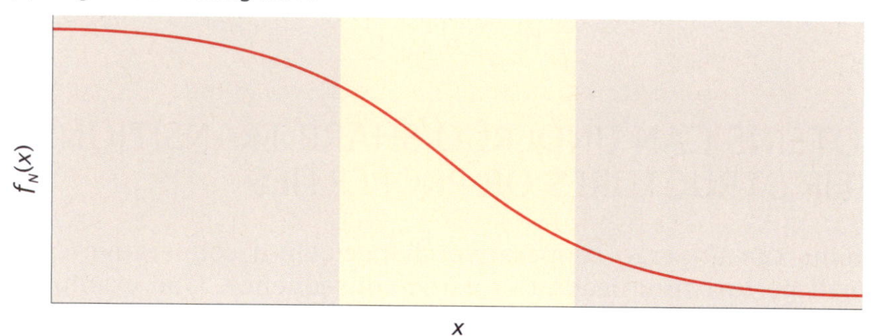

increasing concentrations of denaturant from one beaker to the next (Figure 5.1). The first beaker has no denaturant, the second beaker has a small amount of denaturant, and the third beaker has a lot of denaturant. In the no-denaturant beaker, all the protein molecules are folded into their native structures. In the high-denaturant beaker, all the protein molecules are denatured. The middle beaker has an intermediate denaturant concentration that is at the midpoint of the denaturation transition. What is the state of protein folding in the middle beaker? Figure 5.1A and B show two possibilities: (A) Half of the protein molecules have fully native structures and half of the proteins are fully denatured. Very few of the molecules have structures that are intermediate between native (N) and denatured (D). This transition is *cooperative* or *two-state.* It is characterized by having intermediate populations that are zero or small. (B) If, instead, at the denaturation midpoint, you see a significant population of protein molecules having an intermediate structure between native and denatured, then the transition is called noncooperative or *multistate* because you would need to invoke more than just the two states (N and D) to model it. You call the third state (and others) *intermediate states.* The existence of intermediates—or not—defines the nature of the transition.

What experiments would tell you if a transition is cooperative or noncooperative? Just observing a sigmoidal shape in a denaturation curve is not sufficient. A denaturation curve just tells you the *average* state of the system in each beaker (for example, the average fraction of native contacts, without distinguishing whether any two contacts occur within one chain or in two different chains). Figure 5.1C indicates that such an averaged denaturation curve may have the same sigmoidal shape no matter whether the underlying transition is cooperative or noncooperative. Rather, to learn the nature of the transition, you need to measure the individual subpopulations themselves. To know whether your system is cooperative or not, you need more than just the *population-average* state of the system at a given value of denaturing agent x; you need to know the *distribution of populations*. To prove whether or not your system is cooperative requires some direct measurement of the distinct populations, in this case N, D, and intermediates (I) in each of the beakers. Methods such as mass spectrometry can often distinguish among some subpopulations.

Stable and Unstable States Are Represented on Energy Landscapes

The nature of a transition is evident from an *energy landscape*. To describe landscapes, we first describe *order parameters*. For protein folding, let's define a variable ξ. An order parameter ξ is simply a progress coordinate that tells you the state of the system, in this case from fully denatured to fully folded. $\xi = 0$ means that all the proteins are fully denatured and $\xi = 1$ means that all the proteins are fully folded. An example of an order parameter for folding is the chain's radius of gyration. A native state is compact (has small radius), and a denatured state is expanded (large radius). So, the radius of gyration distinguishes the two states of interest. Or your order parameter might be an experimental quantity that distinguishes the native from denatured populations. A commonly used order parameter is the fraction of contacts in the chain that are native-like. A contact is a close interaction, typically closer than about 4.5 Å, between any pair of atoms belonging to two residues.

Figure 5.2 shows the general relationship between the population $p(\xi)$ of molecules in state ξ and its corresponding free energy $\Delta G(\xi)$:

$$p(\xi) \propto e^{-\Delta G(\xi)/RT}. \tag{5.1}$$

States (ξ) that have the largest populations have the most negative free energies.

Figure 5.3 shows the shapes of two different energy landscapes: (A) a cooperative, two-state transition and (B) a gradual change. The boiling of water is a cooperative process. At exactly the midpoint of boiling, half the water molecules are in low-density steam and half are in high-density liquid water. This is manifested as two dominant populations, or, equivalently, two minima in the free energy. Interestingly, the folding of many small single-domain proteins is cooperative: at the folding midpoint, the molecules in the beaker are either fully folded or unfolded, and very few of the protein molecules are in some intermediate state of folding; see Figure 5.3A. In contrast, Figure 5.3B shows a conceptual alternative of what would be expected if protein folding were a gradual change. This would have a single population, and thus a single minimum in free energy. Here we explore physical models of cooperative and noncooperative and gradual changes in proteins, starting from a simple, classical model, for helix–coil transition.

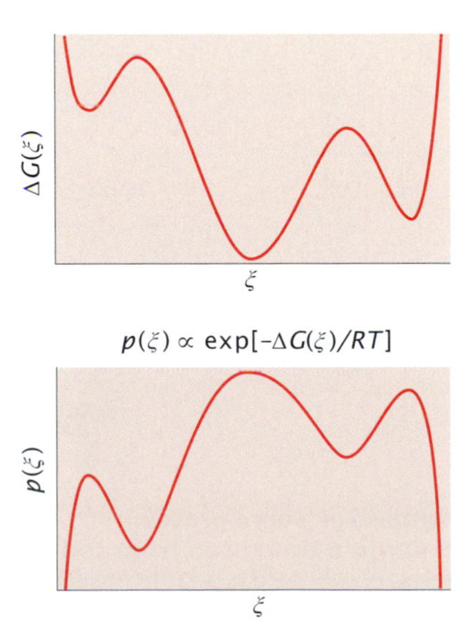

Figure 5.2 Relating populations $p(\xi)$ to free energies, $\Delta G(\xi)$.

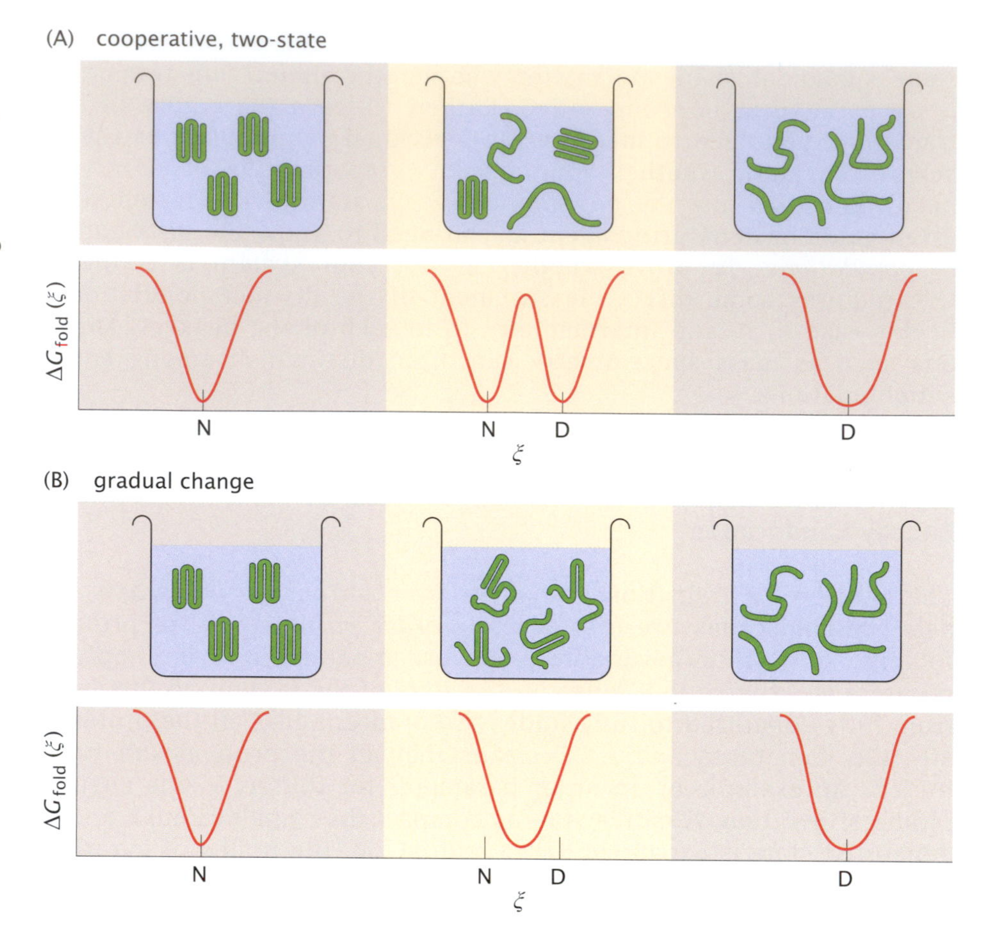

Figure 5.3 How can you tell if a model transition is cooperative? Its energy landscape has a barrier. (A) A two-state transition has multiple free-energy minima and a barrier between those states. (B) For comparison, a gradual change has a free-energy minimum that shifts continually from N, to midway to D, to D, with changing external conditions.

PROTEINS AND PEPTIDES CAN UNDERGO A COOPERATIVE HELIX–COIL TRANSITION

Before considering protein folding, let's look at a simple type of large change called the *helix–coil transition.* Some types of polymers have two experimentally distinguishable states: a *coil* state, which is a large ensemble of disordered conformations, and a *helical* state composed essentially of a single conformation in which the chain spirals in a helix shape (Figure 5.4). Changing the solvent conditions or the temperature can cause such polymers to undergo a sharp transition between the coil state and the helix. What drives this conformational change?

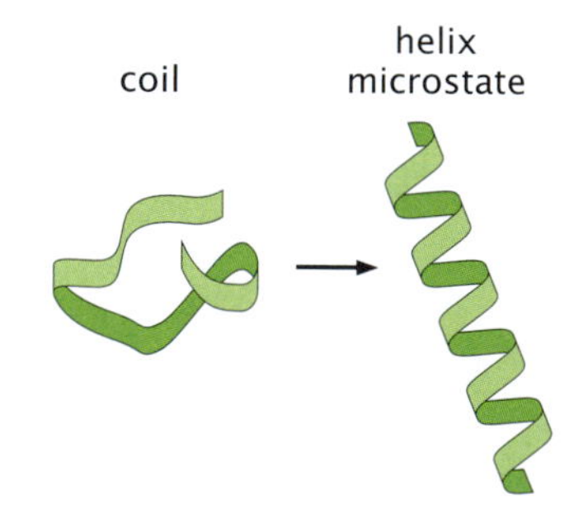

Figure 5.4 Some proteins undergo a transition from coil to helix. A *coil* state is a large ensemble of disordered conformations (many microstates). The helix is a relatively unique conformation (single microstate).

Let's express the conformation of a polypeptide as a one-dimensional binary string of characters: "c" for every unit (for example every residue) that is in a coil conformation and "h" for every residue in a helical configuration. For example, one particular conformation of a 16-mer chain can be expressed as

$$\text{ccchhhhhhhcccccc.} \tag{5.2}$$

Now, our goal is to make a physical model of the helix–coil transition from which we can compute experimentally observable properties. Often, it is easiest to measure properties that are averages over all the molecules in a container. Our aim is to compute averages over probability distributions. Different molecules will have $n = 1, 2, 3, \ldots$ helical units. We want to know the average helicity $\langle n \rangle$, where *average* is defined to be

$$\langle n \rangle = \sum_{n=0}^{N} np(n), \tag{5.3}$$

for a chain having N total units. Now, in order to compute $\langle n \rangle$, we need to know the probability distribution $p(n)$. It is given by

$$p(n) = \frac{\omega(n)}{\sum_{n=0}^{N} \omega(n)}, \tag{5.4}$$

where $\omega(n)$ is the *statistical weight* that accounts for the population of chains that have n units in a helical conformation. A statistical weight is a relative population; it does not have to be normalized to sum to one, as probabilities must. Equation 5.4 shows how probabilities (normalized quantities) are related to statistical weights. A crucial quantity in the denominator of Equation 5.4 is the *partition function* Q,

$$Q = \sum_{n=0}^{N} \omega(n), \tag{5.5}$$

which is the sum of statistical weights over all the possible states of the system, where *state n* in this case refers to the set of conformations having n helical turns. So, our objective in modeling is to find a way to compute $\omega(n)$, the statistical weight for state n of the system.

Now, consider the three levels of statistical weights: (1) We need the statistical weight *for each* h *or* c *residue* in the chain. (2) We need the statistical weight *for each microstate of a whole chain*, that is, for one particular sequence of h and c units. (3) We need to sum the statistical weights over *whatever microstates compose the chain macrostate* of interest. Macrostates usually refer to some experimentally observable state. For example, a denatured state is a macrostate collection of microstates.

(1) Statistical weights for individual h or c units. To begin, we define the statistical weight of each single coil unit, c, to be 1. We are free to choose this one statistical weight arbitrarily because only ratios of statistical weights matter for computing probabilities. Next, we assign a statistical weight of s to each helical residue, h. You can think of s as the equilibrium constant for converting a c unit to an h unit. s depends on the chemical nature of the unit and on the chemical and thermal environment of the chain. The environmental conditions of solvent and temperature sometimes cause an amino acid to prefer the helical conformation. If $s > 1$, it means that you are considering a situation in which helix is more favorable than coil. Or, if the conditions lead to $s < 1$, it means the residue favors the coil state. $s = 1$ means that h and c are equally populated. In reality, different amino acids will have different values of s, but for our purposes here of capturing the essential ideas in the simplest possible model, we take all amino acids to have the same values of s.

(2) Statistical weights for whole chains of combined h and c units. To construct Q, you can reason with the rules of probability, as described in Box 5.1.

Box 5.1 A Useful Aside about Probabilities and Averaging

Recall the two main rules of probability: (1) If states A and B are mutually exclusive and if you want to compute the statistical weight for seeing either state A OR B, then you add: $\omega(\text{A OR B}) = \omega(\text{A}) + \omega(\text{B})$. (2) If states A and B are independent, and if you want to compute the statistical weight for seeing both states A AND B, then you multiply: $\omega(\text{A AND B}) = \omega(\text{A})\omega(\text{B})$. We will add or multiply statistical weights and probabilities accordingly.

Here is a math shortcut that is useful for computing averages. We will have partition functions that take the form

$$Q = 1 + s + s^2 + s^3 + \dots \tag{5.6}$$

In such expressions, the term s^n is the statistical weight of a helix having n turns, and the corresponding probability is

$$p(n) = \frac{s^n}{Q} \tag{5.7}$$

Suppose you want to compute the average number of helical turns, $\langle n \rangle$. Then combining Equations 5.6 and 5.7 with Equation 5.3 gives

$$\begin{aligned} \langle n \rangle &= \sum_n np(n) = \frac{s + 2s^2 + 3s^3 + \dots}{1 + s + s^2 + s^3 + \dots} \\ &= \frac{s}{Q}\frac{dQ}{ds} = \frac{d \ln Q}{d \ln s}. \end{aligned} \tag{5.8}$$

The second line of Equation 5.8 shows that this average can also be obtained by taking a derivative of $Q(s)$ (because $sds^n/ds = ns^n$ and because $d \ln Q = (1/Q)dQ$.

Now we apply this reasoning to perform calculations for different models. First, in order to see the nature of cooperativity, let's start with a model that, by definition, has no cooperativity.

A first model to try: independent units

Consider a model in which each c and h unit is independent of the others. In that case, the partition function over all the microstates will be

$$Q = (1 + s)^N, \tag{5.9}$$

because each unit in a chain can be either c or h, and these are mutually exclusive options for a unit. This accounts for the term $(1 + s)$. Then, you raise $(1 + s)$ to the power N because there are N units in the chain, and because you are seeking the statistical weight that unit 1 and unit 2 and unit 3 ... are each in a particular state. Substituting Equation 5.9 into Equation 5.8 for average fractional helicity gives

$$\frac{\langle n \rangle}{N} = \frac{s}{NQ}\frac{dQ}{ds} = \frac{s(1 + s)^{N-1}}{(1 + s)^N} = \frac{s}{1 + s}. \tag{5.10}$$

Now, suppose you perform a series of experiments in which the solvent or temperature are systematically changed. Each experiment corresponds to a different value of s, the helical propensity. Equation 5.10

for this independent-units model predicts that the average helicity changes gradually as a function of s, not sigmoidally (Figure 5.5).

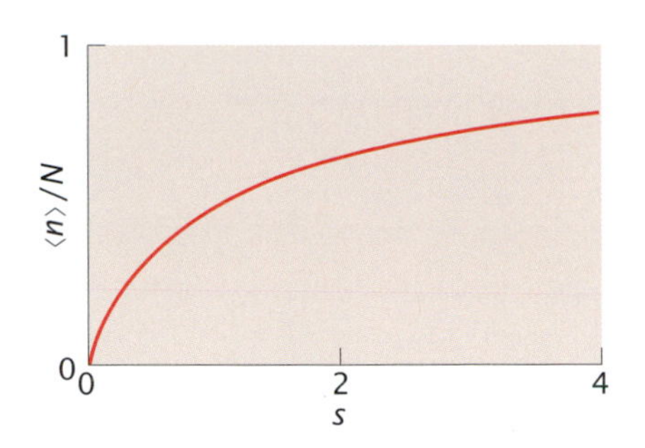

Figure 5.5 If units are independent, the average helicity changes only gradually with helical propensity s. In the independent-units model, if s were caused to vary by changing temperature or solvent, the average helicity $\langle n \rangle / N$ versus s would change only gradually. And the slope of this curve of helicity per monomer would not depend on chain length N (because monomers are independent). But experiments show that transitions become sharper as N gets larger.

This independent-units model does not predict a sigmoidal shape for $\langle n \rangle / N$ versus s and does not predict that this function (helicity per unit) gets steeper with N, both of which are observed in experiments. So, next we consider a model that better captures helix–coil cooperativity.

A second model: two-state model

Now, let's try a different model. In this model, the chain is either all coil (cccc...c), or all helix (hhhh...h), and no other microstate is populated. This model is maximally cooperative, having only two states. If any one unit is in the h state, all units are in that state. Its partition function is

$$Q = 1 + s^N. \tag{5.11}$$

The reasoning here is that you have either all coil, with statistical weight 1^N OR all helix, with weight s^N, and these two chain states are mutually exclusive. Substituting Equation 5.11 into the helicity equation, Equation 5.8, gives

$$\frac{\langle n \rangle}{N} = \frac{s^N}{1 + s^N}. \tag{5.12}$$

This model predicts that $\langle n \rangle / N$ versus s is a sigmoidal function and gets steeper with increasing N (Figure 5.6). These features are qualitatively correct, but quantitatively, this transition is too steep. In experiments, such steepness is found only for chains having $N > 1000$.

So, now consider a third model, due originally to John Schellman, that is intermediate between these two extremes. In the following sections, we use the term "two-state" not just in the previous sense of zero intermediate population, but also for situations where the intermediate populations are relatively small. In the Schellman model, the formation of a helix involves two aspects: initiation and propagation.

The Schellman Model Describes the Helix–Coil Transition as Nucleation Followed by Growth

The Schellman model [1] defines a helix as a stretch of uninterrupted h's. To keep the math simple, suppose that there is, at most, one helical segment (of any length n) in the chain. This *single-helix approximation* is often valid for peptides in solution that are shorter than about 20 amino acids long. Again, the statistical weight of c is 1 and that of h is s. But we now introduce another parameter σ. Whereas s is a *propagation parameter*, for converting one c to one h, σ is a *nucleation parameter*, the equilibrium constant for initiating a helix. That is, every "first" h in a helix is assigned a statistical weight σs. Every subsequent h in a helix is assigned a weight s. Box 5.2 shows how you assign statistical weights to the different chain configurations.

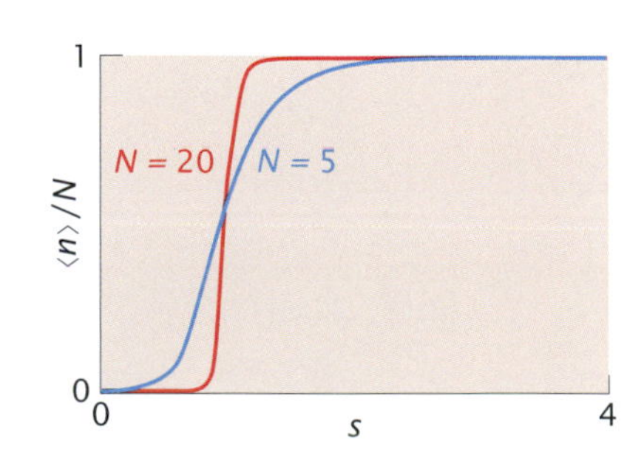

Figure 5.6 The two-state model predicts a sharp increase in helicity with s. But the predicted dependence on N is much steeper than seen in experiments.

Box 5.2 Examples of Statistical Weights

Chain configuration	Statistical weight
ccccccccc	1
ccccchccc	σs
cccchhccc	σs^2
chhcchhcc	$\sigma^2 s^4$
chhhhhhhh	σs^8

You can compute the statistical weight ω_n of all the conformations having a single helix containing n helical amino acids in a row, anywhere in the chain, as

$$\omega_n = (N - n + 1)\sigma s^n, \tag{5.13}$$

relative to a chain that is configured with all its residues in the coil state. The factor of $(N - n + 1)$ in Equation 5.13 counts the number of possible starting locations where an n-mer stretch of helical units $(n > 0)$ can begin in the N-mer chain.

Now, to convert ω_n to a probability $p(n)$, you divide by the sum over all helix lengths, Q_1 (where the subscript 1 indicates a single helix, to distinguish from helix-bundle models below):

$$\begin{aligned} Q_1 &= 1 + \sum_{n=1}^{N} \omega_n \\ &= 1 + \sigma N s + \sigma(N-1)s^2 + \sigma(N-2)s^3 + \ldots + \sigma s^N, \end{aligned} \tag{5.14}$$

where the leading term, 1, is the statistical weight for the coil state. So, the probability that a chain has n helical residues is

$$p(n) = \frac{\omega_n}{Q_1} = \frac{(N - n + 1)\sigma s^n}{Q_1}, \tag{5.15}$$

and the probability of having an all-coil chain is $p(0) = 1/Q_1$.

You can compute the average helicity as

$$\langle n \rangle = Q_1^{-1}[N\sigma s + 2(N-1)\sigma s^2 + 3(N-2)\sigma s^3 + \cdots + N\sigma s^N]. \tag{5.16}$$

Here is how you use the Schellman helix–coil model. You are given the maximum helix length N, the nucleation parameter σ, and the propagation parameter s. Now use Equation 5.14 to compute the partition function Q_1. Then use Equation 5.15 to compute any population $p(n)$ of interest to you, including the population of the fully helical molecule, $p(N)$. To compare with experiments, you want the average helicity $\langle n \rangle$ from Equation 5.16. (Using standard statistical mechanical expressions, you can also take a second derivative of Equation 5.16 to compute the variance in helix length.)

For certain values of the parameters σ and s, the Schellman model predicts two-state cooperativity. Figure 5.7 shows the distributions of helical lengths for three different values of s. In this example, the intermediate states i are much less populated than either the all-coil state

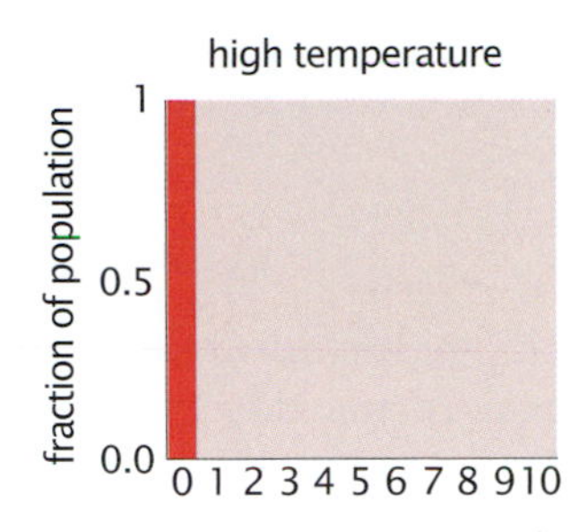

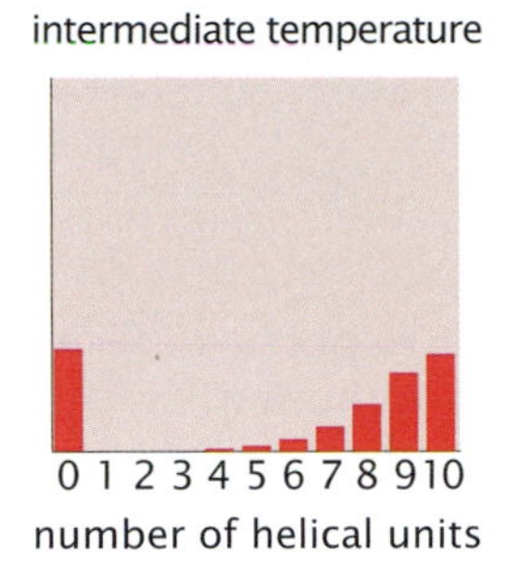

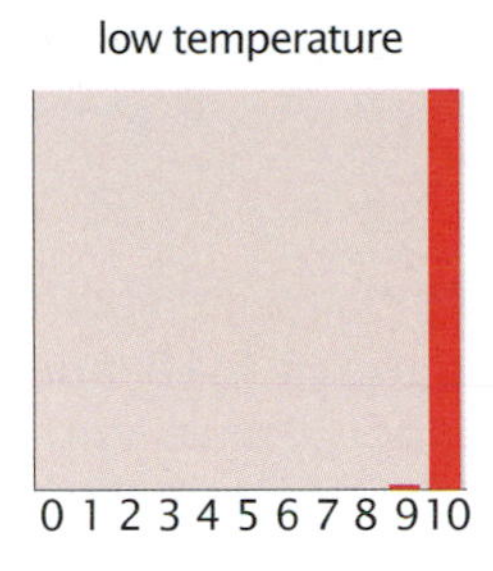

Figure 5.7 Example of a helix–coil transition that is cooperative, having limited populations of intermediate states. Calculations using Equation 5.15 are for $N = 10$, $\sigma = 10^{-4}$, and various values of s. The chain is all-coil under conditions of high temperature ($s = 0.02$) and all-helix at low temperature ($s = 200$). At a middle temperature ($s = 2.5$) (for which Equation 5.14 gives $Q_1 = 3.64$, for example) all the intermediate states are less populated than the coil ($i = 0$) or helical ($i = 10$) states.

or the all-helical state, that is, $p_1 > p_i < p_N$ at the transition midpoint. Two-state behavior occurs when nucleation is difficult (that is, when $\sigma \ll 1$), and when propagation is favored (that is, when $s > 1$). In this case, the coil is favored by its large conformational entropy. The helix is favored by its low energy, due to its multiple hydrogen-bonded units along the chain. The states intermediate between full helix and full coil are less favorable on both counts. In short, when nucleation is difficult, you will find either coil molecules or long helices, but not short helices. It is entropically costly to form the first helical turn, but it becomes energetically favorable to do so if the chain forms a sufficiently large number of helical turns.

Helices Can Be Denatured by Heating

In the last section, we expressed the helix–coil equilibrium as a function of the two parameters σ and s. However, often we will be interested instead in how the helix–coil populations depend on experimentally controllable variables such as temperature or denaturant concentration. The helix–coil equilibrium depends on temperature through $s = s(T)$ and $\sigma = \sigma(T)$. You can express the temperature dependence of the equilibrium coefficients σ and s in terms of their corresponding enthalpy ΔH and entropy ΔS. Doing so gives useful insights into the microscopic physical bases for σ and s. $s(T)$ can be expressed in terms of a free energy $\Delta G(T) = -RT \ln K = \Delta H - T\Delta S$ for the formation of one helical turn:

$$s(T) = e^{-\Delta G/RT} = e^{-\Delta H/RT + \Delta S/R}, \tag{5.17}$$

where ΔH ($= \epsilon_{hb}$, where $\epsilon_{hb} < 0$) is the enthalpy decrease upon forming one helical contact relative to the coil conformation, by forming a hydrogen bond. $\Delta S(T) = -R\ln(z-1)$ is the entropy decrease for extending the helix by one more residue, assuming that each residue has access to z local conformational (or isomeric) states (for example, we could make a simple estimate of $z = 3$ for helical, β-strand, and coiled states). In other words, the entropy of a nonhelical amino acid is $S = R\ln(z-1)$, while that of a helical amino acid is $S = R\ln 1 = 0$, hence the change $\Delta S = -R\ln(z-1)$ accompanying the transition from coil to helix, for each amino acid. R is the gas constant and T is the absolute temperature. The enthalpy change for extending the helix by one residue is favorable (negative) and the entropy change is unfavorable (negative). So, the temperature dependence of the propagation parameter is given by

$$s(T) = \frac{e^{-\epsilon_{hb}/RT}}{z-1}. \tag{5.18}$$

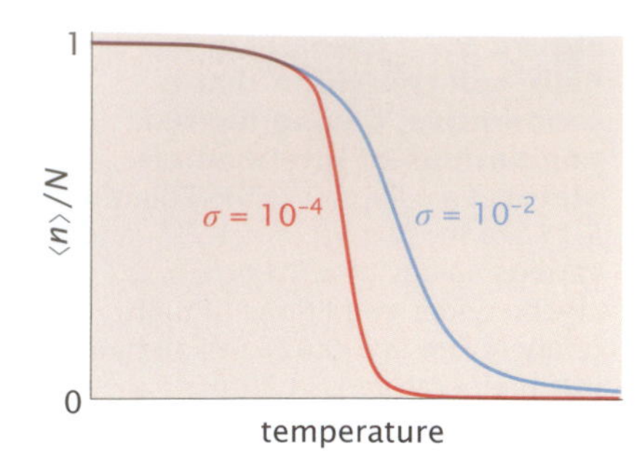

Figure 5.8 The Schellman helix–coil model predicts a sharp thermal transition. This calculation is for a model of $N = 10$ helical bonds, assuming $z = 5$ possible directions for each bond, an energy of $\epsilon_{hb}/R = -500\,K$ for each helical bond, and two different values of σ. The average number of helical residues, $\langle n \rangle$, is calculated from Equation 5.16. The smaller the value of σ, the sharper the transition.

Similarly, nucleation can also be expressed in terms of an enthalpy ϵ_{nuc}:

$$\sigma(T) = \frac{e^{-\epsilon_{nuc}/RT}}{(z-1)^3}. \tag{5.19}$$

Equations 5.18 and 5.19 express $\sigma(T)$ and $s(T)$ in terms of a model of energy and entropy components. These local terms describe how nucleation is entropically disfavored relative to propagation. The cost of nucleation involves the entropy of fixing the conformation in order to form a hydrogen bond between the C=O group of amino acid i and the N–H group of amino acid $i+4$. The first of these four residues can be oriented in any direction; then the next three are fixed, hence the factor $(z-1)^3$. In contrast, the cost of propagation involves the entropy of fixing only one residue. Because of the simplicity of our model, these factors are only approximate, but they capture the physics that nucleation is entropically less favorable than propagation.

Figure 5.8 shows that $\langle n \rangle$ is a sigmoidal function of temperature T: the helix melts out with increasing temperature. It also shows that the transition becomes sharper as σ becomes smaller (that is, when the energetic cost of nucleation is very high).

Figure 5.9 shows that this simple helix–coil theory fits experimental measurements of average helicity as a function of temperature or as a function of denaturant (urea) concentration.

What values of σ and s should you use? Different amino acids have different helical propensities, s. (The value of σ, which ranges from 10^{-2} to 10^{-4} in different models, is thought to be not very dependent on amino acid type). Table 5.1 gives one compilation of experimental values. In general, it is found that alanine has a high propensity to form a helix, while proline is a helix breaker, for example. Sometimes, to explore matters of principle, you use a single value for σ and a single value for s for all amino acids, as we have done earlier. Other times, you may prefer to account for the different s values of the different individual amino acids within a peptide or to go beyond our single-helix approximation. For advanced helix–coil modeling, see Appendix 5A.

PROTEIN FOLDING COOPERATIVITY ARISES FROM SECONDARY AND TERTIARY INTERACTIONS

Now, how should we understand cooperativity in a more complex process such as protein folding? Small proteins are found to fold

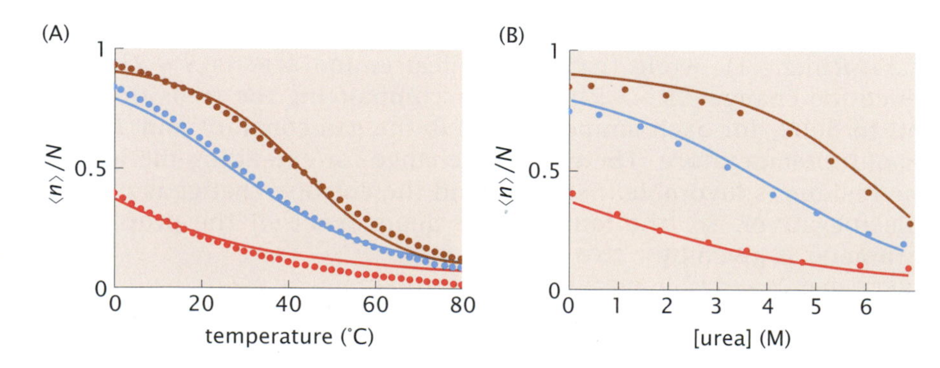

Figure 5.9 The helix–coil model can model helix denaturation by temperature or urea, for chain lengths $N = 50$ (*brown*), $N = 26$ (*blue*), and $N = 14$ (*red*): (A) thermal denaturation data; (B) urea denaturation. For computational details, see [2]. (A, adapted from JM Scholtz, H Qian, EJ York, et al. *Biopolymers*, 31:1463–1470, 1991; B, adapted from JM Scholtz, D Barrick, EJ York, et al. *Proc Natl Acad Sci U S A*, 92:185–186, 1995. Copyright (1995) National Academy of Sciences, USA.)

with two-state cooperativity, that is, with negligible populations of intermediates. Helix–coil theories alone are not sufficient to explain protein-folding cooperativity. For one thing, β-sheet proteins fold cooperatively too. For another thing, most helices in folded proteins are short, yet helix–coil theory says that short peptides should not form stable helices. What are we missing?

Protein collapse theories alone are also not sufficient to explain protein-folding cooperativity. The hydrophobic residues in a protein cause the protein to collapse into a compact structure in water, and this collapse process is abrupt. However, lattice-model studies show that polymer collapse leads to noncooperative transitions, not cooperative ones. Figure 3.9 in Chapter 3 shows that the 6-mer HP model has a significant population of intermediate structures at the midpoint of the denaturation transition. This lack of cooperativity is not because 6-mer chains are too short—longer chains, too, undergo noncooperative collapse.

So, if we can't explain the two-state nature of protein folding by either helix–coil processes alone or collapse alone, what does explain it? Folding cooperativity appears to be due to a combination of secondary and tertiary interactions [3]. When two helices are forming in protein folding, each helix forms with some cooperativity on its own, but, in addition, the packing of the two helices next to each other helps stabilize the two-helix pairing even more. A protein helix is commonly *amphipathic*, meaning that it has a stripe of hydrophobic residues along one side. The hydrophobic stripes of adjacent helices often face each other when those secondary structures pack together.

To illustrate how tertiary interactions between secondary structures can contribute stability and cooperativity, let's focus on helix-bundle proteins. In native helix-bundle proteins, multiple helices (usually three or more) are aligned and packed against each other, side by side, like a bundle of rods. As the helices form cooperatively, they also bundle together, contributing even more stability. Helices help each other to form. Here is a simple model that illustrates the idea.

Table 5.1 Helical propensities of amino acids

Residue	s
Ala	1.54
Arg^+	1.10
Leu	0.92
Lys^+	0.78
Glu^0	0.63
Met	0.60
Gln	0.53
Glu^-	0.43
Ile	0.42
Tyr	0.37–0.50
His^0	0.36
Ser	0.36
Cys	0.33
Asn	0.29
Asp^-	0.29
Asp^0	0.29
Trp	0.29–0.36
Phe	0.28
Val	0.22
Thr	0.13
His^+	0.06
Gly	0.05
Pro	≈0.001

Data measured at 273 K (from A Chakrabartty, T Kortemme, and RL Baldwin. *Protein Sci*, 3:843–852, 1994.)

Helix–Helix Interactions Contribute to Folding Cooperativity in Helix-Bundle Proteins

First, let's model a two-helix bundle assembly. Then, we will model a three-helix bundle protein. The two-helix chain has N monomers. Figure 5.10 shows the most important conformations for our two-helix model. We divide the chain conformations into classes: **(i) Coil state:** The whole chain is fully denatured, having no structure. **(ii) Single-helix state:** The chain is Schellman-like: it can contain a single stretch of helix anywhere and of any length (up to the maximum of N residues long). To keep the math simple, let's approximate each helical turn as having four monomers (rather than 3.6 per turn in α-helices). **(iii) Two adjacent helices are zipped up, but only partially.** The chain has two helices side by side. The two helices have exactly the same number $m = 1, 2, 3, \ldots, M$ of helical turns. We neglect unmatched helix lengths or helices in wrong locations, because their populations will be much smaller. Connecting the two helices is a loop of coil segments. (So, we must have $M < N/2$.) The nonhelical remainder of each chain is coil. **(iv) Native state: Two adjacent helices are fully zipped up together, $m = M$.**

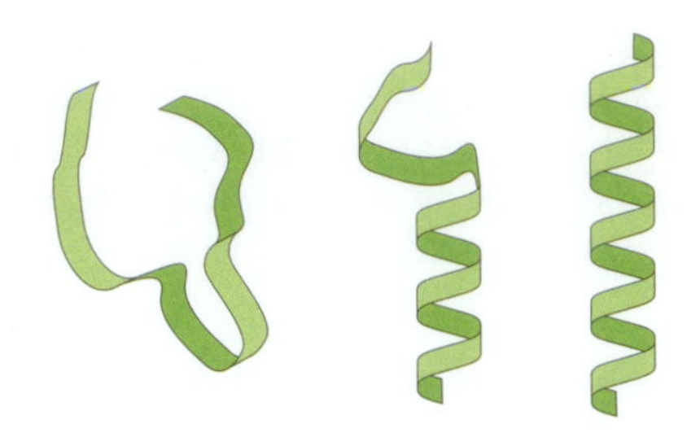

(A) Q_1 (single helix-coil states) (B) Q_2 (two-helix states)

Figure 5.10 Classifying the two-helix-bundle conformations. (A) Three states of the single helix–coil state: No helix (that is, random coil), partially zipped-up helix, and fully zipped helix. (B) Two states of the two-helix-bundle states: partially formed helices having $m = 1, 2, 3, \ldots$ turns each, or native fully formed two-helix bundle having $m = M$ turns of each helix.

Now, we want to calculate the populations of the various protein conformations. Let's construct the statistical weights for these states. First, notice that we have already modeled all the states involving any single helix or coil: the set of conformations on the left side of Figure 5.10. The partition function for these states is just Q_1, given by Equation 5.14 for a chain of length N.

Second, let's count the two-helix-bundle states, shown on the right side of Figure 5.10. For a given zipping state m of the two-helix bundle, the statistical weight is

$$\omega_m = \sigma^2 s^{8m} r^m. \tag{5.20}$$

Here, σ^2 accounts for nucleating both helices, s is the statistical weight for the formation of each helical turn (giving a total of $8m =$ (4 monomers per turn of each helix) $\times$ (2 helices) $\times m$ turns in each helix). And r (> 1) is an equilibrium constant that expresses how much extra stabilization results from each one of the m direct contacts between the adjacent helices. Helix–helix contacts are often hydrophobic and packing interactions. Equation 5.20 is the counterpart of Equation 5.13 for the single-helical chain.

The contribution to the overall partition function from the two-helix bundle conformers of various lengths m (the counterpart of Equation 5.14) is

$$\begin{aligned} Q_2 &= \sigma^2 \sum_{m=1}^{M} s^{8m} r^m \\ &= \sigma^2 (s^8 r + s^{16} r^2 + \cdots + s^{8M} r^M). \end{aligned} \tag{5.21}$$

To compute the population of any of these states, put the statistical weight of that particular state in the numerator, and the sum $Q_{2\text{hb}} = Q_1 + Q_2$ in the denominator. For example, $p(2) = \sigma^2 s^{16} r^2 / (Q_1 + Q_2)$. Then, you can compute averages, such as the average length of the two-helix bundle, using the definition of average, $\langle m \rangle = \sum_{m=0}^{M} m p(m)$.

You can express r in terms of a free energy ϵ_{hh} of forming a helix–helix contact interaction:

$$r = e^{-\epsilon_{\text{hh}}/RT}. \tag{5.22}$$

When there are no helix–helix interactions, $\epsilon_{\text{hh}} = 0$, resulting in $r = 1$. When helix–helix interactions are stabilizing, $\epsilon_{\text{hh}} < 0$ and you have $r > 1$; this is called *positive cooperativity*. In this model, protein folding cooperativity can arise from both the helix–coil contributions (expressed in terms of σ and s) and the helix–helix attractions, expressed by $r > 1$.

Here is how you use this two-helix-bundle model. You are given N (the maximum number of hydrogen bonds that could be formed if the whole chain were a helix); σ and s (the helix–coil nucleation and propagation parameters); r (the equilibrium constant for each favorable helix–helix interaction); and M, the number of amino acids of the maximal-length helix when the protein is in the two-helix-bundle state. (Or, you can begin with ϵ_{hb}, ϵ_{nuc}, and ϵ_{hh}.) Then, you compute the partition function Q_{2hb}, the populations of the states of interest in the model, and averages and variances. Box 5.3 shows how you can generalize this to handle three-helix-bundle proteins, using Q_1 and Q_2, which you have already calculated.

Box 5.3 Three-helix-bundle proteins

Using the same logic as before, you can model the folding of a three-helix-bundle protein. Suppose the dominant three-helix-bundle states are those in which each of the three helices has the same number of turns, m. And suppose that the helices are perfectly adjacent, so there are $3m$ pairwise helix–helix contacts: between helices 1–2, 2–3, and 1–3. Again, take σ as the nucleation constant for each helix, s as the propagation equilibrium constant, and r as the pairwise helix–helix equilibrium constant. Within this simple model, the partition function for the three-helix-bundle states will be

$$Q_3 = \sigma^3 \sum_{m=1}^{M} s^{12m} r^{3m}, \tag{5.23}$$

where $M(< N/3)$ is the maximum number of turns of each helix in the native structure. To compute the population of any state of the three-helix-bundle protein, put the statistical weight of the state you're interested in into the numerator, and put the sum $Q_1 + 3Q_2 + Q_3$ into the denominator. Figure 5.11 shows how this type of simplified model captures the folding–unfolding cooperativity in a three-helix bundle induced by temperature and urea.

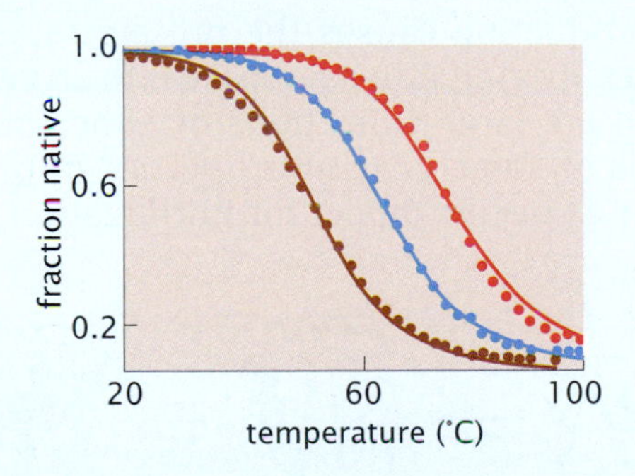

Figure 5.11 The three-helix-bundle model illustrates denaturation by temperature and urea. The thermal denaturation of alpha 3C protein is shown for three concentrations of urea: 2.0 M (*brown*), 1.5 M (*blue*), and 1.0 M (*red*). (Data from JW Bryson, JR Desjarlais, TM Handel, and WF Degrado. *Protein Sci*, 7:1404–1414, 1998; figure from K Ghosh and KA Dill. *J Am Chem Soc*, 131:2306–2312, 2009. Reprinted with permission from American Chemical Society.)

In summary, the negligible populations of intermediate states that are observed in the folding of small proteins can be explained as the product of two types of cooperativity: the individual helix–coil process is cooperative, and helix–helix packing interactions are further stabilizing. The helix zipping factor s combines with the helix–helix interaction factor r to cause the two-helix-bundle state to be more stable than any partially zipped or partially packed states. Whenever $r > 1$, it means that helix–helix interactions help stabilize the bundle states. Three-helix bundles are more stable than two-helix bundles in this model because of the factor of r^3 in the three-helix bundles compared with the factor of only r in the two-helix bundles.

This model is quite simplified. For one thing, we have only enumerated here the "dominant states," not all the possible conformations. Some additional classes of conformations could be included, at the cost of extra mathematical complexity. When we model the two-helix states, we have considered only those in which the helices are packed perfectly together, because the r^m factor says that nonbundled pairs of helices will be less populated. And we have left out the combinatoric factor that would allow two-helix-bundle states to form in non-native locations in the sequence, because those terms are small. The point of simple models such as this is to capture general principles, not the fine details.

PROTEINS CAN ASSEMBLE COOPERATIVELY INTO AGGREGATES, FIBRILS, OR CRYSTALS

Now, let's consider another cooperative process. Proteins can often associate into multiprotein assemblies (Figure 5.12) as a sharp function of protein concentration. Examples include the formation of protein complexes or assemblies, or aggregation, precipitation, crystallization, amyloidogenesis, or fibrillization. For instance, proteins can *crystallize* into symmetrical repeating patterns of native molecules packed next to each other. Crystallization is important because it enables determination of protein structures by X-ray diffraction or scattering. Multiprotein assembly also occurs in *inclusion bodies*: in biotechnology, organisms are often engineered to overproduce a particular protein of interest. Overproduction causes the protein to accumulate in high concentration, often driving it to aggregate into structures called inclusion bodies, which are large collections of either misfolded or native proteins. Let's look at some general properties of protein aggregation before we consider a specific model for fibrillization.

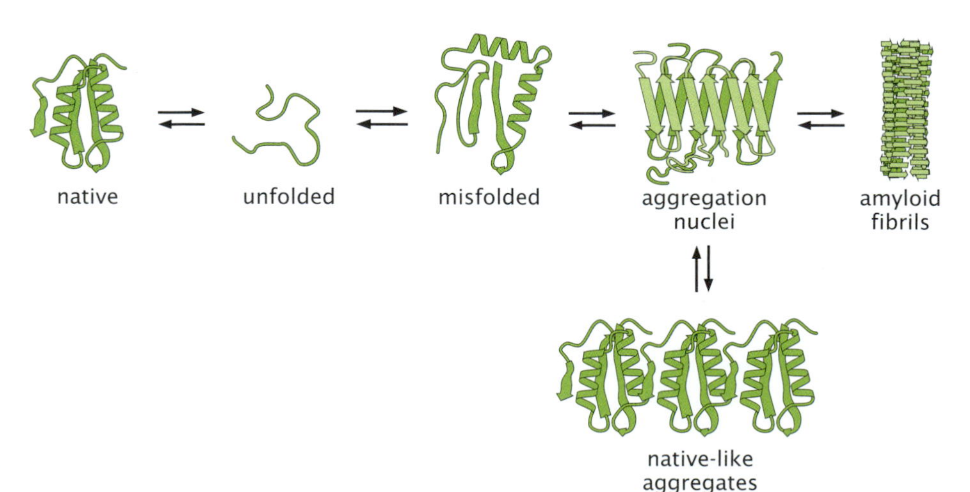

Figure 5.12 Proteins can adopt various states—native, unfolded, misfolded, aggregated, crystalline, or fibrillar, for example.

Attractive Interactions Can Drive Proteins to Aggregate

Protein aggregation processes are not yet well understood. Some are irreversible, so they cannot be studied by equilibrium methods. However, a few guiding principles are known. First, charged proteins repel each other. Proteins have the strongest tendency to associate, aggregate or crystallize when they have no net charge, which occurs when the pH of the solution is equal to the isoelectric point of the protein, called the *pI* (Figure 5.13A) [4].

Second, adding salt favors the aggregation of proteins that have substantial net charge. If two protein molecules have a net charge of the same sign, adding salt shields the charges, weakening the repulsion, and promoting protein processes such as aggregation, fibrillization and crystallization. For example, Figure 5.13B shows that lysozyme solubility decreases as NaCl concentration is increased, meaning that aggregation increases [5].

Third, protein molecules can stick to each other through hydrophobic interactions. Aggregation can occur among proteins that have hydrophobic "sticky" patches, or when interior hydrophobic residues are exposed. For example, heating proteins often unfolds them, at least partially, favoring aggregated states.

Below, we focus on one simple model of just one type of aggregation, namely, fibril formation, chosen because the model is quantitative, agrees with available experiments, and gives a few basic insights.

Amyloid Peptides Can Assemble to Form Fibrils

Some types of proteins, when put into solution at high concentration, will form fibrillar aggregates. A fibril is like dry spaghetti in a package: the many individual chain molecules are stretched out fairly straight, lined up, and packed closely together. As a function of the protein concentration, fibril formation results from a sharp transition: at low concentrations, most of the proteins are free and

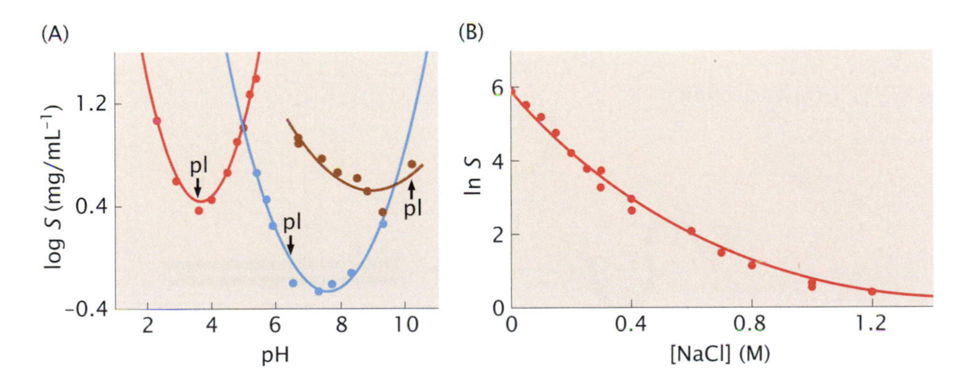

Figure 5.13 A protein's solubility depends on pH and salt concentration. (A) Solubility is minimal near the isoelectric pH. The *red curve* shows the solubility *S* of RNAse Sa (pI = 3.5); the *blue curve* shows the solubility of a mutant of RNAse Sa with pI = 6.4; and the *brown curve* shows a mutant with pI = 10.2. Note that in the case of the wild-type protein, the minimum occurs exactly at pI = pH, where it is slightly shifted in the two mutants, due to other effects (such as dipolar interactions) that perturb the solubility. (B) Proteins are less soluble in high salt concentrations. Adding salt shields net charges on proteins, reducing the repulsions between the proteins, and so reducing protein solubility and facilitating aggregation. This is called *salting out*. In other cases typically not involving charged proteins, increasing salt can increase the solubility, called *salting in*. Data shown are for lysozyme (A, from KL Shaw, GR Grimsley, GI Yakovlev, et al. *Protein Sci*, 10:1206–1215, 2001. B, from E Ruckenstein and IL Shulgin. *Adv Coll Interface Sci*, 123–126:97–103, 2006. With permission from Elsevier.).

independent and monomeric in solution. However, above a particular concentration, the molecules assemble into aggregates. Fibril formation appears to be quite general; it is observed for many different peptides and proteins. Fibrils and soluble oligomers are observed in folding diseases such as Alzheimer's, Parkinson's, Huntington's, and prion diseases. In some cases, a protein that has a normally stable native structure can be perturbed to expose a hydrophobic surface, leading the protein to stick to others, forming aggregates. In other cases, peptides that do not have stable folded structures can aggregate. For example, Aβ is a peptide of 40–42 amino acids that is implicated in Alzheimer's disease. α-synuclein has 140 amino acids, with a highly charged and unstructured 44-residue C-terminus that is implicated in Parkinson's disease. And extended glutamine sequences (for example, more than 40) at the N-terminal segment of the huntingtin protein can lead to aggregation into plaques in Huntington's disease.

Figure 5.14 shows a simple model for the cooperativity of the monomer-to-fibril equilibrium [6]. We are interested in computing the concentration distribution of the different fibril sizes. Let $[A_1]$ represent the concentration of monomeric chains in solution (in units of the number of fibrils per unit volume). $[A_2]$ is the concentration of fibrils containing two protein chains, and $[A_m]$ is the concentration of fibrils containing m protein chains. Let's express our independent variable as $x = [A_1]$ to simplify using standard notation for binding polynomials (see Chapter 4). We now express $[A_m]$ in terms of the concentration x of the isolated chains, as follows.

First, express the concentration of an m-mer fibril as the binding equilibrium of adding one chain to an existing fibril $(m-1)$-mer:

$$K = \frac{[A_m]}{[A_{m-1}]x}, \tag{5.24}$$

where K is the binding constant for adding one protein to a growing fibril. Rearranging Equation 5.24 gives

$$[A_m] = (Kx)[A_{m-1}]. \tag{5.25}$$

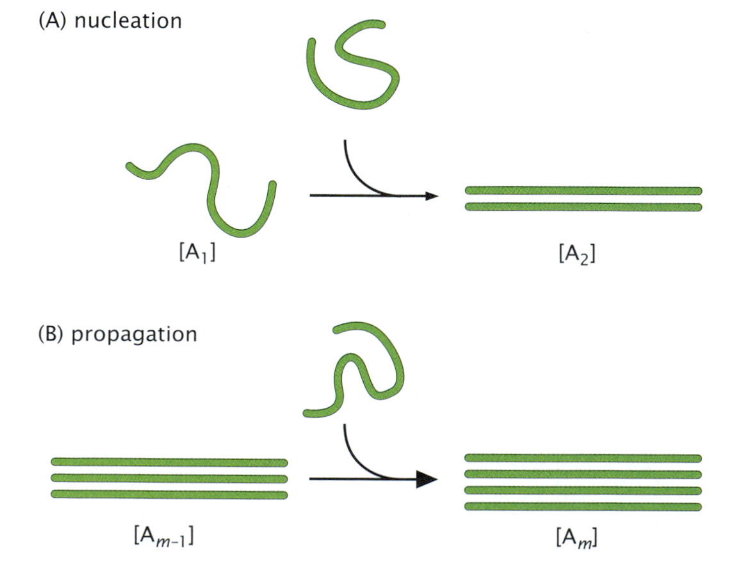

Figure 5.14 Fibril formation can be modeled as two steps: nucleation and propagation. (A) In *nucleation*, the first two chains come together to start the fibril. (B) In *propagation*, additional chains add to the growing fiber.

Equation 5.25 describes the *fibril propagation* step, for growing a fibril from size $m-1$ to size m; see Figure 5.14B. That is, you multiply by a factor of Kx for every chain that adds to the growing fibril. Now, in the same spirit as in helix–coil theory, we also define a *fibril nucleation* step. Let the nucleation equilibrium quantity δ account for intiation when the second protein adds onto the first to begin the formation of the fibril:

$$[A_2] = (Kx)[A_1]\delta = \delta Kx^2. \tag{5.26}$$

Here, (Kx) is the propagation parameter and δ is the nucleation parameter (similar to s and σ in helix–coil theory). $\delta \ll 1$ means that initiating fibril formation is difficult. The difference between our fibrillization model and helix–coil theory is that fibril formation depends on *protein concentration*, whereas helix–coil theory, which describes only the conformations of a single isolated molecule, does not. The cooperativity we model here is a matter of binding equilibrium, not conformational equilibrium.

In order to obtain the concentration $[A_m]$ in terms of x, the concentration of free monomeric protein, you multiply by $(Kx)^{m-1}$ because you have propagated by adding $m-1$ chains, and you multiply by δ because you initiated the fibril to get to the dimer. So, you have

$$[A_m] = \delta x(Kx)^{m-1}. \tag{5.27}$$

You can use Equation 5.27 to compute the relative concentrations of all the species, if you know the value of x. But, before going further, let's switch to another common way of expressing the relative concentrations. Our quantities above, $[A_m]$, are numbers of *m-mer fibrils per unit volume*. Now, we want to count protein molecules, not fibrils. We want c_m, the numbers of *protein molecules in m-mer fibrils per unit volume*. To convert, use $c_m =$ (m proteins per fibril) $\times$ (concentration of fibrils): $c_m = m[A_m] = (\delta/K)m(Kx)^m$. c_1 is the concentration of free protein molecules, c_2 is the concentration of protein molecules in 2-mer fibrils, and so on.

The total concentration of protein in solution is a sum over state concentrations:

$$\begin{aligned} c_{\text{tot}} &= c_1 + c_2 + c_3 + \dots \\ &= c_{\text{monomer}} + c_{\text{fibril}} \\ &= x + \frac{\delta}{K}\sum_{m=2}^{\infty} m(Kx)^m. \end{aligned} \tag{5.28}$$

You can see that Equation 5.28 resembles the helix–coil partition function in Equation 5.14. Both are sums of statistical weights over all the possible states accessible to the system. Both have a nucleation equilibrium constant (σ for helix–coil processes and δ for monomer–fibril assembly). Both have a propagation constant (s for helix–coil processes and Kx for monomer–fibril assembly). Both monomer–fibril assembly and helix–coil transitions are two-state transitions if nucleation is difficult. In both processes, intermediate states are not populated. For helix–coil processes, if σ is small, then the chain will be a coil or a long helix, but not a short helix. For monomer–fibril processes, if δ is small (nucleation is difficult), then most of the proteins will be either in the form of monomers or in big fibrils, but not in small fibrils. A key factor that governs the balance between monomers and fibrils is Kx: if either

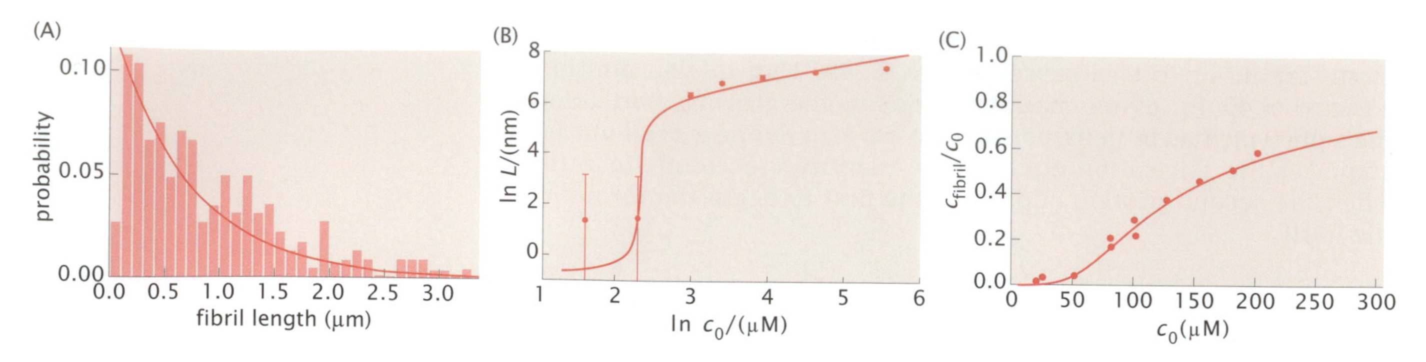

Figure 5.15 Amyloid fibril growth. (A) The length distribution of fibrils in amyloid. The bars show the experimental distribution of lengths of the amyloid peptide α-synuclein, for 30 μM bulk concentration. (B) Amyloid fibrils grow longer sharply as a function of peptide concentration. (C) The concentration of fibrils increases nonlinearly with the concentration of α-synuclein. For computational details, see [6]. (A and B, data from ME van Raaij, J van Gestel, IMJ Segers-Nolten, SW de Leeuw, and V Subramaniam. *Biophys J*, 95:450–458, 2008; C, data from E Terzi, G Hölzemann, and J Seelig. *J Mol Biol*, 252:633–642, 1995.)

K is large or the protein concentration x is large, the system will be mostly fibrillar. Increasing the protein concentration in solution leads to a transition from the monomeric to fibrillar state.

Here are three main conclusions from this model (details are given in Appendix 5B): (1) For a given protein concentration, the distribution $[A_m]$ of fibril lengths m is exponentially decreasing (Figure 5.15A). (2) Adding more protein into the solution leads to a cooperative increase in the *average lengths* of fibrils (see Figure 5.15B). (3) Adding more protein into the solution increases the *concentration* of fibrils, nonlinearly (see Figure 5.15C).

Cooperativity is a common feature of biological mechanisms. For example, viral capsids appear to assemble in all-or-none fashion from many protein molecules at a time. Also, the forces and velocities of molecular motors—which are proteins that move along protein tracks to perform molecular transport or to create forces and flows—are enhanced when multiple motor molecules work cooperatively together. And in *chemotaxis*—the process by which cells move toward food—receptor proteins assemble cooperatively in the membrane to amplify the cell's detection of food signals. In addition, the assembly of protein machines from their component proteins may be cooperative, but little is yet known.

SUMMARY

We have considered various cooperative processes in proteins. Some peptides and proteins undergo a sharp helix–coil transition, from a large ensemble of denatured conformations to a single helical conformation. The classical helix–coil model involves nucleation and propagation. Nucleation is unfavorable, but propagation is slightly favorable, so the chain forms either no helix or long helices. Protein folding in general is also cooperative. Both the secondary and tertiary structures can contribute to stability and cooperativity. In addition, we explored how proteins at high concentrations can associate to form fibrillar assemblies.

APPENDIX 5A: ADVANCED HELIX–COIL THEORIES

Various advanced helix–coil theories allow you to account for the specific sequences of amino acids in the chain and allow you to treat multiple stretches of helix within a chain. Examples are the models of Zimm and Bragg [7], Lifson and Roig [8], and Agadir [9]. Here is the basic idea of the Zimm–Bragg model. Suppose you want to allow for multiple helices in a chain of different s values for each monomer type.

First, define the *generator matrix*,

$$\mathbf{G} = \begin{bmatrix} 1 & \sigma s \\ 1 & s \end{bmatrix}. \tag{5.A.1}$$

The partition function will be

$$Q_N = [1 \quad \sigma s]\, \mathbf{G}^{N-1} \begin{bmatrix} 1 \\ 1 \end{bmatrix}. \tag{5.A.2}$$

$\Bigg($Try it for $N = 2: Q_2 = [1 \quad \sigma s] \begin{bmatrix} 1 & \sigma s \\ 1 & s \end{bmatrix} = 1 + \sigma s + \sigma s + \sigma s^2$, which gives

$$Q_2 = \underbrace{1}_{\text{cc}} + \underbrace{2\sigma s}_{\text{ch+hc}} + \underbrace{\sigma s^2}_{\text{hh}}.\Bigg)$$

Now, for a specific sequence, such as Ala-Trp-Gly, take the values of s for each amino acid from Table 5.1, and choose a value of σ. Then, replace the generic matrix $\mathbf{G}^3$ with the product $\mathbf{G}_{\text{Ala}}\mathbf{G}_{\text{Trp}}\mathbf{G}_{\text{Gly}}$. Each such matrix uses a fixed value of σ and the s that is appropriate for that particular amino acid.

APPENDIX 5B: AMYLOID AGGREGATION THEORY

Here are a few further details of the amyloid fibrillization model described in the text. You can interpret fibrillization equilibria by looking at which concentration terms are large and which are small in Equation 5.28, for particular values of δ, K, and c_{tot}. Note that our independent variable here for protein concentrations is c_{tot}, not x. c_{tot} is the independent variable because you control it; it is the total concentration of protein that you put into the solution. Once you fix the value of c_{tot}, that determines the concentration of free monomer, x. This poses a little computational obstacle: you need to express the protein concentration c_m as a function of c_{tot}, rather than expressing $c_m(x)$ as a function of x. You can do so by solving Equation 5.28 numerically. The solution to this equation gives you the populations of all the fibrillar species of different lengths, either as protein concentrations c or as fibril concentrations $[A]$. From those values, you can compute average quantities, such as described in the next section.

Most Fibrils Are Relatively Short

How many fibrils are long, and how many are short? What fraction $f(m)$ of all fibrils has m peptide molecules per fiber? The relative populations of fibrils of different sizes are given by

$$f(m) = \frac{[A_m]}{\sum_{j=1}^{\infty}[A_j]} = \frac{(Kx)^m}{\sum_{j=1}^{\infty}(Kx)^j} \approx (Kx)^{m-1}(1 - Kx), \tag{5.B.1}$$

where we have used Equation 5.27. The approximation on the right-hand side comes from using the summation relationship $\sum_{m=1}^{\infty} y^m =$

$y(1-y)^{-1}$, which is valid as long as $y < 1$. Equation 5.B.1 predicts that more fibrils are short, and fewer fibrils are long. The fibril fraction $f(m)$ decreases geometrically with m for a fixed peptide concentration. This prediction is compared with experiments in Figure 5.15A.

Here is an important subtle point. On the one hand, $[A_m]$ diminishes with m. On the other hand, the concentration $c_m = m[A_m]$ first increases at small m, reaches a peak, and then decreases at large m. So, fibrils have a well-defined average length, which we compute below.

Fibril *Lengths* Increase Sharply with Peptide Concentration

Adding peptide to the solution can cause a sharp jump in the average size of fibrils. How does the average fibril length change with peptide concentration? You can readily compute the average fibril length as

$$\langle m \rangle = \sum_{m=1}^{\infty} mf(m) = \frac{\sum_m m(Kx)^m}{\sum_m (Kx)^m} = (Kx)(1-Kx)^{-1}, \tag{5.B.2}$$

where we have used Equation 5.B.1 and the additional summation relationship $1 + 2y + 3y^2 + \ldots = (1-y)^{-2}$. An important conclusion is qualitatively expressed in Equation 5.B.2. As x increases to the point that $Kx \to 1$, Equation 5.B.2 predicts that the average fibril size becomes infinite: $\langle m \rangle \to \infty$. It indicates that fibrils undergo a sharp transition in their average lengths. In dilute peptide solutions, fibrils are short. Increasing the peptide concentration leads to a sharp increase in fibril lengths. However, computing the average lengths and transition point accurately requires some numerical computations, for reasons noted previously. Figure 5.15B shows the results of the full calculation, and compares the model with experimental data on α-synuclein [6]. It shows how fibrils undergo a sharp length transition with increasing peptide concentrations in solution.

Increasing the Concentration of Peptides Increases the Fibril Concentration

Now, how does the fibril concentration depend on the amount of protein in solution? At high protein concentrations, each added peptide chain goes into forming fibril. Recall from earlier that for $\delta \ll 1$, our model gives two states for c_m: monomers and long fibrils. So, we can assume that $\sum_{m=2}^{\infty} \approx \sum_{m=1}^{\infty}$ because we are neglecting only a very small population of short fibrils. This allows us to use the relationship $\sum_{m=1}^{\infty} my^m = y(1-y)^{-2}$, which is valid for $y < 1$, where $y = Kx$ in this case. This gives a closed-form expression for the fibril concentration:

$$c_{\text{fibril}} = \frac{x\delta}{(1-Kx)^2}. \tag{5.B.3}$$

Now use Equations 5.28 and 5.B.3 to compute the fraction of protein, f_{fibril}, that is in the form of fibrils:

$$f_{\text{fibril}} = \frac{c_{\text{fibril}}}{c_{\text{tot}}} = \frac{\delta Kx/(1-Kx)^2}{x + \delta Kx/(1-Kx)^2} = \frac{\delta K}{(1-Kx)^2 + \delta K}. \tag{5.B.4}$$

To interpret Equation 5.B.4, note that the quantity Kx ranges from 0 to 1. As $c_{\text{tot}} \to \infty$, $Kx \to 1$. In this limit of high peptide concentration, you can see that (1) $f_{\text{fibril}} \to 1$, meaning that any additional proteins added to solution will go into forming fibrils, and (2) $x \to (1/K)$ (because $Kx \to 1$), meaning that the concentration of monomers reaches a saturation concentration that you can compute from the propagation constant.

REFERENCES

[1] J Schellman. The factors affecting the stability of hydrogen-bonded polypeptide structures in solution. *J Phys Chem*, 62:1485–1494, 1958.

[2] K Ghosh and KA Dill. Theory for protein folding cooperativity: Helix bundles. *J Am Chem Soc*, 131:2306–2312, 2009.

[3] T Bereau, M Bachmann, and M Deserno. Interplay between secondary and tertiary structure formation in protein folding cooperativity. *J Am Chem Soc*, 132:13129–13131, 2010.

[4] KL Shaw, GR Grimsley, GI Yakovlev, et al. The effect of net charge on the solubility, activity, and stability of ribonuclease Sa. *Protein Sci*, 10:1206–1215, 2001.

[5] E Ruckenstein and IL Shulgin. Effect of salts and organic additives on the solubility of proteins in aqueous solutions. *Adv Coll Interface Sci*, 123–126:97–103, 2006.

[6] JD Schmit, K Ghosh, and KA Dill. What drives amyloid molecules to assemble into oligomers and fibrils? *Biophys J*, 100:450–458, 2011.

[7] BH Zimm and JK Bragg. Theory of the phase transition between helix and random coil in polypeptide chains. *J Chem Phys*, 31:526–535, 1959.

[8] S Lifson and A Roig. On the theory of helix–coil transition in polypeptides. *J Chem Phys*, 34:1963–1974, 1961.

[9] V Muñoz and L Serrano. Development of the multiple sequence approximation within the Agadir model of α-helix formation. comparison with Zimm–Bragg and Lifson–Roig formalisms. *Biopolymers*, 41:495–509, 1997.

SUGGESTED READING

Cantor CR and Schimmel PA, *Biophysical Chemistry. Part III: The Behavior of Biological Macromolecules.* WH. Freeman, San Francisco, 1980.

Chiti F and Dobson CM, Protein misfolding, functional amyloid, and human disease. *Annu Rev Biochem.* 75:333–366, 2006.

Dill KA and Bromberg S, *Molecular Driving Forces: Statistical Thermodynamics in Biology, Chemistry, Physics, and Nanoscience*, 2nd ed. Garland Science, New York, 2011.

Flory PJ, *Statistical Mechanics of Chain Molecules.* Wiley-Interscience New York, 1969; new edition by Hanser Publishers, Munich, 1989.

The Principles of Protein Folding Kinetics

CHAPTER 6

THE LEVINTHAL PARADOX MOTIVATED THE SEARCH FOR A PROTEIN FOLDING MECHANISM

In this chapter, we describe the rates and routes of protein folding and unfolding. By *routes*, we mean the time-course of conformational changes that occur as a protein folds. At what stage during the course of the folding process does a protein acquire its secondary structures, its hydrogen bonds, its hydrophobic core, and its tight side-chain packing? Is there a general mechanism that explains commonalities of folding and unfolding routes over different sequences and folds, and as a function of external variables, such as temperature and denaturants? How can a protein "find" its native structure spontaneously and quickly from its unfolded state, sometimes within microseconds? What is the *speed limit* of folding, and what is the physical basis for it?

Why is protein folding kinetics important? For one thing, folding kinetics was central to the historical development of protein science. In 1968, Cyrus Levinthal posed a puzzle that is called the "Levinthal paradox" [1]. Proteins face a "needle-in-a-haystack" problem. The haystack is the astronomically large conformational space that an unfolded protein must search in order to find the needle, its one native structure. If you estimate that there are between $z = 3$ and 8 conformational states for each peptide unit and if there are $N = 100$ amino acids in a protein, then the number of conformations the protein must search (the size of the haystack) is huge, $z^N \approx 10^{50}$–10^{90}. The Levinthal paradox is the question of how a protein can fold to its native structure so quickly, often in microseconds to seconds, in light of this huge search problem.

Levinthal's paradox inspired the following idea: if we could learn the *pathways* of folding, then perhaps we could infer a folding *principle* or *mechanism* by which conformations are explored or ignored, as so many different types of proteins reach their native structures so quickly and directly. Perhaps knowledge of the folding *intermediate states* (partially assembled structures that are formed on the way to the native state) could give us clues about how each type of protein "knows" which vast stretches of its conformational space not to search. Such knowledge could help us improve computer search methods for predicting native protein structures from their amino acid sequences. And knowledge of folding kinetics could help us understand the dynamics of other processes, including ligand binding, allosteric changes in conformation, and mechanistic actions.

FOLDING RATE EXPERIMENTS ARE CAPTURED BY MASS-ACTION MODELS

We begin with the most basic experiment in folding kinetics. You start with a dilute solution of proteins that are unfolded because of the denaturing conditions of the solution. You switch the solution to *folding conditions* by suddenly changing the denaturant concentration, pH, or temperature. You then observe an optical property, such as circular dichroism (CD) or fluorescence, $y(t)$, as a function of time t. You monitor the rate of the increasing population of native molecules in the test tube. This type of experiment is often called *in vitro refolding*, because it takes place in test tubes, not inside the cell, with proteins that were initially unfolded. You can also run the experiment in reverse, unfolding native proteins by jumping the protein solution from native conditions to *unfolded conditions*. You run these experiments with dilute protein solutions, to avoid protein aggregation.

Folding and unfolding processes are usually well described by one or more exponential functions of time:

$$y(t) = A_1 e^{-k_1 t} + A_2 e^{-k_2 t} \tag{6.1}$$

(Figure 6.1). The rate coefficients k_1 and k_2 and the amplitudes A_1 and A_2 can give you insights into the folding process under various conditions.

In this chapter, we describe the insights you get from both *macroscopic* and *microscopic* models. In macroscopic modeling, your goal is

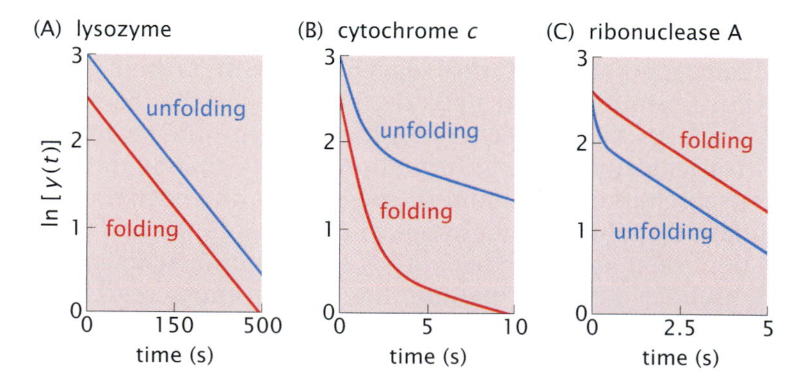

Figure 6.1 Folding and unfolding often follow one- or two-exponential decays. Kinetic relaxation of the amount of native or denatured protein as a function of time after jumping the conditions by light absorbance for (A) lysozyme and (B) cytochrome *c*, and by fluorescence for (C) ribonuclease A. A single exponential is a straight line on these logarithmic plots. The plots show two-state folding and unfolding kinetics for lysozyme; three-state folding and unfolding for cytochrome *c*; and a combination of two-state folding and sequential unfolding for ribonuclease A. (A and B, adapted from A Ikai and C Tanford. *Nature*, 230:100–102, 1971. With permission from Macmillan Publishers Ltd.; C, adapted from TY Tsong, RL Baldwin, and EL Elson. *Proc. Natl Acad Sci USA*, 69:1809–1812, 1972. Copyright (1972) National Academy of Sciences, USA.)

to choose the *mass-action model* that best fits your data. A *mass-action model* is a chosen set of *kinetic states*, such as D, I, and N, and a postulated kinetic scheme of arrows connecting those states (Figure 6.2). Macroscopic modeling simply expresses your experimental results in the shorthand language of these *pathways*. In contrast, microscopic models aim to capture the underlying physics—how the molecular conformations and energies lead to folding rates and routes. Microscopic models aim to give a structural and physical explanation for the time evolution of folding events, and/or the effects of temperature, denaturants, pH, or amino acid sequence on the folding process. First, let's explore macroscopic modeling.

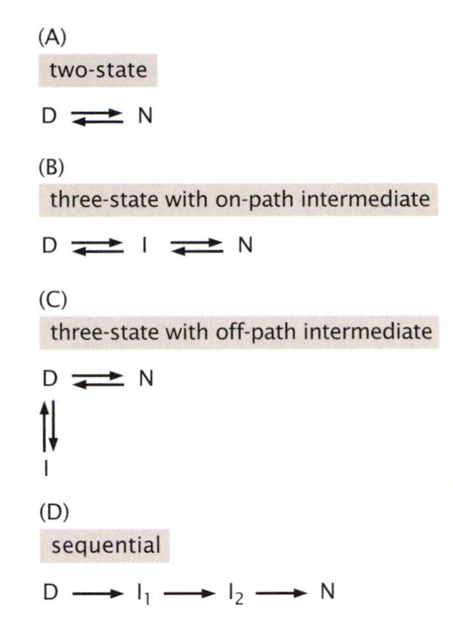

Figure 6.2 Common mass-action models of folding. (A) The two-state model. (B) On-pathway intermediate state. (C) Off-pathway intermediate. (D) A sequential model having multiple intermediates.

Small Proteins Fold Rapidly through Two-State Kinetics

A typical small single-domain protein folds or unfolds in just a single-exponential relaxation process, such as the one shown in Figure 6.1A for lysozyme. Processes in nature that involve single-exponential dynamics are said to follow *two-state* kinetics because the mass-action model you need to explain them requires only two states, in this case D (denatured) and N (native):

$$\mathrm{D} \underset{k_u}{\overset{k_f}{\rightleftharpoons}} \mathrm{N}, \tag{6.2}$$

where k_f and k_u are the folding and unfolding *rate coefficients*,[1] respectively. The time dependence of this process can be described by

$$\begin{aligned} \frac{d[\mathrm{D}]}{dt} &= -k_f[\mathrm{D}] + k_u[\mathrm{N}], \\ \frac{d[\mathrm{N}]}{dt} &= k_f[\mathrm{D}] - k_u[\mathrm{N}], \end{aligned} \tag{6.3}$$

where [D] and [N] are the time-dependent concentrations of the denatured and native states, respectively. If you choose normalized units so that $[\mathrm{D}] + [\mathrm{N}] = 1$, then [D] and [N] are time-dependent *probabilities* or *populations*, $p_{\mathrm{D}}(t)$ and $p_{\mathrm{N}}(t)$, with values ranging from 0 to 1.

There are different ways to solve Equation 6.3. The simplest is to substitute $p_{\mathrm{D}}(t) = 1 - p_{\mathrm{N}}(t)$ and solve for $p_{\mathrm{N}}(t)$. However, that strategy is limited mostly to two-state kinetics. Because we want to explore a broader range of models that apply to different types of dynamics, we use a more general approach, called *master equations*. In this approach, you express the probabilities of the states in a *column vector*

$$\mathbf{P}(t) = \begin{bmatrix} p_{\mathrm{D}}(t) \\ p_{\mathrm{N}}(t) \end{bmatrix} \tag{6.4}$$

and the rate coefficients in a *rate matrix*

$$\mathbf{W} = \begin{bmatrix} -k_f & k_u \\ k_f & -k_u \end{bmatrix}. \tag{6.5}$$

[1] We use the term *rate coefficient* rather than *rate constant*, which is common in the field, because these quantities are not constants. They are usually strong functions of temperature, among other things.

Then, you can express the time evolution of the probabilities as a matrix equation:

$$\frac{d\mathbf{P}(t)}{dt} = \mathbf{WP}(t). \tag{6.6}$$

Appendix 6A gives the procedure for solving master equations. For the earlier two-state case, the native population is

$$p_{\text{N}}(t) = p_{\text{N}}(0)e^{-(k_u+k_f)t} + \frac{k_f}{k_u + k_f}\left(1 - e^{-(k_u+k_f)t}\right), \tag{6.7}$$

and the denatured population is $p_{\text{D}}(t) = 1 - p_{\text{N}}(t)$. The fraction of molecules that are folded converges to the equilibrium value

$$p_{\text{N}}(\infty) = \frac{k_f}{k_f + k_u} \tag{6.8}$$

at long times. Equation 6.7 says that if you jump the conditions at time $t = 0$, you will observe a single-exponential decay to equilibrium with a rate coefficient

$$k_{\text{obs}} = k_u + k_f. \tag{6.9}$$

In one limit, you can start at time $t = 0$ with all molecules unfolded, $p_{\text{N}}(0) = 0$, and apply strong folding conditions, $k_u \to 0$. Then the native-state population will grow exponentially with time, with rate coefficient k_f:

$$p_{\text{N}}(t) = 1 - e^{-k_f t}. \tag{6.10}$$

In the opposite limit, you can start at time $t = 0$ with all molecules fully folded, $p_{\text{N}}(0) = 1$, and apply strong denaturing conditions, $k_f \to 0$. Then the native-state population will disappear exponentially:

$$p_{\text{N}}(t) = e^{-k_u t}. \tag{6.11}$$

Here is how you use Equation 6.7. You adjust the parameters k_f and k_u to achieve the best fit of Equation 6.7 to your experimental data. If your data are more complex than single-exponential dynamics, then you try a different model, typically one that has intermediate states.

The two-state dynamics described earlier is often adequate for small proteins. How should you interpret this exponential time dependence? Does it mean that all the proteins in the test tube are synchronized, with each molecule forming its first helix at the same time that every other protein molecule forms its first helix, for example? Or does it mean that each protein itself folds very quickly, not synchronized with other protein molecules, and that the exponential decay is just reflecting what fraction of the protein molecules are folded or unfolded at a given time? It means the latter. The rate coefficients k_f and k_u are the probabilities per unit time that any particular protein molecule will convert. The *number of molecules* that convert to N at any given time is proportional to how many D molecules are still left unconverted in the test tube at that time. The fewer D molecules there are in solution, the fewer D $\to$ N conversions happen in unit time. This is a concentration effect. This is the basis for the exponential time course in protein folding or unfolding. Box 6.1 shows how you can use this interpretation for determining folding rates from computer simulations.

Box 6.1 You Can Estimate a Protein's Folding Rate from Multiple Short Computer Simulations

Suppose you want to run a computer simulation of folding to determine the folding rate coefficient k_f. (Such simulations are discussed in Chapter 10.) You can run either a few long trajectories or many short ones. You could start the simulation from some unfolded conformation, then run a long trajectory of the protein under folding conditions. Your computer trajectory should be run for at least several multiples of the time $\tau_f = 1/k_f$, the inverse of the folding rate coefficient. Otherwise, you won't have enough data to accurately fit the baseline of your relaxation curves. For example, if the true folding time were $1\,\mu s$, you would need at least $5\text{–}10\,\mu s$ of simulation time. But running long trajectories takes a long time on computers.

Instead, if you have parallel computers, you could run many short trajectories and collect statistics. V Pande showed this on a computer resource called Folding@home. Equation 6.10 shows that for short times ($t \ll \tau_f$), starting from the denatured state, you can approximate the time-dependent population of the native state as

$$p_N(t) = 1 - e^{-k_f t} \approx k_f t. \tag{6.12}$$

Rearranging Equation 6.12 gives

$$\begin{aligned} k_f &= \frac{p_N(t)}{t} \\ &= \frac{1}{t}\left(\frac{\text{number of trajectories in which the protein folds}}{\text{total number of trajectories}}\right). \end{aligned} \tag{6.13}$$

In this way, you can estimate k_f from many short trajectories instead of fewer long ones. You simply count the fraction of folding events you see. This strategy is useful because it is often easier to obtain computer resources for many shorter runs than for fewer long ones. And long trajectories do not necessarily provide adequate sampling of conformational space.

Small globular proteins (shorter than about 100–150 amino acids) usually fold via two-state kinetics. It is remarkable that proteins fold with such simple kinetics, given their heterogeneous and complex molecular structures. We show in the following that any process having single-exponential kinetics can be described as having a *rate-limiting step*, or a *transition state*.

Single-Exponential Kinetics Is Characterized through the Concept of a *Transition State*

Any process having single-exponential dynamics, including two-state protein folding or unfolding, can be described using a concept called *transition state* or *activation barrier*. This concept rests on the fundamental relationship between any equilibrium constant K, such as for folding, and its corresponding free-energy change ΔG_f:

$$\Delta G_f = G_N - G_D = -RT \ln K, \tag{6.14}$$

where RT is the gas constant multiplied by the absolute temperature, and G_N and G_D are the free energies of the native and denatured states, respectively. For two-state kinetics, you can express the equilibrium constant as a ratio of rate coefficients:

$$K = \frac{[N]_{eq}}{[D]_{eq}} = \frac{k_f}{k_u}. \tag{6.15}$$

The first equality is the definition of the equilibrium constant for a two-state process, and the second comes from using equilibrium condition $d[N]/dt = d[D]/dt = 0$ in Equation 6.3.

From substituting Equation 6.15 into Equation 6.14, it follows that

$$\begin{aligned} \Delta G_f &= -RT \ln\left(\frac{k_f}{k_0}\right) - \left[-RT \ln\left(\frac{k_u}{k_0}\right)\right] \\ &= \Delta G_f^\ddagger - \Delta G_u^\ddagger. \end{aligned} \tag{6.16}$$

Equation 6.16 simply relates rate coefficients to free-energy quantities, just as Equation 6.14 expresses *equilibria* in terms of free energies. (We have introduced an *intrinsic rate* constant k_0 into each term on the right-hand side. Mathematically, k_0 is needed to ensure that each argument inside each logarithm is dimensionless. Physically, k_0 represents the maximum speed that either forward or reverse processes can reach, at infinite temperature.) Following immediately from Equation 6.16 is the existence of a third state, represented by ‡, that is neither N nor D, and that has free energy $G^\ddagger$:

$$\Delta G_f^\ddagger = -RT \ln\left(\frac{k_f}{k_0}\right) = G^\ddagger - G_D \tag{6.17}$$

and

$$\Delta G_u^\ddagger = -RT \ln\left(\frac{k_u}{k_0}\right) = G^\ddagger - G_N. \tag{6.18}$$

The state ‡ is called the *transition state* (TS) or *activated state*, and $G^\ddagger$ is its free energy. Rearranging Equation 6.17 gives the folding rate coefficient in terms of the barrier free energy:

$$k_f(T) = k_0 \exp\left(-\frac{\Delta G_f^\ddagger}{RT}\right). \tag{6.19}$$

The state ‡ can be regarded as a bottleneck; it usually has a positive free energy, $\Delta G_f^\ddagger > 0$, meaning that both the forward and backward rates of the system are slower than the speed limit, that is, $k_f/k_0 < 1$ and $k_u/k_0 < 1$. So, $G^\ddagger$ represents a hilltop point on a *reaction-coordinate diagram* (Figure 6.3). A reaction diagram is a plot of the free energies of the various states along the way between the two end states. Reaction-coordinate diagrams convey the information that the forward and reverse rates can be expressed in terms of free energies, and that there is a sum relationship between the free energies: $\Delta G_f = \Delta G_f^\ddagger - \Delta G_u^\ddagger$. Any process involving single-exponential kinetics is interpretable in terms of transition-state barriers. The importance of Equation 6.19 is in describing an experimentally observable quantity, the folding rate coefficient, in terms of another quantity, the free-energy barrier, that expresses underlying driving forces.

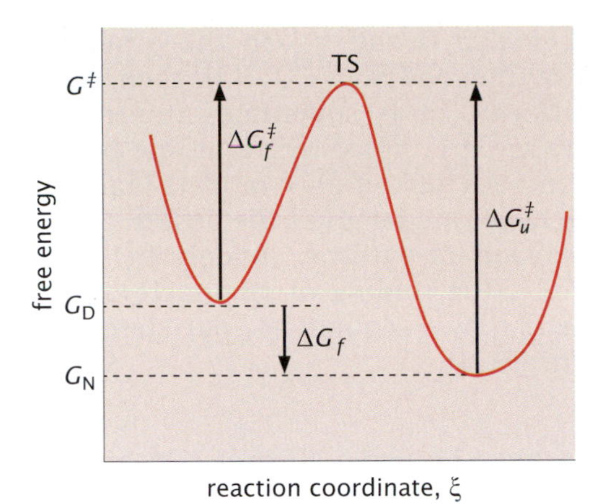

Figure 6.3 A reaction-coordinate diagram is a one-dimensional representation of the relationship between equilibrium and rate quantities: $\Delta G_f = \Delta G_f^{\ddagger} - \Delta G_u^{\ddagger}$. This diagram conveys this relationship on a diagram showing a real or fictitious *reaction coordinate* ξ, which represents the progress of the reaction.

Here's how the transition-state concept is used for proteins. First, it implies a possible structural basis for rate bottlenecks. Is there a particularly "challenging" structure of the protein that causes it to be the slow step in folding? However, protein folding is different than small-molecule chemistry, where the transition-state concept is traditionally applied. The making and breaking of covalent chemical bonds involves large energies, usually tens of kilocalories per mole, whereas the making and breaking of noncovalent bonds, as in protein folding, involves much smaller energies, in the range of $RT = 0.6\,\text{kcal mol}^{-1}$ at $T = 300\,\text{K}$. So, transition states for protein conformations are *thermal ensembles*, not single molecular structures.

Second, the transition-state equation, Equation 6.19, is also useful for interpreting the effects of temperature on kinetics. If you observe that $\ln k_f(T)$ is proportional to $1/T$, it implies that k_0 and $\Delta G_f^{\ddagger}$ are constants, independent of temperature. In this case, called *Arrhenius kinetics*, increasing the temperature increases the speed of a process. Arrhenius kinetics is often observed for protein *unfolding* and for conformational changes of small protein pieces, such as helices and β-hairpins. However, Arrhenius kinetics are sometimes not observed for protein *folding*: for ultrafast-folders, heating doesn't speed up folding. We give examples later in this chapter.

What is the speed limit k_0? The speed limit of gas-phase chemical reactions is $k_0 = RT/h = 0.16$ conversions per picosecond at room temperature (h is Planck's constant). But protein-folding kinetics is not comparable to that of chemical-bond-forming reactions in the gas phase. Folding entails a protein sloshing around in solvent. The shortest time for protein folding has been estimated to be around $1\,\mu\text{s}$, so for protein folding, $k_0 \approx 10^6\,\text{s}^{-1}$ [2]. This also happens to be approximately the maximum speed that a 10- or 20-mer peptide can form a helix in solution.

Large Proteins Typically Fold via Multi-Exponential Kinetics

Large proteins tend to fold slowly, and with multi-exponential rates. By definition, the two-state model will not fit a multi-exponential process. To fit two exponentials, you must invoke at least three states, typically native, denatured, and a *kinetic intermediate* state. There are two main types of three-state models. In the model $\text{D} \rightleftharpoons \text{I} \rightleftharpoons \text{N}$, the folding intermediate state I is called *on-pathway* because the symbol appears in-line between D and N. On the other hand, if your data may be better fit by the model, I $\rightleftharpoons$ D $\rightleftharpoons$ N, then the intermediate state I is called *off-pathway*, or is sometimes referred to as a *misfolded intermediate*, because it is not directly in-line between D and N. You choose between on-path and off-path models (and possibly other models) based on whichever model gives the best fit to the data. And, when neither on-pathway or off-pathway models fits, then try a different model. For example, the folding of ribonuclease A (see Figure 6.1C) has been explained [3] by a sequential, four-state model, similar to the one shown in Figure 6.2D.

What Is the Difference between a Kinetic Intermediate State and a Transition State?

The expression D $\rightarrow$ TS $\rightarrow$ N resembles the expression D $\rightarrow$ I $\rightarrow$ N. What is the difference between a kinetic intermediate state I and a transition state TS? Intermediates are defined as states that become substantially populated during a kinetic process, while transition states do not. Think of the kinetic states D, N, or the intermediates, $I_1, I_2, \ldots I_m$, as metaphorical "buckets" that can fill up and empty out. In this metaphor, the folding process begins with a full bucket D and an empty bucket N. Folding ends when the N bucket is full and D is empty. Now consider a situation in which there is an intermediate bucket (labeled either I or TS) interposed between D and N. Bucket D pours into this intervening bucket, which has a hole in its bottom, so it empties into bucket N. The difference between an intermediate and a transition state is the size of the hole in the middle bucket. As the D bucket pours into an intermediate I, the I bucket leaks out into the N bucket only slowly, so the I bucket fills up significantly throughout the middle of the kinetic process, before it drains out at later times. In contrast, think of TS as a bucket that has a large hole in its bottom, so that it never fills up very much (Figure 6.4A). Even if the D bucket is pouring into the TS bucket rapidly, the TS bucket empties into the N bucket as fast as it is being filled, so TS never becomes very full. In short, I states reach significant populations and observable kinetic features, while TS states have only small populations, so they are consistent with only two observable kinetic states.

Our buckets are simply a metaphor for the *eigenvectors* of a master equation; see Appendix 6A. Once you know the dynamics of a model, you can compute its eigenvalues and eigenvectors. Think of one eigenvalue (a rate) and an eigenvector (a set of changes in the populations of states reflecting the inflows and outflows of the different states)

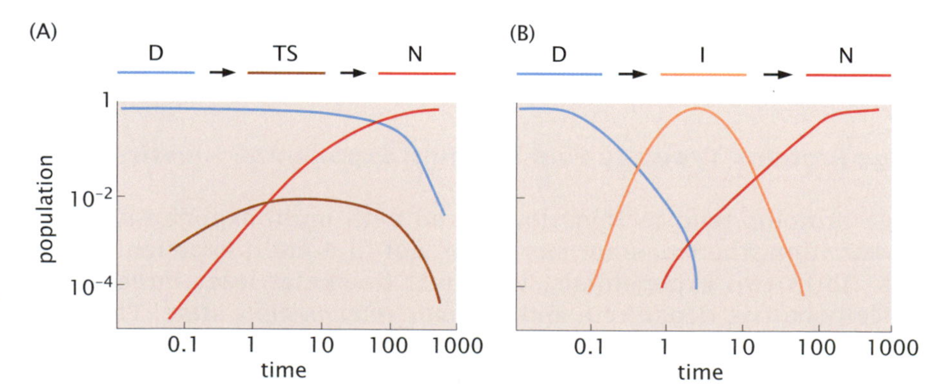

Figure 6.4 Transition states have small populations. Kinetic intermediates have larger populations. The plots show populations of states versus time since the initiation of folding. (A) The transition state (TS) in D $\rightarrow$ TS $\rightarrow$ N is only transient, never reaching a high population, and not contributing an observable kinetic phase. (B) The intermediate state (I) in D $\rightarrow$ I $\rightarrow$ N has a detectable population during the folding process, leading to an observable kinetic phase (see [4]).

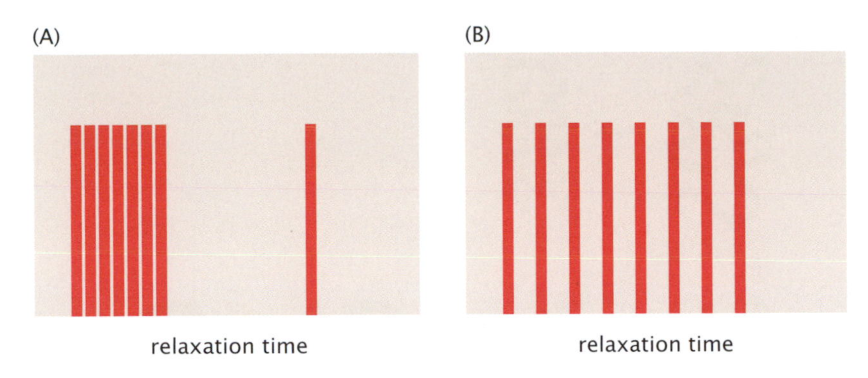

Figure 6.5 A gap in the eigenvalue spectrum implies apparent two-state kinetics. The bars indicate the different relaxation times of multi-exponential processes. Different processes have characteristic spacings. (A) When the eigenvalue spectrum has a single slowest process that is well separated from other (faster) processes, the slowest exponential will be the most prominent observable, since faster processes will equilibrate over that timescale. (B) When the eigenvalues are not well separated, the kinetics are more complicated and multi-exponential.

as describing one "generalized process" or metaphorical "generalized bucket". Solving a master equation gives a sum of such generalized independent processes. Each process has its own timescale (eigenvalue). Each process is a sum of different amounts of flows to and from the different states. Eigenvalues and eigenvectors are the quantitative manifestation of the bucket idea. They express a given relaxation process as a sum of contributions from different modes of relaxation. Each reaction *topology* (that is, particular arrangement of states and arrows) leads to some eigenvalue spectrum. Single-exponential kinetics is observed when an eigenvalue spectrum has a gap between the slowest rate and all others (Figure 6.5A). In multi-exponential kinetics, there is no such gap between eigenvalues (see Figure 6.5B).

RATE MEASUREMENTS GIVE INSIGHTS INTO THE PATHWAYS OF PROTEIN FOLDING

What are the folding *routes*? When do different structures form during the folding process? Different experimental probes will report different information. Suppose you have two probes, one that reports helix formation and another that reports chain collapse. In principle, watching the time dependence of both probes during a folding experiment of a protein could tell you the relative order of the collapse and helix-formation events of that protein. But here's the challenge. The small proteins that are commonly studied are two-state folders; that is, they fold with only a single kinetic phase. So, without other information, all you can say about two-state folders is that "everything happens within a single kinetic event." If you want to create a more detailed narrative about folding pathways, you need independent information or you need to study multistate folders, where you can characterize the kinetic intermediate states.

Figure 6.6 shows three hypothetical folding pathways. In one, folding happens as a single event. Another possibility is that two helices form separately at about the same time, and then the two fully formed helices dock together. A third option is that one helix forms first, providing a surface onto which the second helix can form. And, there are many other options. The search for folding mechanisms has driven

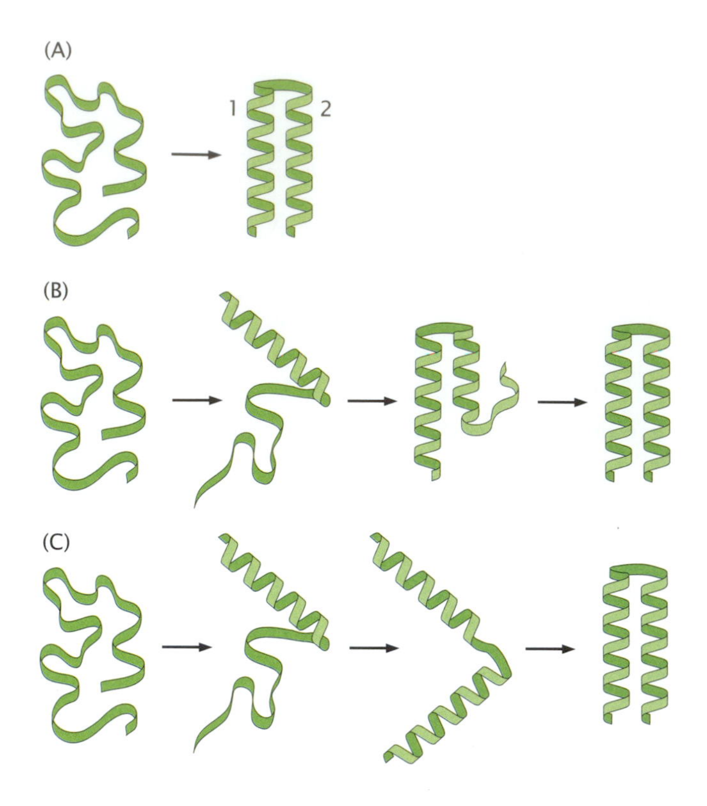

Figure 6.6 Different possible folding routes. Different kinetic routes from the denatured state (disordered, on the left) to the native state (helices, on the right). (A) Two-state process: it happens all at once. (B) First, helix 1 forms, which serves as a template, assisting the formation of helix 2. (C) Helix 1 forms independently of helix 2, then they come together.

important advances in experimental methods, some of which are described in this chapter.

One well-studied kinetic intermediate state involves *proline isomerization*. Prolines can interconvert between two different isomeric states, *cis* and *trans* peptide bonds. In the native structure, the proline interconversion is slow. In the denatured protein, the proline will be in a Boltzmann equilibrium between the two isomeric states. Upon folding, each of the protein's non-native prolines must convert to whatever is that proline's native-state isomer. So, folding has two kinetic phases. In the slow phase, the protein starts with the wrong proline isomer. That folding is slow because prolines are slow to isomerize. In the fast phase, the protein begins in conformations that are already in the correct isomeric state.

Another kinetic intermediate is found in the folding of the bovine pancreatic trypsin inhibitor protein (BPTI), whose native structure has three disulfide bonds among six Cys residues. Interestingly, the disulfide bonds in BPTI do not form in a systematic increasingly native-like series of events. Some wrong disulfide bonds form transiently first, then they are undone, then the correct disulfides finally form. These incorrect disulfides have been called *on-pathway misfolded states*.

Other proteins, too, can pass through non-native states on their folding routes. The folding kinetics of lysozyme shown in Figure 6.1A indicates a single exponential, when probed by a single method. However, by using multiple methods, it has been shown that the two domains of hen lysozyme behave differently during folding. The α-domain folds rapidly. Once the α-domain is formed, the β-domain then folds. The protein overshoots in the fast process, indicating that helix formation probably goes too far, and leads to non-native contacts, before being rescued by the slower β-sheet formation step. The folding of hen lysozyme thus appears to involve multiple paths and considerable complexity [5]. β-lactoglobulin also folds with multiphase kinetics. β-lactoglobulin is a β-barrel comprised of nine antiparallel β-strands, one major α-helix, and four short helices. In folding, the molecule first forms an α-helical structure transiently before adopting its native β-barrel structure.

Mutational Studies Can Probe Folding Pathways

You can get insights into folding routes by studying mutated proteins. A method called *Φ-value analysis* [6] gives information about which parts of the protein fold slowly and which fold more rapidly. In Φ-value analysis, you mutate a single amino acid at a time in the protein. You measure both the folding rate coefficient $k_{f,\,\mathrm{wt}}$ of the wild-type protein and the folding rate coefficient $k_{f,\,\mathrm{mut}}$ of the mutant. Equation 6.17 indicates that the *change* in the transition-state barrier free energy resulting from the mutation will be given by

$$\Delta\Delta G_f^{\ddagger} = -RT\ln\left(\frac{k_{f,\,\mathrm{mut}}}{k_{f,\,\mathrm{wt}}}\right). \tag{6.20}$$

To perform Φ-value analysis, you must also measure the change in protein stability, $\Delta\Delta G_f$, caused by the mutation, or equivalently the change in equilibrium constant of the mutant, $K_{f,\mathrm{mut}}$, relative to that of the wild type, $K_{f,\mathrm{wt}}$. The Φ value for folding is defined as the ratio

$$\Phi_f = \frac{\Delta\Delta G_f^{\ddagger}}{\Delta\Delta G_f} = \frac{\ln(k_{f,\,\mathrm{mut}}/k_{f,\,\mathrm{wt}})}{\ln(K_{f,\,\mathrm{mut}}/K_{f,\,\mathrm{wt}})}. \tag{6.21}$$

Φ is a number that is usually between 0 and 1. Figure 6.7 shows how Φ values are interpreted in terms of transition states.[2] Observing $\Phi_f = 0$ means that your mutation had no effect on the folding rate. Observing $\Phi_f = 1$ means that your mutation affected the folding free-energy barrier as much as it affected the folding stability. So, mutational results

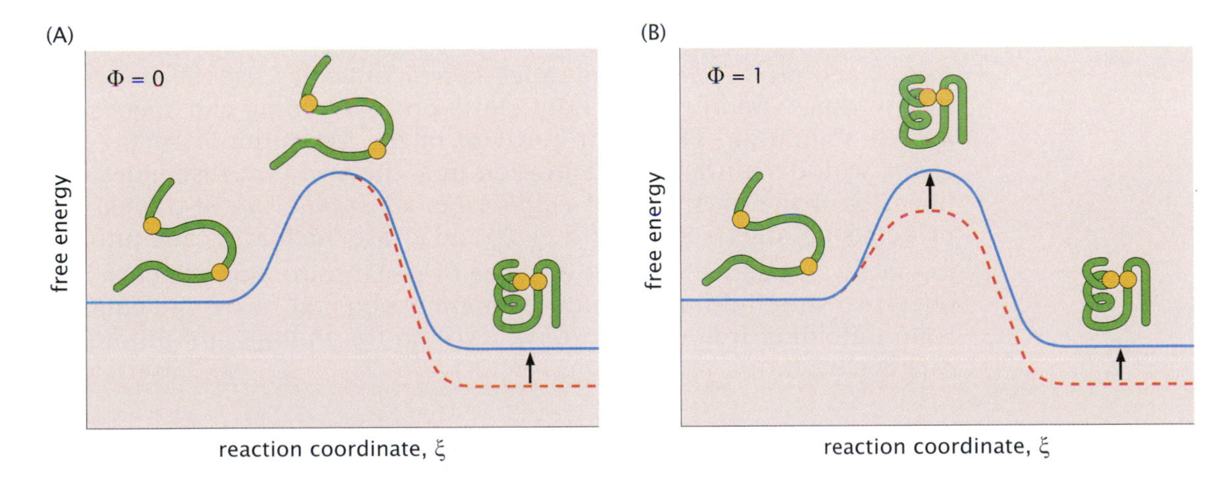

Figure 6.7 A Φ value is often interpreted in terms of a structure of the transition state. (A) $\Phi = 0$ means that the mutation site has a denatured-like structure when it is in its transition state. (B) $\Phi = 1$ means that the mutation site has a native-like structure when it is in its transition state. The *blue solid curve* is for wild type, and the *red dashed curve* is for mutant.

[2] We have described here the *folding* Φ value, Φ_f; you can also define a corresponding *unfolding* Φ_u.

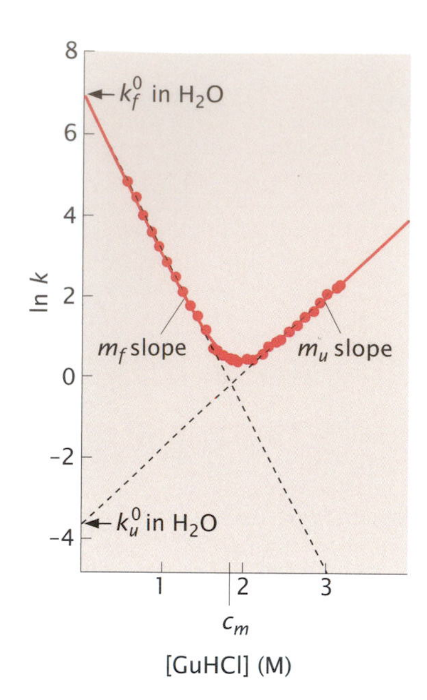

Figure 6.8 A chevron plot shows how denaturants affect folding kinetics. The *left straight* line shows that reducing the denaturant concentration speeds up folding. The *right straight* line shows that increasing denaturant speeds up unfolding. c_m is the denaturation midpoint. Extrapolation of the lines to zero denaturant gives the folding and unfolding rate coefficients k_f and k_u in water. The results shown here are for the acyl-CoA-binding protein. (Adapted from KL Maxwell, D Wildes, A Zarrine-Afsar, et al. *Protein Sci*, 14:602–616, 2005.)

are often interpreted as follows. If you see $\Phi_f = 0$ for your mutation, it means that residue in the chain has not yet reached its folded structure when the rest of the chain is passing through the transition state, while $\Phi_f = 1$ means that residue is native when the chain passes through its folding transition state.

If you find nonzero $\Phi > 0$ at multiple sites, then the transition state is called *diffuse*, meaning that many different amino acids in that protein are involved in controlling the folding rate. If, instead, you find $\Phi \approx 0$ broadly throughout the protein, the transition state is called *polarized*, meaning that just a few of the protein's amino acids are controlling the folding rate. For example, chymotrypsin inhibitor 2 has a diffuse transition state, and ubiquitin has a polarized transition state.

Denaturants Can Change Folding Rates: The Chevron Plot

Adding denaturants, such as guanidinium hydrochloride (GuHCl) or urea, can slow down folding and speed up unfolding. The effects of denaturants on folding rate coefficients are represented using a *chevron plot*. A chevron plot shows the logarithm of the observed relaxation rate on the vertical axis versus the denaturant concentration on the horizontal axis (Figure 6.8). The term *chevron* refers to the plot's V shape (or inverted-V shape if you plot the logarithm of folding time instead). The left branch of the V describes the folding rates you see when the system is jumped to low denaturant concentrations, telling you about the folding process. The right branch of the V describes jumping to high denaturant concentrations, telling you about the unfolding process.

For proteins that fold with two-state kinetics, the logarithm of folding and unfolding rate cooefficients are linear functions of denaturant concentration $[d]$ (see Figure 6.8); that is,

$$\begin{aligned} \ln k_f &= \ln k_f^0 + m_f[d], \\ \ln k_u &= \ln k_u^0 + m_u[d]. \end{aligned} \tag{6.22}$$

$m_f < 0$ and $m_u > 0$, which are called *kinetic m-values*, are the slopes of the two lines. Adding denaturant slows protein folding and speeds up protein unfolding. The linear behavior of the logarithm of *rate coefficients* with denaturant concentration in a chevron plot resembles the linearity of the logarithm of *equilibrium constants* (the m-values) for protein stabilities (see Chapter 3). k_f^0 and k_u^0 are the folding and unfolding rates, respectively, in the absence of denaturant. For proteins that obey two-state folding/unfolding kinetics, you can relate the equilibrium unfolding free energy $\Delta G_u = -\Delta G_f$ to the folding and unfolding rate coefficients:

$$\Delta G_u = RT \ln\left(\frac{k_u}{k_f}\right) = RT \ln\left(\frac{k_u^0}{k_f^0}\right) + (m_u - m_f)[d]. \tag{6.23}$$

At the denaturation midpoint (that is, at c_m in Figure 6.8), you have $k_f = k_u$.

Why are chevron plots V-shaped? Recall from Equation 6.9 that the observed relaxation rate coefficient is the sum of folding and unfolding rate coefficients: $k_{obs} = k_f + k_u$. At low denaturant concentrations, the protein is native and k_u is small, so $k_{obs} \approx k_f$, and therefore the left arm reports mostly on the folding process. At high denaturant concentrations, the protein is denatured and k_f is small, so $k_{obs} \approx k_u$ (see Figure 6.8). Therefore, you get two straight lines when you plot the logarithm of k_{obs} versus denaturant concentration, giving a shape like the letter V. Kinetic m -values for several small proteins are given in Maxwell et al. [7].

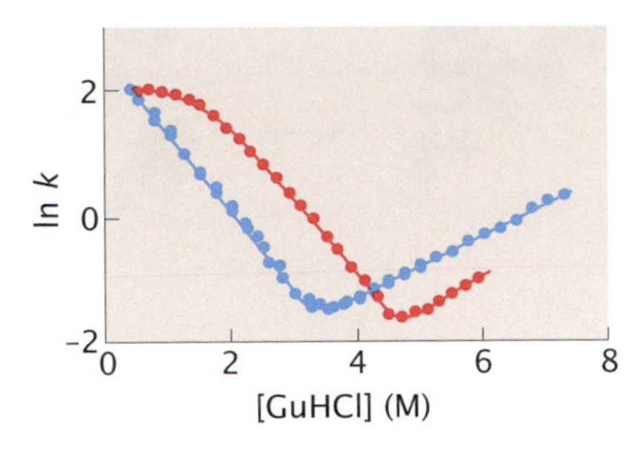

Figure 6.9 Adding salt speeds up folding and slows down unfolding. Chevron plots for the folding of ribosomal protein S6, in the absence (*blue*) and the presence (*red*) of sodium sulfate. The rollover (*red*) indicates that folding reaches maximum speed at low denaturant concentrations. The salt stabilizes the native protein. (Adapted from DE Otzen and M Oliveberg. *Proc Natl Acad Sci USA*, 96:11746–11751, 1999. Copyright (1999) National Academy of Sciences, USA.)

A V-shaped plot indicates that a protein has two-state folding kinetics. When folding is more complex than a single exponential, a chevron plot is not V-shaped. Figure 6.9 shows *rollover*, where one arm of a chevron plot is curved or forms a plateau, rather than a straight line. Rollover indicates the presence of a folding intermediate or a kinetic trap: further strengthening the folding conditions (moving to the left on the *x*-axis of the chevron diagram) doesn't speed up folding. Rollover means that a stronger external driving force cannot overcome some internal speed limit of the protein.

You can combine mutations with chevron plots. To do this, you first measure the folding and unfolding rates of your protein in a series of different denaturants, to make a chevron plot for your wild-type protein. Then you make a mutation at a particular amino acid site in the protein. Now, you make another chevron plot for the mutant. Plot the two chevrons on the same figure. Figure 6.10 shows two limiting cases: (A) In one case, the mutation changes the folding arm of the chevron plot, but not the unfolding arm. This indicates the mutation site contributes to folding-rate control. (B) In the other case, the mutation changes the unfolding arm, but not the folding arm. This indicates the mutation site is not rate-controlling for folding.

Here is a key implication from chevron studies on two-state proteins: the main folding mechanism (the slowest step) is independent of the starting state of the protein. The folding arm of a chevron plot is determined fully by the final state of the solution to which the protein is jumped. The folding arm does not depend on the initial state of the protein prior to the jump in conditions. You could have started the folding process from highly denaturing conditions, where the protein would have no partially folded structure. Or you could have started the folding process from weakly denaturing conditions, where the protein might have had considerable partial structure. It doesn't matter. The slowest relaxation time doesn't change as a function of different initial conditions.

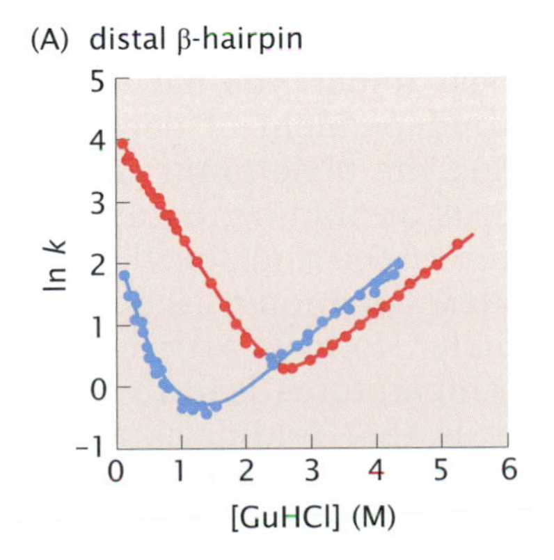

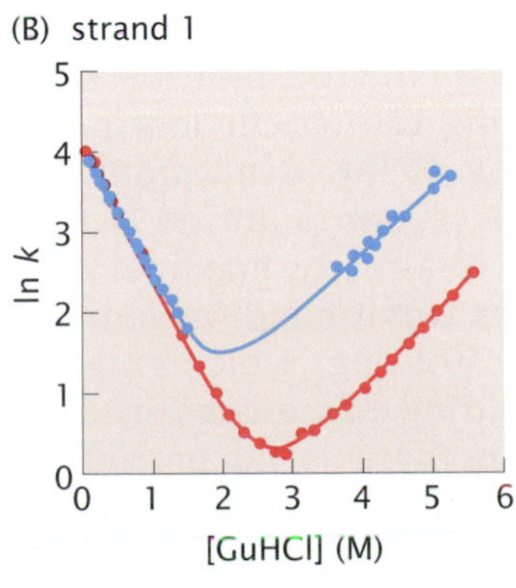

Figure 6.10 Some mutations change folding rates; some change unfolding rates. (A) Mutations in the "distal" β-hairpin of the SRC kinase SH3 domain change only the folding arm of the chevron plot. (B) Mutations in the first β-strand change only the unfolding arm. Wild type is shown by *red* and mutant by *blue*. (Adapted from DS Riddle, VP Grantcharova, JV Santiago, et al. *Nat Struct Biol*, 11:1016–1024, 1999. With permission from Macmillan Publishers Ltd.)

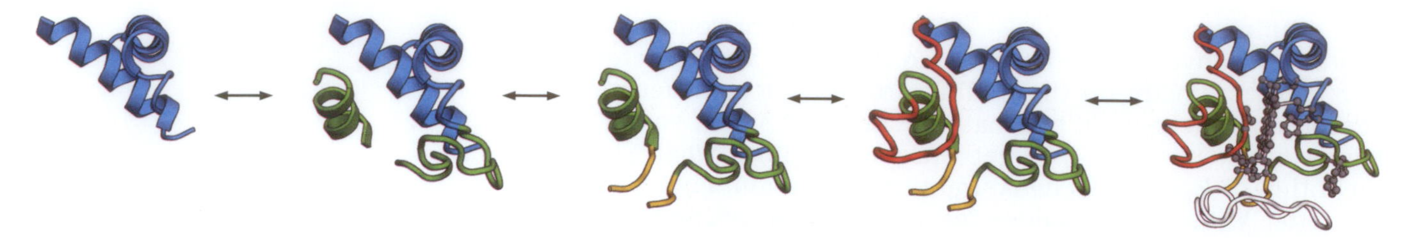

Figure 6.11 Cytochrome *c* folds in units of foldons. The foldon units are the pair of two terminal helices that form a tertiary contact (*blue*); a central, so-called 60's helix and loop (*green*); a two-stranded β-sheet connected to the green regions (*yellow*); and two loops (*red* and *white*). (Adapted from SW Englander, L Mayne, and MMG Krishna. *Q Rev Biophys*, 40:287–326, 2007. With permission of Cambridge University Press.)

Whole Secondary Structures Often Fold as a Unit

Proteins tend to fold in units of *motifs* or secondary structures, rather than as individual amino acids. Structural folding units are called *foldons* or *partially unfolded forms*. Foldons have been observed using *hydrogen exchange* (HX). HX can measure either equilibrium fluctuations, or properties of folding kinetics. In equilibrium HX, you have a series of protein solutions with increasing amounts of denaturant. Each amino acid has an amide proton. Prior to the experiment, you replace those amide protons with deuterium atoms. In solution, the deuterium atoms on the protein will exchange with the hydrogen atoms in the surrounding solvent water, with an equilibrium constant that depends on how exposed those amide groups are to the water. It is found that in a series of increasing amounts of denaturant, whole secondary structures are often protonated at once. In kinetic HX, you transiently pulse the conditions so as to capture the protonation state of the protein during a particular time interval of folding. Both experiments show that secondary structures can fold rapidly relative to other folding events. Figure 6.11 shows the sequence of stabilities of the individual structural elements in cytochrome *c*.

Some proteins fold so fast that they appear to have essentially no free-energy barrier. Called *ultrafast folders*, they can fold on timescales of tens of microseconds.

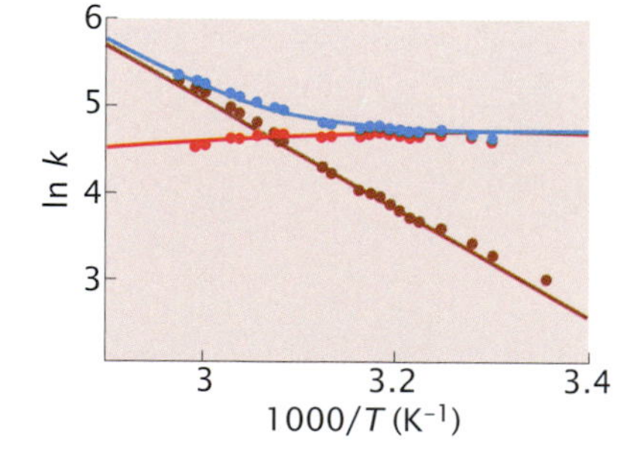

Figure 6.12 Evidence of barrierless folding: folding rate coefficients (*red*) are almost independent of temperature (for the engrailed homeodomain protein). In contrast, unfolding rate coefficients (*brown*) follow normal Arrhenius behavior; see Equation 6.B.12. The observed rate coefficient (*blue*) is the sum of folding and unfolding rates. (Model fits from K Ghosh, SB Ozkan, and KA Dill. *J Am Chem Soc*, 129:11920–11927, 2007. Copyright (2008) National Academy of Sciences, USA.)

Ultrafast Folders Shed Light on the Speed Limits to Protein Folding

How could you determine if a process is barrierless? First, a barrierless process is fast. Second, the absence of a kinetic barrier means you don't have two distinguishable states, so you won't get two-state (single exponential) kinetics. Barrierless processes can have complex kinetics. Third, in a chevron plot for a barrierless folder, you may observe that the folding rate coefficient becomes independent of denaturant concentration at low denaturant. Lowering the denaturant concentration beyond a certain point no longer speeds up folding, because the folding rate is already maximal (see Figure 6.9). And fourth, barrierless processes won't have Arrhenius kinetics. Equation 6.19 describes how two-state folding, which entails a barrier, follows Arrhenius kinetics. When folding has a barrier, higher temperatures lead to faster rates. But when there is no barrier, it means that folding is already happening at the maximum possible rate, so increasing the temperature will not speed it up further. Figure 6.12 shows the case of a protein for which temperature does not accelerate folding. In ultrafast folders, $k_f(T)$ is approximately constant with temperature. Interestingly, however, the unfolding of ultrafast folders usually does obey Arrhenius kinetics. These temperature dependencies are explained by the Zwanzig–Szabo–Bagchi (ZSB) model (see Appendix 6B).

In the remainder of this chapter, we switch from experiments and macroscopic models to microscopic modeling of how folding processes are encoded in protein structures. The macroscopics aim to capture experimental data in terms of kinetic macrostates, such as N, D, and I_1, I_2, ... Macroscopic modeling is not intended to explain the physical basis of folding, or why one protein folds differently than another, or how the different conditions of solvent or temperature speed up or slow down folding, or to predict folding rates. For the latter questions, we now turn to *microscopic* models of kinetics.

HOW DO PROTEINS FOLD SO FAST? THEY FOLD ON FUNNEL-SHAPED ENERGY LANDSCAPES

Let's return to Levinthal's question: how can proteins fold so quickly? And how are small proteins able to do this, independently of their amino acid sequence, independently of starting denaturing conditions, and no matter what denatured conformation the chain begins in? These questions are answered by recognizing the shape of a protein's *folding energy landscape*. An energy landscape is a mathematical function $G(x_1, x_2, x_3, \ldots, x_L)$ of independent variables $x_1, x_2, x_3, \ldots, x_L$. A protein's independent variables x_i are the chain conformational degrees of freedom, a total of $L = 3n - 6$ of them for a system of n atoms, excluding the six external (three rigid-body translational and three rigid-body rotational) degrees of freedom. These variables may be described in terms of many geometric features, such as bond angles, bond lengths, intermolecular distances, and locations and orientations of water molecules, for example. For protein folding, the function G is the free energy.[3] G is a high-dimensional surface, because it depends on these many degrees of freedom of the protein. What is the nature of this function G? What is the shape of the surface of G as a function of the variables x_i?

Let's consider some possibilities. It could be that a protein's folding energy landscape might have the shape of a golf course in a high-dimensional space (Figure 6.13A). That is, G might be perfectly flat everywhere except for a highly localized well, representing the native structure, where the free energy must be lower than all the other states (since the native state is stable) under native conditions. A golf-course landscape would mean that all conformations of the chain have identical internal free energy, except for the native state, which has a lower internal free energy.

However, statistical mechanical theories developed in the 1980s showed that folding free energy landscapes are not shaped like golf courses. They are shaped like funnels [8, 9, 10]; see Figure 6.13B and

[3]The y-axis on an "energy" landscape is more correctly a free energy, called the *internal free energy*. The internal free energy combines all the energies and entropies, except for the chain conformational entropy. After all, each point on the landscape represents a different chain conformation. The entropies that are included in the internal free energy are those that involve the solvent degrees of freedom.

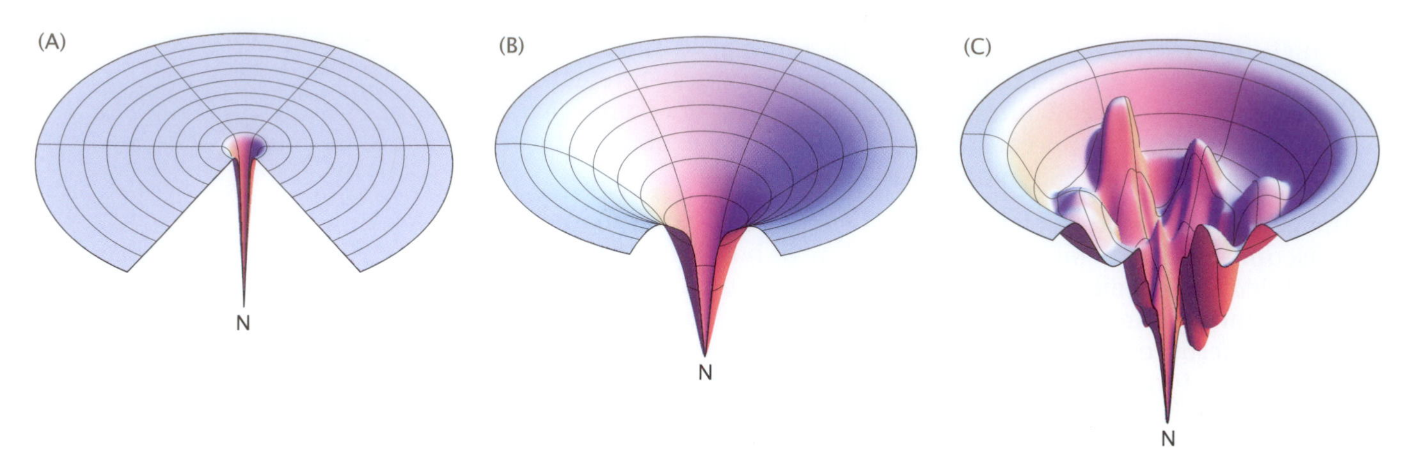

Figure 6.13 Energy landscapes help to visualize protein folding processes. (A) A "golf-course" energy landscape is flat everywhere except for the minimum of the folded state. As in the Levinthal paradox, folding on such a landscape would take a very long time, because the native structure can only be found by random diffusional searching of the entire large conformational space. (B) Protein folding is more appropriately described by a funnel energy landscape, where every random conformational step leading downhill in energy (y-axis) also leads to a reduction of the further conformational search, illustrating how protein folding can be so fast. (C) A rugged energy landscape indicates the kinetic traps (local minima) that a typical protein encounters while folding.

Chapter 3. Energy landscapes for folding are large and open at the top and small and confined at the bottom. The funnel shape comes from the statistical mechanical density of states of protein molecules. There are many open chain conformations that, collectively, have a high entropy and few compact native-like states of low free energy. It is easy to see how this rationalizes Levinthal's puzzle. The thermodynamic principle of free-energy minimization means that you can think of the tendency toward equilibrium metaphorically as a ball rolling down a hill. If you roll a ball randomly on the golf landscape, it would take a long time to find a hole on a metaphorical high-dimensional landscape, implying that folding would be very slow. But if you roll a ball randomly on a funnel, even a very high-dimensional one, the ball will roll downhill, finding its way to the bottom, no matter where it starts on the funnel. This indicates both how folding can be so fast and also how the native structure can be reached from any of the huge number of different denatured microstates.

What Do You Learn from Folding Funnels?

First, funnels explain how a solution of protein molecules, all starting from different denatured microstates, can reach the same native structure, and rapidly. Most proteins, irrespective of amino acid sequence, fold on funnel-shaped landscapes because they collapse from the many unfolded states to the one or few native structures. Funnels also explain kinetic heterogeneity, namely that different individual chain conformations reach the native structure through different microscopic folding trajectories, possibly at different rates (Figure 6.14). This folding heterogeneity is at the microstate level, not the macrostate level. To learn about dynamic heterogeneity at this microscopic level, you would need to observe more than just the average folding rate coefficient k_f. You need to measure how much folding flows through different possible microscopic routes. This is becoming possible through single-molecule experiments, which can see individual trajectories of individual molecules. In addition, simple models give a quantitative answer to the Levinthal paradox, of how proteins fold so fast—in milliseconds, rather than millions of years. Appendix 6B describes the ZSB

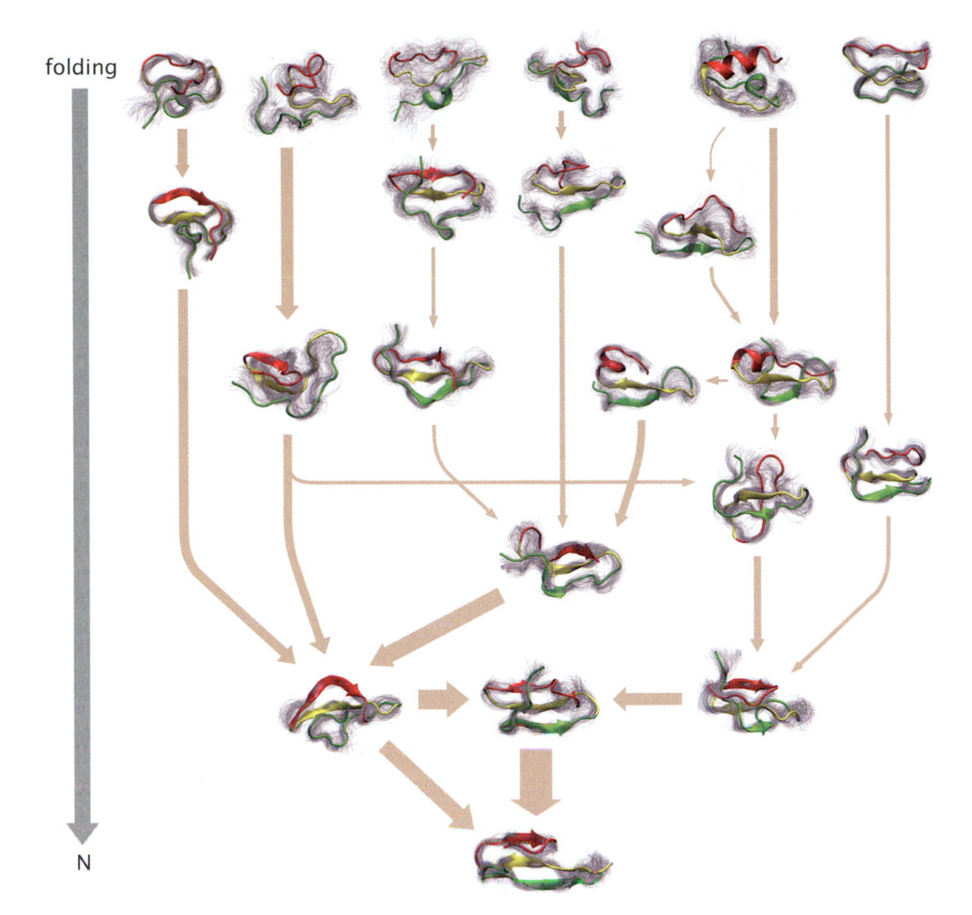

Figure 6.14 Microscopically, the chain folds via many routes. This figure shows a molecular dynamics simulation of the folding of the Pin WW domain. The thickness of the arrows indicates the relative frequencies of the different conformational transitions seen in the simulation. (Adapted from F Noe, C Schutte, E Vanden-Eijnden, et al. *Proc Natl Acad Sci USA*, 106:19011–19016, 2009.)

Model, which shows how protein folding speeds can arise even from quite shallow slopes of folding funnels. Typical protein folding speeds are attained when individual native-like interactions are more favorable than non-native interactions by only 1–2 kcal mol^{-1}.

Another class of models, called *Go models*, which are named after their originator N Go [11], have given useful insights about folding funnels and the folding routes of individual proteins [12, 13]. In a Go model, you assign energetically favorable interactions only to native contacts. In that way, you can turn the true energy landscape into a smooth funnel, to explore the most efficient routes to native states. Another major model is the spin-glass model, which gives a simple approximate way of describing the bumpiness features of folding energy landscapes (see Appendix 6C).

Moreover, cartoons of folding funnels can convey useful insights. For example, Figure 6.13A shows a *golf-course* landscape, illustrating the early expectations of an infinitely slow random search; Figure 6.13B shows a *smooth funnel,* indicating fast folding; Figure 6.13C shows a *bumpy funnel,* indicating small kinetic traps of the type that can reduce folding speeds.

There is experimental evidence for funneled-landscape folding. For example, D Barrick et al. [14] have studied *repeat proteins*, molecules that have multiple units of small foldable peptides. Each repeat unit can fold individually, but their folding rates and equilibria also depend on cooperative interactions among the units. The folding of these proteins occurs through parallel processes because of the many equivalent repeating modular subunits (Figure 6.15).

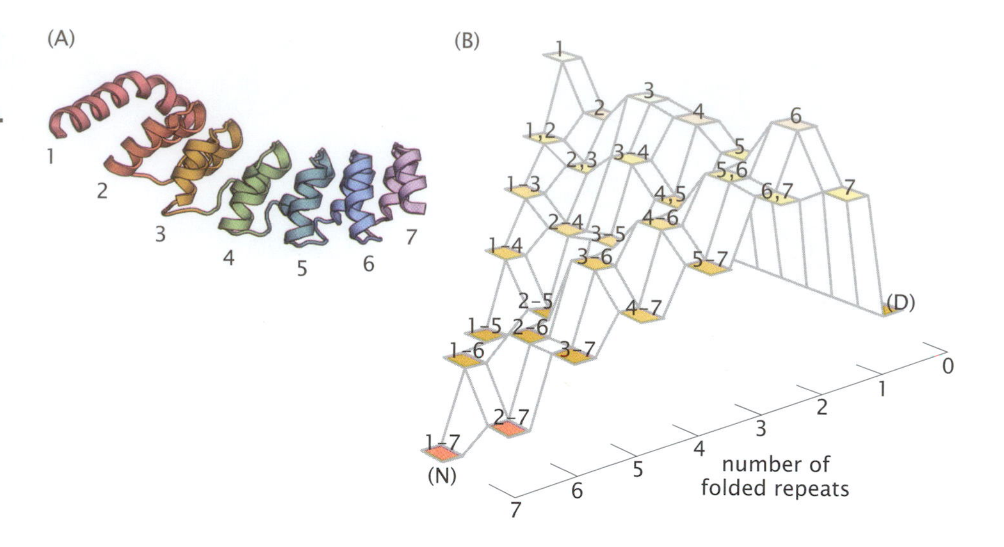

Figure 6.15 A repeating-domain protein lends itself to detailed measurement, showing a funnel-shaped folding landscape. (A) The Notch ankyrin protein has seven repeat units. Folding can be measured in individual domains. (B) The energy landscape of folding states shown as a function of the number of folded repeats and their location. (Adapted from E Kloss, N Courtemanche, and D Barrick. *Arch Biochem Biophys*, 469:83–99, 2008. With permission from Elsevier.)

DIFFERENT PROTEINS CAN FOLD AT VERY DIFFERENT RATES

Protein folding speeds depend not only on mutations, temperature, and denaturants. Protein folding speed also depends on a protein's native structure. How does a protein's sequence and structure encode its folding rate and route? Which conformations are explored and which are not? Is there a general folding mechanism? That is, is there a single narrative for the sequence of structural events that happens that applies across a broad spectrum of different types of proteins? Here, we describe some observations that relate protein structures to folding speeds.

Knowing a protein's native structure can help you predict its folding speed. Figure 6.16 shows how folding speed depends on features of the native structure. Folding speeds vary over orders of magnitude. Figure 6.16A shows that folding rates correlate with the *absolute contact order* (ACO) of native structures, a measure of the "localness" of a protein's native contacts [15]. Proteins having more local contacts and fewer nonlocal contacts in their native structures tend to be faster folders. α-helical proteins tend to fold faster than β-proteins. A similar correlation in Figure 6.16B shows that the more secondary structures a protein has, the slower the protein folds, on average [16]. Why should adding secondary structures slow down the folding process?

The elemental unit of folding kinetics is the foldon. Folding appears to occur by motifs of individual secondary structures. Even though individual secondary structures of a native protein are not stable by themselves, they are more stable than alternative structures and can be further stabilized by assembling together into tertiary structures.

The dominant folding paths form structures in the order *local-first, global-later*. Upon jumping from unfolding to folding conditions, a polymer molecule can only perform localized conformational sampling within small sections of the chain on the earliest timescales—for example, a turn of a helix or a β-strand pair. Forming nonlocal contacts requires larger searches and longer times. On intermediate timescales,

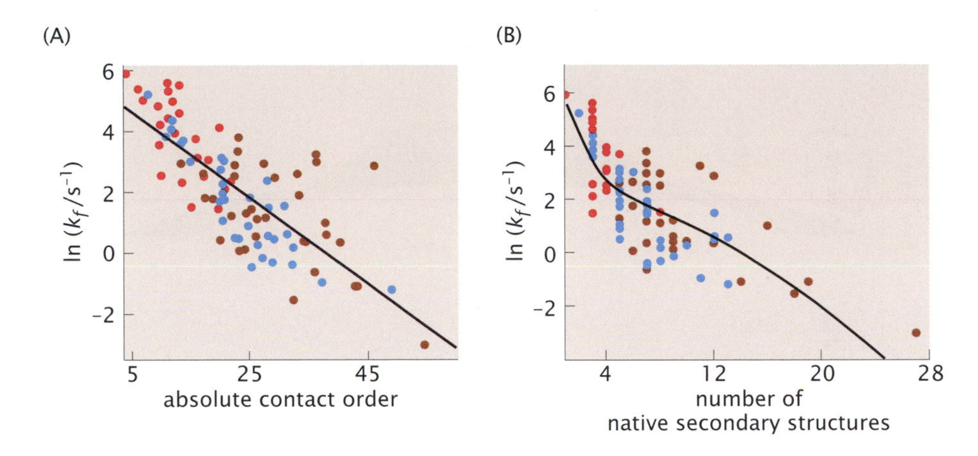

Figure 6.16 Folding rates depend on structural features. (A) Folding rates correlate with the *absolute contact order* (ACO) of a protein's native structure. The ACO measures the average "nonlocalness" of the protein's contacts. Proteins that are mostly helical fold faster, and proteins that are mostly β sheets fold slower. ACO $= (1/C_N)\sum \Delta C_{ij}$, where C_N is the total number of native contacts and $\Delta C_{ij} \equiv |j - i|$ is the *contact order* between residues *i* and *j*, the separation between them along the sequence. (B) Folding rates are slower for proteins having more secondary structures. *Red* shows α-helical proteins, *blue* β-proteins, and *brown* αβ-proteins. (From GC Rollins and KA Dill. *J. Am. Chem. Soc.*, 2014, 136 (32), pp 11420–11427, 2014. Reprinted with permission from American Chemical Society.)

a local piece of chain can *grow* or *zip* into bigger local structures, such as full helices or β-strand pairs. Over the slowest timescales, secondary structure pieces have time to come together and *assemble* into the protein's tertiary structure. Local-first, global-later (also called "Zipping & Assembly" [17]) can rapidly find the native structures of proteins without searching whole conformational spaces [18]. For example, experiments show that inserting polyglycine loops of increasing lengths, which increases the conformational search, also slows the folding process [19].

Figure 6.17 shows a proposed folding mechanism, called the *Foldon Assembly Model* [16]. First, one foldon forms somewhere in the chain. It forms rapidly but in low population. A second foldon then forms by assembling onto the first one. This double-foldon assembly has an even lower population than its predecessor, the single foldon. The process continues, with additional secondary structures assembling onto the growing framework, with diminishing populations (because each helix added is a step uphill in free energy). The final step is the formation of the native structure, which has high population (low free energy), because of additional packing and stabilizing interactions when the protein achieves its native state. Just prior to reaching the native structure, the protein passes through its folding transition state. The landscape is volcano shaped: uphill at first, then downhill at the very end (Figure 6.18).

The Foldon Assembly Model is a general folding mechanism: (i) Secondary structures are not stable alone. (ii) Tertiary interactions help stabilize them. (iii) Two-state proteins have a free-energy barrier between D and N (hence, single-exponential kinetics). (iv) The slowest

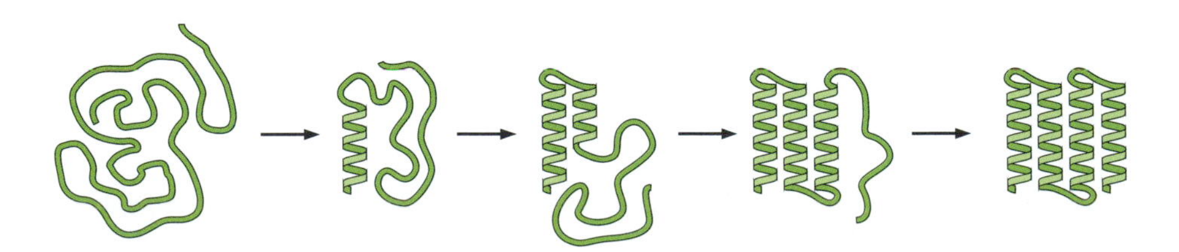

Figure 6.17 Foldon Assembly Model. Under native conditions, a secondary structure flickers on and off—mostly off. But sometimes when it's on, another secondary structure flickers on too, adjacent to the first and stabilized by it. By accretion of secondary structures, the later stages of folding are "less unstable" than the earlier stages, until a fully stable native structure forms at the last step.

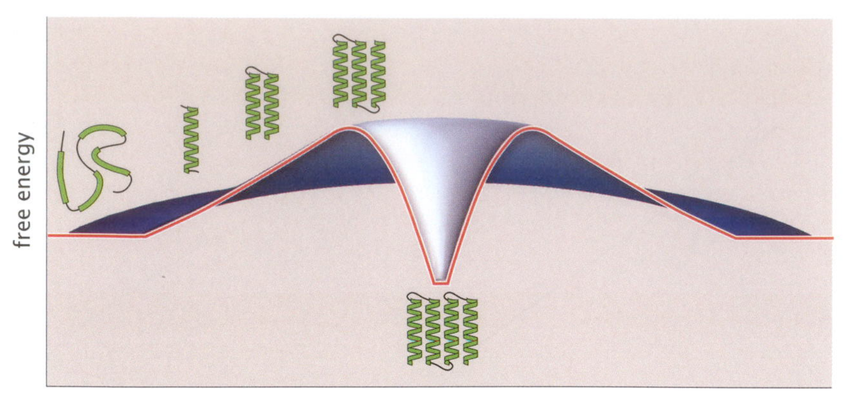

Figure 6.18 The volcano-shaped folding energy landscape shape of the Foldon Assembly Model. Folding is sequential: foldon 1 forms, then foldon 2 adds to it, then foldon 3, etc. These steps are all uphill in free energy, because each individual secondary structure is unfavorable. Only the final step to the full native state is downhill in free energy.

relaxation time should be independent of its starting denatured conformation (see the earlier section in this chapter on chevron plots). The starting state of a protein only affects the fast processes. The slow processes are independent of whether the chain starts from fully denatured or from some other conformation having some residual initial structure. (v) Folding becomes slower with increasing numbers of secondary structures (see Figure 6.16). This model gives a general mechanism, applicable to two-state proteins, over many different folds, and estimates the folding rates from known quantities such as helix–coil and tertiary propensities.

SUMMARY

Measurements of protein folding and unfolding rates give insight into the sequences of folding events and how proteins can fold so quickly, despite the complexities and diversities of their sequences and native structures. Folding rates can depend strongly on temperature, the concentration of denaturants, and on the protein's size, as found on chevron plots, by Φ-value analysis, hydrogen exchange, and other experiments. Small proteins often fold through single-exponential kinetics, while larger proteins often fold through multi-exponential kinetics (involving kinetic intermediates). Master equations provide a general way to capture this phenomenology. Microscopic modeling explains the speed of protein folding in terms of funnel-shaped energy landscapes. The ZSB Model, described in Appendix 6B, shows how a relatively small energy bias (funneling) is sufficient to explain the high speeds of protein folding. In the Foldon Assembly Model, foldons form quickly, then assemble onto an increasingly native structure, for two-state proteins. These simple models give insights into how a protein folds relatively quickly and directly into its native structure.

APPENDIX 6A: MASTER EQUATIONS DESCRIBE DYNAMICS

Master equations are used to describe the time evolution of systems of populations. They treat a very general class of dynamics called *Markov processes,* in which the state of the system at a given time step is fully determined by the state at the previous time step. Master equations resemble classical chemical kinetics, with the fundamental difference that master equations model the probabilistic behaviors of microscopic states, rather than the deterministic behaviors of average properties, such as concentrations. So, the general approach below can be used to describe macroscopic kinetics, or microscopic stochastics, depending on whether the rate coefficients are found simply by fitting to experimental data (a macroscopic model) or whether they are generated by some underlying physical model of a protein's substructures (a microscopic model). Given a model of n different states (either macroscopic or microscopic), and given the set of arrows and states that constitute the topology of the reaction scheme, you can express the time-dependent populations of the states in terms of a vector

$$\mathbf{P}(t) = \begin{bmatrix} p_1(t) \\ \vdots \\ p_n(t) \end{bmatrix} \qquad (6.A.1)$$

and a *rate matrix*

$$\mathbf{W} = \begin{bmatrix} k_{11} & k_{12} & \cdots & \\ k_{21} & k_{22} & \cdots & \\ \vdots & \vdots & \ddots & \\ k_{n1} & k_{n2} & \cdots & k_{nn} \end{bmatrix}, \qquad (6.A.2)$$

where k_{12} represents the microscopic rate of transition from state 2 to state 1, and the diagonal elements are given by the negative sum of all nondiagonal terms in the same column, that is

$$k_{ii} = -\sum_{j,\, j \neq i} k_{ji}. \qquad (6.A.3)$$

As such, the ith diagonal element represents the rate of escape or efflux from state i and all elements in the same row represent the transition/influx into the state i.

The dynamics is described by the set of differential equations:

$$\frac{d\mathbf{P}(t)}{dt} = \mathbf{W}\mathbf{P}(t), \qquad (6.A.4)$$

known as the *master equation*. The solution to Equation 6.A.4 can be expressed as

$$\mathbf{P}(t) = e^{\mathbf{W}t}\mathbf{P}(0), \qquad (6.A.5)$$

where $\mathbf{P}(0)$ is the initial population, $\mathbf{P}(t)$ at time $t = 0$. How can you evaluate $\exp(\mathbf{W}t)$ and solve Equation 6.A.5? You can do this by first *diagonalizing* the matrix $\mathbf{W}$. Diagonalizing a matrix means that you transform $\mathbf{W}$ to a different matrix $\mathbf{\Lambda}$, in which only the diagonal elements are nonzero:

$$\mathbf{\Lambda} = \begin{bmatrix} \lambda_1 & 0 & 0 & \cdots & 0 \\ 0 & \lambda_2 & 0 & \cdots & 0 \\ 0 & 0 & \lambda_3 & \cdots & 0 \\ \vdots & \vdots & \vdots & \ddots & \vdots \\ 0 & 0 & 0 & \cdots & \lambda_n \end{bmatrix}. \qquad (6.A.6)$$

The quantities λ_i are called the *eigenvalues*. Note that by definition the rank of the transition matrix is $n-1$, and therefore its eigenvalue decomposition yields $n-1$ nonzero eigenvalues, and one zero eigenvalue (for example, $\lambda_1 = 0$). The remaining eigenvalues are all negative, and their absolute values represent the frequency of different processes contributing to the overall dynamics. You can perform this transformation by multiplying $\mathbf{W}$ by a matrix $\mathbf{U}$:

$$\mathbf{\Lambda} = \mathbf{U}^{-1}\mathbf{W}\mathbf{U}. \tag{6.A.7}$$

Here $\mathbf{U}^{-1}$ is the inverse of $\mathbf{U}$; that is, it satisfies the equation $\mathbf{U}\mathbf{U}^{-1} = \mathbf{U}^{-1}\mathbf{U} = \mathbf{1}$, where $\mathbf{1}$ is the identity matrix. We show next how to compute the matrices $\mathbf{\Lambda}$ and $\mathbf{U}$. Once you have those matrices, you can compute the full dynamics from

$$\mathbf{P}(t) = \mathbf{U}e^{\mathbf{\Lambda}t}\mathbf{U}^{-1}\mathbf{P}(0). \tag{6.A.8}$$

Note that the product $\mathbf{U}e^{\mathbf{\Lambda}t}\mathbf{U}^{-1}$ represents the time-dependent conditional probability, or transition probability matrix $\mathbf{C}(t)$, the ijth element of which represents the probability of transition from state j to state i at time t. This is a sum of exponentials.[4] You can see this by breaking it into its components:

$$p_i(t) = \sum_k \sum_j U_{ik}e^{-\lambda_k t}[\mathbf{U}^{-1}]_{kj}p_j(0) = \sum_k A_{ik}e^{-\lambda_k t}, \tag{6.A.9}$$

where A_{ik} is the amplitude of the kth relaxation mode of state i and λ_k is the relaxation time of that mode. Note that one relaxation mode has zero eigenvalue (for example, $\lambda_1 = 0$), which, as $t \to \infty$, defines the equilibrium probability of the individual states. That is, the equilibrium probability is found from Equation 6.A.9 upon substituting $t = \infty$, to obtain

$$p_i(\infty) = \sum_j U_{i1}[\mathbf{U}^{-1}]_{1j}p_j(0). \tag{6.A.10}$$

To apply Equation 6.A.8, you must first determine the diagonal matrix $\mathbf{\Lambda}$. Begin with Equation 6.A.7. Multiply it on the left by $\mathbf{U}$ to get

$$\mathbf{U}\mathbf{\Lambda} = \mathbf{W}\mathbf{U}. \tag{6.A.11}$$

You can break the matrix $\mathbf{U}$ into its vector components:

$$\mathbf{U}\mathbf{\Lambda} = \begin{bmatrix} u_{11} & u_{12} & \cdots \\ u_{21} & u_{22} & \cdots \\ \vdots & \vdots & \ddots \\ u_{n1} & u_{n2} & \cdots \end{bmatrix} \begin{bmatrix} \lambda_1 & 0 & 0 & \cdots & 0 \\ 0 & \lambda_2 & 0 & \cdots & 0 \\ 0 & 0 & \lambda_3 & \cdots & 0 \\ \vdots & \vdots & \vdots & \ddots & \vdots \\ 0 & 0 & 0 & \cdots & \lambda_n \end{bmatrix}$$
$$= \begin{bmatrix} \mathbf{u}_1\lambda_1 & \mathbf{u}_2\lambda_2 & \mathbf{u}_3\lambda_3 & \ldots & \mathbf{u}_n\lambda_n \end{bmatrix} \tag{6.A.12}$$

[4]Here is the derivation of Equation 6.A.8 from Equation 6.A.5: the symbolic notation $e^{\mathbf{W}t}$ can be expressed as the series

$$e^{\mathbf{W}t} = \mathbf{1} + \mathbf{W}t + \tfrac{1}{2}\mathbf{W}^2t^2 + \ldots$$

You can express the right-hand side of this equation instead in terms of $\mathbf{\Lambda}$ by transforming each term. Notice that $\mathbf{I} = \mathbf{U}\mathbf{U}^{-1}$ and $\mathbf{W} = \mathbf{U}\mathbf{\Lambda}\mathbf{U}^{-1}$ and $\mathbf{W}^r = \mathbf{U}\mathbf{\Lambda}^r\mathbf{U}^{-1}$ (for any power r, since $\mathbf{\Lambda}$ is a diagonal matrix) so you can express this equation instead as

$$e^{\mathbf{W}t} = \mathbf{U}\left(\mathbf{1} + \mathbf{\Lambda}t + \tfrac{1}{2}\mathbf{\Lambda}^2t^2 + \ldots\right)\mathbf{U}^{-1} = \mathbf{U}e^{\mathbf{\Lambda}t}\mathbf{U}^{-1}.$$

where

$$\mathbf{u}_i = \begin{bmatrix} u_{1i} \\ u_{2i} \\ u_{3i} \\ \vdots \\ u_{ni} \end{bmatrix}.$$

So Equation 6.A.11 becomes

$$\mathbf{u}_j \lambda_j = \mathbf{W}\mathbf{u}_j \tag{6.A.13}$$

Equation 6.A.13 is called an *eigenvalue* equation. To solve it for the values of λ_i, rearrange it into the form

$$(\mathbf{W} - \lambda_j \mathbf{I})\,\mathbf{u}_j = \mathbf{0}. \tag{6.A.14}$$

You can solve this by computing the determinant

$$\det(\mathbf{W} - \lambda_j \mathbf{I}) = 0, \tag{6.A.15}$$

so

$$\det \begin{bmatrix} k_{11} - \lambda_1 & k_{12} & k_{13} & \dots & k_{1n} \\ k_{21} & k_{22} - \lambda_2 & k_{23} & \dots & k_{2n} \\ k_{31} & k_{32} & k_{33} - \lambda_3 & \dots & k_{3n} \\ \vdots & \vdots & \vdots & \ddots & \vdots \\ k_{n1} & k_{n2} & k_{n3} & \dots & k_{nn} - \lambda_n \end{bmatrix} = 0. \tag{6.A.16}$$

Solving this equation gives the eigenvalues λ_i. Once you have the eigenvalues, solve for each $\mathbf{u}$ using Equation 6.A.14 (see the example in Box 6.2). This gives the full dynamical description of a system having a given rate matrix $\mathbf{W}$. Box 6.2 gives a worked example for two-state dynamics. For the more general case of a system consisting of an ensemble of (n) microstates, there are standard computer packages that use this approach to compute eigenvalues and to solve the full dynamics.

Box 6.2 Master Equations Describe Two-State Folding Dynamics

Let's apply the master-equation approach to two-state kinetics for the two states D and N. The master equation is

$$\frac{d\mathbf{P}(t)}{dt} = \mathbf{U}\boldsymbol{\Lambda}\mathbf{U}^{-1}\mathbf{P}(t), \tag{6.A.17}$$

where $\boldsymbol{\Lambda}$ is the diagonal matrix of the eigenvalues of the matrix $\mathbf{W}$ and $\mathbf{U}$ is the matrix of its corresponding eigenvectors. Now, solve for the full time dependence of the populations using $\boldsymbol{\Lambda}$ and $\mathbf{U}$:

$$\mathbf{P}(t) = \mathbf{U}e^{\boldsymbol{\Lambda}t}\mathbf{U}^{-1}\mathbf{P}(0), \tag{6.A.18}$$

where $e^{\boldsymbol{\Lambda}t}$ is a shorthand notation that describes a diagonal matrix composed of the elements $e^{\lambda_i t}$, and λ_i is the ith eigenvalue of $\mathbf{W}$ ($i = 1, 2$ in this case). For two-state dynamics, the rate matrix $\mathbf{W}$ is

$$\mathbf{W} = \begin{bmatrix} -k_f & k_u \\ k_f & -k_u \end{bmatrix}. \tag{6.A.19}$$

Now find the two eigenvalues λ_1 and λ_2 that satisfy the eigenvalue equation

$$\mathbf{W}\mathbf{u}_i = \lambda_i \mathbf{u}_i, \quad (6.A.20)$$

where $\mathbf{u}_i$ is the ith eigenvector ($i = 1, 2$ in the present case). You do this by solving the characteristic equation

$$\det(\mathbf{W} - \lambda\mathbf{I}) = 0, \quad (6.A.21)$$

where $\mathbf{I}$ is the identity matrix of order 2. You can alternatively express this equation in terms of the elements of the matrices:

$$(k_f + \lambda)(k_u + \lambda) - k_u k_f = 0$$
$$\Longrightarrow \lambda^2 + (k_f + k_u)\lambda = 0. \quad (6.A.22)$$

Solving Equation 6.A.22 gives two values, $\lambda_1 = 0$ and $\lambda_2 = -(k_f + k_u)$, which can be put into a diagonal matrix of eigenvalues,

$$\Lambda = \begin{bmatrix} \lambda_1 & 0 \\ 0 & \lambda_2 \end{bmatrix} = \begin{bmatrix} 0 & 0 \\ 0 & -(k_f + k_u) \end{bmatrix}. \quad (6.A.23)$$

Now, let's determine the eigenvectors. For the eigenvalue corresponding to $\lambda_1 = 0$, the eigenvector is $\mathbf{u}_1 = [u_{11} \quad u_{21}]^T$. Application of Equation 6.A.20 gives

$$\begin{bmatrix} -k_f & k_u \\ k_f & -k_u \end{bmatrix} \begin{bmatrix} u_{11} \\ u_{21} \end{bmatrix} = 0. \quad (6.A.24)$$

Note that these two equations are not independent; that is, you are free to choose one of the components of $\mathbf{u}_1$, and the second is then defined by the relation $k_f u_{11} = k_u u_{21}$. So one possible solution is

$$\mathbf{u}_1 = \begin{bmatrix} k_u \\ k_f \end{bmatrix}. \quad (6.A.25)$$

Similarly for $\lambda_2 = -(k_f + k_u)$, you'll find that the eigenvector is

$$\mathbf{u}_2 = \begin{bmatrix} -1 \\ 1 \end{bmatrix}. \quad (6.A.26)$$

In this case, the two components of u_2 are related by $u_{22} = -u_{21}$. So, we are free to arbitrarily choose $u_{21} = -1$. Now, assemble the eigenvectors into a matrix:

$$\mathbf{U} = [\mathbf{u}_1 \ \ \mathbf{u}_2] = \begin{bmatrix} k_u & -1 \\ k_f & 1 \end{bmatrix}. \quad (6.A.27)$$

We also need the inverse of $\mathbf{U}$, which is obtained by solving $\mathbf{U}^{-1}$ $\mathbf{U} = 1$. The result is

$$\mathbf{U}^{-1} = \frac{1}{k_u + k_f} \begin{bmatrix} 1 & 1 \\ -k_f & k_u \end{bmatrix}. \quad (6.A.28)$$

Finally, to obtain the time-dependent populations $\mathbf{P}(t)$, insert $\mathbf{\Lambda}$ (Equation 6.A.23), $\mathbf{U}$ (Equation 6.A.27), and $\mathbf{U}^{-1}$ (Equation 6.A.28) into Equation 6.A.18, to get

$$\begin{bmatrix} P_{\mathrm{D}}(t) \\ P_{\mathrm{N}}(t) \end{bmatrix} = \begin{bmatrix} k_u & -1 \\ k_f & 1 \end{bmatrix} \begin{bmatrix} 0 & 0 \\ 0 & e^{-(k_u+k_f)t} \end{bmatrix} \begin{bmatrix} \frac{1}{k_u+k_f} & \frac{1}{k_u+k_f} \\ -\frac{k_f}{k_u+k_f} & \frac{k_u}{k_u+k_f} \end{bmatrix} \begin{bmatrix} P_{\mathrm{D}}(0) \\ P_{\mathrm{N}}(0) \end{bmatrix}. \quad (6.A.29)$$

Performing these matrix multiplications gives $P_{\mathrm{D}}(t)$ and $P_{\mathrm{N}}(t)$. This is Equation 6.7 in the main text.

APPENDIX 6B: THE ZWANZIG–SZABO–BAGCHI MODEL SHOWS HOW FUNNELS ACCELERATE FOLDING

How does a funnel-shaped landscape explain fast folding? To explore this, let's express a protein's folding equilibrium and kinetics in a highly simplified way. Represent a chain of L amino acids as a one-dimensional string of symbols

$$\text{nnndddnnnnnndddn}\ldots,$$

where each n indicates that a particular residue is in its folded native-like conformation, and each d indicates that a particular residue is in an unfolded non-native conformation. This resembles the way we treated the helix–coil transition in Chapter 5.

Here, we describe the Zwanzig–Szabo–Bagchi (ZSB) model, which shows how Levinthal's paradox is resolved by energy funneling [20, 21]. Let $m = 1, 2, 3, \ldots L$ represent the number of "mistakes" d (that is, native contacts not yet made) in the string. So, $m = 0$ represents the folded native state; $m = 1$ means the molecule has all n's except for one d somewhere in the string; and $m = L$ means that the molecule has no correct (that is, folded native) pieces of structure. m represents a simple one-dimensional "reaction coordinate" for folding.

Now, let's model the energies of a funnel landscape. The larger the number of mistakes in a given conformation, the higher is the energy of that conformation. To keep the model simple, we first suppose that the energy $U(m)$ is a linear function of the number of mistakes, $U(m) = m\epsilon$, where $\epsilon > 0$ is the energy cost of each mistake. Figure 6.B.1 is the energy landscape of the ZSB Model.

However, we need one more ingredient for the ZSB Model. To capture the two-state nature of folding equilibria, we also need an *energy gap*. We assume that the native state is further stabilized by an energy $\epsilon_0 > 0$. Figure 6.B.1 shows this linear energy funnel plus native well.

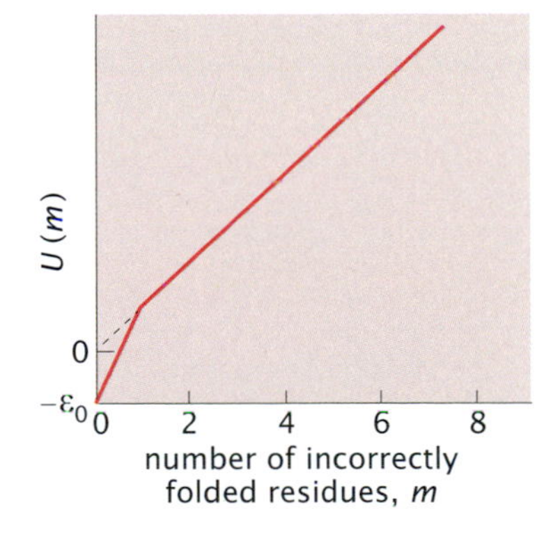

Figure 6.B.1 ZSB funnel energy landscape. In the ZSB Model, the energy increases linearly with m, the number of non-native contacts ("mistakes"). Each mistake costs an energy ϵ. The step from $m = 0$ (native) to $m = 1$ is steeper, with slope ϵ_0. This defines the funnel shape of this landscape, projected onto a one-dimensional axis of m.

The Equilibrium Properties of the ZSB Funnel Landscape Model

Before looking at the folding kinetics, let's consider the equilibrium predictions of the ZSB Model. The equilibrium probability $P_m(\text{eq})$ that a chain has m mistakes is given by

$$P_m(\text{eq}) = \frac{L!}{m!(L-m)!} K^m Q^{-1}, \tag{6.B.1}$$

where K is the Boltzmann factor,

$$K = (z-1)e^{-\beta\epsilon}. \tag{6.B.2}$$

K is an equilibrium constant per residue that gives the energetic disadvantage ϵ and the entropic advantage $z - 1$ of converting each residue from n to d. z is the total number of rotational isomers, so $z - 1$ is the number of isomers in wrong states, and $\beta = (RT)^{-1}$. The combinatorial factor in Equation 6.B.1 counts the number of ways you can arrange m d's and $(L - m)$ n's in a one-dimensional string. Q is the partition

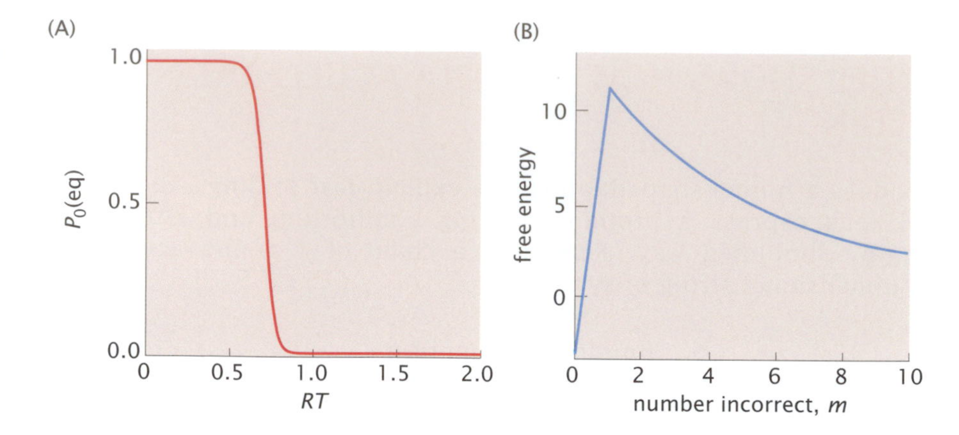

Figure 6.B.2 The ZSB Model predicts two equilibrium states in protein folding. (A) The plot of the native population versus temperature is sigmoidal, indicating cooperativity. (B) There are two minima in the energy versus the number of mistakes, m, corresponding to the native state ($m = 0$) and a denatured state where approximately half of the amino acids are unfolded or misfolded.

function, the sum over the statistical weights of all the states:[5]

$$Q = e^{\beta \epsilon_0} + \sum_{m=1}^{L} \left(\frac{L!}{m!(L-m)!} K^m \right) = e^{\beta \epsilon_0} + (1+K)^L - 1. \tag{6.B.3}$$

The binomial expression $(1+K)^L$ would account for all the possible states of the system if the native state were defined as the state of zero energy. However, our model defines an additional stabilization for the native state, to represent the high level of cooperativity observed in two-state proteins. The term $e^{\beta \epsilon_0} - 1$ subtracts the term from the series for $m = 1$ and adds an additional term to give the corrected weight of the extra-stabilized native state.

The equilibrium populations of the native (folded) state ($m = 0$, here called P_N rather than P_0) and the *first excited* state ($m = 1$) are

$$P_N(\text{eq}) = \frac{e^{\beta \epsilon_0}}{Q} \quad \text{and} \quad P_1(\text{eq}) = \frac{LK}{Q}. \tag{6.B.4}$$

Figure 6.B.2A shows that the native (folded) state population P_0 (see Equation 6.B.4) undergoes a sharp transition with temperature. Thus, the ZSB Model predicts that folding involves a two-state equilibrium. You can see this from the free energy $G_m = -RT \ln P_m(\text{eq})$ as a function of the *order parameter m*. Figure 6.B.2B gives the free energy calculated by substituting $P_m(\text{eq})$ from Equation 6.B.1, and shows that the model predicts two stable states, denatured and native, with a barrier in between (that is, the definition of two-state equilibrium).

The ZSB Model Relates Landscape Shape to Folding Speed

Now, let's compute the folding and unfolding kinetics given by the ZSB Model. Express the kinetics in terms of a master equation

$$\frac{dP_N(t)}{dt} = k_0 P_1(t) - k_1 P_N(t), \tag{6.B.5}$$

where $P_N(t)$ is the native population as a function of time, $P_1(t)$ is the population of the first-excited state, k_0 is the rate coefficient for the folding transition to the native state from the first-excited state (1 →

[5]The last equality in Equation 6.B.3 follows from the binomial relation $(x+y)^L = \sum_{m=0}^{L} \{L!/[m!(L-m)!]\} x^{L-m} y^m$, where $x = 1$ and $y = K$.

N), and k_1 is the rate coefficient for the unfolding (N $\to$ 1). We want to obtain the folding and unfolding rate coefficients k_f and k_u, so we want to convert Equation 6.B.5 into the two-state form

$$\frac{dP_N}{dt} = k_f P_D(t) - k_u P_N(t). \tag{6.B.6}$$

To solve this equation, we first express k_1 in terms of k_0, a given rate coefficient. The principle of detailed balance applied to Equation 6.B.5 asserts that at equilibrium we must have $k_0 P_1(\text{eq}) = k_1 P_N(\text{eq})$. Combining this with Equation 6.B.4 gives

$$k_1 = k_0 \frac{P_1(\text{eq})}{P_N(\text{eq})} = k_0 L K e^{-\beta\epsilon_0}. \tag{6.B.7}$$

Second, express $P_1(t)$ in terms of $P_N(t)$. We assume that the system is in rapid equilibrium among all unfolded denatured states ($m = 1, 2, 3, \ldots, L$). So, we take the time-dependent population of state 1 to be proportional to the time-dependent population of the set of denatured states (D) ($m > 0$):

$$\frac{P_1(t)}{P_D(t)} = \frac{LK/Q}{Q_D/Q}$$
$$\Rightarrow P_1(t) = \frac{LK}{Q_D}[1 - P_N(t)], \tag{6.B.8}$$

where $Q_D = (1 + K)^L - 1$ is the partition function for the denatured state (that is, the full partition function excluding the extra native term) given by Equation 6.B.3. We have also used $P_D(t) = 1 - P_N(t)$. Now, substitute Equations 6.B.8 and 6.B.7 into Equation 6.B.5 to get

$$\frac{dP_N}{dt} = k_0 LK\left[\frac{1}{Q_D}P_D(t) - e^{-\beta\epsilon_0}P_N(t)\right]$$
$$= \left(\frac{k_0 LK}{Q_D}\right)P_D(t) - (k_0 LK e^{-\beta\epsilon_0})P_N(t). \tag{6.B.9}$$

Comparison with Equation 6.B.6 shows that $k_f = k_0 LK/Q_D$ and $k_u = k_0 LK e^{-\beta\epsilon_0}$. In Boxes 6.3 and 6.4, we compute the folding times on a golf course and funnel landscape, respectively.

Box 6.3 Folding on a Golf-Course Landscape Would Be Very Slow

How fast would proteins fold if their energy landscapes were flat, shaped like a golf course? Suppose you have a protein of $L = 100$ amino acids. Let's take $k_0 = 10^{-6}$ s (see page 135) and $z = 4$, which are reasonable estimates. For a flat energy landscape, you have a slope $\epsilon = 0$ and $K = z - 1$ (see Equation 6.B.2). So, $Q_D = z^L - 1 \approx z^L$. Then, Equation 6.B.13 gives the folding time as

$$\tau_f = \frac{z^L}{k_0 L(z-1)}$$
$$= \frac{4^{100}}{(10^6\,\text{s}^{-1})(100)(3)} \approx 5 \times 10^{51}\,\text{s}, \tag{6.B.10}$$

which is astronomically large. If proteins were to fold on golf-course energy landscapes, it would take figuratively "forever."

Box 6.4 Folding on a Smooth Funnel Landscape Is Fast

Figure 6.B.3 shows that even small biases of energy toward the native state can speed up folding enormously. Suppose the energy advantage of forming a native contact is just $\epsilon = 2\,\text{kcal mol}^{-1}$ ($3.34RT$). For $z = 4$ and $L = 100$, this gives $K = (z-1)e^{-\beta\epsilon} = 0.106$, so $Q_D = (1+K)^L - 1 \approx 2.4 \times 10^4$. Then

$$\tau_f = \frac{Q_D}{k_0 LK} = \frac{2.4 \times 10^4}{(10^6\,\text{s}^{-1})(100)(0.106)} \approx 2.3 \times 10^{-3}\,\text{s}. \qquad (6.B.11)$$

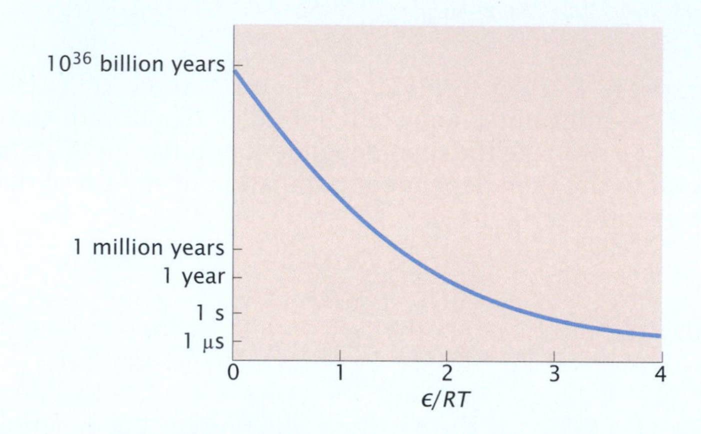

Figure 6.B.3 The time required for a protein to fold depends on its energy bias, ϵ/RT. This shows how a protein can fold rapidly, in fractions of a second, because of the funnel shape of its energy landscape, even if the energy bias is relatively small, $\epsilon \approx 2RT$. (Adapted from R Zwanzig, A Szabo, and B Bagchi. *Proc Natl Acad Sci USA*, 89:20–22, 1992.)

Comparing this folding time of 2 ms with that in Box 6.3, you see that funneling accelerates folding by more than 50 orders of magnitude, given only a bias of $-2\,\text{kcal mol}^{-1}$ when residues are in native-like versus non-native-like conformations. Figure 6.B.3 shows that the predicted folding speed has a strong dependence on the slope of the energy landscape.

The ZSB Model Explains the Unusual Temperature Dependences of Ultrafast Folders

Recall from Figure 6.12 that the folding rates of ultrafast folders cannot be increased by temperature; they are already folding at maximum speed. However, the *unfolding* kinetics of ultrafast folders often follows Arrhenius kinetics (see Figure 6.12). The ZSB Model explains both features. First, the temperature independence of $k_f(T)$ comes from a balance of two quantities, $K(T)$ and $Q_D(T)$, in Equation 6.B.11. Second, it also explains the Arrhenius kinetics of unfolding. Substitute the definition of $K(T)$ from Equation 6.B.2 into $k_u = k_0 LK(T)e^{-\beta\epsilon_0}$ deduced from Equation 6.B.9, and take the logarithm to get

$$\ln k_u(T) = \text{constant} - \frac{\epsilon + \epsilon_0}{RT}, \qquad (6.B.12)$$

which is the Arrhenius law (if ϵ and ϵ_0 are constants, as we have assumed), consistent with experiments (see Figure 6.12).

Equation 6.B.9 with Equation 6.B.6 shows that the folding rate coefficient k_f, or its inverse, the folding time τ_f, is

$$k_f = \frac{1}{\tau_f} = \frac{k_0 LK}{Q_D}, \tag{6.B.13}$$

as given in the main text. And the unfolding rate coefficient is

$$k_u = k_0 LKe^{-\beta\epsilon_0}. \tag{6.B.14}$$

The Foldon Assembly Model is a Folding Mechanism Variant of the ZSB Model

The Foldon Assembly Model is a variant of the ZSB Model; for details, see [16]. But, in short, here is how the Foldon Assembly Model differs from the ZSB Model, Equation 6.B.9, for the specific example of a four-helix-bundle protein. The native population is $P_N(t)$, and

$$\frac{dP_N}{dt} = \left(\frac{4k_0 K_2^3 K_3^3}{Q_D}\right) P_D(t) - \left(\frac{4k_0 K_2^3 K_3^3}{Q_N}\right) P_N(t). \tag{6.B.15}$$

From Equation 6.B.15, we get the rate coefficient for forming the native structure as $4k_0 K_2^3 K_3^3 Q_D$. The factor of 4 accounts for the combinatorics that any one of the four helices can form first. k_0 is the intrinsic rate coefficient for forming a helix. The factor K_2^3 is the equilibrium coefficient for forming the three helices of the precursor to the native state, as if they were independent. However, the helices are not independent. Their interaction is treated by the factor of K_3^3, which accounts for the pairwise tertiary interactions among the helices (hence the subscript 3). Each helix is surrounded by three other helices, each of which stabilizes the first by a factor $K_3 > 1$.

APPENDIX 6C: PROTEIN FOLDING FUNNELS CAN BE BUMPY: THE SPIN-GLASS MODEL

Spin-glass models, adapted from the physics of glassy materials to proteins by PG Wolynes, JN Onuchic, Z Luthy-Schulten, EI Shakhnovich, D Thirumalai, and others, capture essential aspects of the bumpiness of protein folding funnels [22]. Consider a probability distribution $p(U)$ of the protein chain conformations as a function of their energy U. U is the *internal free energy*, the sum of all the intermolecular interactions, due to hydrogen bonding, charge and steric interactions, and hydrophobic and solvation interactions, of a single given chain conformation. We are interested in how the chain entropy $S(U)$, representing how many chain conformations have that energy, depends on U. Focus on the most highly populated denatured state, for which the ensemble-average energy is $\langle U \rangle$. Assume that the distribution around that most-probable denatured state is Gaussian (Figure 6.C.1):

$$p(U) = \frac{1}{\sqrt{2\pi\delta_U^2}} e^{-(U-\langle U \rangle)^2/2\delta_U^2}. \tag{6.C.1}$$

where δ_U^2 is the variance of the energy fluctuations (so δ_U has units of energy and is the standard deviation), representing the width of the Gaussian distribution and characterizing the bumpiness of the landscape for a given protein or foldamer.

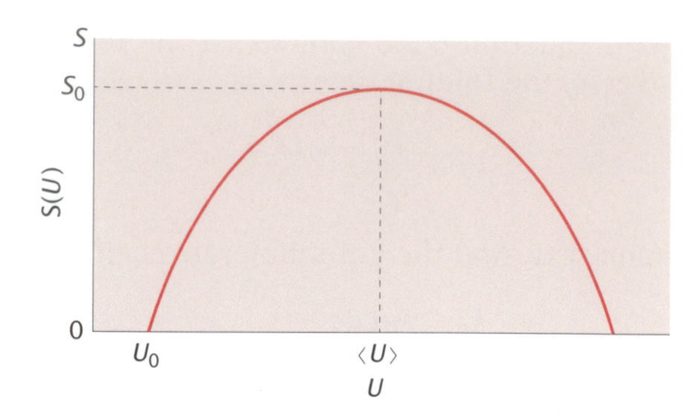

Figure 6.C.1 The Gaussian energy distribution of the spin-glass model. The entropy $S(U)$ is a parabolic function of energy U. This surface is characterized by two slopes, which are inverses of the glass temperature T_g (a measure of bumpiness) and the denaturation temperature T_f (a measure of protein stability).

Substitute $p(U)$ from Equation 6.C.1 into the Boltzmann distribution expression $S(U) = S_0 + k \ln p(U)$, to get

$$S(U) = S_0 + \frac{k(U - \langle U \rangle)^2}{2\delta_U^2}, \tag{6.C.2}$$

where S_0 is the entropy of the dominant denatured state. The central idea of this model is that proteins have *glass-like* states. In a glass, as the system moves to lower and lower energies, it reaches a *kinetic trap*, that is, a lowest possible energy U_0, at which there is only a single microstate, so the entropy of that state is zero: $S(U_0) = 0$. This is called the *entropy catastrophe*. The system cannot easily reach even lower energies, to achieve the global minimum in energy, the crystalline state. Applied here to proteins, the idea is that a folding protein can get caught in a kinetic trap, resembling a glassy state, which slows its folding progress toward its native state. The native state has global minimum energy U_N, but the kinetic trap energy is higher, $U_0 > U_\mathrm{N}$. Our aim is to compute the temperature of the entropy catastrophe, T_g, called the *glass transition temperature*.

To find the energy, $U = U_0$, of the entropy catastrophe trap, substitute $S(U_0) = 0$ into Equation 6.C.2 to get

$$\langle U \rangle - U_0 = \delta_U \sqrt{\frac{2S_0}{k}}. \tag{6.C.3}$$

(We have taken only the positive root because our interest is in the left side of the $S(U)$ curve, $T_g > 0$.) Now, thermodynamics gives a fundamental relationship [23] for computing the temperature if you know the function $S(U)$:

$$\frac{1}{T_g} = \left.\frac{\partial S}{\partial U}\right|_{U_0} = \frac{k(U_0 - \langle U \rangle)}{\delta_U^2}, \tag{6.C.4}$$

where we have evaluated this function at the glass point $U = U_0$. Finally, for comparison, the folding temperature occurs where $T_f \approx \Delta U / S_0$, where $\Delta U = \langle U \rangle - U_\mathrm{N}$, so we can compute a dimensionless ratio

$$\frac{T_f}{T_g} = \frac{\Delta U}{\delta_U} \sqrt{\frac{2k}{S_0}} \approx \frac{\Delta U}{N\delta_U}. \tag{6.C.5}$$

In the last step, we have approximated the denatured state entropy as $S_0 = k \ln z^N$, where z is the number of rotamers per backbone bond and N is the number of chain bonds, and we have left out constants of order unity.

The main points that emerge from such spin-glass models are that (1) proteins can have glass-like kinetic traps that slow the progress of folding; (2) the bumpiness of an energy landscape can be approximated as δ_U, a standard deviation of a Gaussian function; (3) kT_g defines an energy scale for fluctuations, which is of order δ_U/N; and (4) T_f/T_g is a useful dimensionless quantity for comparing polymer folding speeds (T_f/T_g is smaller for slower-folding molecules.)

REFERENCES

[1] C Levinthal. Are there pathways for protein folding? *J Chim Phys PCB*, 65:44–45, 1968.

[2] SJ Hagen, J Hofrichter, A Szabo, and WA Eaton. Diffusion-limited contact formation in unfolded cytochrome *c*: estimating the maximum rate of protein folding. *Proc Natl Acad Sci USA*, 93:11615–11617, 1996.

[3] PS Kim and RL Baldwin. Structural intermediates trapped during the folding of ribonuclease A by amide proton exchange. *Biochemistry*, 19:6124–6129, 1980.

[4] SB Ozkan, KA Dill, and I Bahar. Computing the transition state populations in simple protein models. *Biopolymers*, 68:35–46, 2003.

[5] CM Dobson, PA Evans, and SE Radford. Understanding how proteins fold: the lysozyme story so far. *Trends Biochem Sci*, 19:31–37, 1994.

[6] AR Fersht, A Matouschek, and L Serrano. The folding of an enzyme. I. theory of protein engineering analysis of stability and pathway of protein folding. *J Mol Biol*, 224:771–784, 1992.

[7] KL Maxwell, D Wildes, A Zarrine-Afsar, et al. Protein folding: defining a "standard" set of experimental conditions and a preliminary kinetic data set of two-state proteins. *Protein Sci*, 14:602–616, 2005.

[8] KA Dill. Theory for the folding and stability of globular proteins. *Biochemistry*, 24:1501, 1985.

[9] JD Bryngelson and PG Wolynes. Spin glasses and the statistical mechanics of protein folding. *Proc Natl Acad Sci USA*, 84:7524–7528, 1987.

[10] PE Leopold, M Montal, and JN Onuchic. Protein folding funnels: A kinetic approach to the sequence–structure relationship. *Proc Natl Acad Sci USA*, 89:8721–8725, 1992.

[11] N Go. Theoretical studies of protein folding. *Annu Rev Biophys Bioeng*, 12:183, 1983.

[12] V Muñoz and WA Eaton. A simple model for calculating the kinetics of protein folding from three-dimensional structures. *Proc Natl Acad Sci USA*, 96:11311–11316, 1999.

[13] PC Whitford, JK Noel, S Gosavi, et al. An all-atom structure-based potential for proteins: Bridging minimal models with all-atom empirical forcefields. *Proteins*, 75:430–441, 2009.

[14] E Kloss, N Courtemanche, and D Barrick. Repeat-protein folding: New insights into origins of cooperativity, stability, and topology. *Arch Biochem Biophys*, 469:83–99, 2008.

[15] KW Plaxco, KT Simons, and D Baker. Contact order, transition state placement and the refolding rates of single domain proteins. *J Mol Biol*, 277:985–994, 1998.

[16] GC Rollins and KA Dill. General mechanism of two-state protein folding kinetics. *J Am Chem Soc*, 136:11420–11427, 2014.

[17] SB Ozkan, GA Wu, JD Chodera, and KA Dill. Protein folding by zipping and assembly. *Proc Natl Acad Sci USA*, 104:11987–11992, 2007.

[18] VA Voelz and KA Dill. Exploring zipping and assembly as a protein folding principle. *Proteins*, 66:877–888, 2007.

[19] AR Viguera and L Serrano. Loop length, intramolecular diffusion and protein folding. *Nat Struct Biol*, 4:939–946, 1997.

[20] R Zwanzig, A Szabo, and B Bagchi. Levinthal's paradox. *Proc Natl Acad Sci USA*, 89:20–22, 1992.

[21] R Zwanzig. Simple model of protein folding kinetics. *Proc Natl Acad Sci USA*, 92:9801–9804, 1995.

[22] JN Onuchic, Z Luthey-Schulten, and PG Wolynes. Theory of protein folding: The energy landscape perspective. *Annu. Rev. Phys. Chem.*, 48:545–600, 1997.

[23] KA Dill and S Bromberg. *Molecular Driving Forces: Statistical Thermodynamics in Biology, Chemistry, Physics and Nanoscience, 2nd ed.* Garland Science, New York, 2011.

SUGGESTED READING

Fersht A, *Structure and Mechanism in Protein Science: A Guide to Enzyme Catalysis and Protein Folding*. WH Freeman, New York, 1998.

Gruebele M, Dave K, and Sukenik S, Globular Protein folding *in vitro* and *in vivo*. *Annu Rev Biophys*, 45:233–251, 2016.

Proteins Evolve

CHAPTER 7

Protein molecules can change through biological evolution. As organisms evolve, their proteins are subjected to random mutations, occasionally resulting in changes to their native structures, physical properties, and/or biological actions. Such variations can become encoded into the genome, so descendant proteins can differ from ancestral proteins. On the large scale, modified biomolecules sometimes lead to new species, as when humans diverged from apes. This chapter describes how protein sequence, structure, and function change through evolution.

PROTEINS CHANGE THROUGH EVOLUTIONARY PROCESSES

The principles of biological evolution were first elucidated by Charles Darwin in his book *On the Origin of Species*, published in 1859, based on studies of plants and animals. Darwin's central principle, which now provides the foundation for understanding living systems, is that biological evolution is a process of *descent with modification*. The term "descent" means that modern organisms arose from earlier (that is, ancestral) organisms. The term "with modification" means that modern organisms are not exactly the same as their ancestors. Descent with modification means that all living systems are related to each other. This relatedness is captured in *phylogenetic* or *evolutionary* trees (**Figure 7.1**). The *roots* of the tree represent the most ancient organisms, the *nodes on the branches* represent their ancestors, and the *leaves* (end nodes) represent modern organisms. On the one hand, it was not surprising, even in Darwin's day, that organisms might arise from their ancestors, since this is so readily apparent from watching the birth of children from their parents. On the other hand, it was not so evident prior to Darwin that one *species* of organism could arise from another. Speciation events are much rarer and more difficult to observe.

Darwin's principle of descent with modification has powerful implications at the molecular and cellular levels. It is consistent with the observation that the types of biomolecules found in cells, such as DNA, RNA, proteins, and lipids, are universal throughout biology. The

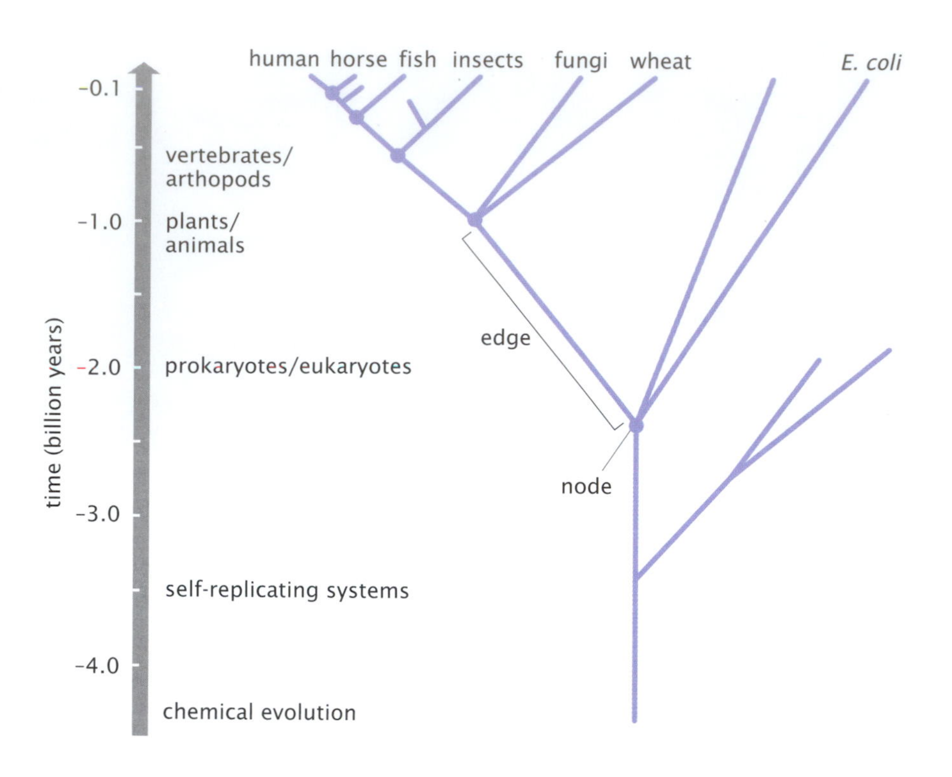

Figure 7.1 Phylogenetic trees show evolutionary relationships. Tip nodes at the top represent today's species. Edges indicate ancestry relationships between parent and daughter species. This left bar indicates the timescale, from the earliest forms of life to the present. (Adapted from MO Dayhoff, RM Schwartz, and BC Orcutt. *Atlas of Protein Sequence and Structure*, 5:345–352, 1978.)

molecules that provide the core functions in one organism also provide the core functions in other organisms. It means that every living system is related to every other. It also means that you can learn about the functioning of one type of cell or organism by studying other organisms. As discussed in the following chapter, this provides a powerful basis for understanding biology and disease, and for discovering new drugs.

What Are the Mechanisms of Evolutionary Change?

Even a thousand years before Darwin, people were breeding plants and animals to have certain traits. Such breeding is called *artificial selection*. Darwin proposed that normal biological evolution was driven by a similar process, which he called *natural selection*. The underlying biochemical basis for evolutionary change is *random mutation and natural selection* of biopolymers: (a) DNA molecules undergo random mutation. Mutations in *coding regions* of the DNA may lead to altered amino acid sequences of proteins.[1] Mutations are not directed toward end results; they just happen randomly. (b) If a mutation happens to confer a *fitness* advantage to the organism, that mutation becomes stably embedded within the population, and propagates into the next generation. In this way, advantageous mutations are preserved, and disadvantageous ones

[1]The regions of DNA that code for proteins or RNA molecules are called *genes* or *coding regions*. Protein-encoding genes are composed of a series of codons (units of three nucleotides), each coding for an amino acid. There are $4^3 = 64$ possible codons (given that the DNA has 4 types of nucleotides: A, T, G, and C), for 20 different types of amino acids, that is, not all nucleotide mutations in the DNA necessarily give rise to an amino acid change. Those DNA mutations that do not result in amino acid change (or result in similar amino acid substitutions) are called *silent mutations* or *synonymous mutations*. Much of the DNA in genomes of higher organisms is in *noncoding regions*, which do not code for proteins. They may have some functional importance, but this is not yet well understood.

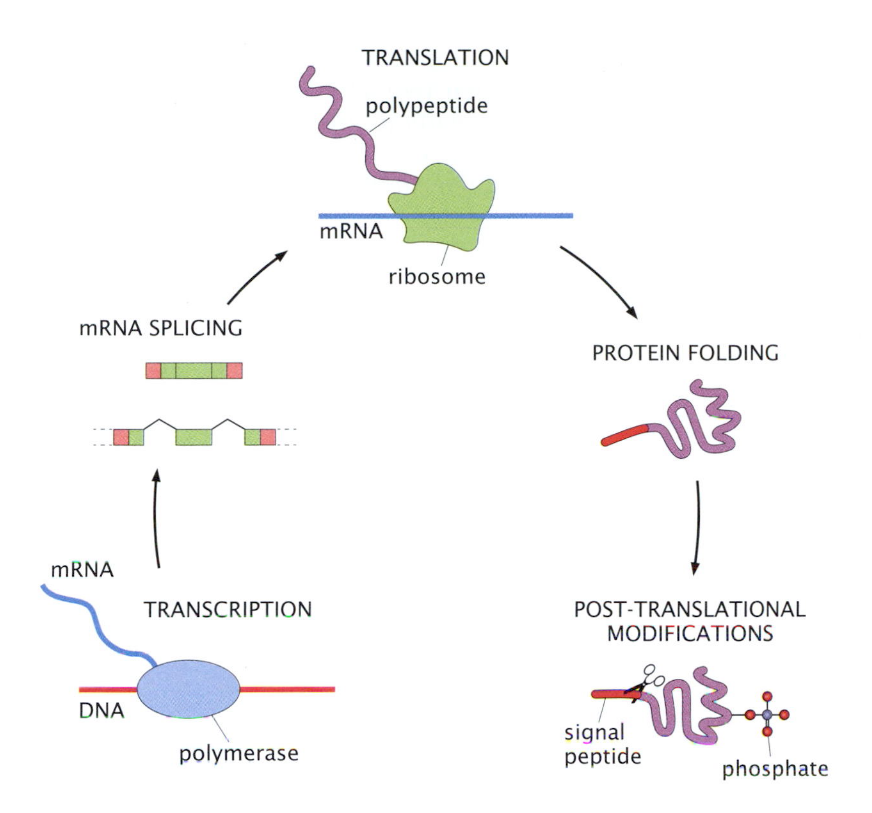

Figure 7.2 Protein modifications can occur at different stages. Errors can occur in transcription, mRNA splicing, translation, folding, or post-translational modifications. Such errors can alter cellular fitness, ultimately impacting genome evolution. (Adapted from DA Drummond and CO Wilke. *Nat Rev Genet*, 10:715–724, 2009. With permission from Macmillan Publishers Ltd.)

are lost. In addition, some biological changes happen by *genetic drift*; these are just random changes that are neither advantageous nor disadvantageous for the survival of the organism, so they are not driven by the forces of natural selection. The distinction is that natural selection is affected by changes in evolutionary pressure (that is, changes in environmental factors that affect the relative fitnesses of different organisms), whereas genetic drift is not.

Mutations can be *deleterious*, *neutral*, or *beneficial* for the fitness of the organism. Mutations can affect a protein's binding affinity, its mechanism of action, its stability, or its propensity to form aggregates, or they can cause mistakes in protein synthesis (Figure 7.2).

Homologs Are Proteins with Similar Functions and a Common Ancestor

The term *homolog*, introduced by R Owen in 1843, expresses the idea that two biological entities are evolutionarily related to each other. A human heart is a homolog of a cow's heart. The two hearts have similar structures and similar functions, so it is reasonable to infer that they are related through evolutionary descent from a common ancestral origin. The bone structure in your hand resembles the bone structure in a dog's paw and a bat's wing, implying that all three anatomical structures are related by evolutionary descent. Similarly, two proteins are *homologous* to each other, or *homologs*, if they have the same function and they are descended from a common ancestor.

Evolution is a process we don't often observe directly, so we must draw inferences indirectly. Let's distinguish between properties that we can *measure* versus properties that we *infer*. How do we know if two entities have a common ancestor when we didn't see the process and can't reproduce it? For proteins, three terms are used for sequence comparisons: *sequence identity*, *sequence similarity*, and *sequence homology*. Sequence identity refers to the count of how many amino acids are identical between two sequences (of nucleic acids or amino acids) when they are optimally aligned. You can measure it, for instance using the *Hamming distance*, which is the count of the number of *mismatches* between two sequences (Box 7.1).

Box 7.1 The Hamming Distance Counts the Mismatches between Strings of Symbols, Such as Amino Acid Sequences

The Hamming distance is the number of unit mismatches between the two sequences (Figure 7.3). Sequence 2 differs from sequence 1 at the three amino acid positions highlighted in light red, so the Hamming distance between sequences 1 and 2 is $d_H(1,2) = 3$. Now compare sequences 1 and 3. Sequence 3 differs from sequence 1 at 10 positions (7 additional ones highlighted in blue), so $d_H(1,3) = 10$. And the Hamming distance between sequences 2 and 3 is $d_H(2,3) = 7$.

(A)

```
1. LVRKVAEENGRSVNS
2. LVAKVAEENKRSTNS
```

(B)

```
1. LVRKVAEENGRSVNS
3. MVAYVGEDHKRCTIS
```

Figure 7.3 Evaluation of Hamming Distance for Pairwise Alignment of Sequences. The alignment of two different sequences (labeled 2 and 3) are shown in (A) and (B), both against sequence 1. Mismatches are highlighted in *red*.

You can also measure the *sequence similarity* between two sequences. To estimate the similarity of protein sequences, instead of just counting the number of times one type of amino acid is substituted for another, you also account for the frequency you expect for such substitutions, based on some metric. Substitution frequencies can reflect the physicochemical similarities between pairs of amino acids. For example, valine and leucine are similar to each other since they are both hydrophobic and aliphatic, while valine and glutamate are dissimilar. By using some measure of how similar one amino acid is to another, you compute how similar one sequence is to another. Like sequence identity, similarity is a numerical quantity that you can define for any two sequences, without introducing assumptions about how those proteins evolved or how they are functionally related.

Sequence homology is different than *sequence similarity*. Sequence homology is not a number you can measure; it is an answer to a yes-or-no question. Do two sequences share a common ancestor? If "yes," those two sequences are said to be homologous. If "no," those two sequences do not have sequence homology. To say that two sequences are homologous just means that they are related evolutionarily. The determination of common ancestry is inferred from knowledge of evolutionary trees, which is obtained based on whatever existing data are available—biomolecule sequences or functional or anatomic similarities, for example. To prevent confusion, you should avoid using the term sequence homology when you mean sequence identity or similarity.[2]

[2]In this book, we largely conform to this rule. However, we sometimes use the term "homology modeling" to refer to computational comparisons of sequence similarities because it is such a common term in the field.

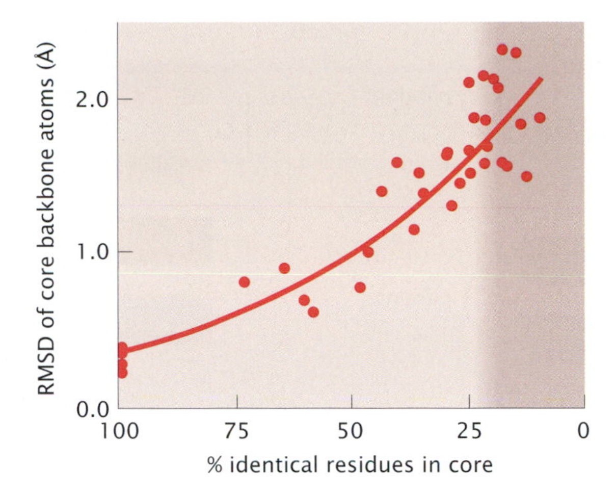

Figure 7.4 When protein sequences are similar, their native structures are often similar. RMSD is the root-mean-square deviation of the two structures when they are optimally superimposed. (Adapted from P Aloy, H Ceulemans, A Stark, and RB Russell. *J Mol Biol*, 332:989–998, 2003. With permission from Elsevier.)

When Two Protein Sequences Are Similar, Their Structures and Functions Are Usually Similar Too

When one amino acid sequence is similar to another sequence, it is usually found that their native structures are similar to each other, and that their biological functions are similar (Figure 7.4). Ninety percent of proteins that have more than 30% sequence identity are structurally homologous to each other [1].

What do we mean by protein *function*? A unified vocabulary of protein functions is given by the Gene Ontology (GO) project [2, 3], which provides a website and search tools for describing the actions, roles, and mechanisms of protein molecules. It defines function in three ways: (a) a protein's biological process, (b) a protein's cellular localization, and (c) a protein's molecular mechanism. In this language, cytochrome *c*'s biological process is oxidative phosphorylation and the induction of cell death, its cellular localization is within the mitochondrial matrix, and its molecular mechanism is identified as an oxidoreductase. GO provides a common language for annotating functions of proteins throughout genomes. As shown in Figure 8.3 in Chapter 8, two proteins having greater than 35% sequence identity are likely to perform similar functions.

There Are Three Types of Homolog Relationships: *Orthologs*, *Paralogs*, and *Xenologs*

It is useful to define different types of evolutionary relationships between proteins. As noted earlier, two proteins *A* and *B* are called homologs if they perform the same function and have sequence similarity, and share a common ancestor. Even so, *A* can be related to *B* in different ways. You and your brother share a common ancestor. You and your cousin also share a common ancestor. But your genetic relationship with your brother is different from your genetic relationship with your cousin. To distinguish the different types of homologous relationships, Walter Fitch in 1974 coined the terms *ortholog* and *paralog* (Figure 7.5). Two proteins *from different species* that have similar sequences and functions are *orthologs*. Two proteins *from the same species* that have similar sequences and functions are *paralogs*. An example of a pair of orthologs is a protease in corn and a protease in humans that have the same mechanism and similar amino acid

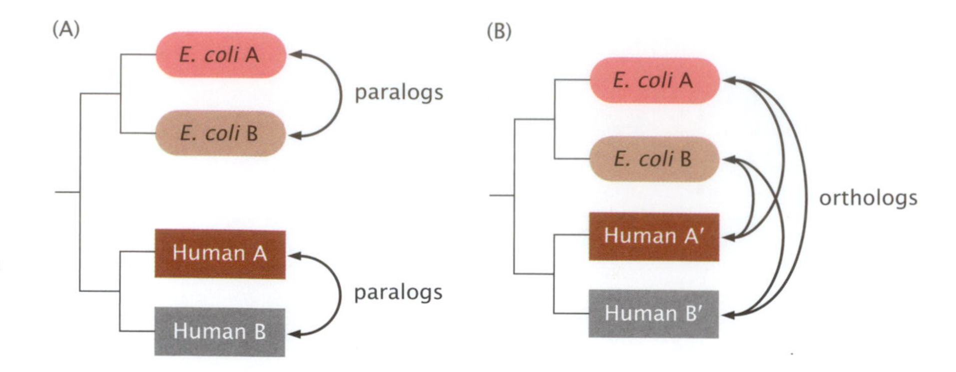

Figure 7.5 Paralog and ortholog relationships. (A) *E. coli* A and *E. coli* B are two different protein sequences that are paralogs, say arising from a gene duplication. Likewise, human sequences A and B are paralogs. (B) Human A′ and Human B′ are two different orthologs to *E. coli* A and to *E. coli* B: they are similar proteins in different organisms that evolved from a single common ancestor.

sequences. An example of a pair of paralogs is two different proteases in humans having the same mechanisms and similar amino acid sequences. The prefix "para" indicates the two proteins are operating in parallel within the same species. The 478 different kinase proteins in humans are paralogs; they operate on different substrates, but they perform the same biochemical function of phosphorylation and share a similar catalytic domain [4].

In paralogs, one protein is assumed to arise from the other by *gene duplication*. When a gene has undergone evolutionary duplication in some type of cell, one copy of that gene can continue to perform its usual function in that cell line, and that copy remains subjected to the same forces of natural selection as before. The other copy of the gene now becomes superfluous, and no longer needs to respond to the same forces of natural selection. So, the new protein is now free to evolve further to perform other functions for the organism.

Or two proteins may be *xenologs* of each other. "Xeno" means foreign or alien. A xenolog is a protein that entered the cell from outside during some evolutionary event. Think of a virus that attacks a bacterium. A gene from the virus might be implanted into the bacterial DNA, and then become stable there. The viral gene then propagates into future generations within the bacterial genome, along with the bacterial DNA. In this example, a viral protein may be a xenolog to a bacterial protein that resembles it. The process by which one gene enters a foreign cell and becomes stably incorporated into its genome is called *horizontal gene transfer*, HGT. HGT is a very important evolutionary process, particularly in lower organisms, like bacteria and fungi. HGT is facilitated under conditions of membrane disruption, or DNA fragmentation, or when cells are stressed or unstable.

The HGT mechanism complicates the idea of Darwinian "descent" because HGT implies that some evolutionary relationships are not representable on a tree-like graph in which an organism fully descends from a single ancestor. When a virus infects a bacterium, it may insert its DNA into the bacterial genome, and the bacterial genome then no longer originates from a single bacterial ancestor. HGT is a process that can accelerate evolution.

Gene Duplication Can Explain the High Symmetries in Protein Structures

Some proteins are highly symmetrical. Figure 7.6A shows the β-trefoil protein; it has threefold symmetry. It is a single polypeptide chain composed of three repeating subsequences. What is the evolutionary origin

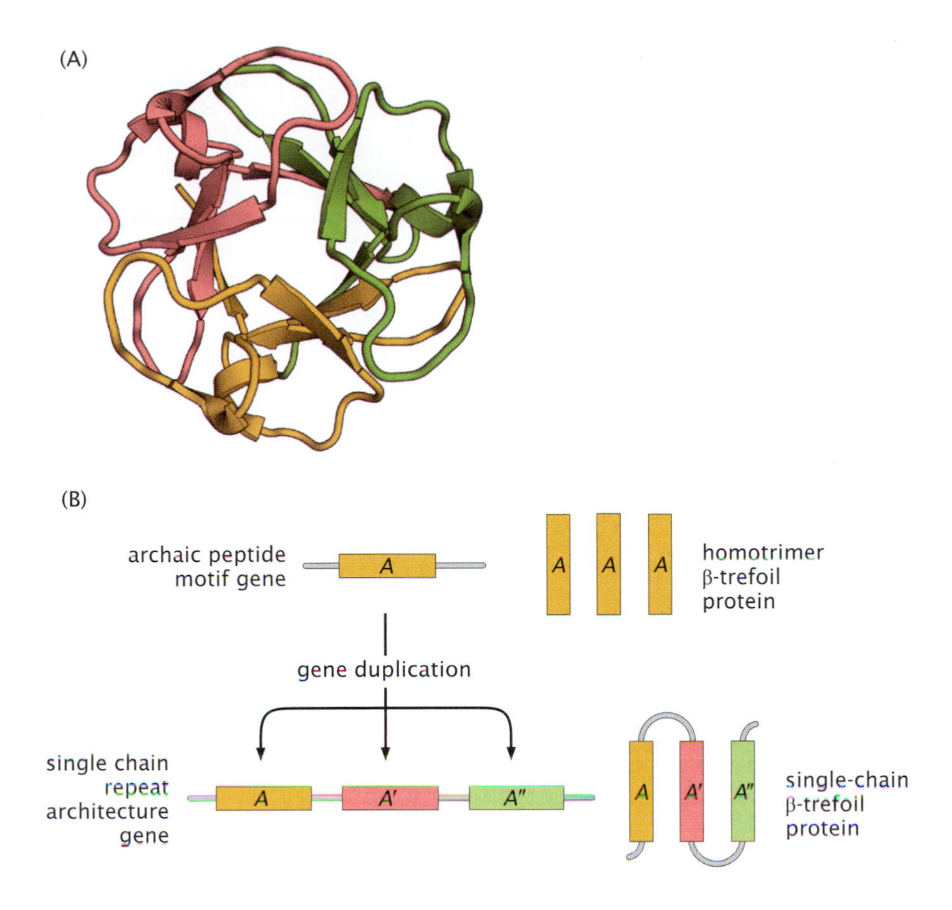

Figure 7.6 How might internal symmetries arise in proteins? Here's an example in the conserved-architecture model. (A) A β-trefoil fold has threefold symmetry. (B) Its evolutionary origin is an ancestral peptide *A*, which does not fold by itself, but binds to two identical copies, which can fold and associate into a noncovalent homotrimer. The gene for *A* now duplicates into two similar peptides *A′* and *A″* that are adjacent genes in the DNA. The three genes fuse leading to a single polypeptide chain, having internal sequence symmetry, that can fold on its own. (Adapted from J Lee and M Blaber. *Proc Natl Acad Sci USA*, 108:126–130, 2011.)

of such symmetries in proteins? Figure 7.6B gives an explanation in terms of the *conserved-architecture model.* First, evolution produces a peptide *A*, which is just a single subunit that is too small to fold on its own. *A* has the property that it can bind (noncovalently) to two other identical *A* peptides, and when it does, the three chains associate with each other and fold into a noncovalent three-chain homotrimer. If the gene encoding the protein *A* that assembles into the homotrimer is duplicated to two similar peptides *A′* and *A″*, the proteins that are encoded by these three genes could now still assemble into heterotrimers. Or, those three genes *A*, *A′*, and *A″* may be neighbors in the DNA sequence and become fused at the DNA level, leading to a covalently linked chain of the three similar subsequences that can now fully fold as a single chain. Gene duplication, divergence (see the following section), and recombination are three essential mechanisms for the expansion of the protein repertoire [5] in more complex organisms.

Evolutionary Processes Can Be *Convergent* or *Divergent*

Comparing present-day protein homologs from different organisms can give insights into their ancestry. Evolution may be convergent or divergent (Figure 7.7).

Often in nature, two different proteins can perform the same biological function. For example, trypsin and elastase are two different serine proteases. Both break down other proteins by hydrolyzing peptide bonds in the backbone. Both proteins use the same catalytic mechanism to convert their substrate (which, in this case, is a peptide bond) into their product (in this case, a broken peptide bond). They are examples

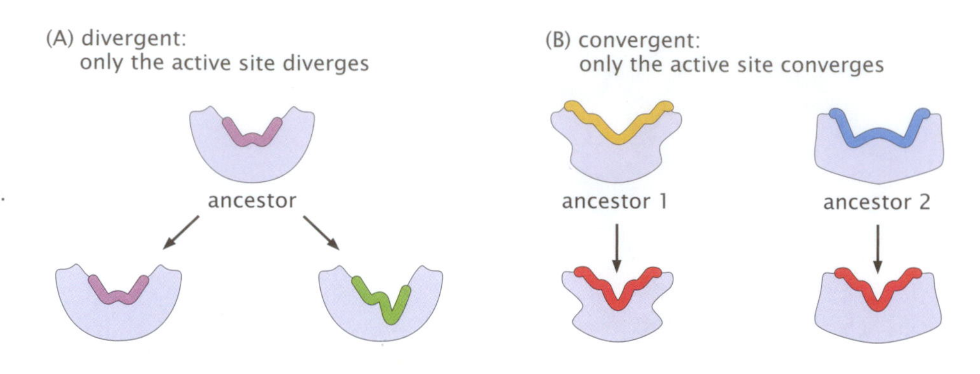

Figure 7.7 Divergent versus convergent evolution. (A) Divergence: Two proteins can have different mechanisms or actions because they diverged from the ancestor (indicated by similar sequences and possibly similar folds). (B) Convergence: Two present-day proteins can have similar functions (because of similar active sites or binding or regulatory sites), despite having low sequence identity with each other, indicating that they originated from different ancestors.

of *convergent evolution*: two proteins that share biological functionalities or mechanisms can arise from different ancestral proteins [6]. Convergent evolution has been found in serine proteases, aspartyl proteases, aldolases, and topoisomerases [7]. Serine proteases require a *catalytic triad*, a serine, histidine, and aspartic acid that are always arranged in the same spatial configuration to catalyze the hydrolysis of peptide bonds, to break down proteins (see Figure 2.4 in Chapter 2). Yet, such catalytic triad structures can arise within completely different amino acid sequences and different native folds. Figure 7.8 shows two carbonic anhydrases, in which a single molecular function is performed by two different chain folds.

In contrast, in *divergent evolution*, two very different protein functionalities can arise from a single ancestral protein. Figure 7.9 shows how many functionalities different protein folds can have. Members of the TIM-barrel family of protein structures perform 16 different known functions, for example. The figure shows that in a database of 229 folds, a protein has a single known function in 175 cases, two known functions in 38 cases, and several functions in a few cases. Sometimes, protein functionalities are *adaptive*: the protein that performs one function when it's in one conformation can perform a different function when it's in a different conformation. For example the *OB fold* binds to different single-stranded DNA segments, but the protein can change conformation to recognize different binding features of different DNA molecules.

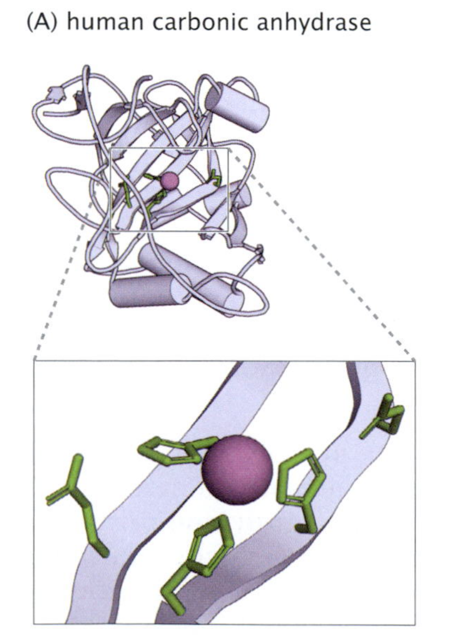

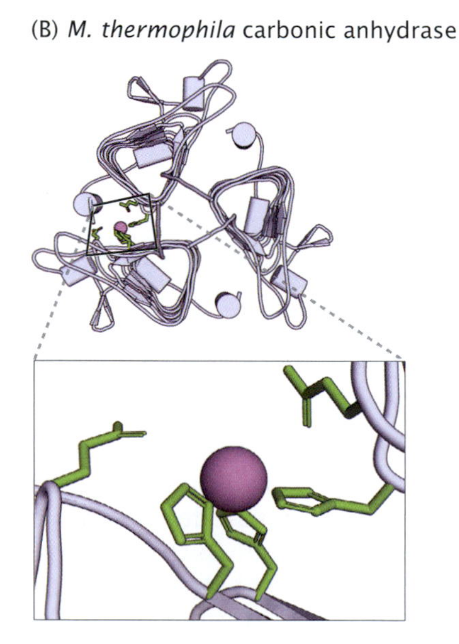

Figure 7.8 Convergent evolution—two different native folds perform the same function. (A) The human carbonic anhydrase has an α/β roll fold, while (B) the carbonic anhydrase from *Methanosarcina thermophila* has a β-helix topology. The *M. thermophila* protein, having no significant sequence identity to any carbonic anhydrase, represents a distinct class of carbonic anhydrases. (Adapted from H Hegyi and M Gerstein. *J Mol Biol*, 288:147–64, 1999. With permission from Elsevier.)

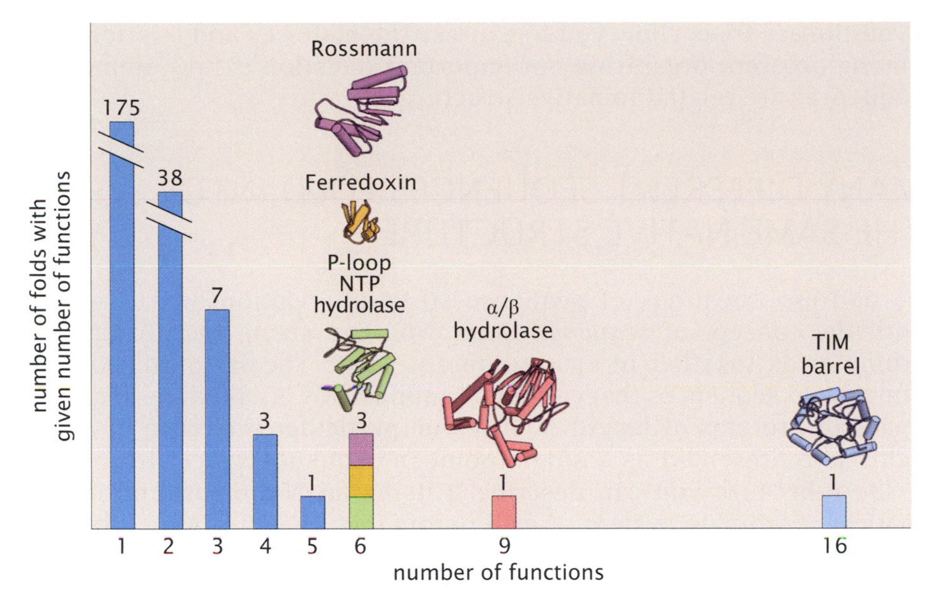

Figure 7.9 How many protein folds have how many different functions? The TIM-barrel is a fold that has 16 known functions (far right). There are 175 folds that perform a single function each (far left). And, ferredoxin, the Rossman fold, and the β-turn hydrolase folds are known to perform six functions each. (Adapted from M Gerstein. *Nat Struct Biol*, Structural Genomics Suppl:960–963, 2000. With permission from Macmillan Publishers Ltd.)

How can you determine convergence or divergence (see Figure 7.7)? Compare two homologous proteins. This can be determined from sequence- and structure-matching, together with comparison of Gene Ontology terms. Proteins are divergent if they have similar sequences and different mechanisms. Proteins are convergent if they have different sequences and similar mechanisms. Of course, proteins could diverge sufficiently over long evolutionary periods that there is no remaining resemblance of any property—either sequence, structure, or function. When relationships are too distant, you cannot draw any inferences about their evolutionary relatedness.

Phylogenetic Trees Show Evolutionary Relationships between Organisms or Sequences

In Darwin's day, evolutionary relatedness was established by comparing the *morphologies* (anatomic structures) between two organisms, such as their bones, teeth, wings, or leaves. However, since the 1960s, more quantitative biomolecular sequence-based methods have emerged for constructing evolutionary trees.

Figure 7.1 shows an evolutionary tree. Each node represents one organism, whether known or unknown. The edges between nodes (called *branches*) have different lengths. The branch length represents the relative evolutionary distance between two organisms. The longer the branch length, the more distant two organisms are from each other. To construct such a tree, you first need to define a *metric* for determining the evolutionary distance between two species, molecules, or sequences. Recall from Box 7.1 that the *Hamming distance* is such a metric. The bigger the Hamming distance, the greater is the difference between two sequences.

Chapter 8 describes how to construct an evolutionary tree for a protein, based on sequences found in an evolutionary lineage. You first line up the different amino acid sequences using *alignment methods.* Then you compute the sequence similarity score between pairs of sequences. The tree is then constructed by drawing each edge to have a length proportional to the corresponding sequence similarity score. Like family trees, evolutionary trees allow you to express the histories and relationships among different organisms. An important question is how amino acid sequences are related to native structures.

MANY DIFFERENT SEQUENCES FOLD INTO THE SAME NATIVE STRUCTURE

To address questions of sequence–structure relationships, we start with the concept of *sequence space*. You can string the 20 different amino acids together in many different ways. The set of all possible amino acid sequences that contain N amino acids is called the sequence space of proteins of length N. Any one particular sequence of amino acids is represented as a single point in sequence space. It is called a *space* because you can describe it using an N-dimensional mathematical coordinate system of grid points (Figure 7.10). Along the first coordinate axis, call it x_1, you have 20 points, each representing one of the 20 amino acids. Along the second axis, x_2, you also have 20 points representing the amino acids. All N axes, up to x_N, are labeled with such points. Axis x_1 represents which amino acid appears at position number 1 in the sequence; axis x_2 represents the amino acid appearing at position number 2, in the sequence, and so on. Figure 7.10 shows how one point in the sequence space represents one amino acid sequence, say $(x_1, x_2, x_3, \ldots) = (\text{Ala, Gly, Trp}, \ldots)$.

How big is sequence space? If each amino acid in the chain is chosen from an M-letter *alphabet* ($M = 20$ for the natural amino acids), then there will be M possible points along each of the N axes. So, sequence space has M^N grid points, representing all the possible different sequences of that chain length. For example, consider a protein of chain length $N = 100$ amino acids from $M = 20$ amino acids. The number of different sequences of that length will be $20^{100} = 1.3 \times 10^{130}$. The concept of sequence space helps to reason about evolution.

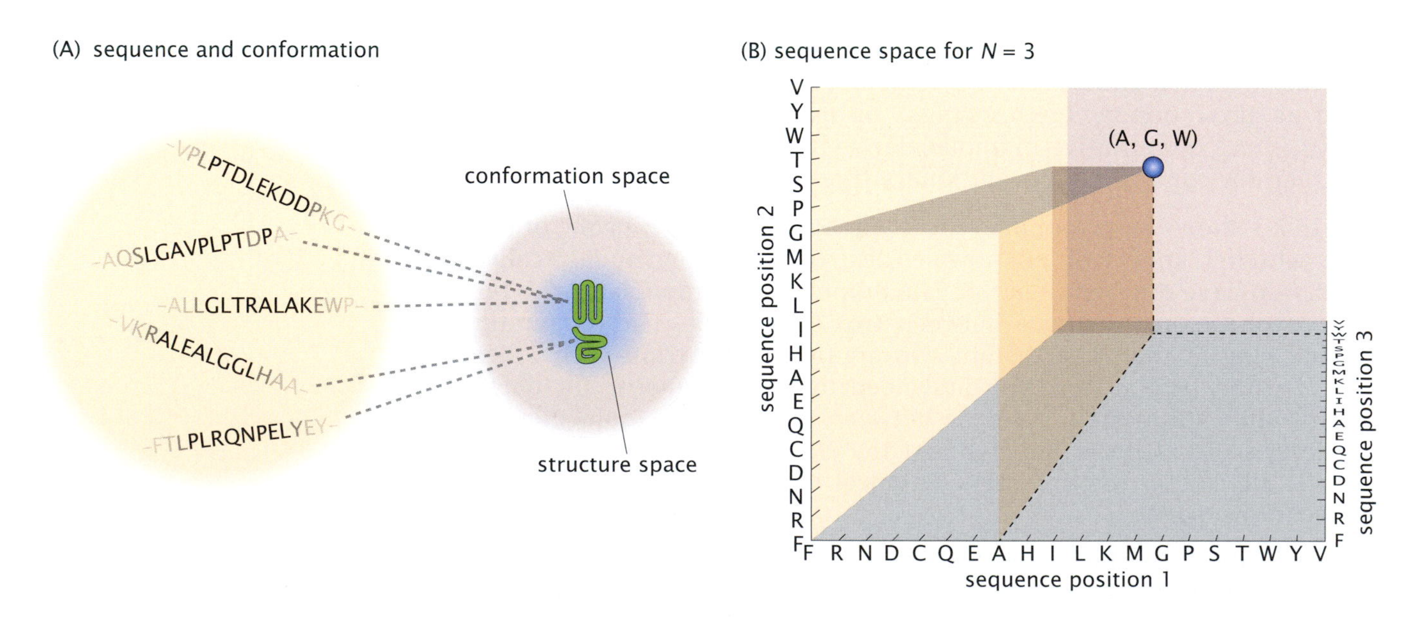

Figure 7.10 What is sequence space? How does it map to structure space? (A) Different sequences can fold to different structures (*green* icons). Many different sequences can fold to a particular native structure. (B) Sequence space. Each axis has ticks for each of the 20 amino acids. The number of axes is the chain length N. So, one point in this N-dimensional space represents one amino acid sequence. Shown here is the sequence AGW for a three-residue chain in such a space.

During evolution, proteins are subjected to *mutational changes*, such as the swapping of one amino acid for another or the loss or gain of amino acids. You can think of the mutational changes in evolution as the hopping or diffusion from one grid point to another in sequence space.

A protein also has a *conformational space*, the collection of all possible conformations of a chain molecule of a given length (see Figure 7.10). Each point in the conformational space represents one chain conformation. For a protein with N residues, each containing two rotatable bonds (ϕ and ψ angles) per residue with z rotational conformations for each, an estimate of the size of conformational space is z^{2N}, for large N. The space of conformations is further restricted by protein-like properties such as high packing density, hydrophobic core formation, and secondary structure propensities. Native-like compact conformations constitute a small fraction of conformational space. Most of conformational space is a large collection of disordered and denatured conformations.

The HP Model Gives Insight into Sequence–Structure Mapping

What is the relationship between the sequences and structures of proteins? We get some insights from studies of reduced-alphabet models, such as the HP model of Chapter 3. For HP chains having $N = 16$ monomers on two-dimensional square lattices ($z = 3$ accessible directions at each step), the sequence space contains $2^{16} = 65,536$ sequences (using $M = 2$ for two residue types, H and P), and the conformational space of all compact and noncompact self-avoiding walks (excluding those conformers related by symmetry) has 802,075 conformations. The number of compact conformations is much smaller: a 16-mer has only 69 fully compact conformations that fit on a 4×4 lattice, for example. So, you can study all possible conformations of all possible sequences by computer using HP proteins on lattices. Despite their simplifications, such models capture some basic principles of sequence-structure mappings.

HP modeling shows that mapping sequences to native structures is *many-to-one* (see Figure 7.10A). That is, many different amino acid sequences may fold to the same native structure. This information is useful for understanding the *plasticity* or *robustness* of proteins to *mutation*. If you replace the amino acid at one position in the chain sequence with a different amino acid, it is called a *single-site mutation*. The natural protein is called the *wild-type* protein, and the mutated version is called the *mutant*. A single-site mutation corresponds to a step from one point to a nearby point in sequence space, just as a frog might jump from one lily pad to the next in a pond. Oftentimes, changing from one sequence S to another sequence S' causes no change in the lowest energy structure to which the protein folds: both sequences fold to the same native structure N. The set of all such sequences that fold to a given native structure is called the *neutral set* or *convergence set*.

A protein structure is called *designable* if it has a large convergence set. That is, a protein structure is designable if many different sequences fold to it. A designable structure is highly robust to amino acid change: various mutations will not cause a change in the protein's native structure. HP modeling indicates that protein structures should generally be highly tolerant of single amino acid changes.

Protein Structures Are Tolerant of Single Amino Acid Changes

Before the 1970s, it was thought that a protein could not tolerate many mutations without destroying its structure. The native structures of proteins were perceived as fragile, that is, fairly rigid and immutable. In those days, studies of protein mutations were rare, and mutations were only known at random locations within a protein. In 1978, in work that transformed protein science, Michael Smith invented *site-directed* or *single-site* or *point* mutagenesis, a much more targeted approach in which it became possible to replace one amino acid with another in a protein's sequence. For that work, he shared the 1993 Nobel Prize. There are now huge numbers of sequences. For example, more than 50,000 sequences are known of kinases alone. Studying sequences can tell you much about what substitutions are tolerated.

It is now clear from decades of extensive studies of targeted mutations that native structures of proteins are plastic and robust, highly tolerant of point mutations. That is, native structures are encodable in many different sequences; proteins have large convergence sets. For example, in T4 lysozyme [8] and the ROP protein, whole hydrophobic cores have been replaced by other hydrophobic residues.

How can a single native structure be encoded by different sequences? Box 7.2 shows two different HP sequences that fold to the same native

Box 7.2 A Given Native Fold Can Be Encoded by Different Sequences

In the HP model, the native state is determined by the maximum number of HH contacts. In Figure 7.11, you see two different HP sequences for which the native structure is identical. Here, two different sequences fold to the identical native structure even though the numbers of native HH contacts differ by one. In general, for longer chains and more fine-grained models, a single native structure can be encoded by a very large number of different sequences.

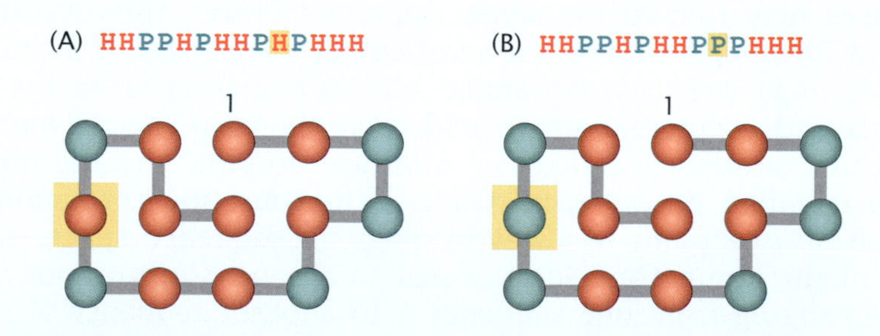

Figure 7.11 These two different HP sequences (A) and (B) encode the same native fold. The sequences differ by the one position indicated by the *yellow box*. The *orange spheres* represent hydrophobic (H) residues and the *green spheres* polar (P) residues. (Adapted from HS Chan and KA Dill. *J Chem Phys*, 95:3775–3787, 1991. With permission from AIP Publishing.)

structure. The fact that many different sequences can encode the same native fold is illustrated in Figure 7.12. It shows the *inverse folding problem* or *protein design problem*. In this type of problem, you are given a particular compact target structure. You want to design a sequence for which that particular structure is the most stable conformation. To design it, you first draw the chain backbone onto the lattice in the native configuration that you want. At this stage of this design process, each amino acid bead does not yet have an identity in terms of H or P. It is just a blank chain. You then design the sequence by "painting" H or P onto each residue position according to some recipe. In this way, you will have taken a target structure and will have designed a sequence that will fold to it.

In the HP model, the design recipe is relatively simple. For amino acid positions that are in the core of the structure, you assign H-type monomers, to make a hydrophobic core. For sites on the surface of the protein structure, it mostly doesn't matter whether you assign an H or a P, because surface interactions don't much affect the stability. In essence, by designing a sequence that has mostly H monomers in the core and mostly P monomers at the surface, you are doing the *positive design* of creating a stable hydrophobic core. And, by putting P monomers on the surface, you are doing some *negative design*, ensuring that other (wrong) chain folds are less stable. Of course, designing real proteins requires consideration of other factors too.

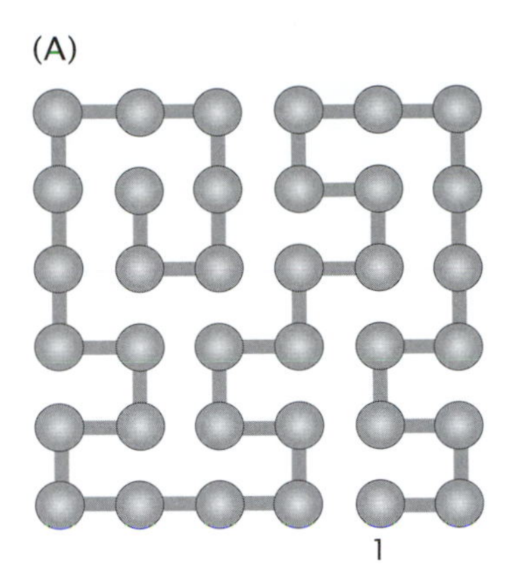

(B)

PPPHHPPPPPPHHHHHH-
HHPPPPPHHPPPPPPHHHH

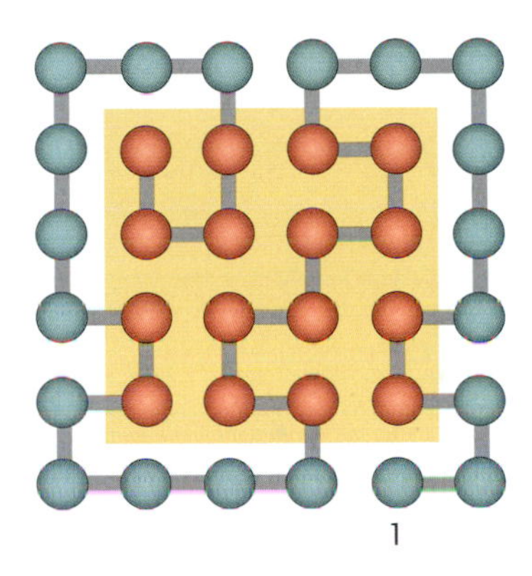

Figure 7.12 How to design an HP sequence to have a particular native fold. (A) Start with a target native structure you want a sequence to fold into. (B) Now, "paint" H onto each core site and P onto each surface monomer. This gives an HP sequence that will fold into the compact structure (A) as a state with very low free energy. Then, take sequence (B) and perform a conformational search to make sure there are no alternative (wrong) conformations of lower free energy to which sequence (B) prefers to fold. This procedure designs sequences with good hydrophobic cores.

Sequence Space Is Filled with Sequences that Collapse to Compact Protein-Like Folds

Now, let's explore the nature of random sequences. Will arbitrary amino acid sequences fold up into native-like structures? Imagine a fictitious beaker of water into which you insert one copy of every possible amino acid sequence of a given chain length. This imaginary *sequence-space soup* would contain a uniform representation of the whole sequence space. Sequence-space soup is a useful concept for addressing questions about the structures and properties of random amino acid sequences.

You might imagine that sequence-space soup is a bleak, uninteresting place with chains that are mostly open, unfolded, and unstructured—bearing essentially no resemblance to folded proteins. On the contrary, HP modeling reveals that sequence-space soup is relatively protein-like (Box 7.3). The typical molecule in sequence-space soup is compact and has much secondary structure. This result follows from the facts that (1) about half of the 20 amino acids are hydrophobic, so any given sequence will have a considerable number of hydrophobic residues, so (2) the chain will be driven to ball up in water into compact structures having a hydrophobic core, and (3) the resultant compactness will help stabilize specific hydrogen-bonded secondary structures that are compatible with those particular compact chain folds.

Box 7.3 What Is the Probability that a Random Sequence Is Protein-Like?

What is the probability of reaching into a random soup of sequences and pulling out one sequence that folds into a protein-like structure? Consider the sequence-space soup of all proteins having $N = 100$ residues. The number of sequences is 20^{100}. Now, consider two different calculations. First, the probability of finding one particular *sequence* is

$$p_{\text{one sequence}} = \frac{1}{20^{100}} \approx 10^{-130}. \tag{7.1}$$

Second, the probability of finding *any* sequence having an HP pattern that folds to a particular *structure* is

$$p_{\text{one structure}} = \frac{1}{2^{100}} \approx 10^{-30}, \tag{7.2}$$

where we take an alphabet of 2 instead of 20 since the folding code is mainly binary (H and P). So, finding a sequence having the right patterning of hydrophobic and polar amino acids needed to specify a particular native fold, by random search, is 100 orders of magnitude more likely than finding a particular sequence (Figure 7.13). In addition, you may not even need exactly the right HP patterning on the surface of a protein, since protein surfaces are generally much more tolerant to mutations. These binary-code arguments are qualitatively consistent with experiments involving the core mutations of lambda repressor [9, 10] and also with sequence randomization experiments [11]. Such arguments show why it is not so very improbable for natural proteins to arise from random amino acid sequences, for example in early evolution.

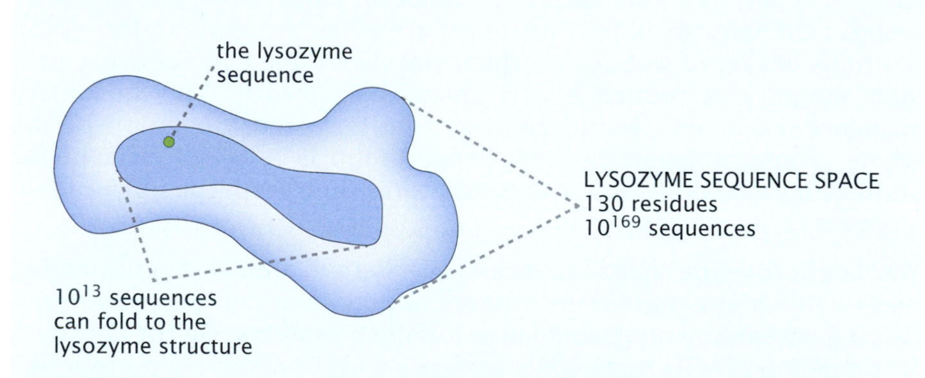

Figure 7.13 What is the probability of finding lysozyme in a soup of all possible sequences? Lysozyme is $N = 130$ residues long. So, in a soup of all the possible $20^{130} = 10^{169}$ sequences, the probability of finding the *one sequence* of lysozyme is essentially zero. But the probability of finding any sequence *that has the right hydrophobic–polar pattern* to fold into lysozyme's structure is much higher. It is around $(1/2)^{(130/3)} = 10^{-13}$, assuming that there are approximately $N/3$ residues in the core, which ought to be hydrophobic. (Adapted from KA Dill. *Protein Sci*, 8:1166–1180, 1999.)

Proteins and Other Polymers Can Be Designed to Fold to Stable Structures

In recent years, it has become possible to engineer proteins and other polymers to fold to desired structures. By using the principles that (a) proteins are constructed from hydrogen-bonded secondary structures, (b) the cores of proteins are mostly hydrophobic and mostly well packed, and (c) stable protein structures rarely have buried charges, taken together with the insights that are available from the many native structures in the Protein Data Bank, several research groups have created human-designed proteins. Figure 7.14 shows *Top7*, a protein designed and studied by the group of D Baker [12]. Top7 was a key advance in two ways: (a) it folds to its targeted structure, and (b) this structure had not been seen in nature before. Proteins have also been designed that can catalyze reactions [13].

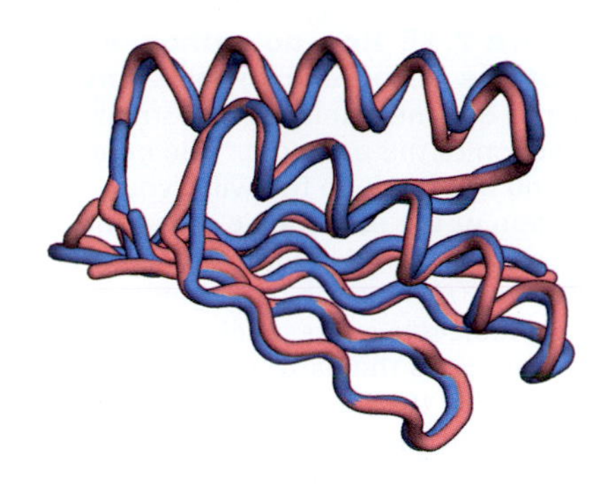

Figure 7.14 Proteins can be designed to fold into prespecified native conformations. Baker and colleagues designed an amino acid sequence to fold to the novel model conformation shown in *blue*. Then, they produced the protein using molecular biology techniques, crystallized it, and determined its structure (*orange*). The designed sequence was thus proven to fold into the target structure. (Adapted from B Kuhlman, G Dantas, GC Ireton, et al. *Science*, 302:1364–1368, 2003. With permission from AAAS.)

Proteins are not the only types of polymer that can fold into specific structures. Some RNA molecules fold into compact unique structures. Polymers that can fold into specific stable structures are called *foldamers* [14]. Polymers called *peptoids*, which have a nonbiological *N*-substituted glycine backbone, have been designed to fold into helical bundles [15].

EVOLUTION IS NOT AN ABSTRACTION. IT'S REAL. IT'S HAPPENING NOW

Evolution is in action all around us. It happens during the lifetime of a cell or organism. Evolution is not necessarily slow. A cancerous tumor is an example of *somatic evolution*, changes of cell properties that happen in the body within the lifetime of the organism. Parasitic worms, infectious viruses, and benign and pathogenic bacteria all evolve and adapt within the bodies of people and animals. In directed evolution (see page 177), laboratory cycles of mutation and natural selection are used to create new proteins of commercial importance over timescales of months to years.

Drug Resistance Is an Example of Evolution in Action

Drug-resistant pathogenic organisms constitute a major biomedical problem. Some drug-resistant bacteria, such as MRSA (methicillin-resistant *Stapholococcus aureus*), sometimes called *superbugs*, are resistant to many different antimicrobial drugs at the same time, unaffected by the main pharmaceuticals in our modern arsenal. Here's one way that drug resistance arises. First, consider a drug that treats a disease. The drug binds to a critical protein of a pathogenic organism such as a bacterium, killing that organism. The problem is that the drug may kill some of those pathogenic cells, but may not kill others, simply due to the natural diversity of the population. As the population replicates in the presence of the drug, the drug-resistant cells can now multiply. In this way, resistant organisms come to dominate the population over many replication cycles. Some evolution is relatively fast. HIV and bacterial pathogens evolve rapidly to become resistant to drugs, often on a global scale, in less than five years. Compare that with the 10- to 20-year period typically required for a drug company to discover and develop a new drug in the first place, and it's easy to see the magnitude of the drug-resistance problem.

How does a protein evolve resistance to a drug? Think about a pathogenic organism; call it *wt* (wild-type). Think about a protein in that pathogenic organism that has a normal function of binding to its natural ligand *A′*; see the top row of Figure 7.15. Think about a drug that binds to the same protein, blocking the natural action of ligand *A′*; see the middle row of Figure 7.15. The top and middle rows summarize the pathogen's normal action and how the drug inhibits the pathogen.

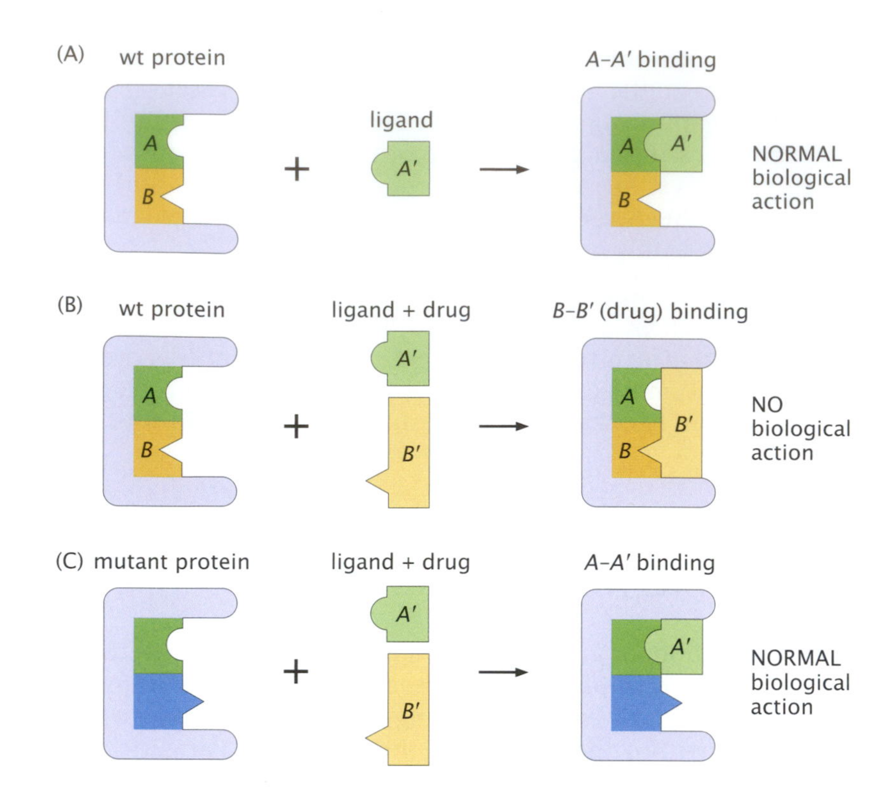

Figure 7.15 How do pathogenic proteins become drug-resistant? Here's one mechanism. (A) Normal function of the pathogen: The natural ligand (*A′*) binds to the wild-type protein of the pathogenic organism, causing its normal biological action. (B) Drug (*B′*) kills the pathogen: A drug molecule successfully outcompetes the natural ligand, preventing the ligand's action on the pathogenic protein, killing the pathogenic organism. (C) Pathogenic protein mutates, resisting the effects of the drug: The pathogenic protein becomes mutated, altering its binding site, preventing the drug from binding, and re-allowing the pathogenic protein's normal action.

Over time, the pathogenic organism can become resistant to the drug by evolving a mutation in the protein. The bottom row of Figure 7.15 shows a mutation in the protein that disrupts the drug binding, restoring the pathogenic protein's normal biological action. This shows one way that organisms can evolve to be drug resistant; and there are other ways to become drug resistant.

Molecular Clocks: Some Evolutionary Changes Proceed at a Constant Rate

Different types of proteins tend to evolve at different rates. However, for a given protein and a given amino acid position in that protein, the rate of mutation is often approximately constant over time, even over different organisms. In 1962, Zuckerkandl and Pauling termed this constancy of mutation rate the *molecular clock hypothesis* [16]. For example in globins, mutation and natural selection cause a change of about one amino acid per site per billion years. One protein diverges from another at a constant rate over evolutionary time. **Figure 7.16** shows the evidence for the molecular clock hypothesis. The *x*-axis of Figure 7.16 gives the time from the present day back to when two species diverged, at times that have been determined by paleontology. The *y*-axis gives the Hamming distance between the two sequences, that is, the number of amino acid differences between them. The figure shows that the number of amino acid changes in different proteins is approximately a linear functions of the time since the proteins diverged evolutionarily.

Figure 7.16 also shows that the molecular clock ticks at different speeds for different proteins. Different amino acid positions within a given protein can evolve at a different rate than other amino acid positions, so the molecular clock can tick at different speeds for different positions within a protein [17]. Some sites in proteins are *conserved*,

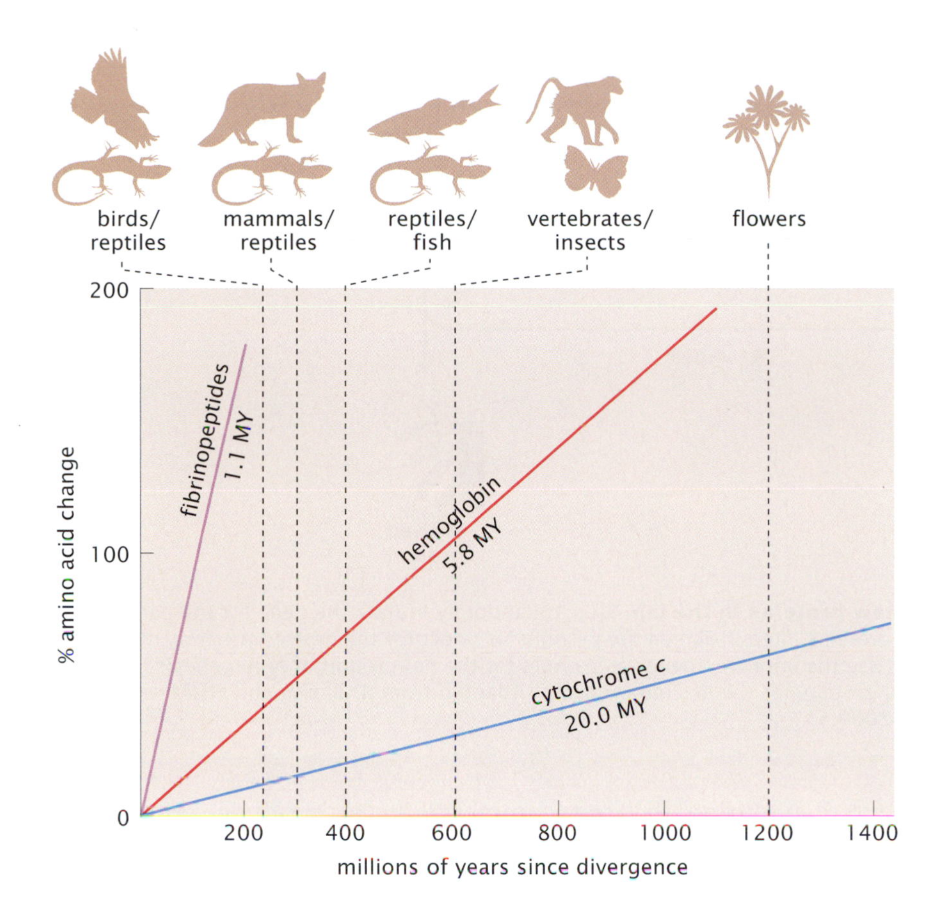

Figure 7.16 Proteins change at constant rates, like "evolutionary clocks." For each given type of protein, the number of amino acid changes grows linearly at a constant rate. The rate is the same for different organisms. The clock rate depends on the type of protein. For example, fibrinopeptides evolve faster than hemoglobins, which are faster than cytochrome *c*. (From R Phillips, J Kondev, J Theriot, and HG Garcia *Physical Biology of the Cell.* 2nd ed. Garland Science, New York, 2012.)

meaning that the amino acids at those sequence positions change relatively slowly. The rate of evolution can depend on physical properties, such as the residue's location (buried versus exposed) or secondary structure. For example, exposed and disordered regions (such as turns and coils) undergo faster substitutions compared with buried or structurally regular regions (that is, helices and sheets). And protein regions that have much conformational flexibility tend to evolve faster. This is because those regions can accommodate neutral mutations without disrupting the protein's function, whereas mutations at buried or tightly packed regions can destabilize its structure, affecting its function. See Figure 8.17 in Chapter 8.

Directed Evolution Is a Way to Improve Proteins in the Laboratory

Directed evolution is a procedure for modifying proteins in the laboratory by performing rounds of mutation and selection in test tubes. Suppose you want to modify a protein to make it more thermally stable or to have improved catalytic activity. You start with the natural protein of interest, then you introduce mutations randomly into different copies of that protein. You then perform an assay or *screen* on the property of interest to find which mutations have improved the property of the protein. You select out those improved variants, amplify them, and then perform further rounds of mutation and selection [18] (Figure 7.17). An advantage of this approach is that, like biological evolution, it does not require knowledge-based design.

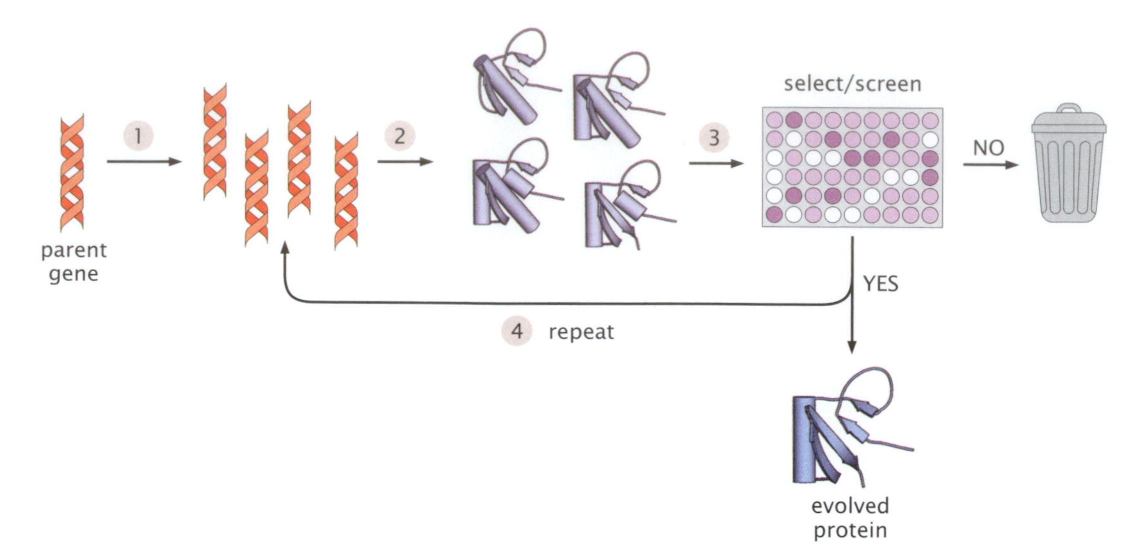

Figure 7.17 Directed evolution leads to new proteins in the lab. Step 1: Randomly mutate the gene for the parent protein to create a library. Step 2: Produce the mutant proteins. Step 3: Screen the proteins to select for the desired property. Reject mutants that don't show improvement. Step 4: Use the improved genes as parents for the next round of mutagenesis and screening. Repeat the process until the evolved protein has the desired property. (Adapted from JD Bloom and FH Arnold. *Proc Natl Acad Sci USA*, 106(Suppl 1):9995–10000, 2009.)

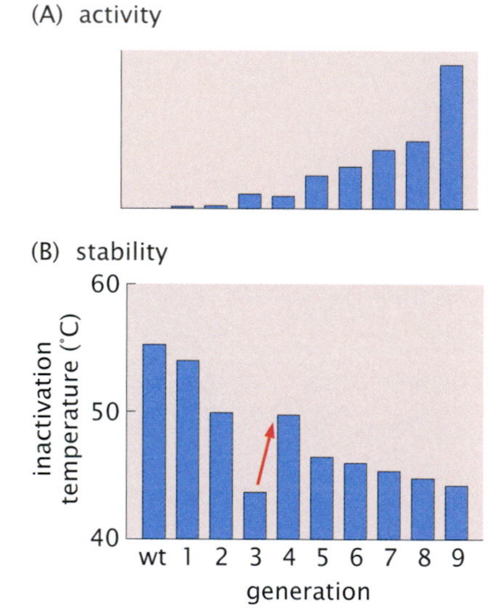

Figure 7.18 Directed evolution systematically gives better enzyme properties. (A) Wild-type cytochrome P450 began by having fatty acid hydroxylase activity. In nine rounds of directed evolution, mutants were selected with increasing propane monooxygenase (PMO) activity. (B) In the first four rounds, protein stability decreased. Then, proteins were selected for better stability (*red arrow*), following which the activity improved. (A, adapted from JD Bloom and FH Arnold. *Proc Natl Acad Sci USA*, 106(Suppl 1):9995–10000, 2009. B, Adapted from R Fasan, YT Meharenna, CD Snow, TL Poulos, and FH Arnold. J Mol Biol, 383:1069–1080, 2008. With permission from Elsevier.)

Directed evolution is used commercially to produce proteins with advantageous industrial properties, such as increased expression levels, enzyme activities for detergents, stabilities at different temperatures or pH values, catalytic rates, and others. Some examples of successes with directed evolution are the 1000-fold increase in the half-life of subtilisin BPN′ in the absence of Ca^{2+} [19], a 1000-fold increase in substrate specificity of β-galactosidase [20], the 10^5-fold increase in the activity of aspartate aminotransferase for β-branched amino and 2-oxo acids [21], and the conversion of a dimeric chorismate mutase into a monomeric form [22].

Figure 7.18 shows the directed-evolution conversion of a cytochrome P450 fatty acid hydroxylase into a propane monooxygenase through nine generations of mutation and selection. The evolved enzyme has good activity over a broad range of temperatures. This evolution led to 20 mutations near the heme-binding pocket of the enzyme, reshaping the substrate-access pathway. It illustrates how proteins can be engineered to have desired properties when subjected to nonbiological laboratory selective pressures. Directed evolution also teaches us about natural evolution. Here are two of the key lessons learned so far.

First, directed evolution experiments show that random mutation and selection are a remarkably effective combination for traversing sequence space to reach desired properties of proteins. Even when the beneficial changes are relatively infrequent, there are almost always good evolutionary routes, involving neutral or beneficial mutations, for achieving goals of directed evolution. The implication is proteins rarely need to pass through states of lesser fitness on their way to becoming more fit. Evolution can avoid kinetic traps, just as protein folding can, because the dimensionality of the space being searched is so vast that there are invariably some routes that are relatively fast and direct.

Second, directed evolution experiments show that when proteins are stable, this assists in providing abundant evolutionary routes. If a protein is marginally stable, then many of the possible mutations would lead to its unfolding, not to an improved protein. Those evolutionary pathways are dead ends. But if a protein is sufficiently stable, then it has more evolutionary exit routes—fewer of the possible mutations would cause unfolding. Consider the following experiment. Start with an amino acid sequence that folds to a given native structure. Next, find its neutral net—all the sequences that differ from the wild-type sequence by at least a single mutation that still fold to the same structure. Now walk around that neutral net to find the sequence that has the maximum mutational stability, that is, the sequence that can tolerate the largest number of single mutations and still fold to the correct structure. Remarkably, it is found in the HP model that those *most highly connected* sequences also tend to have the highest folding stabilities [23, 24]. This implies that sequence space (like conformational space) has funnel-like properties: protein evolution will favor proteins having the highest neutral-space connectivity, and those will also have the greatest conformational stability. Said differently, unstable proteins have few evolutionary routes to better properties because most mutations will unfold unstable proteins. In contrast, stable proteins have many evolutionary routes to better properties because stable proteins can tolerate a few destabilizing mutations without unfolding.

SUMMARY

Protein sequences evolve through mutation and selection. Most single mutations do not change a protein's structure or function: native structures are fairly robust to small changes in sequence. But more extensive mutations can lead to changes in protein structure, dynamics, aggregation propensity, and physical properties, as well as biological mechanisms and functions. Evolving proteins lead to evolving organisms, sometimes leading to very large transformations, such as new biological species. Typically, if two proteins have similar sequences, they also have similar structures and similar functions. Such correlations are the basis for the methods described in Chapter 8: by detecting amino acid sequence similarities, you can draw inferences about the structures and functions of proteins.

REFERENCES

[1] B Rost. Twilight zone of protein sequence alignments. *Protein Eng*, 12:85–94, 1999.

[2] M Ashburner, CA Ball, JA Blake, et al. Gene Ontology: Tool for the unification of biology. The Gene Ontology Consortium. *Nat Genet*, 25:25–29, 2000.

[3] The Gene Ontology Consortium. Gene Ontology Consortium: Going forward. *Nucleic Acids Res*, 43:D1049–D1056, 2015.

[4] G Manning, DB Whyte, R Martinez, et al. The protein kinase complement of the human genome. *Science*, 298:1912–1934, 2002.

[5] C Chothia, J Gough, C Vogel, and SA Teichmann. Evolution of the protein repertoire. *Science*, 300:1701–3, 2003.

[6] EV Koonin and MY Galperin. *Sequence–Evolution–Function Computational Approaches in Comparative Genomics.* Kluwer Academic, Norwell, MA, 2003.

[7] RF Doolittle. Convergent evolution: the need to be explicit. *Trends Biochem Sci*, 19:15–18, 1994.

[8] BW Matthews. Structural and genetic analysis of protein stability. *Ann Rev Biochem*, 62:139–160, 1993.

[9] JU Bowie, JF Reidhaar-Olson, WA Lim, and RT Sauer. Deciphering the message in protein sequences: tolerance to amino acid substitutions. *Science*, 247:1306–1310, 1990.

[10] WA Lim and RT Sauer. Alternative packing arrangements in the hydrophobic core of lambda repressor. *Nature*, 339:31–36, 1989.

[11] S Kamtekar, JM Schiffer, H Xiong, et al. Protein design by binary patterning of polar and nonpolar amino acids. *Science*, 262:1680–1685, 1993.

[12] B Kuhlman, G Dantas, GC Ireton, et al. Design of a novel globular protein fold with atomic-level accuracy. *Science*, 302:1364–1368, 2003.

[13] L Jiang, EA Althoff, FR Clemente, et al. *De novo* computational design of retro-aldol enzymes. *Science*, 319:1387–1391, 2008.

[14] SH Gellman. Foldamers: A manifesto. *Acc Chem Res*, 31:173–180, 1998.

[15] BC Lee, RN Zuckermann, and KA Dill. Folding a nonbiological polymer into a compact multihelical structure. *J Am Chem Soc*, 127:10999–11009, 2005.

[16] E Zuckerlandl and LB Pauling. Molecular disease, evolution, and genetic heterogeneity. In M Kasha and B Pullman, editors, *Horizons in Biochemistry*, pp 189–225. Academic Press, New York, 1962.

[17] EP Rocha and A Danchin. An analysis of determinants of amino acids substitution rates in bacterial proteins. *Mol Biol Evol*, 21:108–116, 2004.

[18] JD Bloom and FH Arnold. In the light of directed evolution: pathways of adaptive protein evolution. *Proc Natl Acad Sci USA*, 106(Suppl 1), 2009.

[19] SL Strausberg, PA Alexander, DT Gallagher, et al. Directed evolution of a subtilisin with calcium-independent stability. *Biotechnology*, 13:669–673, 1995.

[20] JH Zhang, G Dawes, and WP Stemmer. Directed evolution of a fucosidase from a galactosidase by DNA shuffling and screening. *Proc Natl Acad Sci USA*, 94:4504–4509, 1997.

[21] T Yano, S Oue, and H Kagamiyama. Directed evolution of an aspartate aminotransferase with new substrate specificities. *Proc Natl Acad Sci USA*, 95:5511–5515, 1998.

[22] G MacBeath, P Kast, and D Hilvert. Redesigning enzyme topology by directed evolution. *Science*, 279:1958–1961, 1998.

[23] E Bornberg-Bauer and HS Chan. Modeling evolutionary landscapes: mutational stability, topology, and superfunnels in sequence space. *Proc Natl Acad Sci USA*, 96:10689–10694, 1999.

[24] JD Bloom, MM Meyer, P Meinhold, et al. Evolving strategies for enzyme engineering. *Curr Opin Struct Biol*, 15:447–452, 2005.

SUGGESTED READING

Alberts B, Johnson A, Lewis J, Morgan D, Raff M, Roberts K, and Walter P, *Molecular Biology of the Cell*, 6th ed. Garland Science, New York, 2015.

Marsh JA and Teichmann SA, Structure, Dynamics, Assembly and Evolution of Protein Complexes. *Annu Rev Biochem*, 84:551–575, 2015.

Page RDM and Holmes EC, *Molecular Evolution: A Phylogenetic Approach*, Blackwell Science, Oxford, 1998.

Patthy L, *Protein Evolution*, 2nd ed. Wiley-Blackwell, New York, 2008.

CHAPTER

8 Bioinformatics: Insights from Protein Sequences

COMPARING AMINO ACID SEQUENCES GIVES INSIGHT INTO PROTEIN STRUCTURE AND FUNCTION

Bioinformatics methods allow you to compare and analyze sequences of proteins or nucleic acids to learn about their functions. Given an amino acid sequence, you often want to find a similar sequence in a database. These are called *string searches*. Sometimes, you want to align and compare two or more sequences, and measure the distance between them. By measuring distances between sequences, you can construct evolutionary trees, or infer biological functions or binding sites, or learn what drugs might bind to those proteins. By looking at sequence positions that are conserved through evolution, you can draw inferences about protein stability or function. And, if you are designing a drug to disable a pathogenic organism, you can compare the pathogen protein with human proteins, to avoid mistakenly drugging human proteins, causing side effects of drugs.

The first experimental determination of a protein's complete amino acid sequence was the sequence of insulin, published in 1951 by F Sanger. For that work, he was awarded the Nobel Prize in Chemistry in 1958. In the 1960s, M Dayhoff assembled the first datasets of protein sequences in a published database, called the *Atlas of Protein Sequence and Structure*. She developed the first algorithms to cluster sequences into *families* and *superfamilies* based on their sequence similarities. Our knowledge of protein structures grew from 2 in 1960 to 120,000 in 2016 (more than 100,000 of which were determined by X-ray crystallography, 10,000 by NMR, and a few thousand by cryo-electron microscopy). Sequence databases grew even faster (Figure 8.1). Examples are GenBank, which is a database of publicly available DNA sequences [1] and UniProtKB (the Universal Protein KnowledgeBase), which is a database of protein sequences [2]. Figure 8.1 also displays the growth in the total number of protein superfamilies based on the CATH classification database, where protein domains are grouped into superfamilies when there is sufficient evidence that they diverged from a common ancestor [3].

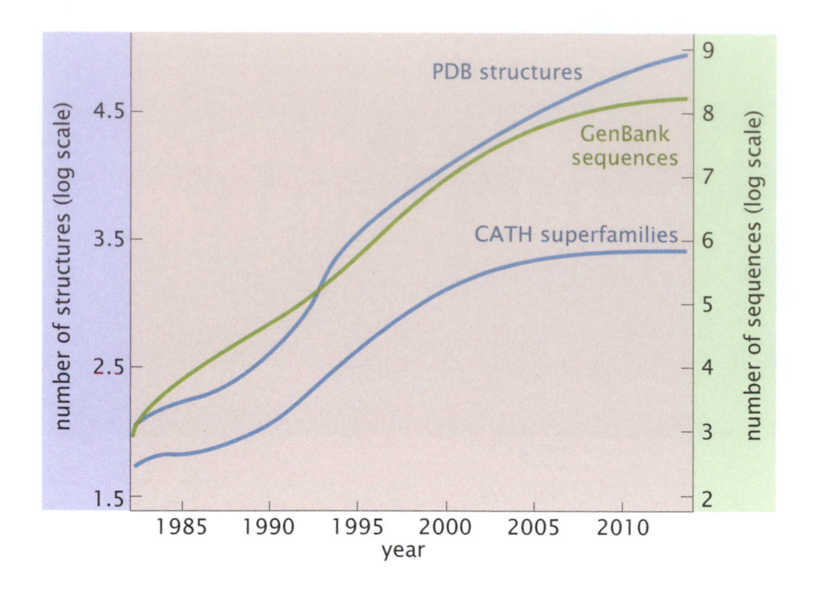

Figure 8.1 The numbers of known sequences and structures have grown rapidly. Sequences from GenBank (*right* ordinate). Structures from the PDB (*left* ordinate). However, the number of new protein superfamilies is reaching a plateau. CATH database version v4.1 [3] contains 2737 superfamilies.

Sequences Change through Evolutionary Mutations

Evolution leads to amino acid changes in protein sequences. There are three types of sequence changes originating from changes in DNA: mutations, insertions, and deletions (Figure 8.2). In a *mutation*, one amino acid gets swapped for another; in a *deletion*, an amino acid is dropped; and in an *insertion*, an extra amino acid is added, as a parent sequence evolves into a daughter. *Indel* is a term that refers to cases where either an insertion or deletion has occurred.

Proteins Having Similar Sequences Usually Have Similar Structures and Functions

If two proteins have similar amino acid sequences, they probably also have similar native structures and similar biological functions and mechanisms of action. In fact, structure diverges more slowly than sequence, and many sequence changes may be accommodated by the same fold, as illustrated in Figure 7.10 in Chapter 7. So, you can use similarities and differences in amino acid sequences to draw inferences in biology. This is the subject of *bioinformatics* and *comparative modeling*, wherein you infer various properties of biomolecules by comparing their sequences.

When two sequences are only slightly similar to each other, they are said to lie in the *twilight zone* of sequence similarity. Sequences that are very similar are in the *reliable zone*. And, sequences that are so different that sequence–function relationships are unpredictable are in the *midnight zone*. The twilight zone is where sequences are 20–35% identical (Figure 8.3). Two proteins with greater than 35% sequence identity are quite likely to have similar structures and perform similar functions.

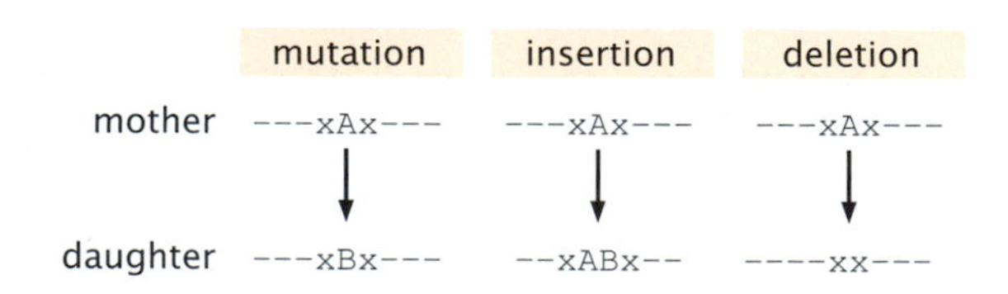

Figure 8.2 Different types of mutational changes: substitutions, insertions, and deletions.

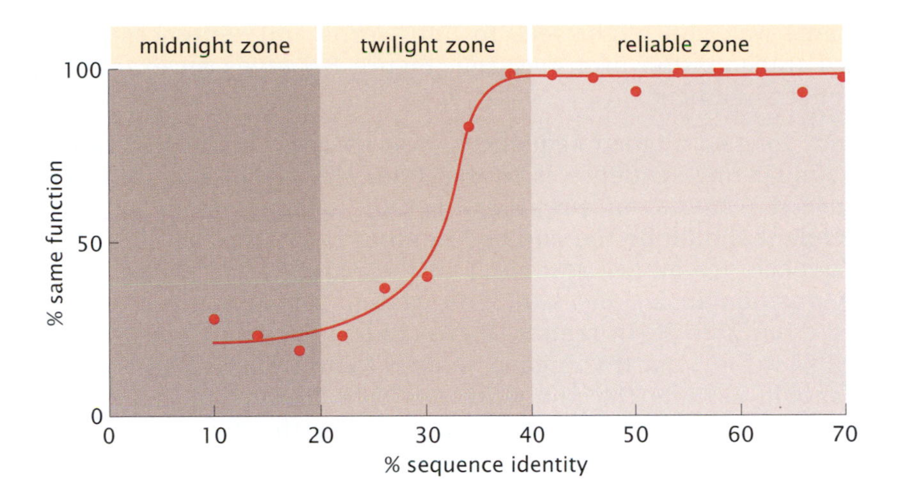

Figure 8.3 Two proteins that have similar sequences usually have similar functions. High sequence identity (reliable zone) implies two proteins have the same function. But sometimes proteins have the same function even when they don't have high sequence identity (midnight zone). In between (twilight zone), the predictability of function from sequence is low. (Adapted from CA Wilson, J Kreychman, and M Gerstein. *J Mol Biol*, 297:233–249, 2000. With permission from Elsevier.)

HOW DO YOU DETERMINE THE RELATEDNESS BETWEEN SEQUENCES?

Figure 8.4 shows two different pairs of parent–daughter sequences. How related is each daughter to its parent? Which mother–daughter pair has the "closer relationship"? The pair on the left has one mutation. The pair on the right has one deletion, one mutation, and one insertion. So, is the pair on the left more closely related? Not necessarily. In order to measure the relatedness, you need a *metric*, that is, a method for *scoring* how much any given mutation, insertion, or deletion is "worth." The degree of relatedness between two sequences depends on how you choose to rank the importance of any particular type of change. It follows that there is a certain arbitrariness in choosing a scoring system. Here are the steps for determining the similarities between two sequences.

First, to make comparisons between sequences, you must *align them* with each other (see page 186). Aligning sequences means superimposing the two sequences to find commonality among their substrings, in

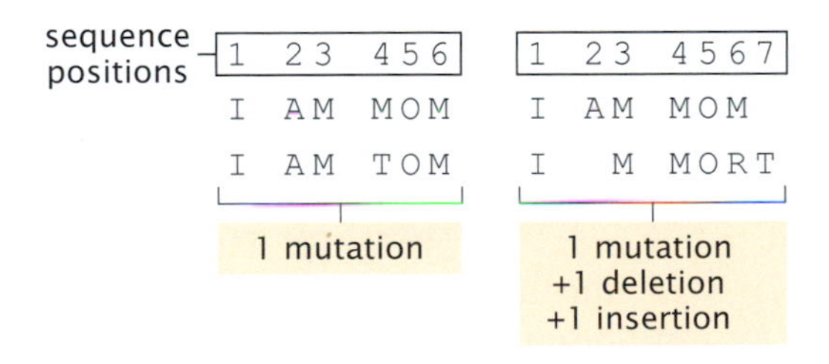

Figure 8.4 A parent sequence can evolve to different daughters. The *left* mother–daughter pair of sequences has one mutation. The *right* mother–daughter pair has one deletion (at position 2, of A), one mutation (at position 6, from M to R), and one insertion (of T, at position 7). However, to determine which mother–daughter pair is more closely related, you need to do more than count them; you also need to score the importance of each modification.

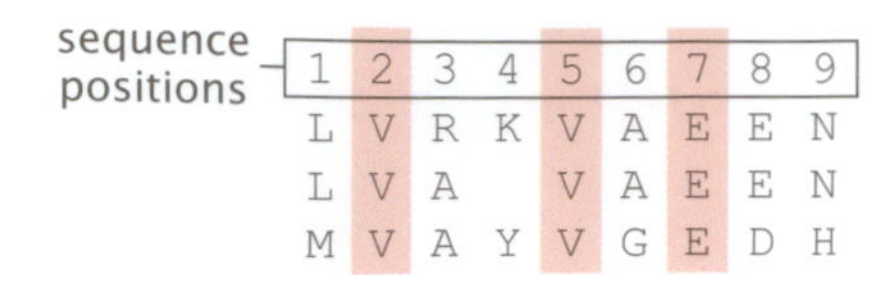

Figure 8.5 Aligned sequences have identical residues at some positions. In this alignment of three sequences, there is an identical residue at sequence positions 2, 5, and 7. Alignment sometimes requires skipping over one site (a deletion), in this case at site 4 of the middle sequence.

the same way you might seek to find words and sentences in a corrupted transmission of a message. Figure 8.5 gives an example of a sequence alignment.

Second, for two aligned sequences, you assign a score that indicates how similar one sequence is to the other. To assign a score, we usually use the *parsimony principle*, that the changes from ancestor to descendant should be the minimal number consistent with differences in sequence. If you see residue X in the parent and residue Y at the same position in an ancestor, you assume only one mutation happened from X to Y. In reality, there could have been many changes, $X \rightarrow Z \rightarrow W \rightarrow Y$, but if you have no direct knowledge of such changes, you should assume the minimum possible change. In addition, you assign some penalty score for *gaps* in either the parent or ancestor, that is, for indels. Different algorithms score gaps in different ways, depending on the gap length and other properties. One expression that is commonly used to assign a penalty $g(k)$ for a gap of k residues is

$$g(k) = -a - kb, \tag{8.1}$$

where a is a constant representing the *gap-initiation* cost and b is the *gap-extension* cost, charged equally for each amino acid in the gap. Equation 8.1 says that all gaps have an intrinsic initiation cost, a, and that the total penalty becomes more costly in linear proportion to the size of the gap. Usually you set $b < a$, to indicate that the penalty for opening a gap is larger than for extending it by an additional amino acid. It is common to use values such as $a = 11$ and $b = 1$ to encode the general insight that two sequences are evolutionarily more closely related if they have fewer indels that are long than if they have more indels that are short.

The Hamming distance (see Box 7.1 in Chapter 7) is simple, just making a binary determination for each position: either the two sequences exactly match, or don't match, at that site. In practice, scoring systems are usually more refined than this. For example, you could score the swap of one amino acid for another based on a physical scale, such as their hydrophobicities or charges or sizes (see Figure 1.5 in Chapter 1). Leucine commonly swaps for valine, for example, because they are both hydrophobic and about the same size. In contrast, cysteine and glutamic acid rarely swap, because their physical properties are different from each other.

Some Amino Acids Swap More Often than Others

When you compare multiple related proteins, you see different distributions of amino acids at different sites. We want to determine the swap frequencies of amino acids. To do that, you can make a database by first pairing up two proteins that you know are homologs and that have high sequence similarities. Then, you align their sequences. You do this for many such pairs of proteins. Then, you count the numbers of times that you find an amino acid of type a at position i in sequence A and an amino acid of type b at the same position i in its homolog sequence A'. Call that frequency $p_{ab}(i)$. Next, in order to turn this into a useful measure of swapping, you perform two mathematical operations. First, you make a correction for what would be found by random chance. Let q_a be the natural frequency of occurrence of amino acid of type a (the number of a-type amino acids divided by the number of all types, across a large set of proteins; see the final column of Table 1.1 in Chapter 1). Then, define the *log-odds* score $s_{ab}(i)$ for sequence position i as

$$s_{ab}(i) = \log \frac{p_{ab}(i)}{q_a q_b}. \tag{8.2}$$

When $p_{ab}(i)/(q_a q_b) = 1$, it means that amino acid types a and b swap at position i at the same rate that would be expected by chance; then $s_{ab}(i) = 0$. When $p_{ab}(i)/(q_a q_b) > 1$, it means that amino acid types a and b swap more often than by random chance (so $s_{ab}(i) > 0$), and when $p_{ab}(i)/(q_a q_b) < 1$, it means that they swap less often than by chance (so $s_{ab}(i) < 0$). The reason for taking the logarithm in Equation 8.2 is because it makes the math more convenient for scoring full amino acid sequences, not just single positions. The total log-odds score for comparing two full sequences A and A' is the sum of $s_{ab}(i)$ over all sites i. This approach, and variants of it (using large sets of proteins and taking averages), are the bases for amino acid *substitution matrices* (Box 8.1), which can assign scores to find the best sequence alignments. A substitution matrix that is commonly used in protein-sequence alignments is *BLOSUM* [4].

Box 8.1 Examples of Substitution Matrices: BLOSUM and PSSM

BLOSUM (BLOck SUbstitution Matrix) was developed to improve upon comparisons of evolutionarily distant sequences. BLOSUM62 is shown in Figure 8.6 (62 indicates that the scores are derived from comparing sequences with 62% identity or less).

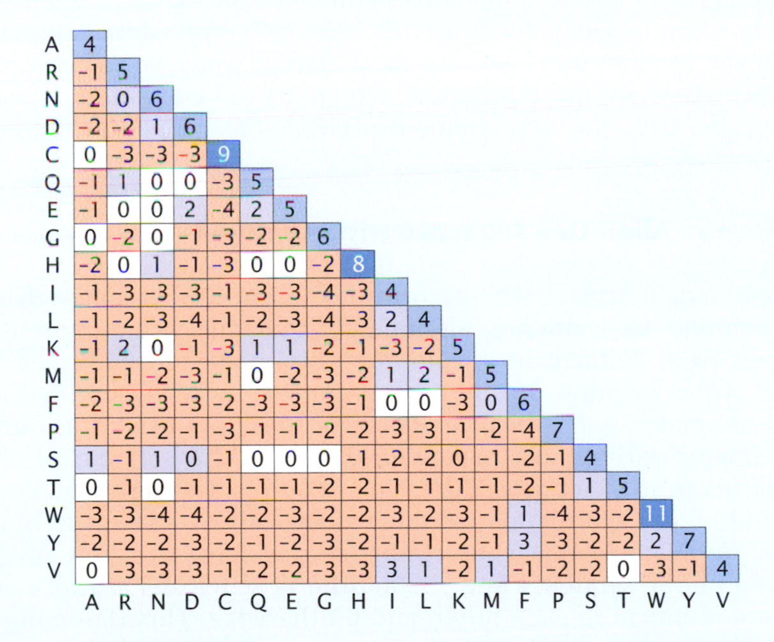

	A	R	N	D	C	Q	E	G	H	I	L	K	M	F	P	S	T	W	Y	V
A	4																			
R	-1	5																		
N	-2	0	6																	
D	-2	-2	1	6																
C	0	-3	-3	-3	9															
Q	-1	1	0	0	-3	5														
E	-1	0	0	2	-4	2	5													
G	0	-2	0	-1	-3	-2	-2	6												
H	-2	0	1	-1	-3	0	0	-2	8											
I	-1	-3	-3	-3	-1	-3	-3	-4	-3	4										
L	-1	-2	-3	-4	-1	-2	-3	-4	-3	2	4									
K	-1	2	0	-1	-3	1	1	-2	-1	-3	-2	5								
M	-1	-1	-2	-3	-1	0	-2	-3	-2	1	2	-1	5							
F	-2	-3	-3	-3	-2	-3	-3	-3	-1	0	0	-3	0	6						
P	-1	-2	-2	-1	-3	-1	-1	-2	-2	-3	-3	-1	-2	-4	7					
S	1	-1	1	0	-1	0	0	0	-1	-2	-2	0	-1	-2	-1	4				
T	0	-1	0	-1	-1	-1	-1	-2	-2	-1	-1	-1	-1	-2	-1	1	5			
W	-3	-3	-4	-4	-2	-2	-3	-2	-2	-3	-2	-3	-1	1	-4	-3	-2	11		
Y	-2	-2	-2	-3	-2	-1	-2	-3	-2	-1	-1	-2	-1	3	-3	-2	-2	2	7	
V	0	-3	-3	-3	-1	-2	-2	-3	-3	3	1	-2	1	-1	-2	-2	0	-3	-1	4

Figure 8.6 The BLOSUM62 similarity matrix. Each matrix element in this matrix is the log-odds ratio (see Equation 8.2) for substituting one particular amino acid for another. The matrix is symmetrical, so only half of it is shown. The off-diagonal elements describe the propensity to swap two different types of amino acids. Positive numbers mean that those swaps happen more often than by chance, while negative numbers indicate they are rarer than by chance. The diagonal terms indicate the degree of conservation of each type of amino acid. For example, tryptophan, cysteine, and histidine are the most highly conserved. (From S Henikoff and JG Henikoff. Proc Natl Acad Sci USA, 89:10915-10919, 1992. Copyright (1992) National Academy of Sciences, USA.)

PSSM (Position-Specific Substitution Matrix). Earlier, we described methods in which substitution frequencies only depend on the types a and b of amino acids, and not on their particular location in the protein sequence. However, in general, substitution frequencies of amino acids among homologous proteins can differ from one site in the sequence to the next (and therefore in the protein's native structure). PSSMs are matrices that account for each residue position i individually.

TO COMPARE SEQUENCES, YOU START WITH GOOD ALIGNMENTS

To determine the similarity between two sequences, you need an *alignment method*. In practice, alignment and scoring are performed simultaneously, because you want to find the alignment with the highest score. A sequence alignment is a procedure for establishing the best position-by-position correspondence (according to some scoring system), by sliding one sequence relative to another, taking account of indels. You can perform a *global alignment*—that is, the best possible alignment taken over the entire length of the two sequences—as shown later. Or, you can perform a *local alignment*, meaning that you make optimal alignments of local stretches of the chain, without paying attention to the rest of the sequence. By combining the local alignments of all possible fragments, you can obtain an overall alignment. It is much faster to compute, and is sufficient for many problems, particularly when two sequences are not very different. Global alignments are slower to compute; they are most useful when you compare two sequences that are quite different, which are called *remote homologs*. Next, we describe the *Needleman–Wunsch method* of global alignment.

How Do You Align One Sequence with Another?

The Needleman–Wunsch (NW) algorithm [5] is an application of *dynamic programming* to sequence alignment. Dynamic programming was invented by R Bellman at a Cold War think-tank in 1953. The basic idea of NW is to build up the best alignment by using optimal alignments of smaller subsequences. Because these are scored as sums of the segments without any dependences across the boundaries of segments, this is an exact procedure. This requires scoring them. Usually, you would use a scoring matrix, such as BLOSUM62. However, in order to have an example that is as simple as possible, let's just assign to residue i in one sequence and j in another a score of 1 if those amino acids are identical to each other, and 0 otherwise. This Hamming-type scoring will provide the number of identical amino acids between the two sequences. So, here, for the purpose of illustration, our scoring scheme amounts to nothing more than just counting the number of identically matching residues in a given alignment.

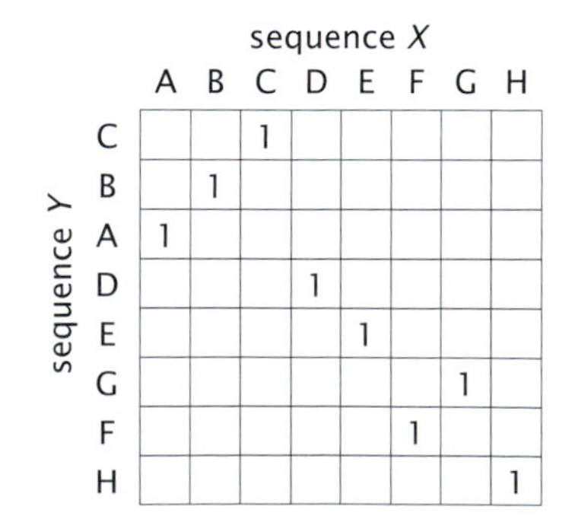

Figure 8.7 In this match matrix, 1's indicate sequence positions having the same amino acid in both sequences. The top left corresponds to the N-terminus and the bottom right the C-terminus.

Let us consider the two sequences: sequence X, ABCDEFGH; and sequence Y, CBADEGFH. The sequences do not need to have the same length, but in this case they do for simplicity. First, we identify which amino acids match identically between the two sequences. All of these are marked by 1's in Figure 8.7.

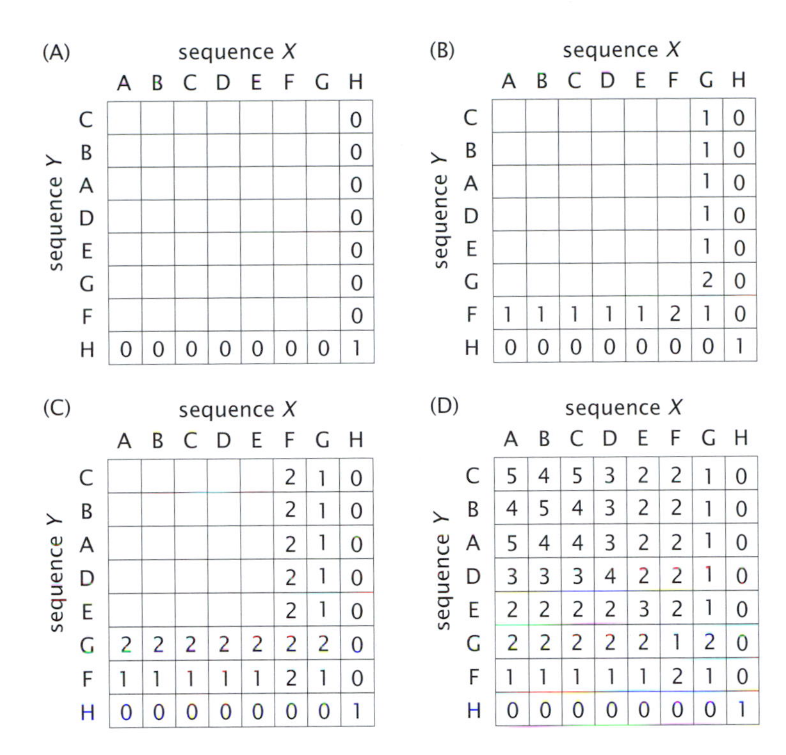

Figure 8.8 Filling in the scoring matrix. (A) The initial step. Start at the C-terminal. Fill in the right column and the bottom row. (B) The second scoring matrix is for the next position toward the N-terminus: fill up the second column from the right and second row from the bottom, accumulating scores as the sites are compared. (C) The third scoring matrix, accumulating matches as sites are compared. (D) A complete scoring matrix, with accumulated scores for all matches stepping from the C-terminus to the N-terminus.

Our goal is to construct a *scoring* matrix **M** that will tell us which alignment of sequences gives the best match. We fill in the elements M_{ij} of the scoring matrix, starting at the bottom right corner and proceeding toward the upper left corner. Each element in turn will be assigned the largest number of identical pairs that can be obtained up until that point, starting from the C-terminal ends of the sequences X and Y. Each matrix element M_{ij} is the best score for the C-terminal portions of the sequences, when i is matched with j, given that the portions ($>i$ and $>j$) of sequences X and Y have been optimally arranged to achieve the highest score.

In Figure 8.8A, we first fill in the last row and last column of the scoring matrix. The only match in that row and that column is in the bottom right corner. So, put 1 in that box and 0's in all the rest.

Then proceed to the next-to-last row and next-to-last column. There is a new match of F with F in this row—*and* a new match of G with G in this column. These are both compatible with matching the terminal H's, so these two elements of the matrix are both assigned the value of 2, meaning that the corresponding C-terminal portion contains two identical matches. All other elements are assigned the value 1 (see Figure 8.8B) because the match of the terminal H's is the only possible match upon pairing any of the other amino acids.

Now, apply the same process for the next position (see Figure 8.8C).

Continuing this process leads you to complete the entire matching matrix (see Figure 8.8D).

The best total score is 5 identity matches. It appears that there are four different ways this score can be achieved, since there are four positions in the final scoring matrix (see Figure 8.8D) having the value of 5. However, a *traceback* procedure reveals that there are more than four sequence alignments that can achieve this score. Figure 8.9 shows the traceback, where you find what these best scoring alignments actually are.

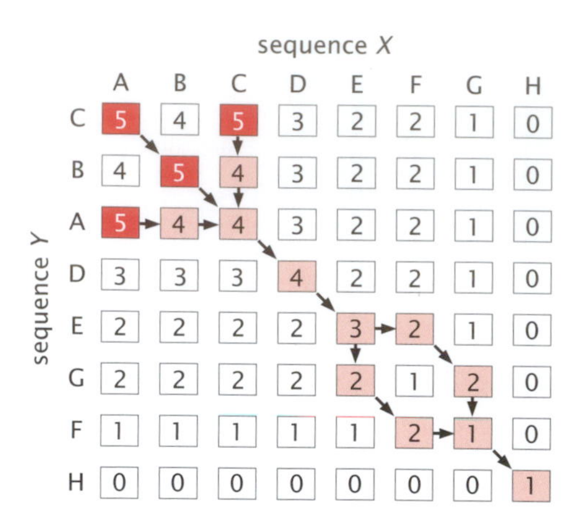

Figure 8.9 Tracing back through the Needleman–Wunsch (NW) scoring matrix to extract the best alignments. The arrows show the traceback to identify the best sequence matches. As you move down or to the right, you place arrows between all of the highest-scoring neighbors. The arrows indicate the paths and the matches leading to the four cases having the highest score, 5. This is a rigorous method because it is a global alignment and includes all combinations of matches, gaps, and insertions.

What are the sequence matches that result from this procedure? Focus on those matches having best scores of 5. Start at the upper left corner corresponding to A(sequence *X*) and C(sequence *Y*). Let's designate them as A(*X*) and C(*Y*), respectively. This value of 5 does not include a new match, since C does not match A, so the score of 5 must have originated from the previous row or column. Indeed, the only adjacent entry from the previous row or column with a value of 5 is from the diagonal entry for matching B(*X*) with B(*Y*). We insert an arrow in Figure 8.9 between these elements. Likewise, we traceback to see where each of the other high values of 5 originated. The rule is that we connect to the highest value either to the right or below, and the selected elements must be equal to or smaller than the starting element. If there is a choice between arrows connecting diagonally and vertically or horizontally, we always choose the diagonal since the other choices would require invoking an indel. In some cases such as E(*X*) matching E(*Y*), you see that the previous best score of 2 is either to the right or below, so this requires inserting arrows pointing to both entries, and the two alternative paths for matches can be seen in the lower right part of the matrix in Figure 8.9.

The traceback matrix gives you the optimal matches. Figure 8.10 shows the two alignments corresponding to the score of 5 for C(*X*), C(*Y*).

Similarly, four additional paths are found starting from the entries A(*X*), A(*Y*) and A(*X*), C(*Y*). Note that matches associated with A(*X*), A(*Y*) and B(*X*), B(*Y*) are identical. This leaves you with two pairs of paths starting at the matched entries B(*X*), B(*Y*) and A(*X*), A(*Y*) shown in Figure 8.11. Overall, this NW matrix leads to a total of six alignments.

All these matches have a score of 5. Which match is the most accurate alignment? You cannot further refine your choice based on the scores that gave you these alignments. But you can pay attention to the points of sensitivity, such as the gap penalties, and these lead directly to the conclusion that alignments 3 and 4 are best. Or, in general, you can use

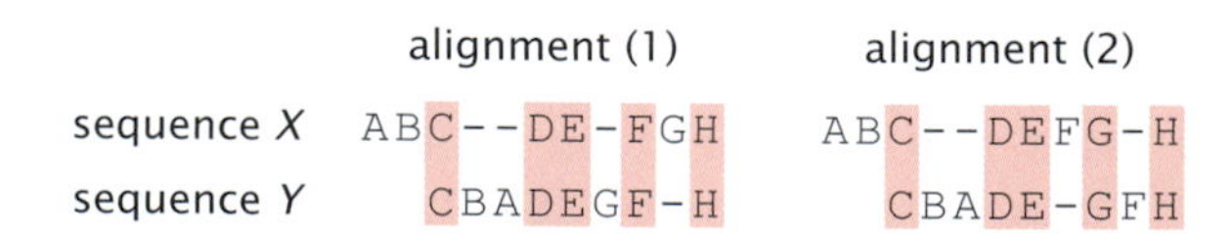

Figure 8.10 Two of the best matches, having scores of 5. *Dashes* designate the gaps in sequence *X* (or the insertion in sequence *Y*), and matching pairs are indicated by vertical *pink* bars.

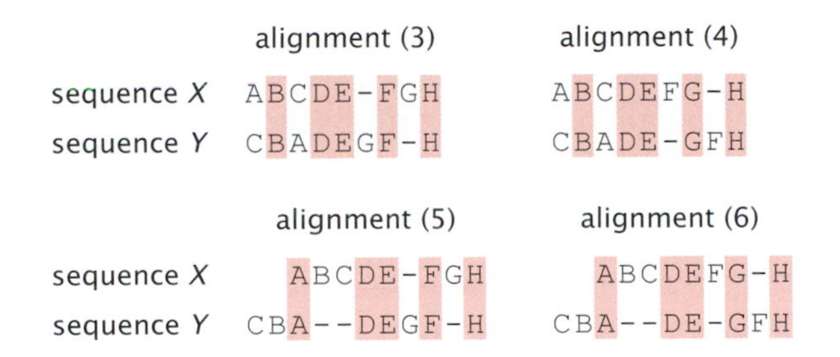

Figure 8.11 The other four best-scoring alignments.

more information about amino acid similarities, from an amino acid similarity matrix (see Figure 8.6) to score the nonidentical matches.

In 1981, Smith and Waterman (SW) developed a variant of the NW algorithm [6]. The SW algorithm was designed to address the problem of matching distantly related sequences. In such cases, there are many regions where the sequences are highly dissimilar. Alignment speeds can be improved by neglecting such regions of low similarity. The SW method performs local (rather than global) sequence alignments. Local alignment avoids low-similarity regions altogether, and just finds good local matches. The main difference from the NW algorithm is that scoring is handled differently in the traceback. For SW, you start at the location in the matrix having the highest score, not the bottom right corner. Then you stop the matching and backtracking when the score remains constant, indicating that any further lengthening of this particular alignment would just produce identically scoring alignments. When you find local alignments in this way, you are guaranteed that they are locally optimal. These collections of locally optimal fragments usually approximate the global alignment. However, it is also possible to end up with short, high-scoring segments that do not necessarily superimpose within the global alignments. Hence, to be guaranteed of the best global alignment, you should use NW.

These classical alignment methods are good at finding even remote homologies, but they are both computationally too slow for large searches. In the following section, we describe the BLAST method (Basic Local Alignment Search Tool), which is much faster and is particularly useful for large searches when there are relatively high sequence similarities. The strategy in BLAST is to compare new (unknown) sequences (*queries*) against all sequences in a database (*targets*) and to find target sequences that closely match the query sequence.

BLAST Uses a Query Sequence to Search a Database for Related Sequences

BLAST is an essential software tool of bioinformatics. You give it your query sequence as input, and it searches large databases to find similar sequences. You can use BLAST to help answer questions such as what organisms have similar sequences, where did a sequence originate evolutionarily, and in what other proteins might you find similar folding motifs. BLAST, which is available online, is a collection of different tools for different purposes [7]. For example, protein blast **`blastp`** compares an amino acid query sequence against a protein sequence database, while **`blastn`** compares nucleic acid sequences against a DNA/RNA database.

BLAST is a local-alignment tool. It detects homologous sequences, not by comparing a sequence in its entirety, but rather by identifying short matches, also called "high-scoring segment pairs" (HSPs). An HSP consists of two sequence fragments (also called "words"), of arbitrarily chosen but equal length, for which the alignment score exceeds a threshold or cutoff score T. For example, suppose that your *query sequence* contains the sequence of letters AIRGF. For **`blastp`** the default "word" size is three letters. In this example, BLAST would search a database of *target sequences* for the words AIR, IRG, or RGF. BLAST locates all common three-letter words between the query sequence and the target sequences in the database. These results are then used to build an alignment. Unlike the SW algorithm, BLAST is heuristic, so it doesn't guarantee the best possible match. The advantage of SW is that it can find remote homologs, where the query has little similarity to target sequences. But BLAST has the advantages of being faster and using less computer memory, so it's better for searching large genomic databases, where global alignment or exhaustive search approaches (such as NW) would simply be too slow.

Here's how you use BLAST (see also Appendix 8A). BLAST starts with your query sequence and breaks it into k-mer parts. This gives you a library of k-mer query words that are shorter than your full query sequence. Now, BLAST uses the query words to search the database (in a process called *scanning*) to find matches. It scores the matches between the query list of k-mer words and the database of k-mer words. These matches form *hits*, which are seeds for growing the sequences (called *extensions*) to see if BLAST can get a longer match between the query word and the database word. BLAST only keeps the query words having the highest scores, that is, those with a score above some threshold value T that you have set. The number of hits depends on the choice of the *threshold score* T for identifying hits. A higher T value yields greater speed, but also an increased probability of missing weak similarities. Thus, there is a tradeoff between speed and sensitivity depending on the choice of T.

Aligning Multiple Sequences Gives More Insight than Aligning Two Sequences

Suppose you have a query protein and you have many sequences of related proteins. You can align all the sequences at the same time. Such *multiple sequence alignments* (MSAs) can be made using methods such as *PSI-BLAST* or *CLUSTAL*, which are available from webservers [8]. MSAs can be useful for extracting information in the twilight zone (see Figure 8.3), because multiple sequences can sometimes average out the noise from single-sequence comparisons. MSAs also give statistical data on sequence positions that are evolutionarily conserved or correlated. PSI-BLAST (Position-Specific-Iterated BLAST) is an extension of BLAST that performs efficient MSAs with gaps. When a BLAST search produces few matches, try PSI-BLAST because it may uncover additional matches. PSI-BLAST generates an "on-the-fly" sequence profile iteratively as a scoring matrix from the specific BLAST search, and continues to generate this matrix upon each reiteration. This is useful for detecting distant sequence homologs.

CLUSTAL aligns three or more sequences efficiently, implementing the NW global alignment algorithm. CLUSTAL is effective as long as the sequences do not contain extensive internal repeat motifs and when the

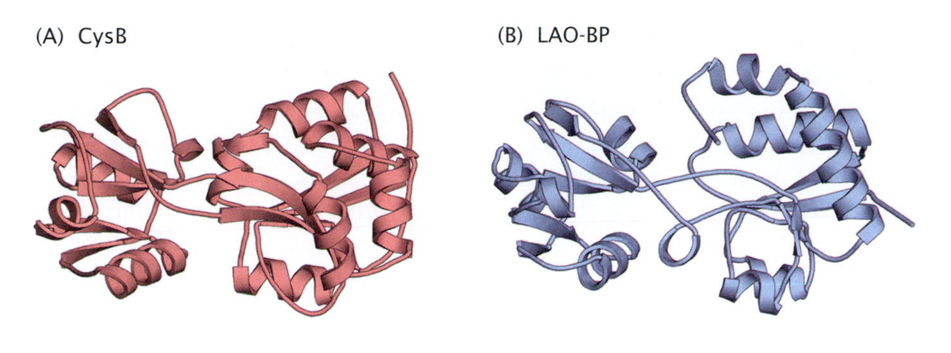

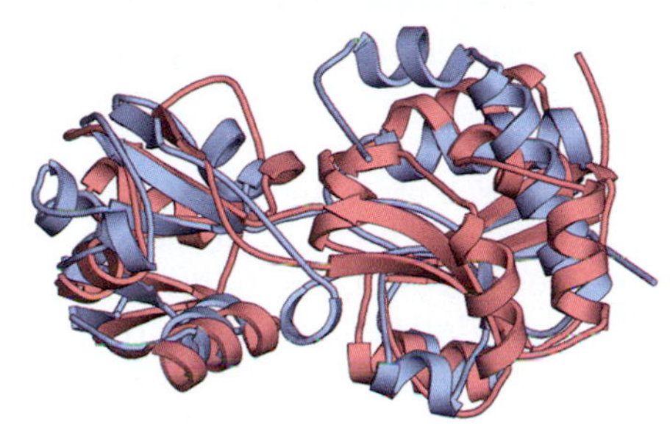

Figure 8.12 Two proteins may have similar native structures, even if they have little sequence identity. The cofactor-binding fragment of CysB (A) and lysine–arginine–ornithine binding protein (LAO-BP) (B) have similar structures, but very little sequence identity (there are only 11 identical residues in common). Superimposing their native structures (C) can help determine how to align their sequences.

sequences are related over their entire length. PSIBLAST works better for sequences where different parts align to different proteins.

You Can Improve Sequence Alignments by Structure Matching

Sometimes two proteins have similar native structures even though they don't have much sequence similarity (Figure 8.12). In such cases, you can superimpose their structures to help you correctly align their sequences. Next, we show how you can use pairwise sequence comparisons to construct evolutionary trees of the relatedness among sets of proteins.

HOW DO YOU CONSTRUCT A PHYLOGENETIC TREE?

Once you have aligned sequences and scored the alignments, you have the basic measure of protein-to-protein distance you need to construct an evolutionary tree (also called a *phylogenetic tree*). An evolutionary tree is simply a graphical representation of a collection of pairwise *evolutionary distances*. There are different methods for constructing phylogenetic trees. Each relies on different assumptions. As noted earlier, we usually first assume parsimony, namely, that sequences changes followed the shortest paths consistent with the observed sequences.

There are two types of trees: *rooted* and *unrooted*. In a rooted tree, the links are all directed arrows pointing from ancestors to descendants, fixing a single root node as the common ancestor. The link lengths in rooted trees can indicate the evolutionary time elapsed from ancestor to descendant. In an unrooted tree, the links are not directed, and no assumption is made about the ancestry or common ancestors. In general, phylogenetic trees have importance both in basic genetics for understanding events such as gene duplications and also for practical problems such as understanding the origins of diseases. The availability of large-scale gene sequencing makes many things possible now.

An example was the determination that West Nile virus in the Western Hemisphere originated in New York and New England.

In Box 8.2, we give an example with the Unweighted Pair Group Method with Arithmetic Mean (UPGMA). This method makes the assumption that there is a constant rate of evolution. It is one of the simplest methods for generating a rooted tree. It produces clusters of sequences connected through nodes. This is one of the earliest ways developed to generate phylogenetic trees, by Sokal and Michener in 1959. It is a general way to obtain a relationship diagram for any set of pairwise data.

Box 8.2 Here's One Way to Generate a Distance-Based Rooted Tree

Let's find the evolutionary tree for a collection of five sequences, which we label *A–E*. We don't need the actual sequences here; we just need to know the distances between them. We will suppose that some distance measure has been defined and that the distances between pairs of sequences are given by the matrix in Figure 8.13A. Note the symmetry: the distance of *A* from *B* is the same as *B* from *A*. So, the matrix is symmetric, and we keep only the part above the diagonal.

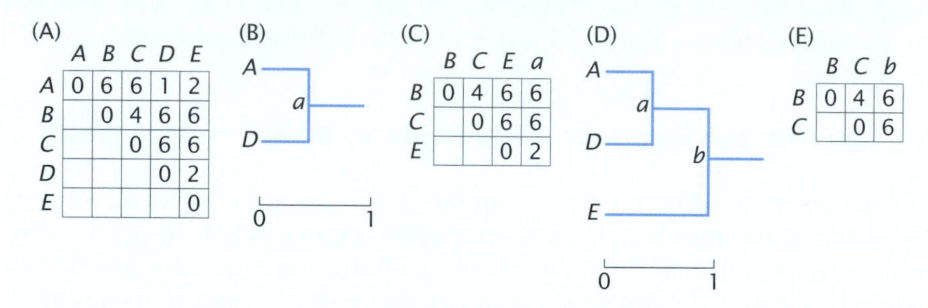

Figure 8.13 First steps in a UPGMA phylogenetic tree generation. (A) Pair distances for 5 sequences. (B) and (C) Choosing the closest pair of nodes to create a new node. This procedure is continued in the next figure.

We begin by finding the pair of sequences that are most similar to each other. In this case, the most similar pair is *A* and *D* (because its corresponding matrix element is the smallest when comparing to all the other off-diagonal terms). The distance between *A* and *D* is 1. Now insert a new node *a* between *A* and *D* having an equal distance of 0.5 to each. Figure 8.13B shows a horizontal axis, on which we mark distances. Distances are taken to be only along the horizontal lines, not the vertical lines. So, the distance from *A* to *D* is $0.5 + 0.5 = 1$, for the sum of the lengths of the horizontal lines.

Now, remove sequences *A* and *D* from the distance matrix (see Figure 8.13C) and include instead the midpoint node *a* located between them. Calculate new distances between the remaining sequences and the node. Note that the distance of *a* from each of the remaining sequences (*B*, *C*, and *E*) is simply the average of the distances of *A* and *D* from *B*, *C*, and *E*.

Within our new matrix, you search again to find the next closest pair. You see that the closest pair of nodes is now *a* and *E*. The distance between nodes *a* and *E* is 2. So, again add a new node *b* between *a* and *E*. The distance between *E* and node *b* is 1. So, this leads to Figure 8.13D.

Now, the shortest remaining distance, which is 4, is between *B* and *C*, and leads to the matrix in Figure 8.13E and the tree in Figure 8.14A.

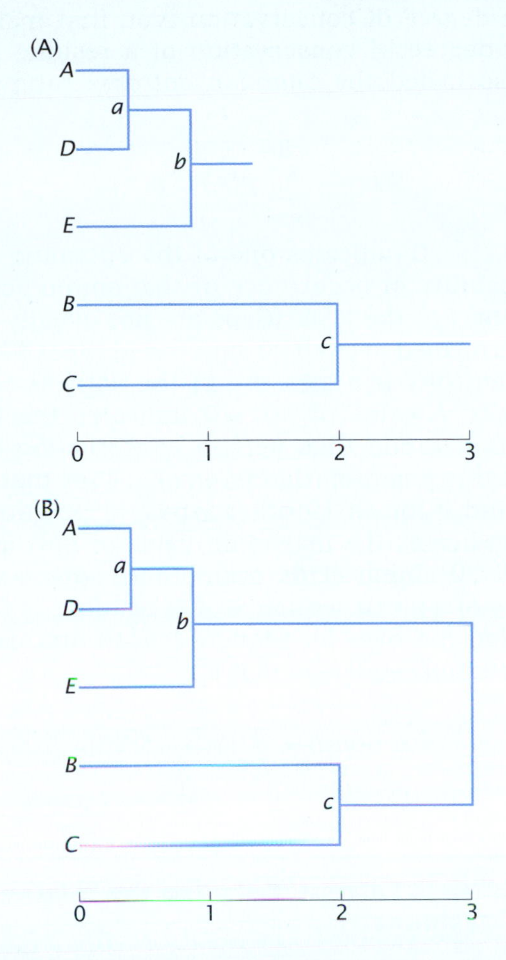

Figure 8.14 Continuation of the phylogetic tree generation from Figure 8.13. (A) New node *c* is placed to connect nodes *B* and *C*. (B) The final phylogenetic tree generated from the pair data in Figure 8.13A.

Finally, you connect the two unconnected branches of the tree. To do this, look at the original distance matrix. We see that the distances between either *B* or *C* and *A*, *D*, or *E* are all 6. So that means that you should place the vertical connecting bar at *x*-axis value 3, yielding the final rooted tree in Figure 8.14B.

You can check that all the distances in this tree are fully consistent with our starting distance matrix (see Figure 8.13).

EVOLUTION CONSERVES SOME AMINO ACIDS AND CHANGES OTHERS

You Can Express the Degree of Residue Conservation Using *Sequence Entropy*

As a protein evolves over time, some of its residues will be more *conserved* than others. An amino acid is conserved if it does not change throughout the evolution of that protein. For example, in a collection that includes a protein and its relatives, if sequence position 26 is always a glycine, it's called conserved. In a protein, amino acid positions that are highly conserved can indicate sites that are important for the action or stability or binding or assembly properties of the protein.

To compute the degree of conservation, you first make an MSA. One measure of the degree of conservation of a residue is the *sequence entropy* $S(i)$ (also called the *Shannon entropy*) for position i in the sequence:

$$S(i) = -\sum_{x_i=1}^{20} p(x_i) \ln p(x_i), \tag{8.3}$$

where $x = 1, 2, 3, \ldots, 20$ indicates one of the 20 amino acid types, and $p(x_i)$ is the probability of occurrence of that amino acid type x at the particular position i of the MSA. (Gaps are not usually included.) This summation is performed over all 20 amino acid types for each sequence position i. The entropy is a measure of the flatness of a distribution function (Box 8.3). A value of $S(i) = 0$ indicates the lowest possible entropy at position i, and thus perfect conservation of a given type of amino acid at this position (that is, $p(x_i) = 1$ for that particular type of amino acid, and 0 for all 19 other types, at position i). A value of $S(i) = \ln 20 \approx 3$ indicates the maximum value of entropy for that site, meaning that all 20 amino acids occur there with equal probability. The total entropy of a given sequence of N residues is found by simply summing Equation 8.3 over all sequence positions, and normalizing relative to the random sequence, that is,

$$S(\text{protein}) = \sum_{i=1}^{N} [S(i) - S^0(i)]. \tag{8.4}$$

Box 8.3 The Sequence Entropy Tells You the "Flatness" of the Distribution of Residues

As shown in Figure 8.15A, if all 20 amino acids occur with equal probability (1/20) at a given protein site, then that position has the maximum possible sequence entropy, $S = 3$. If, as shown in Figure 8.15B, the 20 amino acids occur with different frequencies, the sequence entropy will be smaller, indicating a site that is "more ordered," or "more conserved."

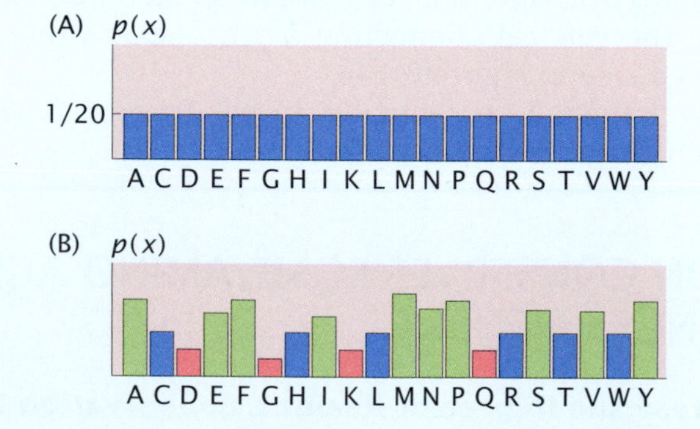

Figure 8.15 Amino acids occur at different frequencies at a given protein site. Higher frequencies are in *green*, average frequencies in *blue*, and lower frequencies in *red*. (A) The flat distribution of frequencies for all amino acids having equal numbers of occurrences. (B) Residue-specific distribution of frequencies of occurrence.

The sum of the terms $S^0(i)$ is equal to $3N$ when the random sequence is defined as having equal probability for any amino acid (that is, $S^0(i) = 3$) at each position i. Alternatively, you can consider the natural occurence probabilities of different types of amino acids as "prior" probabilities to evaluate S^0.

Physical and Biological Factors Affect the Evolutionary Conservation of Amino Acids

Sequence conservation can give various insights. First, it can indicate the role an amino acid plays in a protein's stability or biological action. Figure 8.16 shows that glycine, cysteine, histidine, and tryptophan are among the most conserved residues [9]. This conservation can be justified on a physical basis: Cys is required for disulfide-bridge formation; His is often required for its role in catalysis; Gly can be important for regions of the protein that require extra flexibility or backbone freedom in dihedral angles to make turns; and Trp can be essential where binding to large flat hydrophobic surfaces is required. Hydrophobic residues that usually occupy core regions tend to be conserved. Charged residues with bulky side chains, such as glutamate or lysine, tend to be at protein surfaces and are less conserved.

Second, Figure 8.16 also shows that conserved residues in a protein structure are often located at sites that have the smallest fluctuations, and therefore are the most conformationally constrained. This point is also evident from Figure 8.17A, which shows that amino acids that are in densely packed regions of a protein tend to be conserved. In addition, Figure 8.17B shows that the most conserved residues generally happen to be the most buried residues, irrespective of whether they are polar or nonpolar.

Third, amino acids are sometimes conserved because of the protein's biological mechanism. Sites that are directly involved in specific activities (for example, ligand binding or catalysis) are critical, and cannot be mutated without harming the functioning of the organism. Figure 8.18

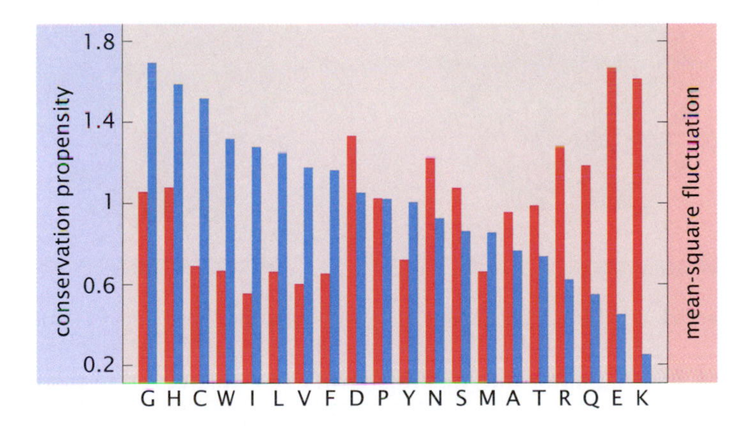

Figure 8.16 Residues that are restricted in their native motions tend to be evolutionarily conserved. (*Blue*) The degree of conservation, based on sequence entropy. (*Red*) The residue motions, as determined by their *mean-square fluctuations*, obtained from their PDB structures. The less motion the amino acid undergoes in the protein's native structure, the greater is its evolutionary conservation. Restriction can arise from being buried in a protein core, or from intrinsic side-chain freedom or other factors. (Adapted from Y Liu and I Bahar. *Mol Biol Evol*, 29:2253–2263, 2012. With permission from Oxford University Press.)

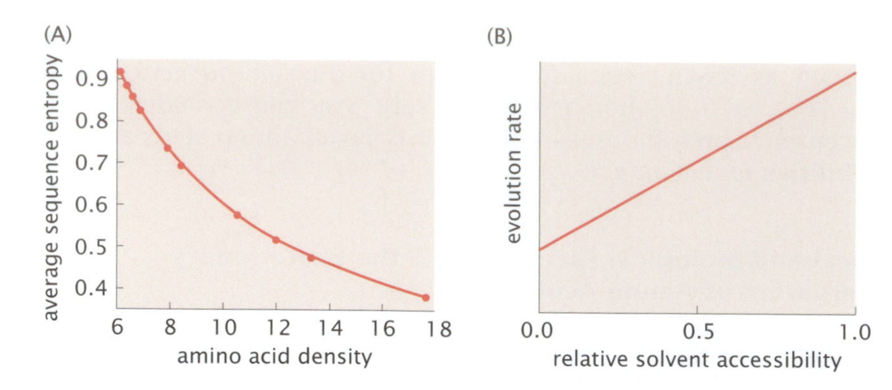

Figure 8.17 Evolution tends to conserve amino acids that are in densely packed and/or buried core regions in a protein. (A) Sequence entropy decreases with increasing amino acid density (such as residue burial in the core). (B) Residues that are more solvent accessible tend to evolve faster, hence are less conserved. (A, from H Liao, W Yeh, D Chiang, et al. *Protein Eng Des Sel*, 18(2):59–64, 2005. With permission from Oxford University Press; B, adapted from EA Franzosa and Y Xia. *Mol Biol Evol*, 26:2387–2395, 2009. With permission from Oxford University Press.)

shows an example: several sites on a DNA glycosylase protein that are found to be highly conserved throughout evolution also happen to be sites that are most important for the protein's biological function, namely, binding to DNA [10]. A webserver called ConSurf allows you to compute and visualize conservation throughout protein structures.

Evolutionary conservation can give you insight into how fast evolutionary processes happen. Appendix 8B gives a Markov model for how fast individual sequence positions change throughout evolution. More conserved sites tend to evolve more slowly than less conserved sites.

Evolutionary Variations Are Sometimes Correlated in the Sequence

Some amino acids *coevolve* with others in the protein. Suppose you make MSAs of a given type of protein from different organisms. Maybe you observe that typically when the 20th residue in the sequence changes, the 46th residue also changes (Figure 8.19). These are called

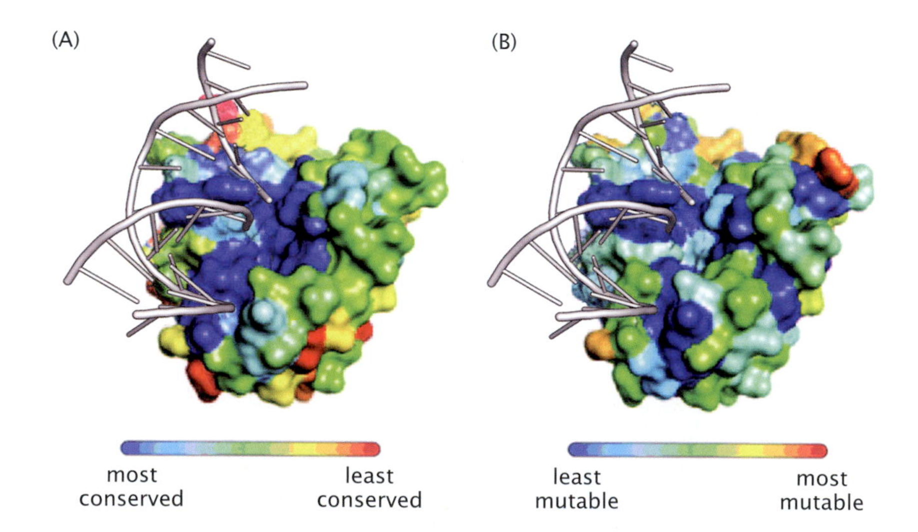

Figure 8.18 Binding and catalytic sites tend to be conserved in evolution. (A) Amino acids in the DNA glycosylase structure bound to DNA (*gray*). The coloring shows evolutionary conservation of the residues (*blue* are most conserved, *red* are most variable). (B) Residue coloring shows the amount of damage that mutagenesis causes to the glycosylase function. The similarity of coloring between (A) and (B) illustrates that evolution conserves sites that are essential to this protein's function. (Adapted from C Pál, B Papp, and MJ Lercher. *Nat Rev Genet*, 7:337–348, 2006; H Guo, J Choe, and L Loeb. *Proc Natl Acad Sci USA*, 101:9205–9210, 2004. Copyright (2004) National Academy of Sciences, USA. Images were generated using ConSurf: H Ashkenazy, E Erez, E Martz, et al. *Nucleic Acids Res*, 38(Web Server Issue):W529–33, 2010.)

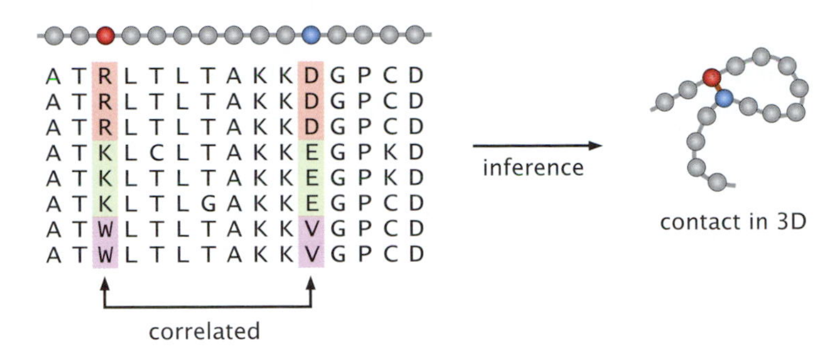

Figure 8.19 Correlated mutations can give information about protein structures. The MSA on the left indicates two sites (columns) that are correlated. When the left residue is *R*, the right residue is *D*, for example. The alignment shows that when one of these positions is changed, there is a compensating change at the other position. Such coevolution patterns can indicate that those positions are neighbors in a protein native structure. This is useful for predicting amino acid contact pairs in protein structures. (Adapted from DS Marks, LJ Colwell, R Sheridan, et al. *PLoS One* 6:e28766, 2011.)

correlated mutations, or *compensatory mutations*. Correlated mutations can give insights into the structure or function of your protein. Two amino acids that are distant in the sequence may fold up to be close to each other and to directly interact in the native structure. They may be packed together in the protein's native structure in such a way that when one residue is replaced with a smaller amino acid, the other residue must be replaced by a larger one, constrained by geometry. And residues that act together in enzyme catalysis are usually conserved. Figure 8.20 shows a network of correlated mutations in a protein that are related to its mechanism.

Mutual Information (MI) Measures the Tendencies of Pairs of Amino Acids to Coevolve

You can measure the degree to which mutations are correlated by using the *mutual information (MI)*, which is a close relative of sequence entropy (Equation 8.3). The mutual information between the ith and jth columns in an MSA is defined as

$$I(i,j) = \sum_{x_i=1}^{21} \sum_{y_j=1}^{21} p(x_i, y_j) \log\left[\frac{p(x_i, y_j)}{p(x_i)p(y_j)}\right], \tag{8.5}$$

where $p(x_i, y_j)$ designates the joint probability of observing both amino acid type x at position i and type y at position j. Here gaps are treated as the 21st amino acid type. High values of $I(i,j) = p(x_i, y_j) \gg p(x_i)p(y_j)$ indicate substitutions that are more correlated. $I(i,j) = 0$ indicates no correlation, or independence; then the joint probability is $p(x_i, y_j) = p(x_i)p(y_j)$. The matrix $I(i,j)$ as a function of residue indices i and j is called a *coevolution map*. A variant of this approach is called *statistical coupling analysis* (SCA) [11]. Because of the many thousands of genome sequences now available, correlation analysis is major way to draw inferences about structure and function (see Chapter 11).

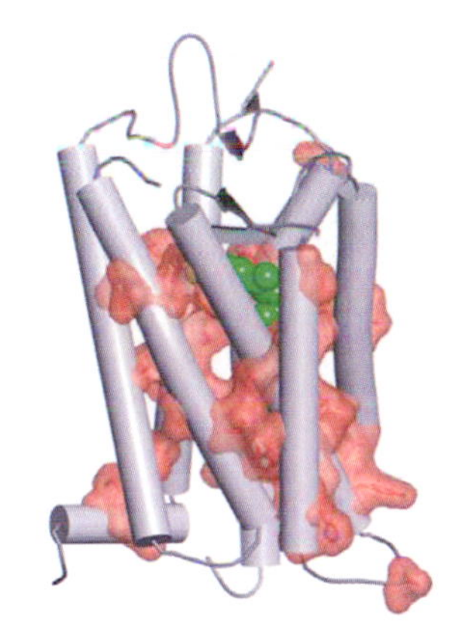

Figure 8.20 The network of correlated mutations in a protein. The *red* regions in bovine rhodopsin describe a connected network of 47 residues, as revealed by correlated-mutation analysis of residue pairs. This network happens to transmit information through the protein. The *retinol* ligand (*green*) triggers signaling through the network when it binds to rhodopsin. (Adapted from GM Süel, SW Lockless, MA Wall, and R Ranganathan. *Nat Struct Biol*, 10:59–69, 2003. Reprinted by permission from Macmillan Publishers Ltd.)

SUMMARY

With bioinformatics methods, you can align two or more sequences with each other and assign scores reflecting their sequence similarities.

If you have a query sequence, you can search large databases to find other similar sequences. You can discover whether certain sequence positions retain one particular type of amino acid more often than others, that is, whether they are *conserved* through evolution. You can also discover whether changes in one sequence position are correlated with changes in another sequence position. These methods are useful for drawing inferences about structure, function, and evolution in proteins. And large-scale sequence comparisons can be used to construct evolutionary trees of life.

APPENDIX 8A: EXAMPLE OF A BLAST RUN

To run the BLAST algorithm, you must first make some choices.

What server? To run **`blastp`**, you can choose one of the public servers NCBI, EMBL-EBI, or ExPASy.

Inputting your sequence. You input your query protein sequence in *FASTA format*. For example, for sperm whale myoglobin (PDB code: 1MBN), the sequence in FASTA format is

```
>1MBN:A-PDBID-CHAIN|SEQUENCE
VLSEGEWQLVLHVWAKVEADVAGHGQDILIRLF KSHPETLEKFDRFKHLKTEAEMKASE
DLKKHGVTVLTALGAILKKKGHHEAELKPLAQSHATKHKI PIKYLEFISEAIIHVLHSRHP
GDFGADAQGA-MNKALELFRKDIAAKYKELGYQG
```

What database do you want to search? There are several protein databases that can be searched by **`blastp`**, including GenBank translations, RefSeq proteins, PDB, PIR, and UniProtKB/Swissprot.

What organisms are of interest? The organisms of interest can be selected or excluded.

Parameter choices. Parameters include how many sequences to return, whether to make adjustments for short sequences, the expectation of a random match, the word size, the choice of what amino acid similarity matrix to use, the gap penalties, and ways to filter or mask certain types of sequences to prevent them from distorting the results.

Interpreting BLAST scores. The output is a set (hit list) of sequence alignments listed in order from lowest to highest expectation (E) value, along with a graphic display color-coded by match quality. E provides a measure of the probability that the output (sequence match) would occur randomly, and thus the lower the E value, the more specific/significant is the match. Significant cases will generally be those having values $E < 10^{-5}$ ($E = 10^{-3}$ would be borderline; $E = 10^{-10}$ indicates a good match).

Figure 8.A.1 shows an example of output from BLAST.

APPENDIX 8B: ESTIMATING EVOLUTIONARY RATES USING A MARKOV MODEL FOR RESIDUE SUBSTITUTIONS

How much time will it take for evolution acting on a sequence to lead to a given distance between two sequences? First, align the two sequences and find the distance between them. Next, make a dynamical model for how fast each amino acid can change from its starting state in one

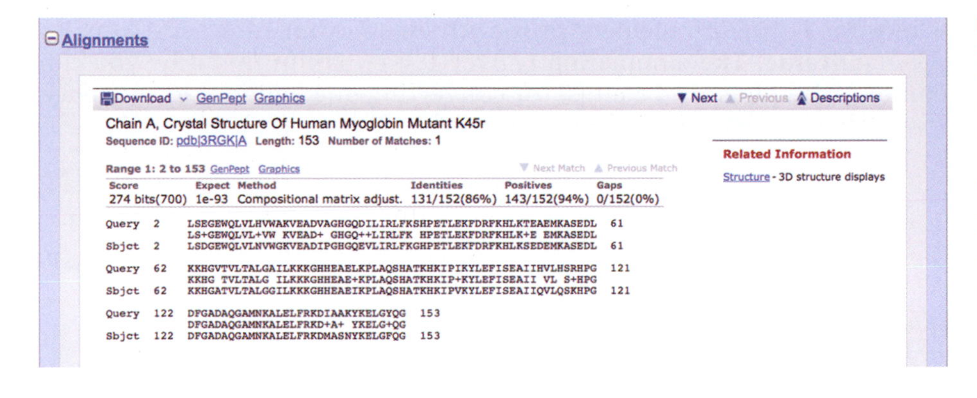

Figure 8.A.1 Example output from BLAST. The query sequence is recombinant human mutant K45R myoglobin (PDB code: 3RGK), and the hit/subject is seal myoglobin (PDB code: 3RGK) in this case. Human myoglobin starts at sequence position 1 and seal myoglobin at position 2, with the middle line showing points of identity or difference. The match is highly nonrandom, indicated by $E = 1 \times 10^{-93}$. In this case, out of the $N = 153$ residues in the sequence, 131 are identical, and similar residues are found for 143 positions. The *middle* line shows the identical residues, and similar residues are marked with +. There is a blank if the residues are dissimilar. (Adapted from M Goujon, H McWilliam, W Li, et al. *Nucleic Acids Res*, 38:W695–699, 2010. With permission from Oxford University Press.)

sequence to its ending state in the other sequence. Using the Markov model described below, you can find the evolutionary time required to reach one sequence from the other.

Here are the details. Suppose you have s sequences, each containing n amino acids. Let h $(1 \leq h \leq n)$ indicate a particular site in the sequence. Let $\mathbf{P}_h$ be the 20-element vector, where each element represents the probability that each particular amino acid occurs at position h. Next, we assume that sequence changes occur according to a Markovian process for the time-dependent substitution of residues [12]. Then you can use a master equation formalism (described in Appendix 6A). To do this, define a transition rate matrix, $\mathbf{W}$ *for each site*. For simplicity, we will omit the site index h in the following, and replace $\mathbf{P}_h$ by $\mathbf{P}$. $\mathbf{W}$ is a 20×20 matrix in which element W_{ab} is the probability that a residue of type b is replaced by a residue of type a, in a single time step Δt, and is a constant over time. You need two more pieces of information to complete the transition matrix $\mathbf{W}$. First, you set the diagonal elements of $\mathbf{W}$ to be

$$W_{aa} = -\sum_{b, b \neq a} W_{ab}, \tag{8.B.1}$$

so the sum of each column equals zero. Second, the off-diagonal terms must satisfy the principle of *detailed balance*:

$$W_{ab} p_b^0 = W_{ba} p_a^0, \tag{8.B.2}$$

where p_a^0 and p_b^0 are the equilibrium probabilities of occurrence of amino acid type a and b, respectively.

The change in the vector $\mathbf{P}$ over time will be given by solving

$$\frac{d\mathbf{P}}{dt} = \mathbf{W}\mathbf{P}(t). \tag{8.B.3}$$

Our goal is to compute the transition (or conditional probability) matrix $\mathbf{M}(t)$ for each site, where the element M_{ab} is defined as the probability of a transition to a residue of type a, given that the residue is type b t time $t = 0$. As described in Appendix 6A, the time-dependent transition probability matrix is given by

$$\mathbf{M}(t) = e^{\mathbf{W}t}, \tag{8.B.4}$$

which says that the matrix elements are given as a sum of exponentials

$$M_{ab}(t) = \sum_j U_{aj} e^{-\lambda_j t} [\mathbf{U}^{-1}]_{jb}, \tag{8.B.5}$$

where U_{aj} is the ath element of the jth eigenvector of $\mathbf{W}$, and λ_j is its jth eigenvalue. The summation is over $1 \leq j \leq 20$. In this way, you can compute the rates (and routes) for converting one amino acid sequence to another at each sequence position h.

REFERENCES

[1] DA Benson, I Karsch-Mizrachi, DJ Lipman, et al. GenBank. *Nucleic Acids Res*, 36:D25–D30, 2008.

[2] The UniProt Consortium. UniProt: a hub for protein information. *Nucleic Acids Res*, 43:D204–D212, 2015.

[3] I Sillitoe, TE Lewis, AL Cuff, et al. CATH: Comprehensive structural and functional annotations for genome sequences. *Nucleic Acids Res*, 43:D376–D381, 2015.

[4] S Henikoff and JG Henikoff. Amino acid substitution matrices from protein blocks. *Proc Natl Acad Sci USA*, 89:10915–10919, 1992.

[5] SB Needleman and CD Wunsch. A general method applicable to the search for similarities in the amino acid sequence of two proteins. *J Mol Biol*, 48:443–453, 1970.

[6] TF Smith and MS Waterman. Identification of common molecular subsequences. *J Mol Biol*, 147:195–197, 1981.

[7] SF Altschul, TL Madden, AA Schäffer, et al. Gapped BLAST and PSI-BLAST: A new generation of protein database search programs. *Nucleic Acids Res*, 25:3389–3402, 1997.

[8] M Goujon, H McWilliam, W Li, et al. A new bioinformatics analysis tools framework at EMBL–EBI. *Nucleic Acids Res*, 38:W695–W699, 2010.

[9] Y Liu and I Bahar. Sequence evolution correlates with structural dynamics. *Mol Biol Evol*, 29:2253–2263, 2012.

[10] C Pál, B Papp, and MJ Lercher. An integrated view of protein evolution. *Nat Rev Genet*, 7:337–348, 2006.

[11] SW Lockless and R Ranganathan. Evolutionarily conserved pathways of energetic connectivity in protein families. *Science*, 286:295–299, 1999.

[12] J Felsenstein. Evolutionary trees from DNA sequences: A maximum likelihood approach. *J Mol Evol*, 17:368–376, 1981.

SUGGESTED READING

Baer CF, Miyamoto MM, and Denver DR, Mutation rate variation in multicellular eukaryotes: causes and consequences. *Nat Rev Genet*, 8:619–631, 2007.

Balaji S and Srinivasan N, Use of a database of structural alignments and phylogenetic trees in investigating the relationship between sequence and structural variability among homologous proteins. *Protein Eng*, 14:219–226, 2001.

de Juan D, Pazos F, and Valencia A, Emerging methods in protein co-evolution. *Nat Rev Genet*, 14:249–261, 2013.

Durbin R, Eddy SR, Krogh A, and Mitchison G, *Biological Sequence Analysis: Probabilistic Models of Proteins and Nucleic Acids*. Cambridge University Press, Cambridge, 1998.

Hopf TA, Colwell LJ, Sheridan R, et al. Three-dimensional structures of membrane proteins from genomic sequencing. *Cell*, 149:1607–1621, 2012.

Marks DS, Hopf TA, and Sander C, Protein structure prediction from sequence variation. *Nat Biotechnol*, 30:1072–1080, 2012.

Mount D, *Bioinformatics: Sequence and Genome Analysis*, 2nd ed. Cold Spring Harbor Laboratory Press, Cold Spring Harbor, NY, 2004.

Zvelebil M and Baum JO, *Understanding Bioinformatics*. Garland Science, New York, 2008.

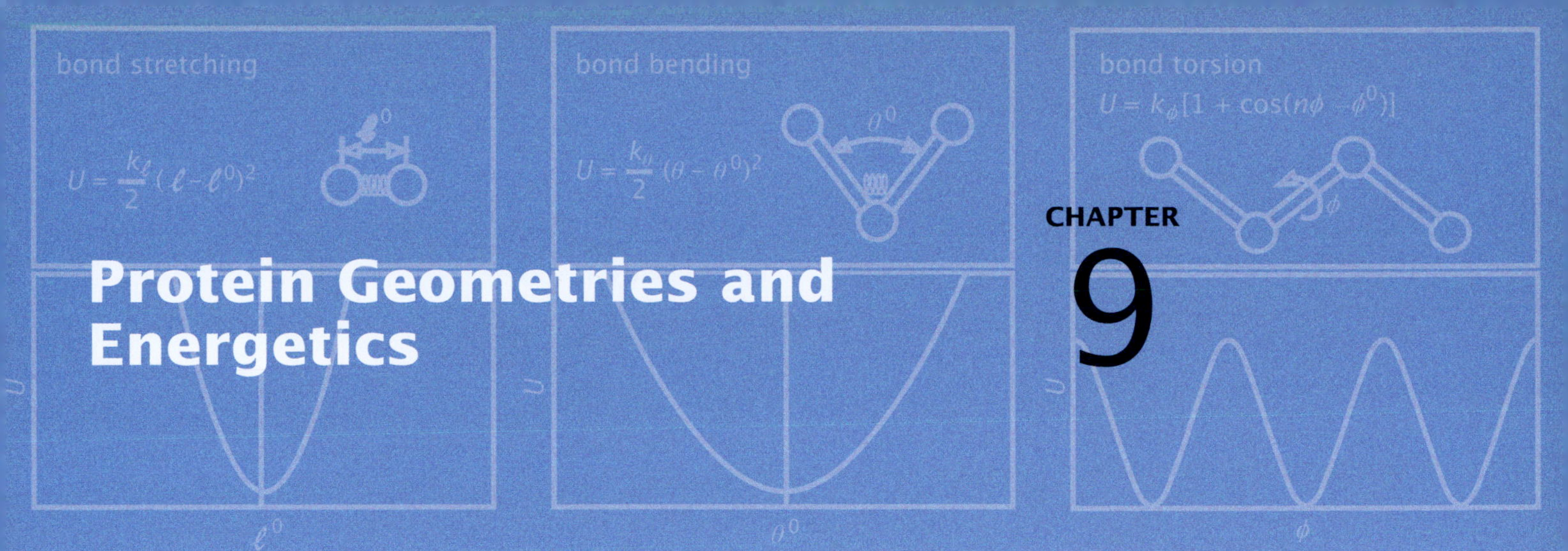

CHAPTER 9

Protein Geometries and Energetics

The structures and actions of proteins derive from their molecular geometries and energies. Physics-based computer simulations can give insights that you cannot get from experiments alone. They fill in the gaps in experimental knowledge. They help you paint the narratives that constitute our understanding of structures, dynamics, and mechanisms in protein science. They harness our knowledge of chemistry and physics to describe—Ångström-by-Ångström and nanosecond-by-nanosecond—the conformational motions, folding routes, binding processes, ensemble distributions, mechanical deformations, collective dynamics, catalytic mechanisms, allosteric changes, and biological mechanisms of proteins. Computer modeling is also used to refine protein structures in conjunction with experiments, and to create visualizations and movies.

For computer modeling of physical structures and actions, you first need a description of protein geometries and interatomic energies. And you need methods for sampling protein conformations and dynamics. We begin with how to describe a protein's molecular structure in the computer.

YOU CAN REPRESENT A PROTEIN STRUCTURE BY ITS ATOMIC COORDINATES

A protein is a chain of interacting *units*, $i = 1, 2, 3, \ldots$, which you may take to be individual atoms or individual amino acids, for example, depending on your modeling objective. Each unit i is represented by its *position vector* $\mathbf{r}_i$ given by the following equation relative to an origin $(0, 0, 0)$:

$$\mathbf{r}_i = \begin{bmatrix} x_i \\ y_i \\ z_i \end{bmatrix} = (x_i, y_i, z_i). \tag{9.1}$$

These are called *Cartesian coordinates*. Any particular set of position vectors for all units of the chain is called a *conformation* or

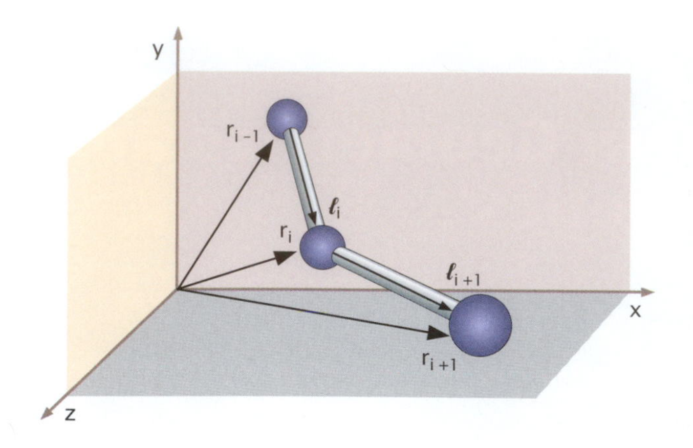

Figure 9.1 A structure can be expressed in *Cartesian coordinates* or *internal coordinates*. In Cartesian coordinates, the position of each unit i (atom or residue, for example) is represented by the vector $\mathbf{r}_i = (x_i, y_i, z_i)$ of its spatial coordinates relative to the origin of a fixed laboratory frame. In internal coordinates, the position of each unit i is represented by the vector ℓ_i relative to the position of the preceding unit $(i-1)$ in the chain.

configuration.[1] Consider a molecule composed of N units.[2] You can define a protein conformation using its set of *position vectors* $\{\mathbf{r}_1, \mathbf{r}_2, \ldots, \mathbf{r}_{N-1}, \mathbf{r}_N\}$ of all the structural units with respect to a given laboratory-fixed reference frame (x, y, z), as shown in Figure 9.1. Coordinates like these for large numbers of proteins are available in the Protein Data Bank (Box 9.1).

Box 9.1 The Protein Data Bank (PDB) Contains the Coordinates of Structurally Resolved Proteins

The PDB, which was created in 1971 [1], is the repository for the atomic coordinates of biomolecular structures. It contains well over 100,000 structures, including proteins unbound, or complexed with other proteins, or with nucleotide chains or small molecules (for example, inhibitors, ligands, ions, water, lipids, or carbohydrates). In the PDB, a protein structure is represented by the Cartesian coordinates of all the experimentally resolved atoms. Each structure has an identification label (ID) and coordinate data in the PDB's standard format.

A piece of a PDB file is shown in Table 9.1 for hemoglobin (PDB ID 1BBB). Hemoglobin is a tetramer. The four subunits are designated as A, B, C, and D (see the "Chain identifier" column). This table shows a section of the PDB file for the amino acids Thr8 and Asn9 of chain A.

Each row in the PDB file shown in Table 9.1 includes the identification of the atom, identification of the amino acid, its subunit identifier, its Cartesian coordinates, its ocupancy fraction, and the X-ray crystallographic B-factors (also called *Debye–Waller* or *temperature factors*). The labels CA, CB, CG, and OD1, refer to C^{α}, C^{β}, C^{γ}, and $O^{\delta 1}$ atoms, where the Roman capital letter is always substituted for the corresponding Greek letter. The B-factor $B_i = (8\pi^2/3)\langle(\Delta r_i)^2\rangle$ is proportional to the mean-square fluctuations of atom i, $\langle(\Delta r_i)^2\rangle$, around the reported mean atomic positions (x_i, y_i, z_i).

PDB files also contain literature references and information about the sequence, secondary structures, hydrogen bonds, disulfide

[1] In this book, we use these two terms interchangeably.

[2] Typically, the hydrogen atom coordinates are not included explicitly, because they are not usually identified by X-ray diffraction, except in very high-resolution structures. Hydrogens are often lumped together with the coordinates of the heavy atoms to which they are attached. In some cases of poor resolution, only the C^{α} atoms are reported.

bridges, the experimental technique (X-ray, NMR, or cryo-EM) used to determine the structure, the crystal space group (for X-ray structures), the number of models (for NMR structures), coordinates of immobilized water molecules, and coordinates of ions and other ligands. Some coordinates may be missing (undetermined, for example, because of large motions or disorder). PDB files provide measures of how well the reported PDB model structure fits the experimental data, and how well it conforms to standard protein characteristics. They also provide information on the sequence position of secondary structures and disulfide bridges. The PDB also contains information on the *biological assembly*, that is, on the biologically functional form (for example, multimeric) of the protein. And the pdb gives translation vectors and rotation matrices for generating symmetrically related subunits in homo-multimeric structures in biological assemblies.

You can visualize the structures specified in the PDB coordinate files using computer-graphics software packages, including PyMOL© [2], Chimera© [3], and VMD© [4], or with JMol©, JSMol©, Simple Viewer®, Protein Workshop, and Ligand Explorer, which are available on the PDB website.

Table 9.1 An example of a PDB file. This shows the information in a PDB file of a part of a protein, including the types and serial numbers of atoms and amino acids, and their atomic coordinates, in units of Ångströms.

Atom record	Atom number	Atom identifier	Amino type	Chain identifier	Residue sequence number	*x* (Å)	*y* (Å)	*z* (Å)	Occupancy	*B*-factor	Element symbol
ATOM	51	N	THR	A	8	15.001	20.876	5.954	1.00	17.71	N
ATOM	52	CA	THR	A	8	16.404	21.205	6.205	1.00	14.67	C
ATOM	53	C	THR	A	8	17.307	20.026	5.883	1.00	15.81	C
ATOM	54	O	THR	A	8	18.345	20.155	5.208	1.00	15.33	O
ATOM	55	CB	THR	A	8	16.606	21.642	7.733	1.00	22.28	C
ATOM	56	OG1	THR	A	8	15.827	22.861	7.909	1.00	30.84	O
ATOM	57	CG2	THR	A	8	18.073	21.891	8.080	1.00	30.82	C
ATOM	58	N	ASN	A	9	16.890	18.864	6.372	1.00	16.01	N
ATOM	59	CA	ASN	A	9	17.599	17.586	6.163	1.00	13.49	C
ATOM	60	C	ASN	A	9	17.735	17.248	4.681	1.00	13.18	C
ATOM	61	O	ASN	A	9	18.793	16.840	4.230	1.00	15.71	O
ATOM	62	CB	ASN	A	9	16.889	16.476	6.923	1.00	12.85	C

As an alternative to Cartesian coordinates, you can represent proteins using *internal coordinates* $\{\mathbf{r}_1, \boldsymbol{\ell}_2, \boldsymbol{\ell}_3, \ldots, \boldsymbol{\ell}_N\}$, which describe the chain configurations by their internal bond vectors $\boldsymbol{\ell}_i$, pointing from atom $i-1$ to atom i (see Figure 9.1), the bond angles, and torsion angles. The length of each bond vector is

$$\ell_i = |\boldsymbol{\ell}_i| = |\mathbf{r}_i - \mathbf{r}_{i-1}|. \tag{9.2}$$

These two coordinate systems contain exactly the same information. However, internal coordinates are often preferred because they are independent of the orientation and position of the structure in space. You often don't need the six degrees of freedom of the absolute location and orientation of the protein in space. With internal coordinates, you can focus on conformations and conformational changes. If you use internal coordinates, you express the positions of all of the atoms

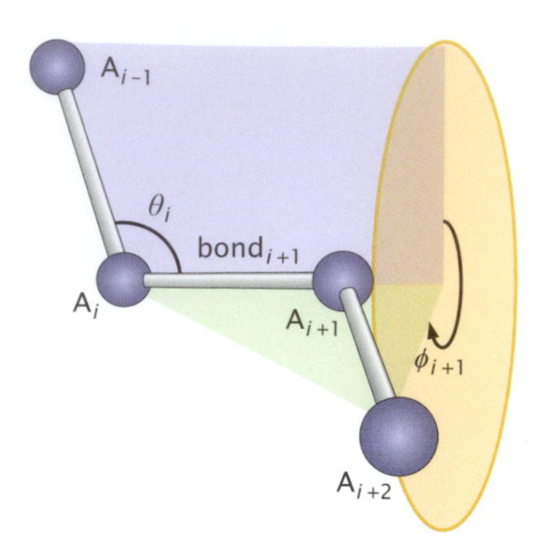

Figure 9.2 Defining the *bond angle* and the *torsion angle*, also called the *dihedral angle*. A dihedral angle depends on the spatial position of four atoms. It is the angle formed by the intersecting planes of successive groups of three atoms. The torsion angle ϕ_{i+1} around bond $i+1$ changes the distance of atom A_{i+2} relative to atom A_{i-1}. The angle θ_i is the bond angle formed by atoms A_{i-1}, A_i, and A_{i+1}, and it affects the distance between atoms A_{i-1} and A_{i+1}.

in terms of their components: their $N-1$ bond lengths $|\ell_i|$, their $N-2$ *bond angles* θ_i, and their $N-3$ *rotational* or *dihedral angles* ϕ_i (Figure 9.2). If you want an even further reduction in the number of degrees of freedom, you can use only the $N-3$ dihedral angles, since bond lengths and bond angles are relatively stiff degrees of freedom.

Box 9.2 shows how to convert from Cartesian to internal coordinates, and Box 9.3 gives an example. You will need this transformation when you manipulate PDB structures, which are all given in Cartesian coordinates. Appendix 9A shows the reverse: how to generate Cartesian coordinates from internal coordinates.

Box 9.2 Here's How to Convert from Cartesian to Internal Coordinates

To convert Cartesian coordinates $\{\mathbf{r}_1, \mathbf{r}_2, \ldots, \mathbf{r}_N\}$ to internal coordinates, first compute the *bond lengths* (for pairs of covalently linked atoms) using Equation 9.2:

$$\ell_i = |\mathbf{r}_i - \mathbf{r}_{i-1}|. \tag{9.3}$$

Then, compute the *bond angles* (from pairs of bonds, or triplets of points) from the dot product of the bond vectors:

$$\theta_i = \theta_i(\mathbf{r}_{i-1}, \mathbf{r}_i, \mathbf{r}_{i+1}) = \cos^{-1}\left[\frac{\boldsymbol{\ell}_i \cdot \boldsymbol{\ell}_{i+1}}{\ell_i \ell_{i+1}}\right]. \tag{9.4}$$

Finally, compute the *torsion angles* ϕ_i of the ith bond. A torsion angle is the angle between the plane containing ℓ_{i-1}, ℓ_i and the plane containing ℓ_i, ℓ_{i+1} (see Figure 9.2). The angle between these two planes is the angle between the normals (which are the perpendicular unit vectors) to the two planes. The normal to the plane defined by the vectors $\boldsymbol{\ell}_i$ and $\boldsymbol{\ell}_{i+1}$ is defined as

$$\mathbf{n}_i = \frac{\boldsymbol{\ell}_i \times \boldsymbol{\ell}_{i+1}}{|\boldsymbol{\ell}_i \times \boldsymbol{\ell}_{i+1}|}. \tag{9.5}$$

A similar expression applies to $\mathbf{n}_{i-1}$. You use the *right-hand rule* to determine the direction of the normal vector: you curl the fingers of your right hand from the first vector of the cross product toward the second (after translating one of them to superimpose their origin), and then your thumb is pointing in the direction of the normal vector.

To compute the dihedral angle, take the dot product between these two normal vectors, in the same way as for finding θ_i in Equation 9.4, that is, $\cos\phi_i = \mathbf{n}_i \cdot \mathbf{n}_{i-1}$. However, in contrast to θ, which varies over the range $0° < \theta < 180°$, the dihedral angle ϕ varies over the full range $[-180°, 180°]$.[3] Note that the inverse cosine gives you an angle, but not its sign. You can determine the sign of this angle by using the sign and the sine function:

$$\text{sign}(\sin x) = \begin{cases} - & \text{if } -180° < x < 0°, \\ + & \text{if } 0° < x \leq 180° \end{cases}$$

$$\Rightarrow \quad \phi_i = \text{sign}(\sin(\mathbf{n}_{i-1} \cdot \mathbf{n}_i)) \cos^{-1}(\mathbf{n}_{i-1} \cdot \mathbf{n}_i). \tag{9.6}$$

[3] A torsion angle of $\pm 180°$ defines the *trans* state (see Figure 9.2).

Box 9.3 Example of Conversion from Cartesian to Internal Coordinates

Here's an example. Consider the two-dimensional configuration of four atoms shown in Figure 9.3.

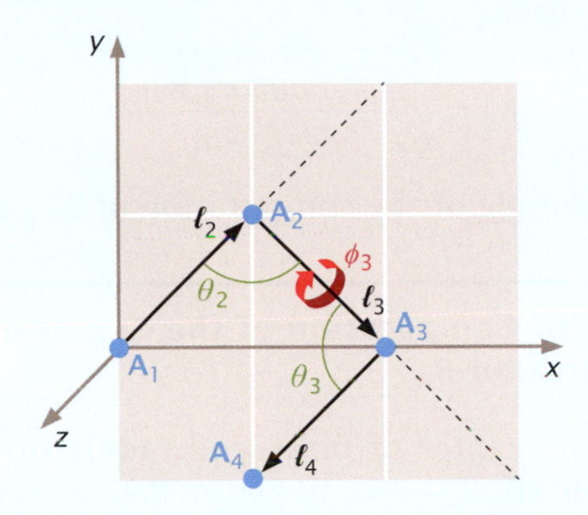

Figure 9.3 Example for converting coordinates. The four atoms are located at $\mathbf{r}_1 = (x_1, y_1, z_1) = (0, 0, 0)$, $\mathbf{r}_2 = (1, 1, 0)$, $\mathbf{r}_3 = (2, 0, 0)$, and $\mathbf{r}_4 = (1, -1, 0)$.

First, compute the bond vectors and their lengths:

$$\boldsymbol{\ell}_2 = \mathbf{r}_2 - \mathbf{r}_1 = \begin{bmatrix} 1 \\ 1 \\ 0 \end{bmatrix}, \quad \boldsymbol{\ell}_3 = \mathbf{r}_3 - \mathbf{r}_2 = \begin{bmatrix} 1 \\ -1 \\ 0 \end{bmatrix}, \quad \text{and} \quad \boldsymbol{\ell}_4 = \mathbf{r}_4 - \mathbf{r}_3 = \begin{bmatrix} -1 \\ -1 \\ 0 \end{bmatrix}.$$

The bond lengths are $\ell_2 = \ell_3 = \ell_4 = \sqrt{2}$.

Then, compute the bond angles:

The bond angle between $\boldsymbol{\ell}_2$ and $\boldsymbol{\ell}_3$ is 90°, as you can see from the following:

$$\cos\theta_2 = \frac{\boldsymbol{\ell}_2 \cdot \boldsymbol{\ell}_3}{\ell_2 \ell_3} = 0 \Rightarrow \theta_2 = 90^\circ.$$

Bond angles are in the range $0 \leq \theta \leq 180^\circ$. Likewise, the bond angle between $\boldsymbol{\ell}_3$ and $\boldsymbol{\ell}_4$ is also 90°:

$$\cos\theta_3 = \frac{\boldsymbol{\ell}_3 \cdot \boldsymbol{\ell}_4}{\ell_3 \ell_4} = 0 \Rightarrow \theta_3 = 90^\circ.$$

Then, compute the torsion angles:

Using Equation 9.5, you see that the normal vectors at positions A_2 and A_3 are parallel and both point inward with respect to the plane of the paper. Also, $\cos(\mathbf{n}_2 \cdot \mathbf{n}_3) = 1$, which implies that $\phi_3 = 0$ (called the *cis* form). This example shows how to compute internal coordinates if you are given the Cartesian coordinates.

From the Coordinates, You Can Compute the Radius of Gyration

A key property of any polymer molecule is its *radius of gyration* R_g. R_g, which is a general property of any set of points in space, is a characteristic measure of size or compactness. It can be measured in X-ray or neutron scattering or other biophysical experiments. To calculate R_g, first calculate the position of the *center of mass* $\mathbf{R}_{\text{cm}}$, defined by the vector

$$\mathbf{R}_{\text{cm}} = \frac{\sum_{i=1}^{N} m_i \mathbf{r}_i}{\sum_{i=1}^{N} m_i}. \tag{9.7}$$

The radius of gyration is the mass-weighted average distance of atoms from the center of mass:[4]

$$R_g = \sqrt{\frac{\sum_{i=1}^{N} m_i (\mathbf{r}_i - \mathbf{R}_{\text{cm}})^2}{\sum_{i=1}^{N} m_i}}. \tag{9.8}$$

Box 9.4 gives an example of the computation of $\mathbf{R}_{\text{cm}}$ and R_g.

Box 9.4 Example of Computation of the Center of Mass $\mathbf{R}_{\text{cm}}$ and the Radius of Gyration R_g

Let's compute the center of mass and radius of gyration for the three-atom configuration shown in Figure 9.4. To keep the math simple, consider a two-dimensional structure and assume all the atoms have the same mass m.

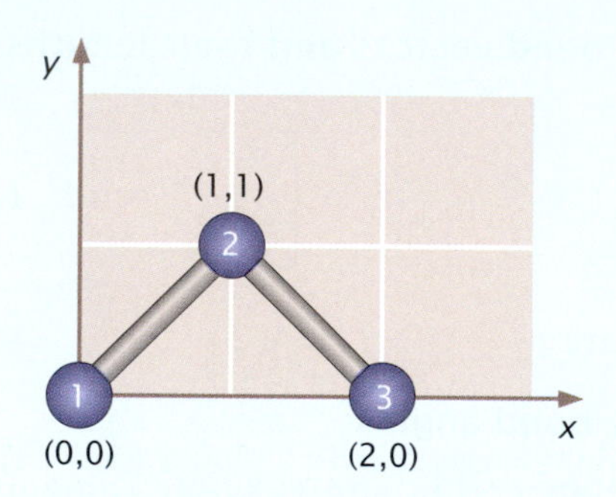

Figure 9.4 Example for computing $\mathbf{R}_{\text{cm}}$ and $\mathbf{R}_g$. A configuration of three atoms located at $\mathbf{r}_1 = (0, 0)$, $\mathbf{r}_2 = (1, 1)$, and $\mathbf{r}_3 = (2, 0)$.

To find $\mathbf{R}_{\text{cm}}$, which is the position vector for the center of mass of this molecule, take the average of the position vectors according to Equation 9.7:

$$\mathbf{R}_{\text{cm}} = \frac{1}{3}(\mathbf{r}_1 + \mathbf{r}_2 + \mathbf{r}_3) = \frac{1}{3}\left(\begin{bmatrix}0\\0\end{bmatrix} + \begin{bmatrix}1\\1\end{bmatrix} + \begin{bmatrix}2\\0\end{bmatrix}\right) = \begin{bmatrix}1\\\frac{1}{3}\end{bmatrix}.$$

Now subtract $\mathbf{R}_{\text{cm}}$ from the position vectors $\mathbf{r}_i$ to get the transformed coordinates (with respect to the center of mass): $\left(-1, -\frac{1}{3}\right)$, $\left(0, \frac{2}{3}\right)$, and $\left(1, -\frac{1}{3}\right)$.

Now, to compute R_g, use Equation 9.8:

$$R_g = \sqrt{\frac{1}{3}\left(\frac{10}{9} + \frac{4}{9} + \frac{10}{9}\right)} = \sqrt{\frac{8}{9}} = \frac{2\sqrt{2}}{3}.$$

[4]The radius of gyration is also related to internal distances within the structure, through the relation $R_g^2 = (2N)^{-2} \sum_{i,j} (\mathbf{r}_j - \mathbf{r}_i)^2$ for identical masses.

How Similar Are Two Protein Structures? Compute the RMSD between Them

Often, you want a single-number measure of how similar or different one protein structure is to another. This information can help you draw inferences about whether proteins are evolutionarily related, whether they perform similar functions, whether they act by similar physical or chemical mechanisms, or whether they bind to the same drugs or metabolites.

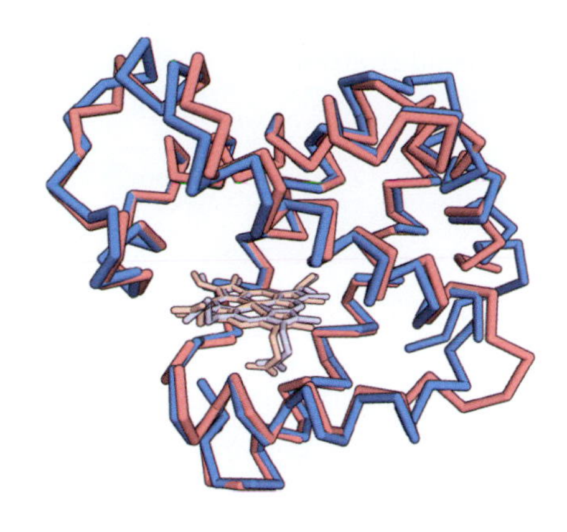

Figure 9.5 Illustrating two proteins that are structurally aligned. This shows two nearly identical myoglobin molecules (*blue*, 1MGN; *light red*, 2NRL), that are structurally aligned. This optimal superposition illuminates the similarities and differences. The differences are small, and are mainly in the loops.

To compare two structures A and B, you first align them and then compute the *root-mean-square deviation* (RMSD) between their Cartesian coordinates. To do this, represent the position vector of the ith atom of protein structure A as $\mathbf{r}_{Ai}$ and the position vector of the ith atom of structure B as $\mathbf{r}_{Bi}$ and repeat this for all atoms ($1 \leq i \leq N$). Now, you must decide which atoms on A correspond to which atoms on B, then make those assignments. Next, find the rigid-body translation and rotation of molecule B that causes the atoms of B to superimpose optimally onto the corresponding atoms of A (Figure 9.5). Details are given in Appendix 9B. Now, designate the resulting (optimally rotated and translated) position vector of the ith atom in B as $\mathbf{s}_{Bi}$. Then, the RMSD between the two structures is defined as

$$\text{RMSD} = \sqrt{\frac{1}{N}\sum_{i=1}^{N}(\mathbf{r}_{Ai} - \mathbf{s}_{Bi}) \cdot (\mathbf{r}_{Ai} - \mathbf{s}_{Bi})}. \tag{9.9}$$

To compute an RMSD, you must use the same number of points in both structures in the same coordinate system; note that Equation 9.9 is in Cartesian coordinates.

TO SIMULATE PROTEIN PHYSICS ON A COMPUTER, YOU NEED A MODEL OF INTERATOMIC ENERGIES

In this and Chapter 10, we describe how to model the physical properties of proteins in the computer. To do this, we need a model for the interatomic interactions among the different possible conformations of the chain. We describe such models as follows.

Molecular Energetics Can Be Described by Atomistic Force Fields

In principle, the starting point for modeling molecular energies is quantum mechanics (QM). However, while QM modeling has been used to study metalloproteins, optical properties (such as the transduction of light in the eye), and enzyme catalysis, it is not practical for most protein modeling. Proteins and their changes are too large and complex for costly QM computations. Instead, the best current alternative—and most common practice—is to use *semi-empirical force fields*. These are widely available in software packages such as AMBER© [5], CHARMM© [6], GROMACS© [7], and NAMD© [8]. In these models, interatomic interactions are treated using a classical mechanics approximation to compute the energy U of a particular chain conformation, as a function of the internal coordinates and the distances r_{ij} between all atoms i and j in the system, of a form such as shown in Figure 9.6.

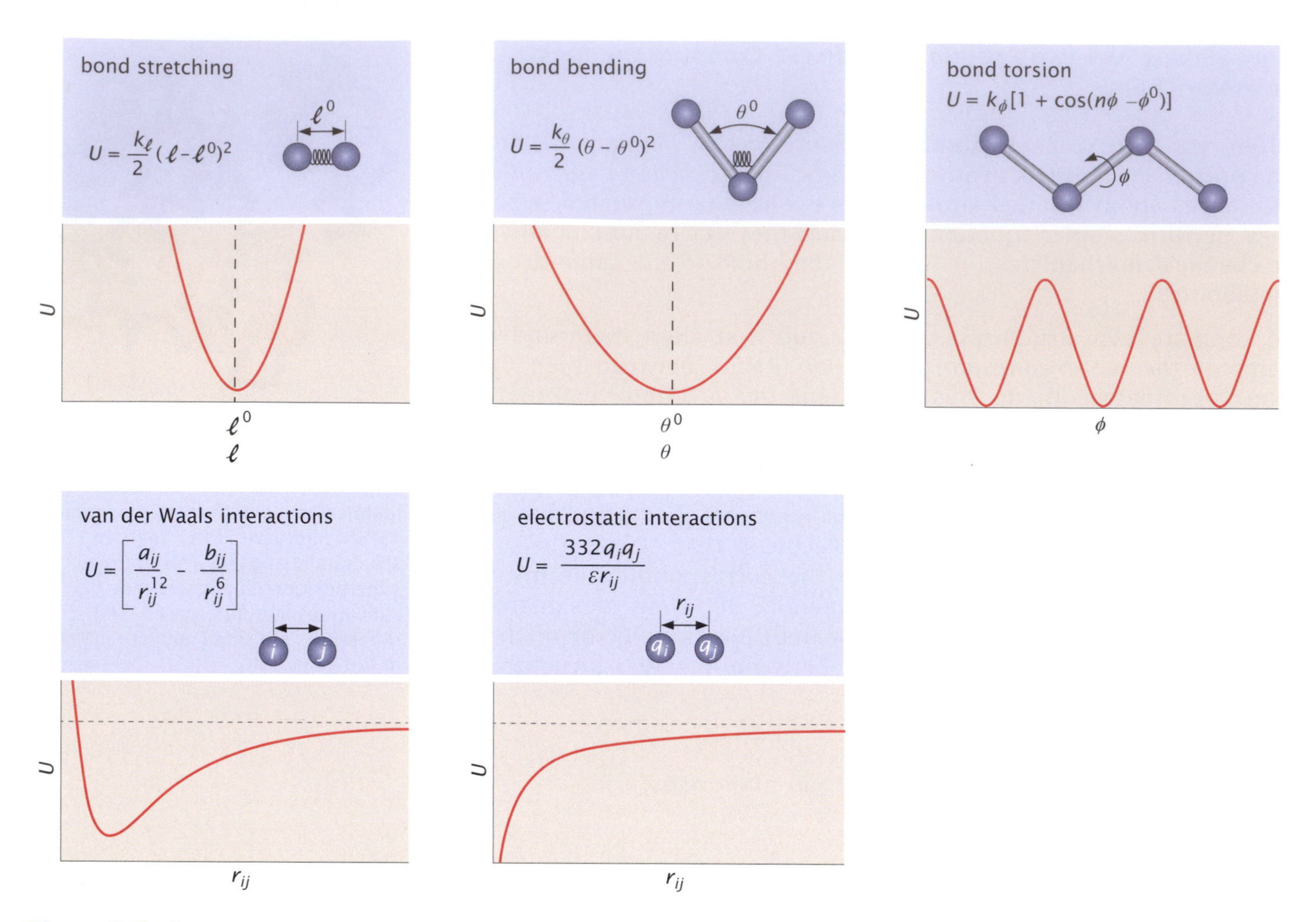

Figure 9.6 The energy components in force fields. Equation 9.10 contains terms for bond stretching, bond angle bending, bond torsions, van der Waals interactions, and electrostatic interactions (shown here as an attraction between opposite charges).

Most current software packages use the following form for the **force field:**

$$U(\mathbf{r}_1, \mathbf{r}_2, \ldots, \mathbf{r}_N) = \sum_{i=2}^{N} \frac{k_\ell}{2}(\ell_i - \ell_i^0)^2 + \sum_{i=2}^{N-1} \frac{k_\theta}{2}(\theta_i - \theta_i^0)^2$$
$$+ \sum_{i=3}^{N-1} \left\{ \sum_{n=1}^{3} k_\phi[1 + \cos(n\phi_i - \phi_i^0)] \right\}$$
$$+ \sum_{i,j} \left(\frac{a_{ij}}{r_{ij}^{12}} - \frac{b_{ij}}{r_{ij}^{6}} + \frac{332 q_i q_j}{\varepsilon r_{ij}} \right), \qquad (9.10)$$

where ℓ_i, θ_i, ϕ_i, and r_{ij} are the variables that express the particular conformation of the molecule in internal coordinates. The superscript 0's refer to constants known from experiments, and n is an index over the stable dihedral conformations, a_{ij} is the Lennard-Jones (LJ) repulsion coefficient between atoms of type i and j, b_{ij} is the LJ attraction coefficient, ε is the effective dielectric constant, 332 is a conversion factor that allows you to calculate the electrostatic potential in units of kcal mol^{-1} if you express the charges q_i and q_j as multiples of e, the charge on a proton[5] (for example, +2 in dimensionless units for Ca^{2+}) and if you express distances in Ångströms. These constants have fixed values for different atom types. Equation 9.10 is for a linear chain, but it is simple to extend it to also include side chains.

Bond Lengths and Bond Angles Are Treated Using Spring-Like Forces

Bond-stretching energies. The first term in Equation 9.10 approximates bond stretching energies using a square-law function $(\ell_i - \ell_i^0)^2$ of the instantaneous (that is, fluctuating) length ℓ_i of bond number i along the chain, taken with respect to its equilibrium length ℓ_i^0. The square law captures the physical principle that there is a most probable (that is, most stable) bond length ℓ_i^0, and a variance, for each given type of chemical bond. The chemical bond length in a protein is assumed to equal that of the same type of bond in a small molecule. This term also captures the principle that if the bond is either stretched or compressed, relative to its most probable length, the energy increases.

The square law is the simplest functional form that captures these properties. It corresponds to a force that is linear with displacement, as in Hooke's law for springs. The importance of this functional form is that each type of chemical bond can be characterized using only two parameters, namely, the equilibrium length ℓ_i^0 and the spring constant k_ℓ, both of which are readily obtained from independent experiments on the chemical bond lengths of small molecules. Square-law energy functions are justified not only by their simplicity, but also because they are the lowest-order terms in Taylor series expansions of more complex models of bond stretching.

Bond-angle energies. Following similar logic, *bond-angle bending* energies are also approximated using a square law. The second term in Equation 9.10 captures the principle that a given chemical bond angle type i has a stable preferred angle $\theta = \theta_i^0$, which can be regarded as an *angular spring*, because the energy increases as a function of angular deviations from this stable value (rather than extensions) with force constant k_θ in either bending direction. Note that the bond stretching is much stiffer than bond-angle bending, which is reflected by the fact that $k_\ell \gg k_\theta$.

Torsional-angle energies. Polymer molecules, including proteins, have torsional degrees of freedom. And the energy barriers between those states can be small, which means that the molecules interconvert relatively readily among those states. Torsional angles are therefore referred to as *soft degrees of freedom*. So, the polymer molecules will have a broad distribution of different conformations. These $N - 3$ torsional degrees of freedom are the main ways protein backbones change their conformations. For proteins, both the backbones and most side chains can adopt different torsional states. In simple polymer chains such as hydrocarbons, the backbone has three prominent torsional states. The third term in Equation 9.10 describes how the energy varies with the torsion angle ϕ, also called the *dihedral angle*. To capture these stable torsional states, the torsional energy is taken to be a periodic function (a cosine). Normally for carbon–carbon bond torsions,

[5] The electron charge $= -1e$, where $e = 4.8 \times 10^{-10}$ esu $= 1.6 \times 10^{-19}$ Coulomb.

there are three stable states, so the different minima are indexed by $n = 1, 2, 3$. Torsional energies are fit with the parameter k_ϕ.

These simple spring-law and cosine models originated in the 1930s and 1940s to describe the vibrational spectra of small molecules using a minimal number of adjustable parameters. The nonbonded terms in Equation 9.10 emerged later, in the 1950s and 1960s, to help predict more complicated conformational properties of molecules. Applications of these force-field models to proteins were pioneered by S Lifson, A Warshel, M Levitt, M Karplus, JA McCammon, H Scheraga, and P Kollman.

van der Waals Interactions Are Short-Ranged Attractions and Repulsions

The last summation in Equation 9.10 describes the *nonbonded* interactions among atoms of types i and j that are close together in space and are not covalently bonded. The atom pairs i and j are included in the summations only if they are separated by at least three or four intervening bonds. Two neutral (that is, uncharged) atoms will weakly attract each other from a distance. They will repel when they come too close. The balance of attraction and repulsion leads to a stable "bond" separation (that is, a noncovalent interaction). This type of interaction is captured using the so-called *Lennard-Jones* (LJ) or *van der Waals* or 6–12 potential. The attraction is short-ranged, decreasing as $1/r_{ij}^6$ with increasing interatomic distance. This attraction arises because an atom has an electron cloud that becomes distorted in the presence of another atom, leading to dipoles that attract. The repulsion is even shorter-ranged, $1/r_{ij}^{12}$, increasing sharply at small distances. Repulsion arises from the Pauli exclusion principle, to avoid overlap of the electron clouds, also referred to as *steric clashes*. This nonbonded energy function requires two parameters, a_{ij} and b_{ij}, which are specific for the atom types of the two interacting atoms i and j.

Note that the LJ parameters depend on types of atoms involved but not on their positions in the chain. The term $1/r_{ij}^{12}$ arose in the early days of modeling when computing was more expensive. Simply squaring the $1/r_{ij}^6$ is fast to compute, has few parameters, and was sufficient at the time. More accurate functional forms are known, but they require more parameters. So, most semi-empirical biomolecule force fields still use the simple 6–12 LJ form.

Charge Interactions Are Modeled Using Coulomb's Law

Some atoms are polar or charged. Charge–charge interactions are described by Coulomb's law (the last term in Equation 9.10). It captures the principle that like charges repel each other, and unlike charges attract. For two charges q_i and q_j separated by a distance $r_{ij} = |\mathbf{r}_j - \mathbf{r}_i|$, the interaction energy is $U = 332 q_i q_j / \varepsilon r_{ij}$ if you express lengths in Ångströms, energy in kcal mol^{-1}, and charges in units of e, that is, multiples of the charge on a proton.

ε is the *dielectric constant*; it captures the principle that the electrostatic interactions between two charges q_1 and q_2 are weakened when those charges are put into a polarizable medium, such as water, which has a high dielectric constant. Two charges produce an electric field that can align the water molecules between them, with the result that intervening waters shield the two charges from each other. The dielectric concept is adapted from macroscale electrostatics and used in force fields on the microscale. While it captures the qualitative differences in charge interactions in different media, it is only an empirical approximation for treating the heterogeneity inside a protein. ε is usually taken to be in the range of 1–12 for charges buried in the nonpolar interior of a protein; $\varepsilon = 1$ in vacuum and 78 in bulk water at room temperature. Ions have full charges. Polar atoms have *partial charges* ($|q| \leq 1$), a fraction of the unit charge on an electron. For example, commonly used charges for peptide backbone atoms are $q = -0.35$ for N, $q = +0.25$ for H, $q = +0.10$ for C^{α}, $q = +0.55$ for C (carbonyl), and $q = -0.55$ for O (Figure 9.7). These charges satisfy the requirement for a neutral amino acid that the net charge sums to zero. The values of partial atomic charges are estimated from quantum mechanical modeling of small molecules.

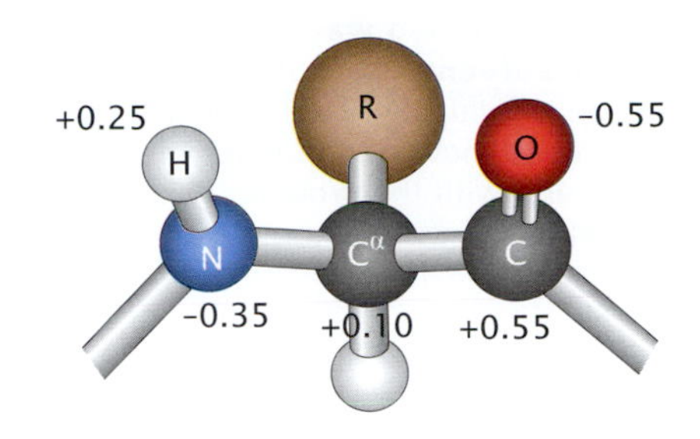

Figure 9.7 The partial charges on the atoms in a peptide unit.

Dipoles. A dipole is an arrangement in which an amount of positive charge is separated by some distance from the same amount of negative charge, for example at the two ends of a bond. The *dipole moment* is the charge on either end of the dipole multiplied by the charge separation. It has units of *Debyes*, D. For example, the dipole moment of an arrangement of two fixed charges in which a single electron and a single proton are separated by 1 Å is

$$\mu = (4.8 \times 10^{-10}\,\text{esu})(10^{-8}\,\text{cm}) = 4.8 \times 10^{-18}\,\text{esu cm} = 4.8\,\text{D}.$$

The peptide group has a dipole moment of 3.5 D. The charges responsible for its dipole moment are shown in Figure 9.7. Water (in the gas phase) has a dipole moment of 1.85 D. Figure 9.8 shows the arrangement of charges in water that gives rise to its dipole moment.

Hydrogen bonds. Hydrogen bonding occurs when a hydrogen-bond donor atom, such as nitrogen, has its hydrogen atom positioned near a hydrogen-bond acceptor atom, such as oxygen, forming the three-atom interaction –N–H···O. Equation 9.10 contains no explicit expression for hydrogen bonds. Force fields typically capture hydrogen bonding as the net electrostatic attraction between the partial charges on the atoms that form hydrogen bonds. This simplification tends toward collinearity of hydrogen bonding, whereas in reality they are sometimes bent.

Solvent Interactions Are a Major Determinant of Protein Conformations

Protein conformational changes are driven not only by a protein's internal atom–atom interactions, but also by the different solvation states of the different conformations. When a protein changes conformation, or binds to a ligand, some of the protein's atoms begin in one medium (say, surrounded by water) and end in a different environment (say, contacting some other part of the protein or the ligand). For example, when a protein chain is unfolded in water, many of its amino acids are solvated by water, but upon folding, those amino acids become buried, away from contact with water. The burial of the hydrophobic amino acids in the protein's core is a major driving force for protein folding. What is the free-energy cost of changing solvation? To account for solvation and desolvation in computational modeling, you typically

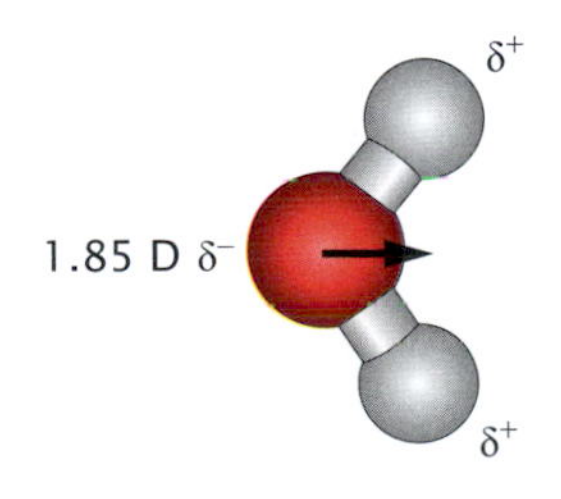

Figure 9.8 The partial charges on a water molecule. Water has partial positive charges on its hydrogens and a partial negative charge on its oxygen. The net charge on water is zero, but water has a dipole moment of 1.85 D (shown by the arrow).

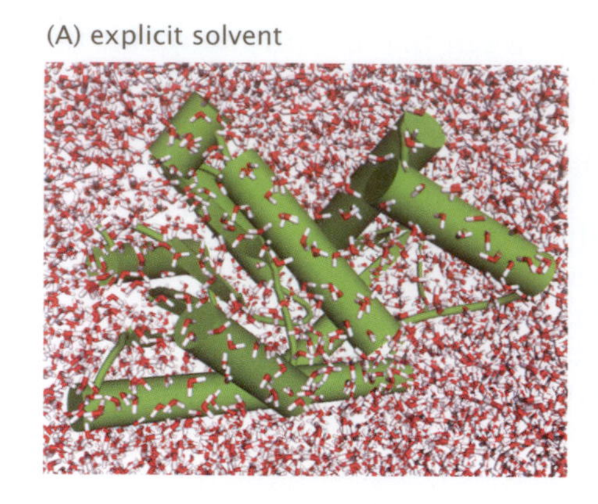

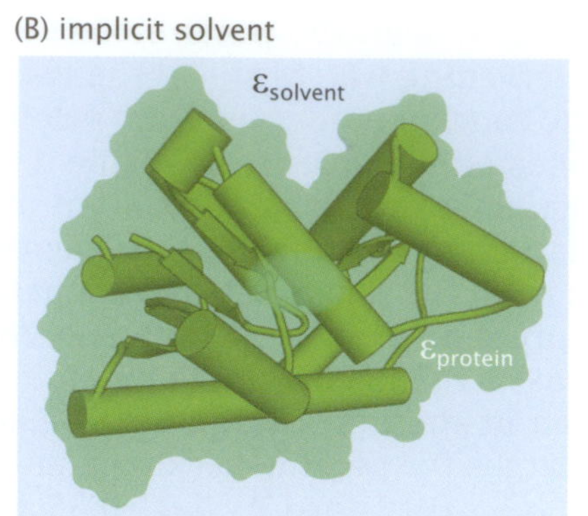

Figure 9.9 Two ways to model the water solvent. (A) Explicit models utilize individual water molecules that are free to move and interact with each other and with the protein. (B) Implicit models suppose that the medium is a continuum having a given dielectric constant.

use either an *explicit-water* or *implicit-water* model, described below (Figure 9.9). In explicit-water modeling, you include atomically detailed water molecules in your simulation. In implicit-water modeling, you treat the solvent surrounding a protein, instead, as a continuum.

Explicit-Water Models Represent Waters as Individual Molecules

Explicit-water modeling treats waters as individual molecules. Popular explicit-water models include SPC (simple point charge), and TIP3P, TIP4P, and TIP5P (transferable intermolecular potentials, where 3P, 4P, and 5P refer to the three, four, and five interaction centers shown in Figure 9.10). In explicit-water models, you represent a protein as being surrounded by thousands of explicit water molecules. Each water molecule interacts both with the atoms of the protein and with other water molecules. Brownian forces cause the waters to jiggle around during a simulation. The number of water atoms far exceeds the number of protein atoms in the simulation box, in order to fully solvate the protein. Explicit-water simulations are regarded as the "gold standard" among semi-empirical models for capturing the physics of solvation as accurately as possible, but they are computationally expensive because of the large number of atom–atom interactions among water atoms and the protein.

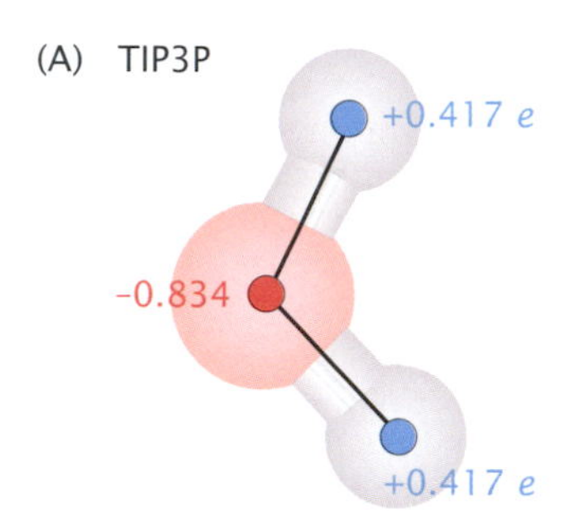

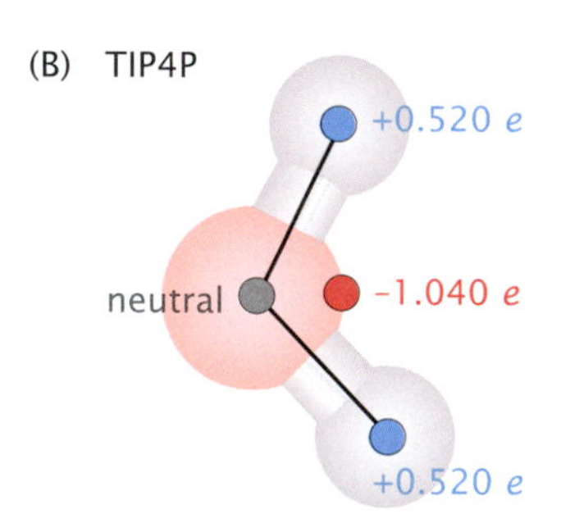

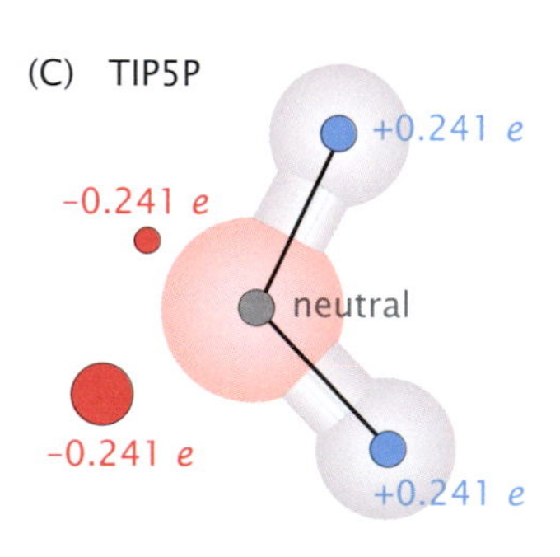

Figure 9.10 Explicit models of water. (A) TIP3P, (B) TIP4P, and (C) TIP5P are explicit models that have three, four, and five point charges, respectively. Negative partial charges are shown in *red*, positive partial charges in *blue*, and neutral points in *gray*. In TIP5P, the two charges on the left are above and below the plane of the water atoms, in symmetric positions.

Implicit-Water Models Treat Water as a Continuous Medium

Implicit-water modeling treats the solvent water as a simple homogeneous continuum (see Figure 9.9), rather than as a sea of individual particulate water molecules. The advantage is that implicit-water computations are fast, so you can study larger molecules or processes. But implicit models sacrifice some accuracy in the physics. To treat the interactions between protein and water, the solvation component in the free energy, $\Delta G_{\text{solvation}}$, has two components:

$$\Delta G_{\text{solvation}} = \Delta G_{\text{nonpolar}} + \Delta G_{\text{electrostatics}}, \tag{9.11}$$

where $\Delta G_{\text{nonpolar}}$ accounts for the short-ranged nonelectrostatic interactions of protein surface atoms and neighboring water molecules, and $\Delta G_{\text{electrostatics}}$ treats the interactions of charges and partial charges on the protein with each other and with the water. $\Delta G_{\text{nonpolar}}$ is often assumed to depend on the surface area A_i of atom type i multiplied by the free energy per unit area of solvating that atom type, γ_i:

$$\Delta G_{\text{nonpolar}} = \sum_i \gamma_i A_i. \tag{9.12}$$

The solvation parameters γ_i are typically derived from experimentally measured free energies of transfer of small molecules from vacuum or oil into water, for example for individual amino acids. The electrostatic component of the solvation free energy in implicit-solvent models is usually treated with either the *generalized Born* or the *Poisson–Boltzmann* models.

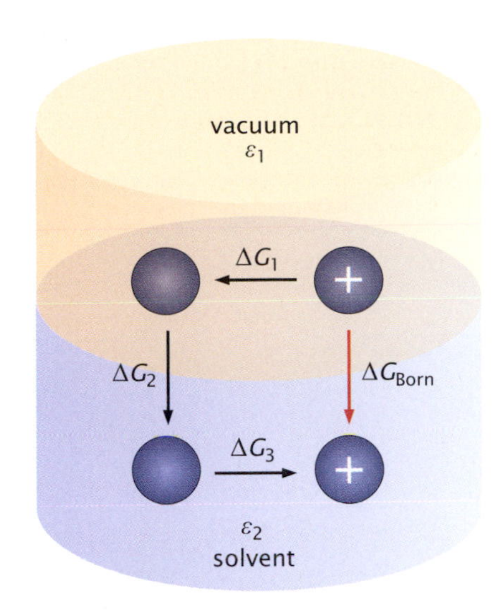

Figure 9.11 Thermodynamic cycle for transferring an ion from one dielectric medium to another. The free energy of transfer of an ion from vacuum to solvent, ΔG_{Born}, is the sum of three terms: ΔG_1 for discharging the charge in vacuum, ΔG_2 for transfering the neutral molecule to the solvent, and ΔG_3 for recharging the ion in the solvent.

The Born model. The *Born model* gives an estimate of the free-energy change ΔG_{Born} that results when a charge moves from one dielectric environment to another. Figure 9.11 shows a sphere of radius a having charge q. The sphere moves from a medium having a dielectric constant ε_1 to a medium having dielectric constant ε_2. The total free energy of transfer is computed as a sum of (1) discharging the charged sphere in its first environment, (2) transferring the neutral sphere from the first to the second medium, and (3) charging up the neutral sphere in the second medium. The Born free energy of transfer is

$$\Delta G_{\text{Born}} = \Delta G_1 + \Delta G_2 + \Delta G_3 = RT\left(\frac{\ell_B q^2}{2a}\right)\left(\frac{1}{\varepsilon_2} - \frac{1}{\varepsilon_1}\right) + \Delta G_2, \tag{9.13}$$

where ℓ_B is the *Bjerrum length* [9].[6] An example of a calculation using the Born model is given in Box 9.5.

Box 9.5 Ions Preferentially Partition More into Water Than into Oil

Let's compute the Born energy for the unfavorable process of transferring an ion from water to oil. To move a sphere of radius $a = 4\,\text{Å}$, having a charge of $q = 1$, from a medium of $\varepsilon_1 = 80$ (water) to $\varepsilon_2 = 4$ (an oil-like medium), Equation 9.13 predicts an increase in free energy of $\Delta G_{\text{Born}} \approx +10\,\text{kcal mol}^{-1}$. One implication of this is that when proteins fold, they will rarely bury charges in their hydrophobic cores, because this free-energy cost is too unfavorable. So, charges are usually on the surfaces of folded soluble proteins, not inside their cores. Another implication is that charged molecules don't readily pass through bilayer membranes, because that would require partitioning into an oil-like environment. Instead, they usually cross membranes through polar channels in membrane proteins.

Now, consider a more general version of this problem in which your molecule is a *collection of multiple spheres*, having charges $q_1, q_2, \ldots$, rather than a single sphere. Again, the molecule is transferred from a medium having dielectric constant ε_1 to a medium having dielectric constant ε_2. In this case, the *generalized Born* (GB) model estimates the electrostatic free energy of transfer as

$$\Delta G_{\text{GB}} = \frac{332}{2}\sum_{i,j}\frac{q_i q_j}{f_{\text{GB}}(r_{ij}, R_i, R_j)}\left(\frac{1}{\varepsilon_2} - \frac{1}{\varepsilon_1}\right), \tag{9.14}$$

[6] $\ell_B = 560\,\text{Å}$ in vacuum at $T = 298\,\text{K}$.

where f_{GB} is a mathematical function of the distances r_{ij} between the ions and their *effective Born radii* R_i and R_j. Charges are in units of e, distances are in Ångströms, and the free energy is in units of kcal mol^{-1}. κ is the Debye–Hückel screening parameter, which depends on the salt concentration in solution. f_{GB} takes different forms in different GB models [10].

The Poisson–Boltzmann Model. The Poisson–Boltzmann (PB) model is an alternative implicit-solvent model for computing approximately the charge–charge interactions, taking into account their dielectric environment. It can also account for the effects of salt ions in the surrounding solution. Here's an overview (for details, see Appendix 9C). A protein molecule in a particular configuration is a three-dimensional constellation of positive and negative charges that is located in a continuum solvent. The solvent may also contain mobile salt ions and counterions that are free to move by diffusion toward or away from the fixed charges on the protein. We want to determine the electrostatic free energy that arises from the sum of each charge interacting with every other charge. This will give you the electrostatic contribution to the conformational free energy. We have already expressed this contribution in Equation 9.10, which sums the charge–charge interactions among the protein atoms. But, for now, we will just focus on this electrostatic contribution alone.

You can compute a three-dimensional contour surface, called the *electrostatic potential* $\Phi(\mathbf{r})$. $\Phi(\mathbf{r})$ looks like a geographic contour map of mountains and valleys. Imagine a hiker at some point on a contour map. A geographic contour map tells you places of high and low gravitational potential. A hiker must perform work to reach a point of high potential, but also can spontaneously fall downhill. The electrostatic potential gives you a similar kind of information, except that it is for electrostatics, not gravitation. The electrostatic potential $\Phi(\mathbf{r})$ is a mathematical surface that describes the energies experienced by mobile charges near a constellation of fixed charges (on the protein in this case). Similar to the mobile hiker on a gravitational potential surface, the high points on the electrostatic potential surface tell you unfavorable places for positive charges (which are favorable places for negative charges). And low points on the electrostatic potential surface tell you favorable places for positive charges (which are unfavorable places for negative charges). In practical terms, you determine $\Phi(\mathbf{r})$ by computing a numerical solution to a partial differential equation, called the *Poisson–Boltzmann equation*, rather than by summing Coulombic interactions, because it is computationally more efficient. Some of the popular software packages that perform grid-based PB calculations for proteins include DelPhi, UHBD, and APBS©. Figure 9.12 shows an example of an electrostatic potential surface around a small protein.

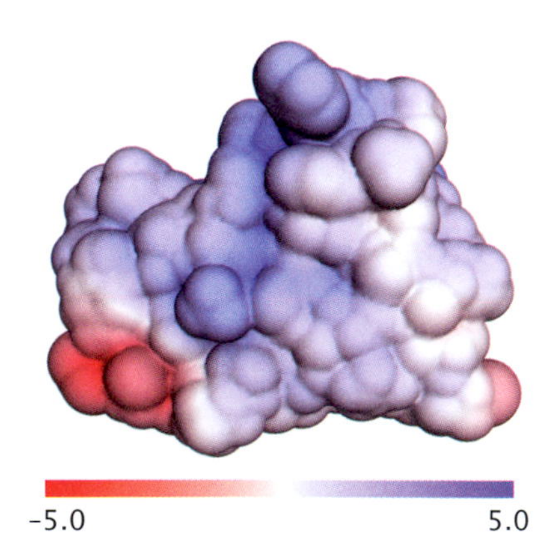

Figure 9.12 The surface of a protein, colored by its electrostatic potential Φ. (*Red*) Negative electrostatic potential (which attracts mobile positive charges). (*Blue*) Positive potential (which attracts negative charges).

Now, to compute the full energy of a particular protein configuration in water, you combine the ingredients of this chapter. A given chain configuration has particular coordinates $\mathbf{r}$. You compute the total intraprotein atom–atom energy $U(\mathbf{r})$ for that configuration using the force-field equation, Equation 9.10. For configuration $\mathbf{r}$, you also compute the nonpolar solvation energy and the electrostatics solvation energy, using either implicit or explicit solvent modeling. You then add the solvation energy to $U(\mathbf{r})$. This gives you the total energy of that chain configuration in water. In Chapter 10, we describe how to simulate a broad range of properties of protein structure, stability, folding, dynamics, and mechanism, by combining such energy modeling with computational methods that systematically step proteins through different possible configurational snapshots.

SUMMARY

To model the structures and physical properties of proteins, you need ways to express and manipulate their atomic coordinates. You also need a model of the interatomic interactions: the bond lengths, angles, and torsions, as well as the van der Waals and electrostatic interactions. In addition, you need to account for how the solvent, typically water, interacts differentially with the different atoms exposed in different chain conformations. The main treatments of water are explicit or implicit models. And the physical quantities they treat are the nonpolar, polar, and charge interactions of the protein with water, as well as—in explicit models—the water–water interactions.

APPENDIX 9A: HOW TO COMPUTE CARTESIAN COORDINATES FROM INTERNAL COORDINATES

Here is how you can convert from internal to Cartesian coordinates. First let's define bond-based local coordinate systems, as in Figure 9.A.1. We use these local frames to transform into the frame appended to the first atom A_1 of the chain. Define the tranformation matrix as

$$\mathbf{T}_i(\theta_i, \phi_i) = \begin{bmatrix} \cos(\pi - \theta_i) & \sin(\pi - \theta_i) & 0 \\ \sin(\pi - \theta_i)\cos(\phi_i - \pi) & -\cos(\pi - \theta_i)\cos(\phi_i - \pi) & \sin(\phi_i - \pi) \\ \sin(\pi - \theta_i)\sin(\phi_i - \pi) & -\cos(\pi - \theta_i)\sin(\phi_i - \pi) & -\cos(\phi_i - \pi) \end{bmatrix}, \tag{9.A.1}$$

which transforms the vectorial quantities that are defined in the bond-based frame $i + 1$ (see Figure 9.A.1) into their representation in the ith frame [11]. The first atom is A_1 and the first bond is ℓ_2. Here the torsion angle is taken to be 0° for the *cis* form. Represent the bond connecting

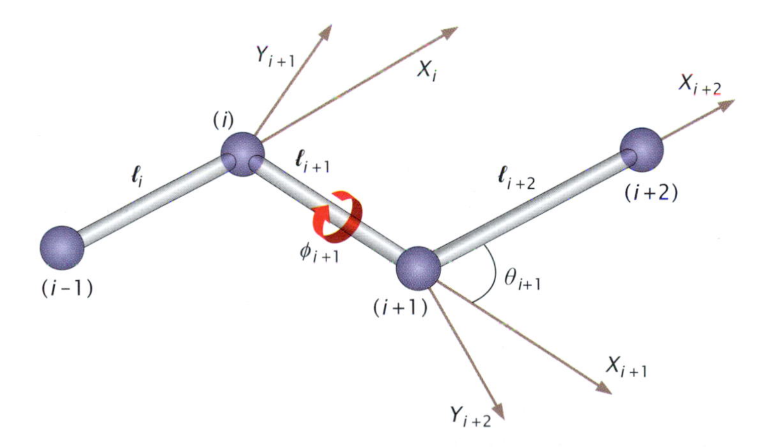

Figure 9.A.1 Definition of bond-based frames for modeling the conformations of chain molecules. This shows a chain segment of four backbone atoms. The ith bond connects atoms $i - 1$ and i along the main chain. Its torsion angle is ϕ_{i+1}. $\pi - \theta_{i+1}$ is the angle defined by bonds $i + 1$ and $i + 2$. The x_{i+1} and y_{i+1} axes of the bond-based coordinate system $x_{i+1}y_{i+1}$ appended to the bond $i + 1$ are shown. y_{i+1} lies in the plane defined by bonds i and $i + 1$, and makes an acute angle with x_i. The z_{i+1} axis, not shown, completes a right-handed coordinate system, that is, it points toward the reader from the plane of the paper, as found by curling your right-hand fingers from x_{i+1} to y_{i+1} and observing the direction your thumb points.

atoms i and $i+1$ as

$$\ell_{i+1}^* = [\ell_{i+1} \quad 0 \quad 0]^T \tag{9.A.2}$$

in the local frame $i+1$. The asterisk indicates that the definition refers to the local frame $i+1$.

Our aim is to compute the Cartesian position $\mathbf{r}_i$ in terms of the internal coordinate matrix $\mathbf{T}_i(\theta_i, \phi_i)$. To do this, you multiply successive matrices of each bond along the chain, to get

$$\mathbf{r}_i = \ell_2^* + \mathbf{T}_2(\theta_2, \phi_2)\ell_3^* + \mathbf{T}_2(\theta_2, \phi_2)\mathbf{T}_3(\theta_3, \phi_3)\ell_4^* + \ldots + \prod_{k=2}^{i-1}[\mathbf{T}_k(\theta_k, \phi_k)]\ell_i^* \tag{9.A.3}$$

for the ith internal position vector $\mathbf{r}_i$, where ϕ_2 is set equal to 180°.

APPENDIX 9B: HOW TO OPTIMALLY SUPERIMPOSE TWO STRUCTURES

We want to align (that is, superimpose) two structures. Figure 9.B.1 shows the initial situation in which two structures A and B are not yet aligned. This example is for two dimensions to make it simpler to follow. Structure A has coordinates $\mathbf{r}_{A1} = (x_{A1}, y_{A1}) = (0, 0)$, $\mathbf{r}_{A2} = (1, 1)$, and $\mathbf{r}_{A3} = (2, 0)$. Structure B has coordinates $\mathbf{r}_{B1} = (x_{B1}, y_{B1}) = \left(1, \frac{7}{4}\right)$, $\mathbf{r}_{B2} = (1, 0)$, and $\mathbf{r}_{B3} = \left(0, -\frac{3}{4}\right)$.

(1) Translate structure B to overlay the mass centers of the two structures:

$$\mathbf{r}'_{Bi} = \begin{bmatrix} x_{Bi} \\ y_{Bi} \end{bmatrix} + (\mathbf{R}_{A,\,cm} - \mathbf{R}_{B,\,cm}), \tag{9.B.1}$$

where

$$\mathbf{R}_{B,\,cm} = \frac{1}{3}\begin{bmatrix} x_{B1} + x_{B2} + x_{B3} \\ y_{B1} + y_{B2} + y_{B3} \end{bmatrix} = \frac{1}{3}\begin{bmatrix} 1 + 1 + 0 \\ \frac{7}{4} + 0 - \frac{3}{4} \end{bmatrix} = \begin{bmatrix} \frac{2}{3} \\ 1 \end{bmatrix},$$

$$\mathbf{R}_{A,\,cm} = \frac{1}{3}\begin{bmatrix} 0 + 1 + 2 \\ 0 + 1 + 0 \end{bmatrix} = \begin{bmatrix} 1 \\ \frac{1}{3} \end{bmatrix}.$$

This operation ensures that the two structures have the same translational state (their mass centers overlap). So the new coordinates

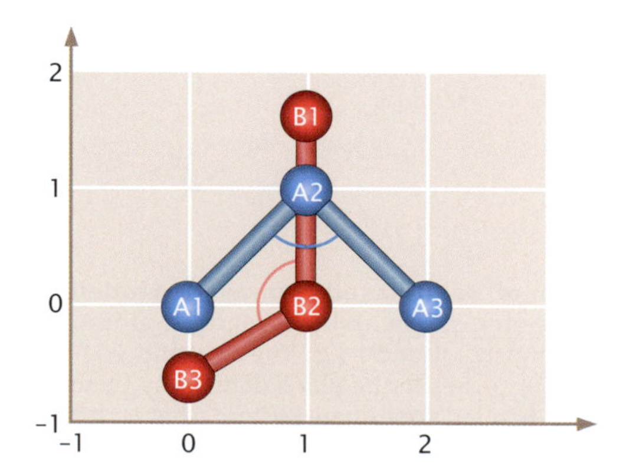

Figure 9.B.1 Example of two starting conformations A (*blue*) and B (*red*) that you want to align. The algorithm in the text rotates and translates one molecule to reach the optimal superposition.

of B′ are

$$\mathbf{r}'_{B1} = \begin{bmatrix} \frac{4}{3} \\ \frac{13}{12} \end{bmatrix}, \quad \mathbf{r}'_{B2} = \begin{bmatrix} \frac{4}{3} \\ -\frac{2}{3} \end{bmatrix}, \quad \mathbf{r}'_{B3} = \begin{bmatrix} \frac{1}{3} \\ -\frac{17}{12} \end{bmatrix}.$$

(2) Rotate B′ to get the optimal orientation relative to A. In order to determine the rotation matrix $\mathbf{T}(B' \rightarrow A)$, you need to describe the orientation of each structure. You do this by determining the three principal axes (in three dimensions), $\mathbf{p}_1$, $\mathbf{p}_2$, $\mathbf{p}_3$, for a given structure. In general, you can find the principal axes from the *singular value decomposition* of the conformation matrix. The conformation matrix is defined by the series of *m* position vectors organized in a $(3 \times m)$-dimensional matrix, for a structure in three dimensions. In this two-dimensional example, the conformational matrix has dimensions of 2×3, and is given by $[\mathbf{r}_{A1}, \ \mathbf{r}_{A2}, \ \mathbf{r}_{A3}]$ for structure A and $[\mathbf{r}'_{B1}, \ \mathbf{r}'_{B2}, \ \mathbf{r}'_{B3}]$ for the translated structure B. Each structure is defined by two principal axes: $\mathbf{p}_{A1}$, $\mathbf{p}_{A2}$ and $\mathbf{p}_{B1}$, $\mathbf{p}_{B2}$.

The principal axes of structure A are defined by the elements of U_A obtained by the singular value decomposition of the coordinate matrix:

$$\begin{bmatrix} x_{A1} & x_{A2} & x_{A3} \\ y_{A1} & y_{A2} & y_{A3} \end{bmatrix} = \mathrm{U_A \Lambda_A} \mathbf{V}_A^T \tag{9.B.2}$$

$$\begin{bmatrix} 0 & 1 & 2 \\ 0 & 1 & 0 \end{bmatrix} = \begin{bmatrix} u_{A11} & u_{A12} \\ u_{AA21} & u_{A22} \end{bmatrix} \begin{bmatrix} \lambda_{A1} & 0 \\ 0 & \lambda_{A2} \end{bmatrix} \begin{bmatrix} v_{A1} & v_{A2} & v_{A3} \\ w_{A1} & w_{A2} & w_{A3} \end{bmatrix},$$

so that

$$[\mathbf{p}_{A1}, \ \mathbf{p}_{A2}] = \begin{bmatrix} u_{A11} & u_{A12} \\ u_{A21} & u_{A22} \end{bmatrix}. \tag{9.B.3}$$

You can calculate the principal axes for structure B′, $\{\mathbf{p}_{B1}, \mathbf{p}_{B2}\}$, using a similar procedure.

With these principal axes defined, the transformation matrix that superimposes structure B′ onto structure A is found from the law of cosines; that is, taking the dot product between these principal axes,

$$\mathrm{T}(B' \rightarrow A) = \begin{bmatrix} \cos(\mathbf{p}_{A1} \cdot \mathbf{p}_{B1}) & \cos(\mathbf{p}_{A1} \cdot \mathbf{p}_{B2}) \\ \cos(\mathbf{p}_{A2} \cdot \mathbf{p}_{B1}) & \cos(\mathbf{p}_{A2} \cdot \mathbf{p}_{B2}) \end{bmatrix}. \tag{9.B.4}$$

The new rotated coordinates of structure B are then evaluated from

$$\mathbf{r}''_{Bi} = \begin{bmatrix} x''_{Bi} \\ y''_{Bi} \end{bmatrix} = \mathrm{T}(B' \rightarrow A) \begin{bmatrix} x_{Bi} \\ y_{Bi} \end{bmatrix} \tag{9.B.5}$$

for $i = 1, 2, 3$. This transformation ensures that the two structures have identical overall rotational state.

APPENDIX 9C: THE POISSON–BOLTZMANN EQUATION TREATS ELECTROSTATIC INTERACTIONS

If you have a three-dimensional constellation of charges, one way to compute the electrostatic free energy among them is just to sum the Coulombic interactions. However, it is often computationally more efficient, instead, to solve a partial differential equation. We do this in two steps below. First, for situations involving no mobile ions, we can solve the *Poisson equation*. Then, if you have mobile salt ions in addition to the fixed charge distribution, you can solve the *Poisson–Boltzmann* (PB) equation. Let's first consider a fixed charge distribution in a continuum dielectric medium, with no salt. Solving Poisson's equation is simpler than summing all Coulombic interactions when you have many charges. To use Poisson's equation, you start with knowledge of the distribution of charges, $\rho(\mathbf{r})$. If charge q_j is located at position $\mathbf{r}_j$ in space, and if you have several such charges, you can write the charge density as

$$\rho(\mathbf{r}) = \sum_j q_j \delta(\mathbf{r} - \mathbf{r}_j). \tag{9.C.1}$$

Here $\delta(\mathbf{r} - \mathbf{r}_j)$ is the delta function, equal to 1 at positions $\mathbf{r}_j$ where there is charge and 0 at positions where there is no charge. Given the knowledge of the position dependence of charge density $\rho(\mathbf{r})$ and of the dielectric permittivity $\varepsilon(\mathbf{r})$, the *Poisson equation* of electrostatics is

$$\nabla \cdot [\varepsilon(\mathbf{r})\nabla\Phi(\mathbf{r})] = -\rho(\mathbf{r}). \tag{9.C.2}$$

where the symbol ∇ is the gradient operator.[7] Solving this second-order differential equation will give you the function $\Phi(\mathbf{r})$ of spatial position, called the *electrostatic potential*.[8] Note that the dielectric constant is allowed to depend on position in this equation, consistent with the possible heterogeneity of the environment. If ε is a constant, the same throughout a given spatial region, then you can express Equation 9.C.2 as the *Laplace equation*:

$$\frac{\partial^2\Phi(\mathbf{r})}{\partial x^2} + \frac{\partial^2\Phi(\mathbf{r})}{\partial y^2} + \frac{\partial^2\Phi(\mathbf{r})}{\partial z^2} = \frac{\rho(\mathbf{r})}{\varepsilon}. \tag{9.C.3}$$

The Poisson equation 9.C.2 treats systems having charges that are fixed in space, such as those of charged side chains in a native protein. Often it is also of interest to consider additional charges that come from mobile ions, produced from dissolved salts like sodium chloride in the solution with the protein. To include these additional mobile ions, we add a Boltzmann distribution for the mobile ions to the Poisson equation. The number of mobile ions (of type *i*) per unit volume within a region of space is given by the Boltzmann distribution

$$n_i = n_i^0 \exp\left(\frac{-q_i\Phi}{kT}\right), \tag{9.C.4}$$

where n_i^0 is the number density of ions in bulk solution, q_i is the charge on the ion, and Φ is the electrostatic potential in that region of space. The Boltzmann distribution specifies that anions accumulate where the potential is positive, and that cations accumulate where the potential is negative.

---

[7]This operator specifies taking the derivative, when it applies to a scalar quantity. It yields a vector of the partial derivatives of the scalar with respect to *x*-, *y*-, and *z*-components.

[8]To solve, you need two boundary conditions: one corresponding to a reference point and another on the protein surface. Typically, the electrostatic potential is set to zero at an infinite distance from the protein as a reference point. The second boundary condition is then set at the protein surface by mapping the partial atomic charges onto this surface.

The Poisson–Boltzmann equation combines the Poisson equation for the effects of the fixed charges with the Boltzmann distribution of the mobile ions to give

$$\nabla \cdot [\varepsilon(\mathbf{r})\nabla\Phi(\mathbf{r})] = -\rho(\mathbf{r}) + \sum_i q_i n_i^0 \exp\left[\frac{-q_i\Phi(\mathbf{r})}{kT}\right]. \quad (9.C.5)$$

The overall charge density at a given position **r** is obtained by adding the charge distribution contributed by the charges of the protein, $\rho(\mathbf{r})$, to that of the ions in the environment. In a 1:1 salt solution where the net ion charge density is zero, the Poisson–Boltzmann equation is commonly given in the form

$$\nabla \cdot [\varepsilon(\mathbf{r})\nabla\Phi(\mathbf{r})] - \kappa^2\varepsilon(\mathbf{r}) \sinh\left[\frac{e\Phi(\mathbf{r})}{kT}\right] = -4\pi\rho(\mathbf{r}), \quad (9.C.6)$$

where e is the charge of a proton, $\varepsilon(\mathbf{r})$ is the dielectric constant (in units where the vacuum dielectric constant is 1), and $1/\kappa$ is the Debye–Hückel length, which gives a measure of how far into solution the electrostatic effects due to a charge extend. κ is given by

$$\kappa^2 = \frac{8\pi N_A e^2 I}{1000\varepsilon(\mathbf{r})kT}, \quad (9.C.7)$$

where I is the ionic strength of the solution measured in mol L^{-1}, and N_A is Avogadro's number. The factor 1000 appears because I is expressed in mol L^{-1}, while κ is in cm^{-1}.

In the case where $q\Phi$ is small (compared with kT) we use $\sinh x \approx x$ (keeping only the first term of the Taylor series expansion) in Equation 9.C.6 to obtain the linearized Poisson–Boltzmann equation

$$\nabla \cdot [\varepsilon(\mathbf{r})\nabla\Phi(\mathbf{r})] - \kappa^2 e\varepsilon(\mathbf{r})\Phi(\mathbf{r})/kT = -4\pi\rho(\mathbf{r}). \quad (9.C.8)$$

For highly charged systems, the large value of q requires the use of the full, nonlinear Poisson–Boltzmann Equation 9.C.5.

In practice, for proteins, the Poisson–Boltzmann equation is usually solved using finite-difference methods. In finite-difference methods, you divide space into a grid. The electrostatic potential is solved only for the grid site centers, and then interpolated for points between them. Usually, the positions of charges do not correspond exactly to grid centers, and the charges (partial or otherwise) are split up, with parts being assigned to the nearest grid centers. Then, on each point, the electrostatic potential is computed iteratively until the equation converges to give $\Phi(\mathbf{r})$. The accuracy of the finite-difference solution depends on the grid size: smaller grids yield more accurate results but require more computational effort.

REFERENCES

[1] HM Berman, J Westbrook, Z Feng, et al. The Protein Data Bank. *Nucleic Acids Res*, 28:235–242, 2000.

[2] The PyMOL Molecular Graphics System. Schrödinger, LLC. https://www.pymol.org/.

[3] EF Pettersen, TD Goddard, CC Huang, et al. UCSF Chimera—a visualization system for exploratory research and analysis. *J Comput Chem*, 25:1605–1612, 2004.

[4] W Humphrey, A Dalke, and K Schulten. VMD—Visual Molecular Dynamics. *J Molec Graphics*, 14:33–38, 1996.

[5] WD Cornell, P Cieplak, CI Bayly, et al. A second generation force field for the simulation of proteins, nucleic acids, and organic molecules. *J Am Chem Soc*, 117:5179–5197, 1995.

[6] BR Brooks, CL Brooks, III, AD MacKerell, Jr., et al. CHARMM: the biomolecular simulation program. *J Comput Chem* 30:1545-1614, 2009.

[7] S Pronk, S Páll, R Schulz, et al. GROMACS 4.5: A high-throughput and highly parallel open source molecular simulation toolkit. *Bioinformatics*, 29:845–854, 2013.

[8] JC Phillips, R Braun, W Wang, et al. Scalable molecular dynamics with NAMD. *J Comput Chem*, 26:1781–1802, 2005.

[9] KA Dill and S Bromberg. *Molecular Driving Forces: Statistical Thermodynamics in Biology, Chemistry, Physics, and Nanoscience*, 2nd ed. Garland Science, New York, 2011.

[10] D Bashford and DA Case. Generalized Born models of macromolecular solvation effects. *Annu Rev Phys Chem*, 51:129, 2000.

[11] PJ Flory. *Statistical Mechanics of Chain Molecules*. Hanser, Munich, 1988.

SUGGESTED READING

Allen MP and Tildesley DJ, *Computer Simulation of Liquids*. Oxford University Press, Oxford, 1989.

Hockney RW, The potential calculation and some applications. In Alder B, Fernbach S, and Rotenberg M, editors, *Methods in Computational Physics*, Volume 9, pp 135–211. Academic Press, New York, 1970.

T_5 T_4 T_3 T_2 T_1

Monte Carlo step

CHAPTER 10

Molecular Simulations and Conformational Sampling

Computer simulations give insights into the structures, transitions, and mechanisms of proteins, in terms of the underlying atomic interactions. Simulations can complement experimental data, by leveraging physical principles, by adding the time dependences, and by giving detailed *narratives* of molecular processes. Experiments are often coarse-grained over time and/or space, and do not provide details atom-by-atom and nanosecond-by-nanosecond. Computer simulations can fill in those details. Knowing the details is like watching a whole movie, rather than just seeing the opening and closing scenes. Simulations can also help with the design of proteins or drugs; they can help you refine protein structures; and they are useful for making movies or visualizations of mechanisms. Reviews are given by the group of Schulten (Perilla et al.) [1], Karplus and Kuriyan [2], and Dror et al. [3].

Computer simulations must search and sample protein motions and conformations. You need to explore different conformations in order to model dynamical processes and to seek stable states, having low free energy. Here's a metaphor. Suppose you want to find the main walking path or deepest valley on a mountainside. It's not likely that you'll find either of them by a static snapshot or by random meandering. However, there are principled strategies that help you find deep valleys or important pathways efficiently, without exhaustively searching everywhere.[1]

In this chapter, we describe the main methods—Monte Carlo and molecular dynamics—for simulating molecular structures and processes. Here is the general problem of conformational searching on a computer (Figure 10.1). You first put the chain into some starting *snapshot* (that is, a single microstate) of interest, making sure to satisfy the covalent connectivities of the atoms. For a given set of energy parameters (that is, a particular force field), you compute the energy U_1 of that initial conformation. Next, you use some procedure to change some or all of the torsion angles and distances. Then, you compute the energy U_2

---

[1] *Searching* refers to seeking particular states, often the states of lowest free energy. *Sampling* refers to collecting sufficient statistics to determine the populations of those states. Often, the words are used interchangeably.

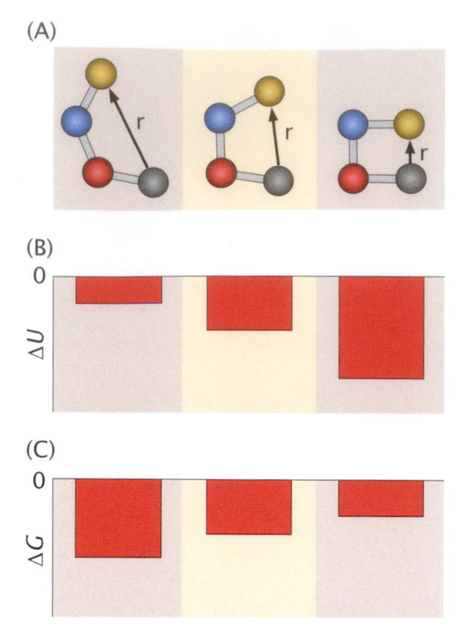

Figure 10.1 Stable conformations result from both low energies and high entropies. To find the important (most populated) states of nature, we seek states of minimum free energy $\Delta G = \Delta U - T\Delta S$, which is a balance of low energy and high entropy. (A) Three conformations along a trajectory. (B) The energy ΔU decreases. (C) The net free energy ΔG increases, in this case because of loss of entropy.

for the new conformation. You continue making incremental changes, leading to a series of conformations $i = 3, 4, 5, \ldots$. Such a sequence of conformational snapshots is called a *trajectory*. Now, in order to predict nature's important states—for example those that are observable in experiments—you need to find the *most populated states*. According to thermodynamics, this means seeking the states having lowest free energy G. And finding states of minimum free energy $G = U - TS$ (where U is the energy, T is the temperature, and S is the entropy) means finding states that have both low energy U (the deepest valleys on energy landscapes) and high entropy S (the widest basins). In addition, you may want to know how a protein works, which can involve conformational fluctuations, cooperative changes, motions and dynamics. Then, you want the most important conformational pathways and the rates of changes along them.

YOU CAN FIND STATES OF LOW ENERGY BY ENERGY MINIMIZATION

To find conformations having low energy, we want a search method that can go downhill on energy landscapes. First, let's distinguish between a *local minimum* and a *global minimum*. A local minimum is the bottom of *any* energy well. A global minimum is the bottom of the *deepest* energy well among all the wells on the whole landscape. Nature's most populated states are predicted by finding the global minima of free energy. We start with a simpler challenge: if you start at a point on a mathematical surface such as B on Figure 10.2), how can you find the nearest stable state A, a local minimum? We describe a method called *energy minimization*.

To Compute Forces, Take the Derivative of the Potential Energy

The principles of molecular motion are described by Newton's law $\mathbf{f} = m\mathbf{a}$, which says that a particle tends to accelerate along the direction of the force it experiences at a given instant. Given an energy function U as in Equation 9.10, you can determine the forces using Newton's law

$$\mathbf{f}_i = m_i\mathbf{a}_i = -\nabla_{\mathbf{r}_i} U, \tag{10.1}$$

where $\mathbf{f}_i$ is the force experienced by the particle (or atom) i due to the presence of the potential energy U. m_i and $\mathbf{a}_i$ are the respective mass and acceleration of particle i. $\nabla_{\mathbf{r}_i}$ designates the gradient operator, that is, partial derivatives of U, such as $\partial U/\partial x_i$, with respect to the x-, y-, and z-components of the position $\mathbf{r}_i$ of the particle i. So, the x-, y-, and z-components of $\mathbf{f}_i$ are found from the partial derivatives of U as

$$\mathbf{f}_i = \begin{bmatrix} f_{ix} \\ f_{iy} \\ f_{iz} \end{bmatrix} = \begin{bmatrix} -\partial U/\partial x_i \\ -\partial U/\partial y_i \\ -\partial U/\partial z_i \end{bmatrix}. \tag{10.2}$$

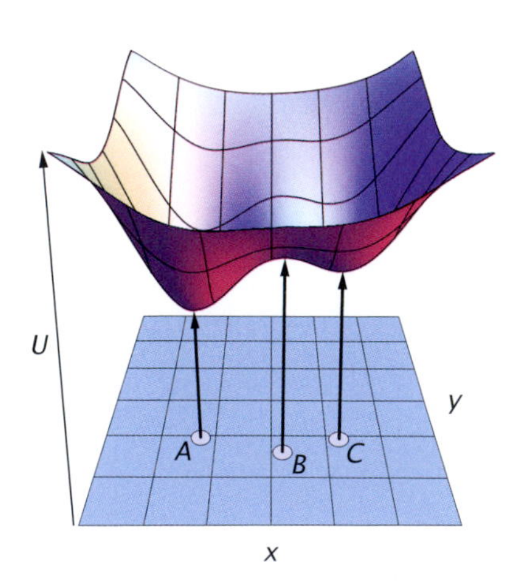

Figure 10.2 An energy landscape, illustrating a local minimum *C*, a global minimum *A*, and a saddle point *B*. The energy $U(x, y)$ is a function of degrees of freedom x and y, such as bond angles and lengths.

The negative sign indicates that the force exerted on the atom i is in the direction that leads to a *decrease* in energy. The system is in mechanical equilibrium when the forces are zero (that is, when $\sum_i \mathbf{f}_i = 0$), which occurs when the energies are at either a local or a global energy minimum. In a molecular dynamics simulation, many different forces act on atom i. In a simulation, you need to determine the gradient of the total potential U, which is the sum of all the forces, in order to get the total force acting on each atom i. Examples of the relationships between forces, displacements, and energy gradients are given in Boxes 10.1 and 10.2.

Box 10.1 Near Energy Minima, the Forces Are Linear, Acting Like Hooke's-Law Springs

Local or global minima on energy landscapes are well bottoms, such as points A and C in Figure 10.2. You can approximate the energy near a well bottom as a function of a coordinate x by a square-law or *harmonic* potential

$$U(x) = \frac{k_s(x - x_0)^2}{2}.$$

This square law describes the energy of a spring, as in Hooke's law, where the force is linear in x, with spring constant k_s (see Equation 10.2):

$$f(x) = -k_s(x - x_0). \tag{10.3}$$

The negative sign indicates that the force acting on the particle is in the direction *opposite* to that of the deformation, $\Delta x = (x - x_0)$. For bond stretching or bending, the force constants are $k_s = k_\ell$ or k_θ, respectively. And $(x - x_0)$ is replaced by $(\ell - \ell_0)$ and $(\theta - \theta_0)$, respectively. The bond stretching constant k_ℓ is large, implying stiff springs.

Box 10.2 Examples of Forces as Gradients of Potentials

(1) The force resulting from a van der Waals interaction

Consider two atoms i and j, separated by a vector $\mathbf{r}_{ij} = \mathbf{r}_j - \mathbf{r}_i$, where $\mathbf{r}_i$ is the position vector of atom i. The corresponding (scalar) interatomic distance is

$$r_{ij} = \left[(x_j - x_i)^2 + (y_j - y_i)^2 + (z_j - z_i)^2\right]^{1/2}.$$

The van der Waals interaction potential is

$$U = \frac{-a_{ij}}{r_{ij}^6} + \frac{b_{ij}}{r_{ij}^{12}},$$

where a_{ij} and b_{ij} are the attraction and repulsion coefficients between atoms i and j. To get the forces acting on the atom i due to this interaction, we want the gradient of U with respect to the change in the position vector $\mathbf{r}_i$ of atom i. So, the derivative with respect to the x-component of $\mathbf{r}_i$ is given by

$$\frac{\partial U}{\partial x_i} = \frac{-6a_{ij}(x_j - x_i)}{r_{ij}^8} + \frac{12b_{ij}(x_j - x_i)}{r_{ij}^{14}}.$$

Repeating this for the y_i and z_i directions, you see that the gradient of the potential with respect to $\mathbf{r}_i$ is

$$\nabla_{\mathbf{r}_i} U = \left(\frac{-6a_{ij}}{r_{ij}^8} + \frac{12b_{ij}}{r_{ij}^{14}}\right)\begin{bmatrix}(x_j - x_i)\\(y_j - y_i)\\(z_j - z_i)\end{bmatrix},$$

or the force acting on atom i due to its van der Waals interaction with atom j is

$$\mathbf{f}_i = \left(\frac{6a_{ij}}{r_{ij}^8} - \frac{12b_{ij}}{r_{ij}^{14}}\right)\mathbf{r}_{ij}.$$

Note that the gradient with respect to $\mathbf{r}_j$ is identical in magnitude, but opposite in sign. Thus the forces exerted by the van der Waals potential on the two atoms are equal but have opposite directions, consistent with the tendency to restore the interatomic distance to its equilibrium value.

(2) The force due to bond stretching or contraction

Now, consider two atoms $i-1$ and i that are covalently bonded and interact through a stretching potential

$$U = \frac{k_\ell}{2}(\ell_i - \ell_i^0)^2,$$

where ℓ_i is the fluctuating (instantaneous) length of the bond and ℓ_i^0 is the equilibrium length for that type of bond. Express ℓ_i in terms of the components of the two position vectors $\mathbf{r}_i$ and $\mathbf{r}_{i-1}$ (see Equation 9.2 in Chapter 9):

$$\ell_i = \left[(x_i - x_{i-1})^2 + (y_i - y_{i-1})^2 + (z_i - z_{i-1})^2\right]^{1/2}.$$

Now, take the derivative with respect to x_i, to get

$$\frac{\partial U}{\partial x_i} = \frac{k_\ell(\ell_i - \ell_i^0)}{\ell_i}(x_i - x_{i-1}).$$

Take the corresponding derivatives for the y_i and z_i directions, to get

$$\nabla_{\mathbf{r}_i} U = \frac{k_\ell(\ell_i - \ell_i^0)}{\ell_i}\begin{bmatrix}(x_i - x_{i-1})\\(y_i - y_{i-1})\\(z_i - z_{i-1})\end{bmatrix}$$

$$= \frac{k_\ell(\ell_i - \ell_i^0)}{\ell_i}\boldsymbol{\ell}_i,$$

where $\boldsymbol{\ell}_i$ denotes the bond vector ($\boldsymbol{\ell}_i = \mathbf{r}_i - \mathbf{r}_{i-1}$). You can compute the forces from these derivatives of potential functions. For example, the vector force $\mathbf{f}_i$ on atom i due to stretching or contracting bond i is given by

$$\mathbf{f}_i = m_i\frac{d^2\mathbf{r}_i}{dt^2} = -\nabla_{\mathbf{r}_i} U = \frac{-k_\ell(\ell_i - \ell_i^0)}{\ell_i}\boldsymbol{\ell}_i$$

and that on atom $i-1$ due to the same stretching potential is $\mathbf{f}_j = -\mathbf{f}_i$. The change in bond length induces two forces, one on atom i and one on atom $i-1$. The force on atom $i-1$ is identical in magnitude to that on atom i, but opposite in sign. Similarly, a change in a bond angle θ_i exerts forces on three atoms, $i-1$, i, and $i+1$ (see the definition of bond angle in Equation 9.4 in Chapter 9). And, a change in the dihedral angle ϕ_i exerts forces on atoms $i-2$, $i-1$, i and $i+1$ (see Figure 9.6 in Chapter 9).

Following Gradients Downhill Leads to States of Low Energy

Following gradients downhill on mathematical surfaces can be done by various methods, including *steepest descent*, *conjuguate-gradient*, and *Newton–Raphson*. In order to determine when a simulation has *converged* to an energy minimum, you check whether the energy at one evaluation is the same as for the previous evaluation, to within some small difference. When the force is sufficiently close to zero in going downhill, you have found a stable state—either a local or global minimum of energy or a saddle point.

There are three main limitations of the energy minimization procedure: (1) It just finds stable states, not the conformational dynamics. (2) It searches only downhill, not uphill. So, it cannot climb over energy hills. It is only likely to find local minima. Protein landscapes are very bumpy, so finding global minima requires methods that can climb over energy barriers. And (3) it finds only states of low energy, not low free energy. That is, it seeks only basins that are deep, while it should seek basins that are also wide. (If state A is one low-energy microstate while B is 27 low-energy microstates, then A can have the lowest energy, while B might have the lowest free energy).

A goal is to understand nature's populated states. To find the most populated states, you must seek the global minima of free energy. A global minimum of the energy is shown in Figure 10.2, labeled A. We now consider search methods that go beyond simple minimization. On the one hand, you need to find the global minimum efficiently (that is, without searching all possible states), because this is computationally expensive, if not impossible, for large biomolecular systems. On the other hand, you also need methods that are *population-preserving*, also referred to as *satisfying detailed balance*. Stated differently, you cannot know whether you are in the deepest valley unless you know the altitudes (energies) of all the valleys in the whole landscape. Your method must be able to determine the relative energies, or relative populations, of the different states. You cannot do this with just any arbitrary sampling method. Just visiting a conformation in a computer simulation does not tell you the population (that is, the probability) of that conformation. To know populations requires knowing free energies, which describe macrostates that have the optimal combination of low energy and high entropy. We need sampling methods that have a *principled basis* for providing proper estimates of the populations. The two main methods are molecular dynamics (MD) and Monte Carlo (MC) simulations.

MOLECULAR DYNAMICS SIMULATIONS SOLVE NEWTON'S EQUATIONS OF MOTION ITERATIVELY

The method of MD simulations was introduced by Alder and Wainwright in 1957 [4] to study the interactions of hard spheres and by Rahman and Stillinger in 1971 to study water [5]. In 1977, McCammon, Gelin, and Karplus [6] were the first to apply MD methods to proteins. Advantages of MD over simple minimization are that MD can cross energy barriers to reach states of low free energy, not just low energy, and that MD helps generate trajectories of conformational motions, which are often important for understanding biomolecular mechanisms of action (**Figure 10.3**).

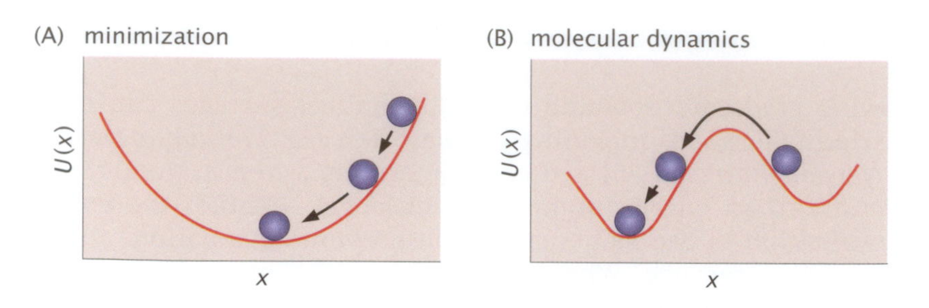

Figure 10.3 Minimization versus molecular dynamics. (A) Minimization is a process that takes only downhill steps on landscapes. (B) In MD simulations, the system follows Newton's law of motion on a landscape, mostly toward lower-energy states, but because it also accounts for battering by random thermal motions, it sometimes goes uphill. So MD can climb over barriers, ultimately reaching closer to globally minimal states. And MD has the advantage that it can capture the dynamics.

Suppose you have a system consisting of N interacting particles (typically atoms). To generate an *MD trajectory*, you use $3N$ differential equations called *Newton's equations of motion*,

$$m_i \frac{d^2\mathbf{r}_i}{dt^2} = \mathbf{f}_i = -\nabla_{r_i} U, \tag{10.4}$$

to evaluate the acceleration $a_i = d^2\mathbf{r}_i/dt^2$ of each atom $1 \leq i \leq N$ at each of many (typically $\sim 10^6$–10^9) sequential time steps. You evaluate the gradient of the potential energy U with respect to the three components of each atom's position vector. U accounts for both intramolecular (bonded and nonbonded) and intermolecular interaction energies. Equation 10.4 shows how the instantaneous forces and accelerations of the atoms are determined as gradients of the potential U. $d^2\mathbf{r}_i/dt^2$ is the second derivative of $\mathbf{r}_i$ with respect to time by the action of the gradient of the potential. The negative sign indicates that each atom accelerates in the direction toward lower potential. For each atom i, the three components of Equation 10.4 are

$$\begin{aligned} m_i \frac{d^2 x_i}{dt^2} &= -\frac{\partial U}{\partial x_i}, \\ m_i \frac{d^2 y_i}{dt^2} &= -\frac{\partial U}{\partial y_i}, \\ m_i \frac{d^2 z_i}{dt^2} &= -\frac{\partial U}{\partial z_i}. \end{aligned} \tag{10.5}$$

Therefore, if you know the instantaneous coordinates $\mathbf{r}(t) = (\mathbf{r}_1, \mathbf{r}_2, \ldots, \mathbf{r}_N)$ at a given time t, you can evaluate the overall potential U, which is completely defined by the atomic coordinates. And you can evaluate the accelerations $\mathbf{a}(t) = (\mathbf{a}_1, \mathbf{a}_2, \ldots, \mathbf{a}_N)$ of all particles. How do you generate an MD trajectory? The two-stage procedure, of initiation and progression, is described next.

How Do You Compute a Molecular Dynamics Trajectory?

Suppose you know all the atomic positions $\mathbf{r}_i(t)$, velocities $\mathbf{v}_i(t)$, and accelerations $\mathbf{a}_i(t)$ at time t. (Knowing the accelerations is equivalent to knowing the forces, because you also know the particle masses.) And suppose you know all previous positions, velocities, and accelerations (at times $t-1$, $t-2$, $t-3$, ..., 1). From that information, we want to compute the positions, velocities, and accelerations for the next instant in time, $\mathbf{r}_i(t+\Delta t)$, $\mathbf{v}_i(t+\Delta t)$, and $\mathbf{a}_i(t+\Delta t)$. You can get the new values by expressing Newton's equations 10.5 in terms of finite differences, and by integrating over the previous time steps. You compute $\mathbf{r}_i(t+\Delta t)$ by numerical integration methods such as the *Verlet* [7] or *leapfrog* algorithms (see Appendix 10A):

$$\mathbf{r}_i(t+\Delta t) = 2\mathbf{r}_i(t) - \mathbf{r}_i(t-\Delta t) + (\Delta t)^2\mathbf{a}_i(t). \tag{10.6}$$

By iterating this procedure for all atoms at each time step, you compute a trajectory. The trajectory can be expressed as a series of $3N$-dimensional vectors, $\mathbf{r}(t)$, at successive time steps. The elements of $\mathbf{r}(t)$ are the instantaneous positions of all atoms i, at time t. Equation 10.6 describes the recurrence equation for each time step. But how do you start the simulation? For the Verlet algorithm, you need the coordinates for two initial times t_0 and $t_0 + \Delta t$. To do this, you choose a starting protein structure, $\mathbf{r}(t_0)$. This is usually a local energy minimum obtained with a short energy-minimization protocol (as described previously). Here, $\mathbf{r}(t_0)$ designates the ensemble of all initial position vectors for all N atoms. Then, in order to assign initial velocities $v(t_0)$ to atoms, you choose randomly from the Maxwell–Boltzmann distribution of the velocities,

$$p(v_{xi}) = \left(\frac{m_i}{2\pi RT}\right)^{1/2} \exp\left(-\frac{m_i v_{xi}^2}{2RT}\right), \tag{10.7}$$

for each component v_{xi}, v_{yi}, and v_{zi} of the velocity vector $\mathbf{v}_i$ for particle i at the given simulation temperature T, where R is the gas constant. This gives you the initial position vector for all atoms, $\mathbf{r}(t_0)$, and the initial velocity vector for all atoms, $\mathbf{v}(t_0)$, and you can evaluate the position vector at the next time step, $t_0 + \Delta t$, using simply $\mathbf{r}(t_0 + \Delta t) = \mathbf{r}(t_0) + \mathbf{v}(t_0)\Delta t$. Let's take $t_0 = 0$. Now, using this conformation vector $\mathbf{r}(\Delta t)$, compute $U(\mathbf{r})$, take the gradient of U to get the forces, and divide by the atomic masses to get $\mathbf{a}(\Delta t)$. This procedure provides all quantities on the right-hand side of Equation 10.6, so you can apply the Verlet algorithm for the position vectors at all successive times.

An MD trajectory is deterministic (not probabilistic). Each successive state is uniquely determined by the instantaneous gradient of the potential. This is in contrast to Monte Carlo trajectories, described later in this chapter, which are fully probabilistic. However, each MD run does depend on the initial velocities, and those can differ. So, you usually generate multiple MD trajectories, with different initial velocities. You should check that the average properties or dominant patterns are consistent and reproducible, over the distribution of initial velocities. To obtain such statistical samples, performing multiple short runs is often preferable to single long runs.

What Time Step Δt Should You Use for MD Simulations?

You must set the size of your time step to be shorter than the system's fastest motions. Otherwise, the motions that occur in that time interval will be so large that they will violate the "small-step" assumption in the numerical integrator, causing unphysical energies and accelerations. For example, bonds could become distorted, leading to high energies and accelerations in the next step, driving your simulation to diverge. Figure 10.4 shows the various timescales of protein motions. For proteins, the fastest motions are bond vibrations. So, MD simulation time steps must typically be around 1–2 fs (where $1\,\text{fs} = 10^{-15}\,\text{s}$).

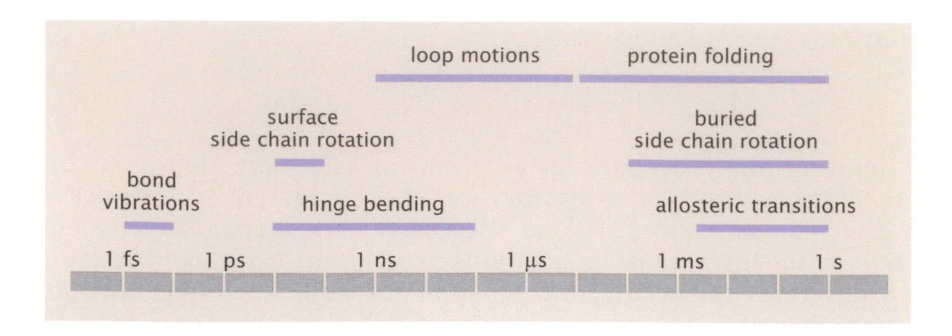

Figure 10.4 The timescales of protein motions. (Adapted from MC Zwier and LT Chong. *Curr Opin Pharmacol*, 10:745–52, 2010. With permission from Elsevier.)

What Is the Computational Time for MD Simulations?

If you want to simulate a typical protein in atomic detail over 1 ms, it requires 10^{12} steps of more than 100,000 atomic position vectors of the protein and surrounding water molecules, after placing the molecule in a simulation box, with periodic boundary conditions (see Appendix 10B). This means that at every single step, one needs to evaluate 10^{10} interactions to calculate the overall energy, and take the derivatives with respect to all N atom positions. This would be prohibitively time-consuming. Instead, there are various ways you can reduce the computational time, including introducing cutoffs that allow you to neglect interactions that act over long distances, or not saving the coordinates at every single step. Even so, atomistic MD simulations can be expensive in computer time and memory, especially for multimeric or multiprotein systems. Here is an estimate of the "real" time cost of generating an MD trajectory for a system of 100,000 atoms. Using 600 processors (CPUs) in a supercomputer (that is, pushing the limits of current computing technology), the computing time would be 500 hours, for generating a trajectory of 1 μs (which would provide statistical data up to ~100 ns). At least two independent trajectories are needed for verifying the reproducibility of the outputs, which means at least 1.5 months of simulations for learning about events occurring within the range $t < 100$ ns. Current software for MD simulations, such as NAMD [8] and Gromacs [9], utilize efficient algorithms that lend themselves to parallel processing to increase efficiency.

How Do You Analyze Trajectories?

Once you have generated a trajectory, you must analyze it. Analyzing it is as important as generating it. A trajectory is a time series of coordinates. The coordinates fluctuate markedly along the trajectory time course. Proper calculations of averages and variances of conformational properties are needed to help you identify important states or modes of motion. In order to get sufficient and accurate statistics, you should either generate multiple trajectories or long individual ones. Getting good averages requires a lot of data and many uncorrelated samples.

Here is how you compute averages from trajectories. Consider the position $\mathbf{r}_{i,f}$ of atom i in a single snapshot, which we identify as a *frame f*, of one such trajectory. You can compute the average position of atom i using

$$\langle \mathbf{r}_i \rangle = \frac{1}{M} \sum_{f=1}^{M} \mathbf{r}_{i,f}, \tag{10.8}$$

where M is the total number of frames. You can compute the instantaneous *fluctuations* around that average, using the quantity

$$\Delta \mathbf{r}_{i,f} = \mathbf{r}_{i,f} - \langle \mathbf{r}_i \rangle. \tag{10.9}$$

The mean-square fluctuation of atom i around its mean position $\langle \mathbf{r}_i \rangle$, averaged over the entire duration of simulations, or over an ensemble of M snapshots, is defined as

$$\langle (\Delta \mathbf{r}_i)^2 \rangle = \frac{1}{M} \sum_{f=1}^{M} (\mathbf{r}_{i,f} - \langle \mathbf{r}_i \rangle)^2, \tag{10.10}$$

and the *root-mean-square fluctuation* (RMSF) is defined as the square root of this quantity, $\langle (\Delta \mathbf{r}_i)^2 \rangle^{1/2}$, which has the same units as your coordinates. Such fluctuations are experimentally measurable. In X-ray crystallographic studies, for example, they are known as *Debye–Waller factors*, also called *B-factors*, given by $B_i = (8\pi^2/3)\langle (\Delta \mathbf{r}_i)^2 \rangle$. RMSFs are time-average quantities that are atom-site-dependent, and are usually plotted as a function of atom site index i as a measure of stability of those sites during simulations.

A related measure of fluctuations for the entire protein (not individual atoms) is the *root-mean-square deviation* (RMSD) in atomic coordinates, often evaluated as a function of time with respect to the initial conformation,

$$\text{RMSD}(t) = \sqrt{\frac{1}{N} \sum_{i=1}^{N} [\mathbf{r}_i(t) - \mathbf{r}_i(0)]^2}. \tag{10.11}$$

You can determine when the RMSD converges during a simulation by seeking a plateau in RMSD(t). Another useful quantity is the RMSD between two conformations A and B:

$$\text{RMSD}(t) = \sqrt{\frac{1}{N} \sum_{i=1}^{N} [\mathbf{r}_i(A) - \mathbf{r}_i(B)]^2}, \tag{10.12}$$

where $\mathbf{r}_i(A)$ and $\mathbf{r}_i(B)$ are the positions of atom i in the two respective conformations, after their optimal superposition.

The next two boxes give examples of what you can learn from molecular simulations of proteins.

Box 10.3 describes how water molecules translocate through a membrane protein called aquaporin. Those simulations show the step-by-step structural details of single-file transport and the energetic factors that drive it.

Box 10.3 What Is the Mechanism of Water Transport through Aquaporin?

One way that osmotic pressures are regulated in cells is through the passive transport of water and other small solutes into and out of cells through a membrane protein called aquaporin. The physiological importance of aquaporin was recognized by the award of the Nobel Prize in Chemistry to Peter Agre in 2003 for his work on the discovery of water channels. Some key questions are: How is transport so selective for water relative to other molecules? Why is water transport so fast? And how does the pore simultaneously prevent unwanted flows of protons?

MD simulations were performed by de Groot and Grubmüller [10] and by Tajkhorshid et al. [11]. Figure 10.5 shows how the water molecules flow through the pore in single file. The simulations show that the water molecules adopt well-defined orientations in the water channel, with the oxygen atoms facing the center of the channel [11]. Highly conserved asparagine–proline–alanine motifs contribute to the selectivity of aquaporin channels against protons by configuring a water molecule to serve strictly as a hydrogen-bond donor.

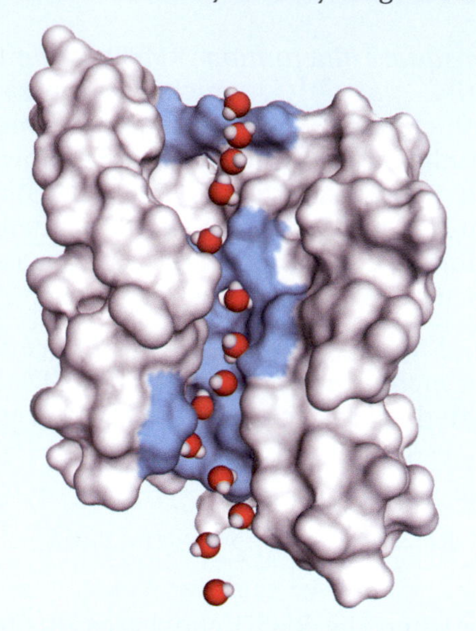

Figure 10.5 Water molecules flow through an aquaporin channel. MD simulations give detailed insights, in this case showing that water molecules (*red* and *white*) flow concertedly, single-file, through the center of the channel (*blue*). The central water coordinates with hydrogen bond donors in the protein, explaining how water is transported while protons are not. (Adapted from BL de Groot and H Grubmüller. *Science*, 294:2353–2357, 2001. With permission from AAAS.)

Sometimes, MD simulations give insights about large conformational changes in biological mechanisms. Box 10.4 describes MD simulations of the activation of a tyrosine kinase. In this case, the conformational landscape obtained from these simulations shows that the passage of the kinase domain from an inactive to an active state involves the opening of an activation loop, succeeded by a salt-bridge switch that stabilizes the active form. Identification of intermediate states can help in the design of allosteric inhibitors that block the activation dynamics. In Chapter 13, we illustrate another allosteric mechanism of inhibition for the same protein family: targeting of both the catalytic site and a myristoylation site at the C-terminus for drugs (see Figure 13.17).

Box 10.4 What Is the Mechanism of the Allosteric Activation of Src Kinase?

Src-family tyrosine kinases are key regulators of immune cell signaling. They are dysfunctional in some cancers. Src kinases have a multidomain architecture with N-terminal SH3 and SH2 domains and a C-terminal tyrosine kinase domain. The kinase domain catalyzes the transfer of a phosphate group from ATP, activating downstream proteins. Two structural elements that play critical roles in activating the kinase domain are the αC-helix on the C-terminal domain, and the activation segment, the A-loop. MD simulations have explored the conformational transition between the active and inactive states of the kinase domain in Src kinases (Figure 10.6). Two intermediate steps have been found. First, the activation loop opens partly from its closed-loop inactive conformation, while the αC-helix stays in its inactive state [12]. Second, the A-loop fully opens in a concerted motion and the αC-helix swings inward to form the active conformation.

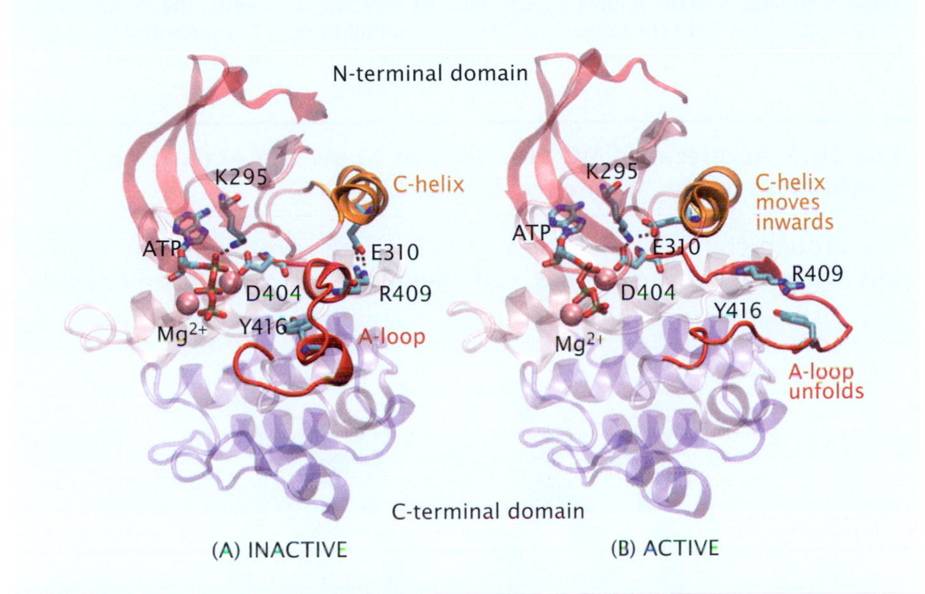

Figure 10.6 MD simulations reveal the activation pathway of the kinase domain of Src kinase. Shown here are the endstates. The transition from (A) inactive to (B) active states involves a concerted action in which the αC-helix moves inward and the A-loop unfolds, leading to the proper placement of the catalytic residues. (Adapted from D Shukla, Y Meng, B Roux, and VS Pande. *Nat Commun*, 5:3397, 2014. With permission from Macmillan Publishers Ltd).

Accelerated Sampling Methods Can Explore Larger Actions and Longer Timescales

You may want to explore larger motions or slower processes than are accessible by simpler MD methods. One way is to use *coarse-grained* models, rather than including full atomistic detail (see Chapters 11 and 12). Another approach is to use *accelerated* sampling methods, such as *metadynamics*, *conformational flooding* [13], *accelerated MD* [14], or others. For example, here is an outline of metadynamics [15]. In metadynamics, you modify the potential function on the fly as you simulate the system. You do this by "pouring computational sand" into the free-energy wells that you have already explored, to fill them up (Figure 10.7). This sand-filling procedure causes the simulation to spend diminishing amounts of time exploring energy wells that it has already explored. This pushes the simulation to search and sample parts of conformational space the simulation hasn't yet seen. Box 10.5 describes how accelerated sampling identifies the activation mechanism of a GPCR.

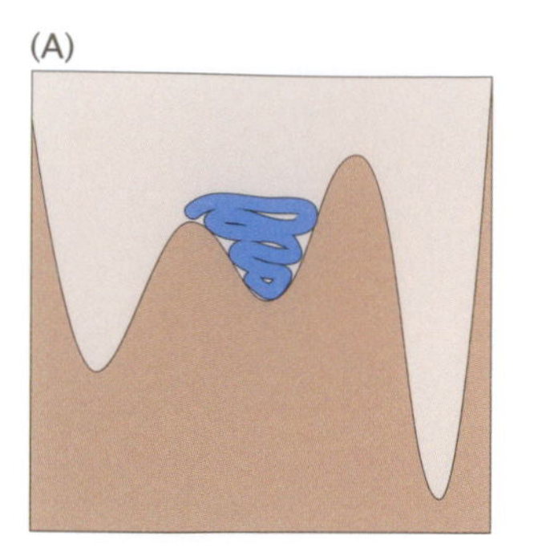

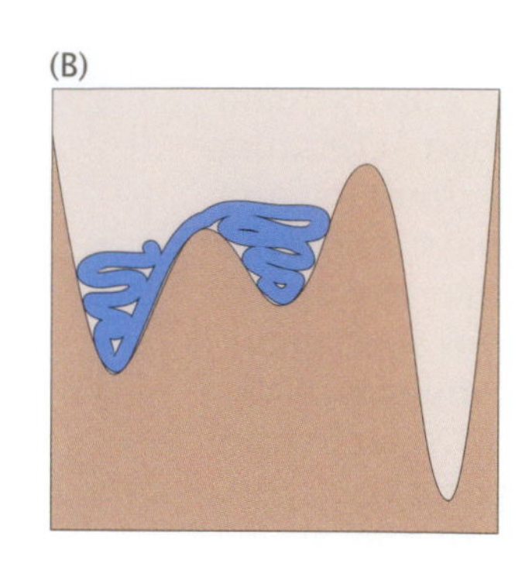

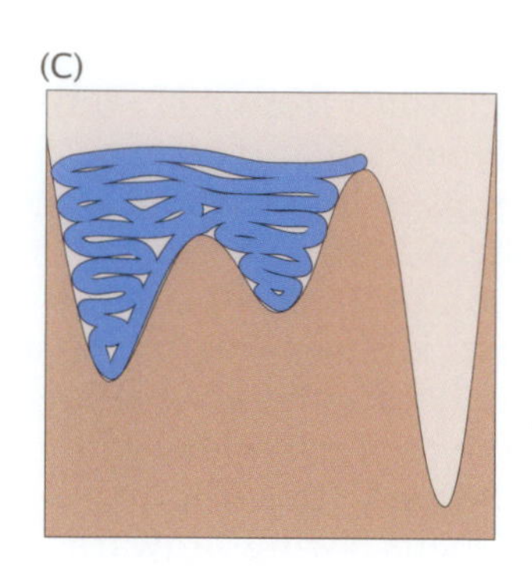

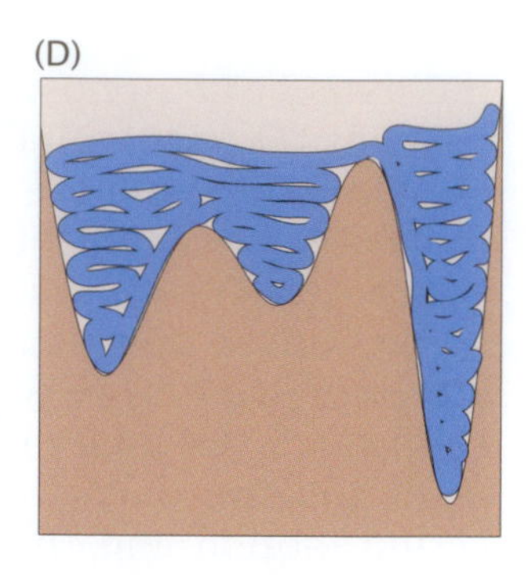

Figure 10.7 Enhanced sampling by "filling wells with computational sand." *Blue* strands represent the trajectories that are sampled – the "computational sand." (A) Initial sampling is in the middle energy well. As sampling proceeds, the middle well fills up with *computational sand.* (B) Sampling now spills over into the left well. (C) Further sampling now fills up both left-hand wells. (D) Sampling now spills over the the rightmost well, reaching states of lower energy than were previously sampled.

Box 10.5 Accelerated MD Simulations Show the Activation Mechanism of the M2 Muscarinic GPCR

G-protein-coupled receptors (GPCRs) are proteins that transmit chemical signals across cell membranes. When hormones or

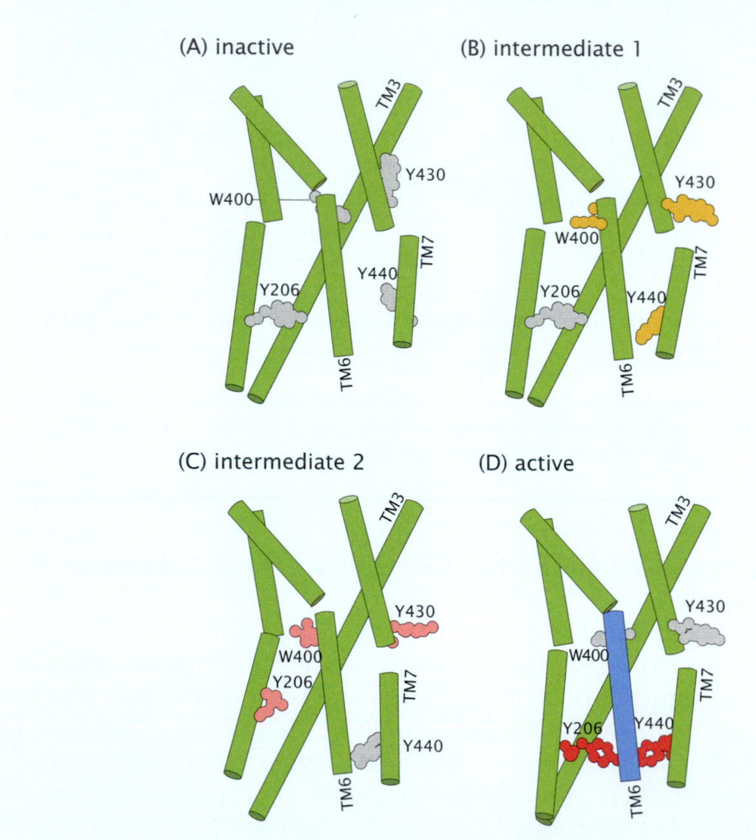

Figure 10.8 A GPCR activation mechanism has two intermediates as seen in accelerated MD simulations. (A) The inactive state. (B) In intermediate 1, the side chains of Trp400, Tyr430, and Tyr440 have moved. (C) In intermediate 2, tyrosine Y206 reorients its side chain to the lipid-exposed side of TM6. (D) In the transition to the active state, Y206 on TM5 and Y440 on TM7 reorient and form a mutual hydrogen bond. The transmembrane helices undergo cooperative reorientations: the cytoplasmic end of TM6 tilts ~6 Å outward away from the helix bundle, and TM5 and TM7 undergo small orientational changes. (Adapted from Y Miao, SE Nichols, PM Gasper, et al. *Proc Natl Acad Sci USA*, 110:10982–10987, 2013.)

neurotransmitters bind to the extracellular parts of a GPCR, the GPCR triggers biochemical responses inside the cell. To study the process of activation of these receptors, JA McCammon et al. used accelerated MD [16]. Figure 10.8 shows the two chemical intermediates along the pathway of activation. Without a ligand bound, the M2 receptor is in a conformational equilibrium between its inactive, intermediate, and active states. Upon activation, three tyrosines and a tryptophan undergo sidechain reorientation, forming a hydrogen bond, and tilting a transmembrane helix away from the bundle of other helices. The opening of the cytoplasmic end of TM6 helps accommodate binding to G-protein.

Stochastic Dynamics Entails Averaging Over Solvent Fluctuations

Yet another strategy for speeding up molecular simulations of proteins is to represent the solvent fluctuations by their average properties, rather than explicitly. In protein solutions, water molecules bombard the protein rapidly in random ways. Solvent bombardment is responsible for hydrodynamic properties such as friction, diffusion, and viscosity. Implicit-solvent modeling captures only equilibrium electrostatics, not these hydrodynamic properties. Explicit-solvent simulations do capture the solvent dynamics, but at great computational expense. In *stochastic dynamics* (SD) simulations, you treat only implicitly the viscous friction of the waters that slow down the protein motions, rather than treating water collisions explicitly. SD treats the effect of solvent Brownian motion as averaged random forces and frictions on the protein.

Two types of stochastic simulations are used for proteins: *Langevin dynamics* (LD) and *Brownian dynamics* (BD). LD uses an equation of motion of the form

$$m_i \mathbf{a}_i = -\nabla_{\mathbf{r}_i} U - \zeta \mathbf{v}_i(t) + \mathbf{f}_i(t). \tag{10.13}$$

In such models, you treat the solvation energetics (which is different from the solvation hydrodynamics) either with an explicit-solvent model or with an implicit-solvent model such as generalized Born or Poisson–Boltzmann. Compared with Equation 10.4, the hydrodynamic effects of solvent in Equation 10.13 are now represented by *stochastic* forces $\mathbf{f}_i(t)$ exerted on each individual atom i of the protein, and by an effective friction coefficient ζ. $\zeta \mathbf{v}_i(t)$ is called the *viscous drag* force. The fluctuating random force represents the same solvent actions that are responsible for its friction and viscosity. The stochastic force $\mathbf{f}_i(t)$ is a random variable with a Gaussian distribution. Its mean is zero, since it is random, and we assume that the force is not correlated in time, so we have

$$\begin{aligned} \langle \mathbf{f}_i(t) \rangle &= 0, \\ \langle \mathbf{f}_i(t) \cdot \mathbf{f}_i(t') \rangle &= 6RT\zeta\delta(t - t'), \end{aligned} \tag{10.14}$$

where the Dirac delta function $\delta(t - t')$ equals 0 if $t \neq t'$, and equals 1 if $t = t'$. In a sufficiently viscous environment (called *overdamped* conditions), you can set the acceleration term (the leftmost term in Equation 10.13) to zero, giving the simpler *Brownian dynamics*:

$$\zeta \mathbf{v}_i(t) = -\nabla_{\mathbf{r}_i} U + \mathbf{f}_i(t). \tag{10.15}$$

At equilibrium, the average of effects of the random force and the frictional dissipation sum to zero, as they should according to Equation 10.4.

In the next section, we switch from MD simulations to equilibrium searching and sampling by Monte Carlo methods.

METROPOLIS MONTE CARLO SIMULATION IS A STOCHASTIC METHOD OF SAMPLING CONFORMATIONS

Monte Carlo (MC) is the name given to a collection of methods that estimate averages and variances by using a series of random samples. Its name derives from casinos in Monte Carlo, Monaco, a favorite gambling spot of an uncle of the mathematician Stanislaw Ulam, the person who developed the method in 1946. Of greatest interest here is the method called *Metropolis Monte Carlo* (MMC), said to be one of the most important advances in theoretical chemistry in the twentieth century. This method is named after the first author of a paper by Metropolis et al. in 1953 [17].

Unlike MD, MC does not give motions or the time dependences of properties. MC is an equilibrium method that seeks states of low free energy, samples the landscape fairly efficiently, and properly preserves populations. But MC has some advantages. First, MC is not limited to femtosecond time steps, so it can be more efficient for conformational searching over many degrees of freedom. Second, MC requires only energies, not their derivatives (which are needed to compute the forces), and this is useful. Some types of energies, such as statistical potentials, are often represented in a simple way: contact or no contact, that is, *square-well potentials*. Such discontinuous potential functions cannot be used in MD because they would give forces that would be infinite. But such potentials are no problem for MC. Another advantage is that MC is not limited to small Newtonian moves at each time step, so MC can often explore conformational space more broadly.

The following subsections show how MC simulations, particularly MMC, are used to compute average properties by efficient sampling of energy landscapes. By *efficient*, we mean that it gives relatively accurate estimates of averages from relatively few sampled points. First, however, we show why random sampling methods are not very efficient.

Here's How to Estimate Averages by Uniform Sampling

Suppose you are given an energy landscape $U(x)$. To keep the math simple, we'll assume that it is a function of only one variable x. The corresponding probability distribution is

$$p(x) = Q^{-1} e^{-\beta U(x)}, \tag{10.16}$$

where

$$Q = \int e^{-\beta U(x)}\, dx \tag{10.17}$$

is the partition function and $\beta = 1/RT$. You can compute the average value of some property, $f(x)$, as

$$\langle f \rangle = \int f(x)p(x)\, dx = \frac{\int f(x)e^{-\beta U(x)}\, dx}{\int e^{-\beta U(x)}\, dx}. \tag{10.18}$$

The problems of interest here are situations where the space of x is so large that you need to estimate this average based on only a relatively small number of samples, say M of them. In such a case, we define

$$\langle f \rangle_{\text{estimate}} = \sum_{i=1}^{M} f(x_i)p(x_i) \approx \langle f \rangle. \tag{10.19}$$

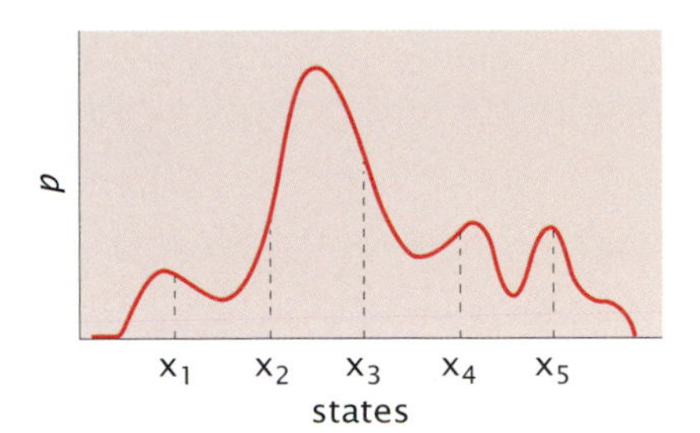

Figure 10.9 Uniform sampling can miss important regions of a distribution. When you want to compute an average $\langle f \rangle$ of some property f (see Equation 10.19) over a probability distribution $p(x)$, you could sample uniformly spaced x values (indicated by the vertical *dashed* lines). But, you would often miss the most important parts of this distribution, where the probability is highest, in this case between x_2 and x_3. So, your estimates of the average $\langle f \rangle$ would have large errors. The MMC method is more efficient and accurate.

In order for Equation 10.19 to give a good estimate of $\langle f \rangle$, you don't want to waste your time sampling places where $p(x)$ is small. Instead, you want to sample values of x where $p(x)$ is sufficiently large that those points would contribute significantly to the estimate in Equation 10.19. Said differently, you want to sample low values of $U(x)$. You want to sample more in the regions of highest population, without wasting much time sampling regions of low population. Figure 10.9 illustrates the problem. The vertical dashed lines in the figure indicate uniform sampling, the simplest approach for the computations in Equation 10.19. According to this procedure, you would apply the following steps:

(a) Sample at uniform intervals: x, $x+\Delta x$, $x+2\Delta x$, $x+3\Delta x$, $x+4\Delta x$, ...

(b) Collect up the values of x for each of these points. We call this "accepting" the value x. (Here, we accept *all* the values. The point of introducing this terminology will become clearer later.)

(c) Weight each accepted value by its appropriate statistical (or Boltzmann) weight to get the probabilities for each point:

$$p_1 = p(x),$$
$$p_2 = p(x+\Delta x),$$
$$p_3 = p(x+2\Delta x),$$
$$\vdots$$

Uniform sampling will give you an estimate of the average according to Equation 10.19. However, it is not generally very useful. For example, Figure 10.9 shows that uniform sampling misses the regions of largest population, such as between x_2 and x_3.

MMC Is an Efficient Method of Sampling Populated States

The Metropolis scheme, which is much more efficient, does not sample at uniformly spaced intervals. Rather, it follows a series of steps that are mostly downhill on the energy landscape, but sometimes uphill.

It can overcome barriers to find valleys having even lower energy, and to ensure sampling that preserves the distribution that it samples. Now instead of accepting all the points, you accept some and reject others, using a Boltzmann decision criterion. The point of the MMC approach is to sample heavily near states that appear to have the highest probabilities, while not wasting much time in less-important regions.

To understand the idea of *Metropolis acceptance probability*, start at step k, where you have conformation x_k. Now, choose randomly from a *move set* and *propose* a new conformation x_{k+1} that the system could take at step $k+1$. Then *accept* or *reject* that proposed move according to the following procedure. Define a quantity R_+ as the ratio of probabilities of those two states:

$$R_+(k) = \frac{p(x_{k+1})}{p(x_k)} = \exp\{-\beta[U(x_{k+1}) - U(x_k)]\}. \qquad (10.20)$$

Since we know the energy function $U(x)$, we can readily compute $R_+(k)$ for any possible step on the landscape. In the Metropolis method, we do not choose just any random step to put into our accumulating sum. Rather, we randomly select a possible move, then we only turn it into an actual move after computing the quantity $R_+(k)$, which tells us whether to accept or reject the potential move. Central to this Metropolis procedure is the acceptance probability a for taking the step from k to $k+1$, defined as

$$a_{k\to k+1} = \begin{cases} 1 & \text{if } R_+(k) > 1, \\ R_+(k) & \text{if } R_+(k) \leq 1. \end{cases} \qquad (10.21)$$

A key part of any MC process is its *move set*, which is a collection of rules for how your system is allowed to transition from one state k to the next, $k+1$. For example, a lattice-model move set is given in Box 10.6. MC is a popular sampling method for coarse-grained and lattice models.

Box 10.7 describes the recipe for performing MMC.

Box 10.6 Example of a *Move Set* for MMC Simulations

Constructing a move set requires some care. First, the move set must be *ergodic*. That is, all parts of the conformational space must be accessible to at least some sequence of moves. Second, some move sets are more efficient than others for accessing the regions of conformational space of interest. You want to choose an efficient move set.

Figure 10.10 shows a move set for MC sampling of the HP lattice model (see Chapter 3). Three types of moves are allowed: (1) rotations of single bonds; (2) rotations of two adjacent bonds; and (3) "crankshaft" rotations of three adjacent bonds.

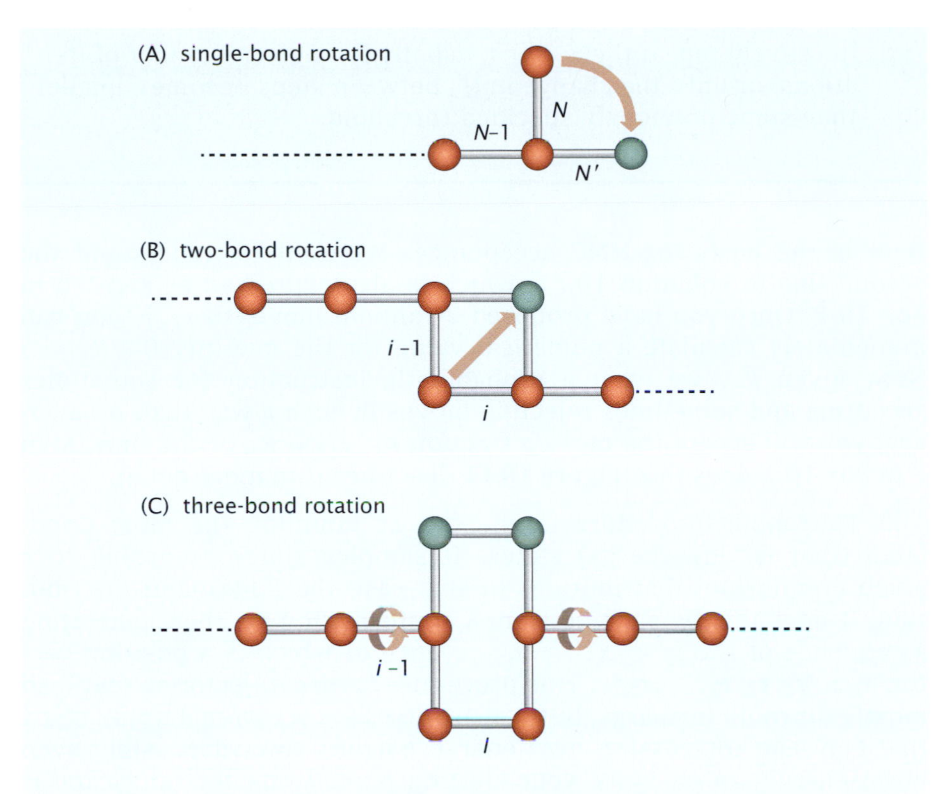

Figure 10.10 Example of a move set for a two-dimensional lattice model: Three different moves are allowed: (A) a single-bond rotation at the end of a chain; (B) a two-bond rotation; and (C) a three-bond "crankshaft" motion.

Box 10.7 The MMC Protocol

Here's how you apply MMC to sampling an ensemble of conformations:

1. Begin with a conformation x_k.
2. Evaluate its energy $U(x_k)$.
3. From a move set, select a random move to a new conformation x_{k+1}.
4. Evaluate the energy of this conformation, $U(x_{k+1})$.
5. Is the proposed conformation x_{k+1} more favorable than conformation x_k? (Is the energy $\Delta U(x_{k+1}) = U(x_{k+1}) - U(x_k)$ negative?)
6. If this proposed move is downhill or level (that is, $\Delta U(x_k + 1) \leq 0$, or $R_+(k) \geq 1$), then accept the move and update the average value $\langle f \rangle$ using Equation 10.19.
7. But if the proposed move is uphill, there are two possibilities: (a) sometimes you accept, and (b) sometimes you reject. Specifically, you accept the move a fraction R_+ of the time. Here is how you do this in MMC simulations:
   (a) Generate a random number η between 0 and 1.
   (b) If $\eta < R_+$, accept this proposed move. If $\eta \geq R_+$, reject this proposed move and add the previous value onto the list one additional time.

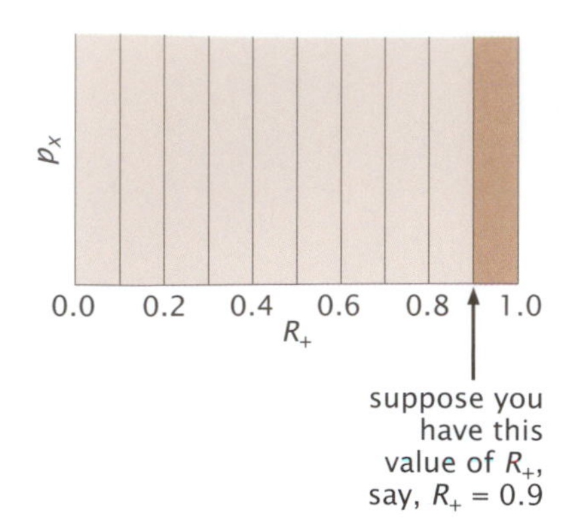

Figure 10.11 How Metropolis acceptance produces proper probabilities. (i) You are given a proposed move. That move has an associated value, say $R_+ = 0.9$. (ii) Now, turn 0.9 into a probabalistic prescription for randomly accepting such a move 90% of the time and rejecting it 10% of the time. (iii) To do that, generate a random number η between 0 and 1 according to the flat distribution shown. (iv) Accept the proposed move whenever $\eta \geq R_+$.

8. Iteratively repeat these steps, usually for a given number of iterations, or until the change in $\langle f \rangle$ between steps becomes smaller than some previously specified threshold.

Here is the basis for MMC acceptance. We want to understand the second line in Equation 10.21, which is also expressed as step 7b in Box 10.7. Once you have proposed a random move, to x_{k+1}, you can immediately calculate a numerical value for the quantity $0 < R_+ \leq 1$. Now, given R_+, we want a probabilistic instruction for sometimes accepting and sometimes rejecting moves in such a way that, on average, you will accept the move a fraction $a_{k \to k+1} = R_+$ of the time. Step 7 in Box 10.7 does this. Figure 10.11 describes it in more detail.

This Metropolis procedure is efficient at sampling the most populated (that is, low-energy) states. It samples states according to a given equilibrium distribution, in this case the Boltzmann distribution. It does this by starting from a conformation x_1, then generating a sequence of states $x_2, x_3, \ldots, x_k, \ldots$, each of which is a position vector, $x_k = \{\mathbf{r}_1, \mathbf{r}_2, \mathbf{r}_3, \ldots, \mathbf{r}_N\}_k$. This procedure favors trajectories that lead mostly downhill in energy, but it also allows for occasional uphill steps that can lead to crossing over energy barriers toward possibly even lower-energy states. Since your starting point x_1 on the landscape is arbitrary, it is not likely to have a low energy. It is only after a sufficiently large number of steps that an MMC procedure is likely to reach low-energy (high-probability) states. In the limit as $k \to \infty$, the MMC procedure approaches the equilibrium distribution for $p(x_k)$ because the acceptance probabilities satisfy a kinetics-like *detailed balance* condition:

$$p_k a_{k \to k+1} = p_{k+1} a_{k+1 \to k}, \tag{10.22}$$

as you can see by combining Equations 10.21 and 10.20.

Finally, there remains the matter of how to obtain a proper average $\langle f \rangle$. If you accept a move, you must *add the accepted value* into your updated average sum $\sum_{i=1}^{k+1} f(x_i)p(x_i)$. But if you reject a move, you must still perform the action of *adding the previous value* into your updated average sum. The reason for this seemingly complicated summing rule is that it samples properly to give you correct averages. Your *proposed moves* sample from the distribution randomly; your *accepted moves* sample the most important parts of the distribution. For example, suppose your landscape has only one state of high probability, $p_{\text{hi}} = 0.5$, and 10 states of low probability $p_{\text{lo}} = 0.05$. Then, to compute a correct average, you need to weight the value of f found in the low-population regions 10 times more heavily than the value you find in the high-population region. This "propose–accept" scheme is a way of getting accurate averages of distributions while sampling them only sparsely.

A virtue of MMC is that all its decisions are based on ratios of probabilities (R_+), not on the absolute probabilities themselves. So, you never need the partition function Q from Equation 10.16. This is important because the determination of Q would require an exhaustive sampling of the entire conformational space, which is computationally very challenging.

There are other acceptance criteria you could have used instead of the one specified by Equation 10.21 that would have also satisfied Equation 10.22 and given the correct probability distribution $p(x)$. However, the advantage of the Metropolis acceptance criterion (Equation 10.21) is that by *always* accepting moves that are downhill in energy, this procedure is much more efficient at finding the important regions of $p(x)$ than other decision procedures would be. Although the MMC procedure gives you trajectories, you should not regard them as physical routes that would be observed in kinetics experiments. Rather, MMC is just a principled way to converge upon a proper equilibrium distribution. And, incidentally, convergence alone does not guarantee that the important states have been sampled. It can only find important states if it has explored those regions.

Box 10.8 illustrates an application of MC simulations, namely, in designing a protein fold that had not been observed in nature before.

Box 10.8 Proteins Have Been Designed by MC Sampling of Sequences

MC sampling has been used to design proteins, by efficiently sampling sequence space. RosettaDesign is a computer server [18] that uses the Rosetta energy function and a Monte Carlo search method to explore possible folded conformations of amino acid sequences [19]. It was used to design a 92-amino-acid sequence that would fold into an $\alpha\beta$ structure. Subsequently, the authors synthesized the sequence and determined its structure by X-ray crystallography. They found that the designed protein, called Top-7, does indeed fold into its intended native structure (Figure 10.12).

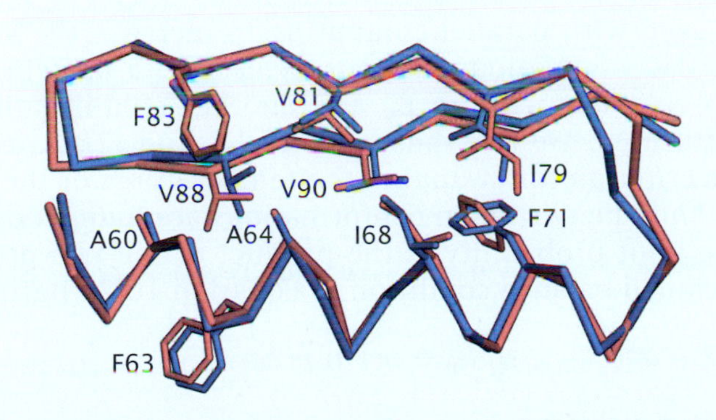

Figure 10.12 A previously unknown protein fold, Top-7, was designed using MC sampling with the Rosetta energy function. (*Blue*) The computationally designed model. (*Orange*) The structure subsequently obtained by crystallography of the synthesized protein indicates a successful design. (From B Kuhlman, G Dantas, GC Ireton, et al. *Science*, 302:1364–1368, 2003. With permission from AAAS.)

ADDITIONAL PRINCIPLES LEAD TO IMPROVED COMPUTATIONAL SAMPLING

In this section, and in Appendix 10C, we describe further enhancements of molecular simulations.

Sampling at Higher Temperatures Allows Broader Exploration of Configuration Space

Varying the temperature of a molecular simulation can lead to more efficient sampling. Let's consider a situation in which we fix the configurational state of the system, but we change its temperature T. We apply a variant of the MMC method. Define $\beta = 1/RT$. Consider a cold temperature β_c and a hotter temperature β_h, and apply Equation 10.22, to get

$$p_c a_{c \to h} = p_h a_{h \to c}. \tag{10.23}$$

The probability ratio is computed as

$$R_+ = \frac{p_h}{p_c} = \frac{e^{-\beta_h U}}{e^{-\beta_c U}} = e^{-(\beta_h - \beta_c)U}. \tag{10.24}$$

To satisfy Equation 10.23, we use the MMC acceptance criterion:

$$a_{c \to h} = \begin{cases} 1 & \text{if } R_+ > 1, \\ R_+ & \text{if } R_+ \leq 1. \end{cases} \tag{10.25}$$

The use of temperature variation is the basis for another important method, called replica exchange, described below.

The Replica-Exchange Method (REM) Is an Efficient Sampling Method

A sampling method called *replica exchange* (or parallel tempering), combines conformational changes and temperature changes into a single simulation to increase the computational efficiency of sampling. The REM is used with parallel computing to increase the sampling of either MC or MD simulations. Suppose you have a protein conformation α_1 at a cold temperature β_c and another conformation α_2 at a hot temperature β_h. We call these states $(1c)$ and $(2h)$. We then seek a Metropolis criterion for swapping the temperatures of the two states to $(2c)$ and $(1h)$. The two chain conformations are independent of each other, so the joint probability is the product of the two probabilities. Thus, the detailed balance condition in Equation 10.22 becomes

$$p(1c)\ p(2h)\ a_{(1c,\ 2h) \to (2c,\ 1h)} = p(1h)\ p(2c)\ a_{(1h,\ 2c) \to (1c,\ 2h)}. \tag{10.26}$$

For the forward direction, we have

$$R_+ = \frac{p(2c)p(1h)}{p(1c)p(2h)} = e^{-\beta_c U_2 - \beta_h U_1 + \beta_c U_1 + \beta_h U_2} = e^{(\beta_h - \beta_c)(U_2 - U_1)}, \tag{10.27}$$

so the acceptance criterion is

$$a_{(1c,2h) \to (2c,1h)} = \begin{cases} 1 & \text{if } R_+ > 1, \\ R_+ & \text{if } R_+ \leq 1, \end{cases} \tag{10.28}$$

where R_+ is given by Equation 10.27 for this swapping process. Replica-exchange swapping is illustrated in Figure 10.13.

In practice, replica exchange performs sampling of multiple different temperatures and attempts to swap only between adjacent temperatures (since these will allow for the largest acceptance ratios) (see

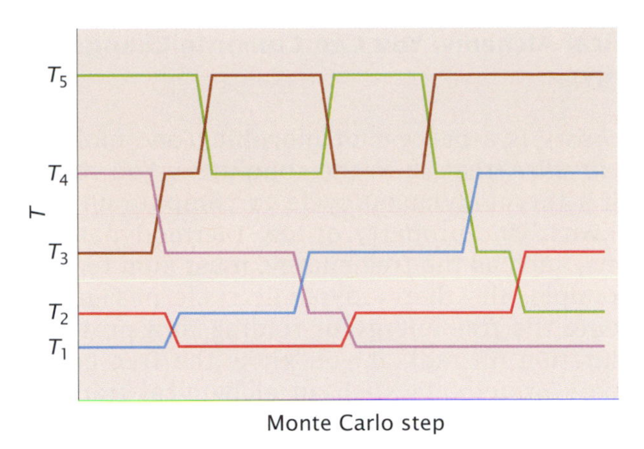

Figure 10.13 The replica-exchange method. In REM, you run many different simulations in parallel (say, by MD), each at a different temperature. Periodically, you query whether two runs at adjacent temperatures can be swapped according to an MMC criterion. You either accept or reject that swap. If you accept, it means that the previously colder simulation is now running at the higher temperature, and vice versa. This procedure is efficient because low-temperature simulations explore small details within local energy wells, while high-temperature simulations can explore much broader reaches of conformational space. Colors correspond to the paths that individual samples traverse.

Figure 10.13). The REM method is highly parallelizable and often very efficient.

Box 10.9 illustrates an application of the REM method to understanding the mechanism of how urea denatures proteins.

Box 10.9 Molecular Simulations Show How Urea Denatures Proteins

One of the most effective ways to denature a water-soluble protein is to add concentrated urea to the protein solution. How does the interaction with urea drive the protein to unfold? It was previously thought that urea unfolds proteins by forming better hydrogen bonds to proteins than water alone can. But A Garcia and coworkers who studied the mechanism of urea–protein interactions using all-atom replica-exchange molecular dynamics (REMD) simulations of the Trp cage protein found a different mechanism [20]. They found that the main action of urea (Figure 10.14) is through the electrostatic and van der Waals forces, not the hydrogen bonds. They also saw that urea interacts preferentially with the side chains, rather than the backbone. Their predicted curves of denaturation by temperature and urea agree well with experiments.

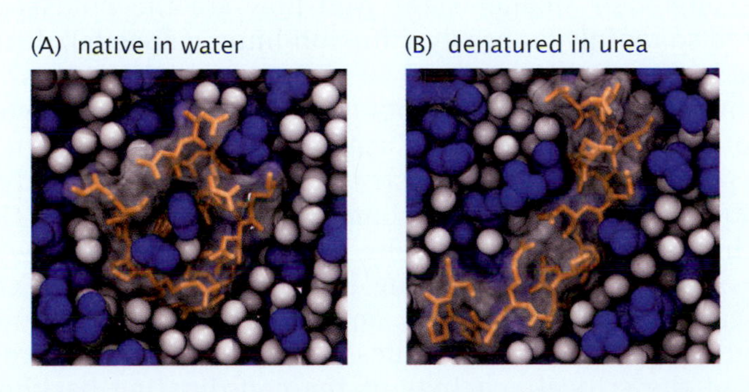

Figure 10.14 The locations of urea (*blue*) and water molecules (*white*) around a small protein called Trp cage (*orange*), obtained from REMD simulations. (A) Water surrounding the native structure. (B) Urea surrounding the denatured structure. (Adapted from DR Canchi, D Paschek, and AE Garcia. *J Am Chem Soc*, 132:2338–2344, 2010. With permission from American Chemical Society.)

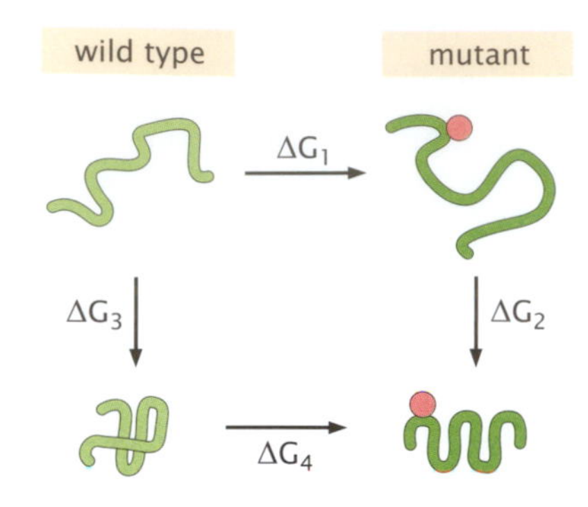

Figure 10.15 A thermodynamic cycle for *alchemical transformation*, in this case a mutation. To predict the stability ΔG_2 of a mutated protein, you can use an alchemical transformation and a thermodynamic cycle. At thermodynamic equilibrium, $\Delta G_2 = \Delta G_3 + \Delta G_4 - \Delta G_1$. So, start with the measured stability ΔG_3 of the wild-type protein. In the computer, you "morph" a mutation into the folded and unfolded chains, to get ΔG_1 and ΔG_4. This alchemical transformation then gives you ΔG_2.

Using Chemical Alchemy, You Can Compute Changes in Free Energy

Chemical alchemy is a process of morphing one molecular structure into another *in silico* (that is, in the computer). You combine chemical alchemy with a thermodynamic cycle to compute an unknown quantity from knowns. One property of any thermodynamic cycle is that state functions, such as the free energy, must sum to zero around the cycle. For example, the thermodynamic cycle in Figure 10.15 shows how to compute the free energy of folding in a protein containing a single-site mutation (in red), if you know the free energy of folding of the wild-type protein. In such an alchemical transformation, the mutated residue is varied continuously, as if you were turning a knob to change its chemical structure and properties gradually. This computational transformation does not correspond to any realizable physical process, but that doesn't matter, because values of state functions only depend on the two end states, not the pathway. The computer makes small changes, which it can do efficiently [21]. The alchemical substitutions are made in both the unfolded and folded states (horizontal steps) to probe the change in free energy due to the amino acid substitution. Their difference ($\Delta G_4 - \Delta G_1$) gives you the free energy difference between the folding free energies of the wild-type and mutant proteins, $\Delta\Delta G_{\text{fold}} = \Delta G_2 - \Delta G_3$.

Box 10.10 illustrates how MD simulations using free-energy-perturbation methods (see Appendix 10C) give insights into how ions pass through a membrane potassium channel called KcsA.

Box 10.10 How Do Potassium Channels Select for Potassium Ions?

Potassium channels are the most commonly observed ion channels in living organisms, present in virtually all types of cells. These are membrane proteins that selectively translocate K^+ ions, and not other ions. This process is essential for maintaining the resting potential (or voltage) across membranes, and regulating the action potentials in neurons. What is the transport mechanism for potassium ions? How is the protein pore selective for potassium ions? For example, why are sodium ions not conducted by potassium channels, despite their smaller size? And how are the potassium ions conducted so rapidly (at nearly diffusion-limited speeds), despite the expected high energetic costs of stripping waters from ions?

MacKinnon and collaborators obtained a crystal structure of a bacterial potassium channel [22], KscA (This work was recognized by the 2003 Nobel Prize in Chemistry.) Different molecular dynamics simulations of KcsA have given somewhat different details. One simulation indicates that the K^+ ions hop through a selectivity filter at four sites, S1–S4, coordinating their interactions with critically spaced carbonyl groups on the amino acids in the channel. The ions alternate with water molecules in single file [23] (Figure 10.16). Because of the critical spacings of the coordinating backbone carbonyl atoms, the channel only conducts K^+, and because of the alternation with water molecules, the system does not need to spend the energy to strip all the waters off the ions. This type of efficient coordination across the selectivity filter would not be achieved if the channeling ions were smaller in size (such as Na^+ ions), explaining the selectivity. A more recent MD study by deGroot et al. indicates a different mechanism, whereby waters are not intercalated between the ions as they flow through the channel [24].

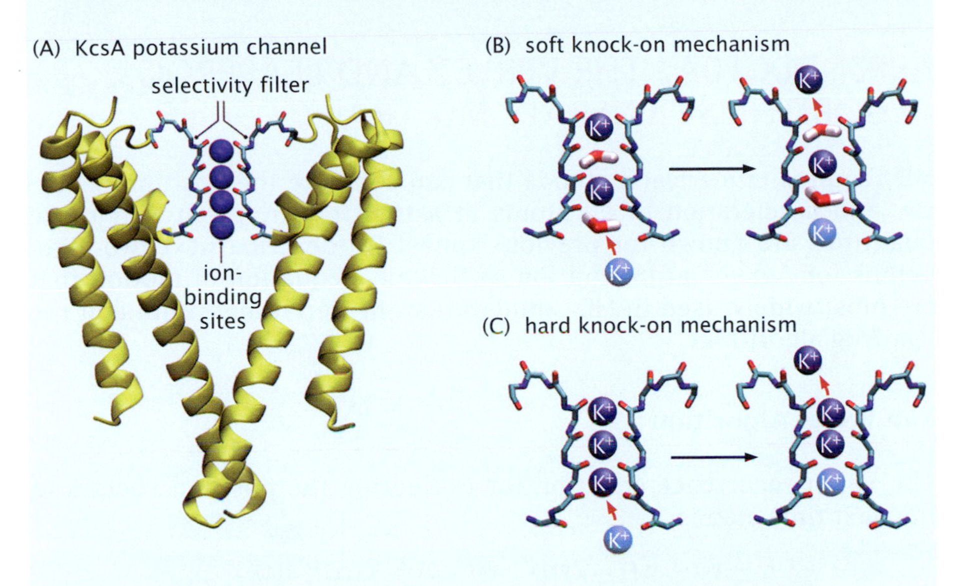

Figure 10.16 Mechanism of potassium translocation through the selectivity filter of potassium channels, from MD simulations, for KcsA. (A) Passage through selectivity filter observed by deGroot and coworkers [24] as a single file of potassium ions (*dark blue* spheres). This contrasts with a previous model (B) where intercalating potassium ions and water molecules were observed to hop from one site to another [22, 25, 26]. In the new Coulomb knock-on model (A and C), the electrostatic repulsion between the ions is the main force driving the hopping from one site to another. (A, from S Bernèche and B Roux. *Biophys J*, 82:772–780, 2002. With permission from Elsevier. B, MS Sansom. *Biophys J*, 78:557–70, 2000. With permission from Elsevier.)

Instead, potassium ions directly interact with each other, and their electrostatic repulsion, or so-called Coulomb knock-on, drives their movement across the selectivity filter, much more efficiently (as required physiologically) than that observed in the presence of intercalating water molecules. The combination of experiments and simulations are providing increasingly deeper understanding of the mechanism of K^+ channeling.

SUMMARY

Molecular simulations of proteins give insights into their mechanisms of actions at the atomic scale. This chapter describes three of the major methods used in biomolecular simulations. First, energy minimization computes the derivatives of an energy function and follows gradients downhill to help identify equilibrium conformations. Second, in molecular dynamics (MD), the computer solves Newton's equations of motion at each time step to generate trajectories representing the time evolution of biomolecular events. MD also has the ability to go uphill, so it can climb over energy barriers to mimic the kinetics and seek states of lowest free energy. Therefore, it is capable of finding states of low energy and states of low free energy. Third, Metropolis Monte Carlo (MMC) is an efficient way to sample complex distributions. It too can surmount energy barriers, and can give proper equilibrium Boltzmann populations, for sufficiently long simulations. Molecular simulations provide narratives that explain protein motions and biological mechanisms in terms of structure at a level of detail that is finer-grained than experiments usually can.

APPENDIX 10A: THE VERLET AND LEAPFROG ALGORITHMS GENERATE MD TRAJECTORIES

MD requires numerical methods that can compute the position, velocity, and acceleration of all atoms at a time $t + \Delta t$, given that those quantities are known for previous times. We describe next two algorithms for numerical integration of Newton's equation of motion that are most widely used in MD simulations: the Verlet algorithm and the leapfrog algorithm.

The Verlet Algorithm

The Verlet recurrence equation for evaluating the position vectors at the next time step is

$$\mathbf{r}(t + \Delta t) = 2\mathbf{r}(t) - \mathbf{r}(t - \Delta t) + (\Delta t)^2 \mathbf{a}(t).$$

To understand where this equation comes from, we can write the position vector at $t + \Delta t$ as a Taylor series expansion around the preceding position vector, $\mathbf{r}(t)$. We also write the position vector for $t - \Delta t$. The resulting series, truncated after the quadratic terms, are

$$\mathbf{r}(t + \Delta t) = \mathbf{r}(t) + \mathbf{v}(t)\Delta t + \tfrac{1}{2}\mathbf{a}(t)(\Delta t)^2 + \ldots,$$

$$\mathbf{r}(t - \Delta t) = \mathbf{r}(t) - \mathbf{v}(t)\Delta t + \tfrac{1}{2}\mathbf{a}(t)(\Delta t)^2 - \ldots .$$

The Verlet equation is obtained by simply summing up these two equations. You see that the velocity term is eliminated, and we obtain the Verlet equation after rearranging the terms of the equality. The velocities at each time step can be evaluated by simply taking the difference of these two Taylor series equations, to obtain

$$\mathbf{v}(t) = \frac{\mathbf{r}(t + \Delta t) - \mathbf{r}(t - \Delta t)}{2\Delta t}.$$

In the Verlet algorithm, you start with two successive position vectors $\mathbf{r}(t - \Delta t)$ and $\mathbf{r}(t)$, and evaluate the acceleration $\mathbf{a}(t)$ from the negative gradient of the potential, computed as a function of $\mathbf{r}(t)$.

The Leapfrog Algorithm

A different way to compute time steps is the leapfrog algorithm, which computes the velocity at the half-time steps: $\mathbf{v}(t \pm \Delta t/2)$. You begin with an initial set of velocities for each atom, drawn from the Maxwell–Boltzmann velocity distribution (Equation 10.7), and initial positions $\mathbf{r}(t)$. The position vectors are used to evaluate the energy at time t, as well as its gradient, which yields the accelerations $\mathbf{a}(t)$ for all atoms.

Then, using $\mathbf{v}(t + \Delta t/2) = \mathbf{v}(t) + \mathbf{a}(t)\Delta t/2$, the position at the next time step is

$$\mathbf{r}(t + \Delta t) = \mathbf{r}(t) + \mathbf{v}(t + \Delta t/2)\Delta t.$$

APPENDIX 10B: PERIODIC BOUNDARY CONDITIONS ARE USED IN MD SIMULATIONS

To solve Newton's equations of motion, which are partial differential equations, you need an initial condition in time and to select a simulation box in which to conduct the simulations. How should you place the boundaries around your system to ensure that your simulation box does not bias the trajectory being generated, and that you maintain the number of particles in your simulation box? It would be computationally cheapest to use a small container around your system, but then your system atoms will have unwanted interactions with the boundary surfaces. If, on the other hand, you use a large container, it will be computationally expensive to simulate the many atoms that are needed to fill it. The solution to these problems is to use *periodic boundary conditions*. You assume several identical copies (replicas) of the simulated system, forming a grid (Figure 10.B.1). The central box should be made sufficiently large that the protein in one box has only negligible interactions with the copy of the protein molecule in the neighboring box. In order to maintain the concentration of small molecules (such as water molecules, ions, or ligands) surrounding the protein, each molecule that crosses the boundary of the central box is assumed to re-enter the same box from the opposite end, as shown in Figure 10.B.1. This way, the total number of molecules is conserved during the course of simulations.

APPENDIX 10C: SOME METHODS FOR ENHANCED SAMPLING

Histogram-Reweighting Methods Let You Predict a Distribution at One Temperature from Another

We describe here how you can compute an average property $\langle f \rangle_h$ at a temperature T_h if you know the average property $\langle f \rangle_c$ at T_c.

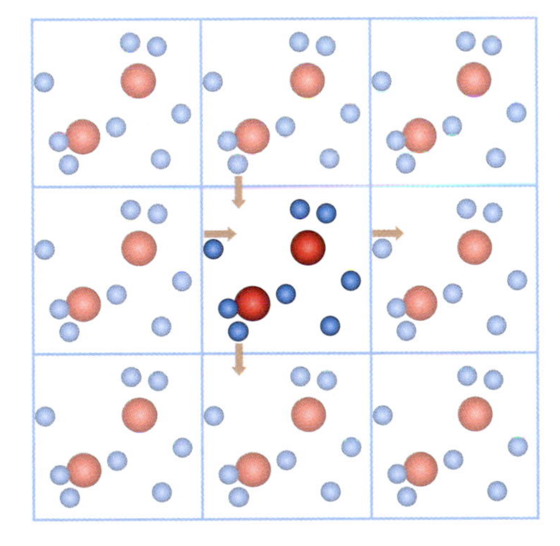

Figure 10.B.1 Periodic boundary conditions are applied by adding replicated copies of the central simulation box as neighboring boxes that surround the central box. To maintain the proper density, each molecule leaving the simulation box re-enters from the opposite end (see the arrows). The important feature is that the number of *blue* waters is preserved, so the concentration is constant. (Adapted from https://protect-us.mimecast.com/s/ 4QneBAFzAd8eTq?domain=isaacs. sourceforge.net, http://isaacs. sourceforge.net/phys/pbc.html.)

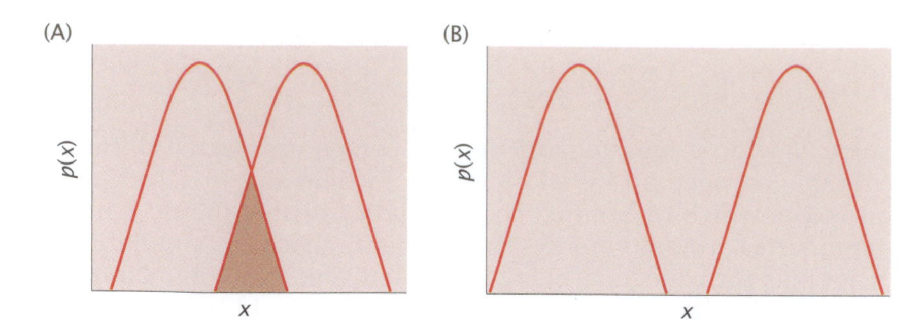

Figure 10.C.1 Histogram reweightings. (A) If you sample two probability distributions having sufficient overlap (*brown* shaded area), then you can reweight by using Equation 10.C.3. (B) If two distributions don't have sufficient overlap, reweighting will not yield good estimates of the other distribution.

The population distributions (*histograms*) for a given conformation x_k having energy U_k at the two temperatures will be

$$\begin{aligned} p_k(h) &= Q_h^{-1} e^{-\beta_h U_k}, \\ p_k(c) &= Q_c^{-1} e^{-\beta_c U_k}, \end{aligned} \tag{10.C.1}$$

where $Q_i = \sum_k e^{-\beta_i U_k}$ is the partition function for $i = h$ or $i = c$.

You can express the new average as

$$\langle f \rangle_h = \frac{\sum_k f_k p_k(h)}{\sum_k p_k(h)} = \frac{\sum_k f_k \left(p_k(h)/p_k(c)\right)}{\sum_k \left(p_k(h)/p_k(c)\right)}, \tag{10.C.2}$$

where the latter equality simply results from dividing both the numerator and denominator by the identical factor of $p_k(c)$. Now substitute Equation 10.C.1 into Equation 10.C.2 to get

$$\langle f \rangle_h = \frac{\sum_k f_k e^{-(\beta_h - \beta_c) U_k}}{\sum_k e^{-(\beta_h - \beta_c) U_k}}. \tag{10.C.3}$$

Equation 10.C.3 describes how to adjust or *reweight* the values of f_k for temperature T_h that you have already obtained from simulations at temperature T_c. The Boltzmann factors $e^{-(\beta_h - \beta_c) U_k}$ are used to reweight these quantities and thus correct for the new temperature. Predicting $\langle f \rangle_h$ from sampling of the distribution at T_c only works if the two probability distributions over U_k have substantial overlap, which means that the temperatures must be sufficiently close together, $\beta_h - \beta_c \approx 0$, as indicated in Figure 10.C.1. If there is not sufficient overlap, then your averages will just be sums over terms that are negligibly small, and this approach will give a poor approximation for $\langle f \rangle_h$ (because you will be sampling low-probability regions near the tails of the distributions). This requires the two temperature to be close to each other. If your temperatures are quite different, you can patch together multiple histograms, each pair of which has good overlap. Patching together different histograms is called the weighted histogram analysis method (WHAM) [27]. WHAM uses all the intermediate state information. The *multistate Bennett acceptance ratio* (MBAR) method uses data from all states to provide predicted free energies of unsampled states [28].

Changes in Free Energy Can Be Computed by Umbrella Sampling, Free-Energy Perturbation, or Thermodynamic Integration Methods

In this section, we consider *transitions* or *changes* from one state to another: differences in free energies, and pathways between states.

Often, you want to sample parts of an energy landscape that are separated by a high energy barrier (Figure 10.C.2A). MMC or MD will only rarely sample and pass over high barriers. So, instead, you can sample such states by introducing an artificial potential that lowers the barrier. You perform simulations on that modified landscape, and then add a correction to account for the effect of your biasing potential after the simulation. This approach is called *umbrella sampling* because the biasing potentials are parabolic, resembling an inverted umbrella (see Figure 10.C.2B).

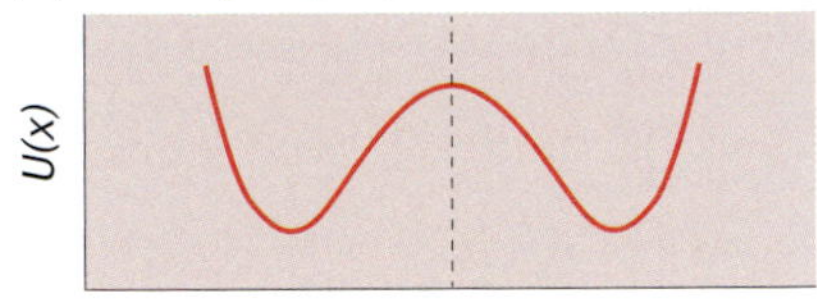

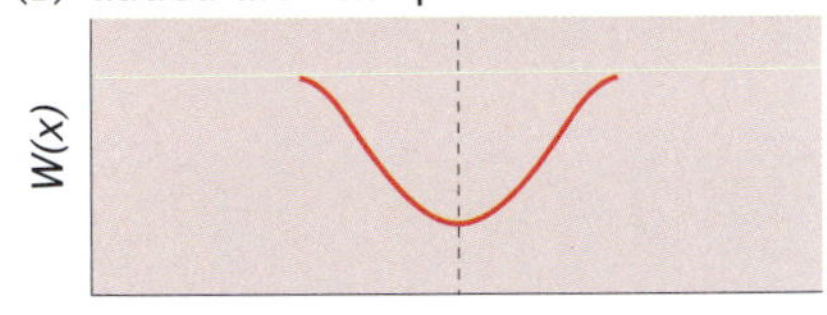

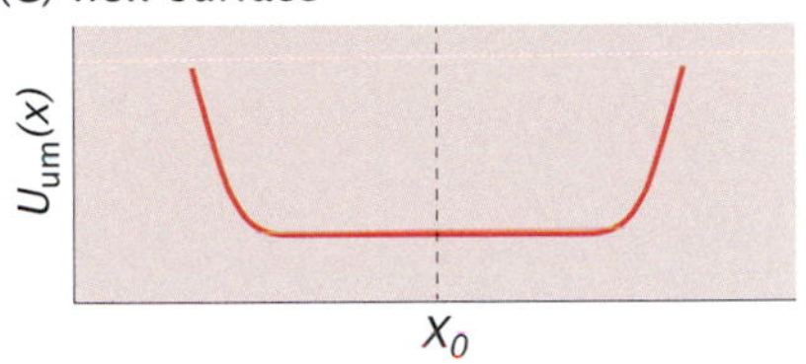

Figure 10.C.2 Umbrella sampling provides a way to sample an energy landscape with high barriers. (A) The actual energy landscape $U(x)$ has a large barrier near $x = x_0$. (B) By adding an artificial umbrella potential $W(x)$, you obtain the biased energy landscape (C) $U_{\mathrm{um}}(x) = U(x) + W(x)$, which is easier to sample more broadly.

The artificial biasing potential in the umbrella sampling method shown in Figure 10.C.2B has the form $W(x) = \exp[-\beta k_s(x - x_0)^2]$. The probabilities of states on the improved surface will have the form

$$p_{\mathrm{um}} = \frac{W(x)e^{-\beta U(x)}}{Q_{\mathrm{um}}}, \tag{10.C.4}$$

where $Q_{\mathrm{um}} = \int W(x)e^{-\beta U(x)}\,dx$. $p_{\mathrm{um}}(x)$ will be the distribution that you will now observe on the modified landscape U_{um} (see Figure 10.C.2C). To see how you can correct for this modification at the end in computing $\langle f \rangle = \sum p(x)f(x)$, first note that

$$\left\langle \frac{f}{W} \right\rangle_{\mathrm{um}} = \sum_k \frac{p_{\mathrm{um}}(x)f(x)}{W(x)} = \frac{\sum_k f(x)e^{-\beta U(x)}}{Q_{\mathrm{um}}} \tag{10.C.5}$$

and

$$\left\langle \frac{1}{W} \right\rangle_{\mathrm{um}} = \sum_k \frac{p_{\mathrm{um}}(x)}{W(x)} = \frac{\sum_k e^{-\beta U(x)}}{Q_{\mathrm{um}}}. \tag{10.C.6}$$

Take the ratio of Equations 10.C.5 and 10.C.6 to see that

$$\langle f \rangle = \frac{\langle f/W \rangle_{\mathrm{um}}}{\langle 1/W \rangle_{\mathrm{um}}} = \sum p(x)f(x). \tag{10.C.7}$$

Equation 10.C.7 shows how you can compute a proper average by correcting for the umbrella bias that allowed you better sampling. Equation 10.C.7 gives you a way to "push" your system from one state to another, over barriers, say along reaction coordinates that are of interest to you. You apply the biasing potential to different positions along the reaction coordinate of interest, to force the molecule to remain close to your trajectory of interest. The output will be a series of histograms along the reaction coordinate, so that the pieces overlap. Then you reweight to correct for the biases of the restraints you applied. Umbrella sampling can give you *potentials of mean force* along pathways or relative free energies of different states.

There are different methods for calculating free energy differences between states, using MC or MD trajectories. Each method has its advantages and disadvantages for different systems or processes. Here are two important ones.

Free-energy perturbation (FEP) is a method that was introduced by RW Zwanzig in 1954 [29] based on statistical mechanical principles, used for evaluating the change in free energy, $\Delta G_{\mathrm{AB}} = G_{\mathrm{B}} - G_{\mathrm{A}}$, associated with a given transition between two conformational states, say $\mathrm{A} \to \mathrm{B}$. The basic equation, known as the Zwanzig equation is

$$\Delta F_{\mathrm{AB}} = -RT \ln \langle e^{-(U_{\mathrm{B}} - U_{\mathrm{A}})/RT} \rangle_{\mathrm{A}}. \tag{10.C.8}$$

Here the term in angular brackets is the average over a simulation performed for state A, and U_B and U_A refer to the internal energies evaluated during the simulation for conformations representative of states A and B. This equation applies only in the neighborhood of A. So, it should be used only when the two states are sufficiently close together in conformational space. These limitations can often be overcome using newer methods, such as the Bennett acceptance ratio (BAR) method.

Thermodynamic integration (TI) is a method in which the chemical identity of a molecule is changed through a continuous nonphysical transformation [30]. For example, you might change a carbon atom into a nitrogen atom by incrementally changing carbon interaction parameters into nitrogen parameters. It is legitimate to use a nonphysical process such as this because we are interested only in the thermodynamic differences, which depend exclusively on the end states, so the pathway of change doesn't affect these differences.

In TI, we need to compute a derivative. The free-energy difference ΔF_{AB} is evaluated by integrating the free energy $F(\lambda) = -RT \ln Q(\lambda)$ of the system using a transition coordinate λ, also called a coupling parameter, which is varied continuously from $\lambda = 0$ (initial state A) to $\lambda = 1$ (final state B) to effect the change being made:

$$\Delta F_{AB} = \int_0^1 \frac{\partial F(\lambda)}{\partial \lambda} d\lambda = -\int_0^1 RT \frac{\partial \ln Q(\lambda)}{\partial \lambda} d\lambda. \qquad (10.C.9)$$

Using $Q(\lambda) = \sum_s \exp[-U_s(\lambda)/RT]$, the free-energy change is rewritten as

$$\Delta F_{AB} = \int_0^1 \frac{k_B T}{Q} \left[\sum_s \frac{1}{RT} e^{-U_S(\lambda)/RT} \frac{dU_s(\lambda)}{d\lambda} \right] d\lambda = \int_0^1 \left\langle \frac{dU(\lambda)}{d\lambda} \right\rangle d\lambda, \qquad (10.C.10)$$

where we replaced $\partial Q/\partial \lambda$ by the term in square brackets. The thermodynamic integration thus reduces to the evaluation of the derivative of $U(\lambda) = U_A + \lambda(U_B - U_A)$ with respect to λ, and its integration over the entire transition path. In practice, this calculation is performed using a series of discrete equilibrium simulations for $\lambda_1 = 0, \lambda_2 = 0.1, \ldots, \lambda_k = 1$. Each simulation samples the neighborhood of a given intermediate state λ_k. The goal is to evaluate the ensemble average of the derivative of the system energy with respect to λ from the configurations in these separate equilibrium runs, then finally evaluating the weighted sum of these derivatives. The weight of each state is also determined from the average potential energy of these equilibrium runs.

REFERENCES

[1] JR Perilla, BC Goh, CK Cassidy, et al. Molecular dynamics simulations of large macromolecular complexes. *Curr Opin Struct Biol*, 31:64–74, 2015.

[2] M Karplus and J Kuriyan. Molecular dynamics and protein function. *Proc Natl Acad Sci USA*, 102:6679, 2005.

[3] RO Dror, RM Dirks, JP Grossman, et al. Biomolecular simulation: a computational microscope for molecular biology. *Annu Rev Biophys*, 41:429–452, 2012.

[4] B Alder and T Wainwright. Phase transition for a hard sphere system. *J Chem Phys*, 27:1208, 1957.

[5] A Rahman and FH Stillinger. Molecular dynamics study of liquid water. *J Chem Phys*, 55:3336–3359, 1971.

[6] JA McCammon, BR Gelin, and M Karplus. Dynamics of folded proteins. *Nature*, 267:585–590, 1977.

[7] L Verlet. Computer "experiments" on classical fluids. I. Thermodynamical properties of Lennard-Jones molecules. *Phys Rev*, 159:98–103, 1967.

[8] JC Phillips, R Braun, W Wang, et al. Scalable molecular dynamics with NAMD. *J Comput Chem*, 26:1781–1802, 2005.

[9] MJ Abraham, T Murtola, R Schulz, et al. GROMACS: High performance molecular simulations through multi-level parallelism from laptops to supercomputers. *SoftwareX*, 1–2:19–25, 2015.

[10] BL de Groot and H Grubmüller. Water permeation across biological membranes: Mechanism and dynamics of aquaporin-1 and GlpF. *Science*, 294:2353–2357, 2001.

[11] E Tajkhorshid, P Nollert, MØ Jensen, et al. Control of the selectivity of the aquaporin water channel family by global orientational tuning. *Science*, 296:525–530, 2002.

[12] D Shukla, Y Meng, B Roux, and VS Pande. Activation pathway of Src kinase reveals intermediate states as targets for drug design. *Nat Commun*, 5:3397, 2014.

[13] H Grubmüller. Predicting slow structural transitions in macromolecular systems: Conformational flooding. *Phys Rev E*, 52:2893–906, 1995.

[14] D Hamelberg, J Mongan, and J McCammon. Accelerated molecular dynamics: a promising and efficient simulation method for biomolecules. *J Chem Phys*, 120: 11919–11929, 2004.

[15] A Laio and FL Gervasio. Metadynamics: a method to simulate rare events and reconstruct the free energy in biophysics, chemistry and material science. *Rep Prog Phys*, 71:126601, 2008.

[16] Y Miao, SE Nichols, PM Gasper, et al. Activation and dynamic network of the M2 muscarinic receptor. *Proc Natl Acad Sci USA*, 110:10982–10987, 2013.

[17] N Metropolis, AW Rosenbluth, MN Rosenbluth, et al. Equations of state calculations by fast computing machines. *J Chem Phys*, 21:1087–1092, 1953.

[18] Y Liu and B Kuhlman. RosettaDesign server for protein design. *Nucleic Acids Res*, 34:W235–W238, 2006.

[19] B Kuhlman, G Dantas, GC Ireton, et al. Design of a novel globular protein fold with atomic-level accuracy. *Science*, 302:1364–1368, 2003.

[20] DR Canchi, D Paschek, and AE Garcia. Equilibrium study of protein denaturation by urea. *J Am Chem Soc*, 132:2338–2344, 2010.

[21] D Seeliger and BL de Groot. Thermostability calculations using alchemical free energy simulations. *Biophys J*, 98:2309–2316, 2010.

[22] Y Zhou, JH Morais-Cabral, A Haufman, and R MacKinnon. Chemistry of ion coordination and hydration revealed by a K^+ channel–Fab complex at 2.0 Å resolution. *Nature*, 414:43–48, 2001.

[23] S Bernèche and B Roux. Energetics of ion conduction through the K^+ channel. *Nature*, 414:73–77, 2001.

[24] DA Köpfer, C Song, T Gruene, et al. Ion permeation in K^+ channels occurs by direct Coulomb knock-on. *Science*, 346:352–355, 2014.

[25] IH Shrivastava and MS Sansom. Simulations of ion permeation through a potassium channel: molecular dynamics of KcsA in a phospholipid bilayer. *Biophys J*, 78:557–570, 2000.

[26] S Bernèche and B Roux. The ionization state and the conformation of Glu-71 in the KcsA K^+ channel. *Biophys J*, 82:772–780, 2002.

[27] S Kumar, JM Rosenberg, D Bouzida, et al. The weighted histogram analysis method for free-energy calculations on biomolecules. I. The method. *J Comput Chem*, 13:1011–1021, 1992.

[28] E Gallicchio, M Andrec, AK Felts, and RM Levy. Temperature weighted histogram analysis method, replica exchange, and transition paths. *J Phys Chem B*, 109:6722–6731, 2005.

[29] RW Zwanzig. High-temperature equation of state by a perturbation method. I. Nonpolar gases. *J Chem Phys*, 22:1420–1426, 1954.

[30] TP Straatsma and HJC Berendsen. Free energy of ionic hydration: Analysis of a thermodynamic integration technique to evaluate free energy differences by molecular dynamics simulations. *J Chem Phys*, 89:5876–5886, 1988.

SUGGESTED READING

Allen MP and Tildesley DJ, *Computer Simulation of Liquids*, Oxford University Press, Oxford, 1989.

Ferrenberg AM and Swendsen RH, Optimized Monte Carlo data analysis. *Phys Rev Lett* 63:1195–1198, 1989.

Frenkel D and Smit B, *Understanding Molecular Simulation: From Algorithms to Applications*, 2nd ed. Academic Press, New York, 2002.

McCammon JA, Harvey SC, *Dynamics of Proteins and Nucleic Acids*. Cambridge University Press, New York, 1988.

Newman MEJ and Barkema GT, *Monte Carlo Methods in Statistical Physics*, Clarendon, Oxford, 1999.

Zuckermann DM, *Statistical Physics of Biomolecules: An Introduction*. CRC Press, Boca Raton, FL, 2010.

Predicting Protein Structures from Sequences

CHAPTER 11

SOME PROTEINS HAVE COMPUTABLE NATIVE STRUCTURES

You can compute the native structures of some proteins from their amino acid sequences. This is important because knowing a native structure is the starting point for understanding a protein's dynamics, its binding partners, and its biological mechanism of action, and for designing drugs. Computational structure prediction can fill a big gap, since the number of known protein structures is fewer than one-thousandth of the number of known amino acid sequences. We describe here some of the methods for computing protein structures.

First, let's define some terms. The *target* is the protein whose structure you want to predict. The sequence of the target protein is called the *query sequence*. In *comparative modeling*, also called *homology modeling*, you predict the target's structure based on your knowledge of some other protein, called the *template* protein. The principle is that if two proteins have similar sequences, they usually have similar structures. So, if you can find a good template protein—a protein that has a sequence that is similar to the target, as well as having a known structure—then you can infer much about the structure of your target protein. In situations when you can't find a good template, an alternative is to see how your sequences fit known structures using a method called *threading*. Yet another strategy, which currently applies only to the smallest simplest proteins, is to fold a protein from scratch using molecular simulations (see Chapter 10). Other useful computational tools can help you guess the types and locations of secondary structures, the side-chain and loop configurations, and which residue pairs may be in contact in the native structure. We first describe comparative modeling.

COMPARATIVE MODELING IS A MAIN TOOL FOR STRUCTURE PREDICTION

Here Is How Homology Modeling Works

Figure 11.1 shows the general procedure for homology modeling. First, using the methods of Chapter 8, search the Protein Data Bank (PDB) to find sequences that are similar to that of your target. These are potential templates. Align the sequence of the template with the target protein. Next, attach the atoms of the target's side chains and loops, assigning them initial coordinates. Then, apply methods such as energy minimization, molecular dynamics, or Monte Carlo to tweak the coordinates (changing the backbone and side-chain angles and the residue–residue distances), sample different conformations, and seek those conformations that are most native-like, by some scoring criterion. You can score structures by physical free energies or by the "nativeness" of a protein structure found using *statistical potentials*, described in this chapter. You can follow up by checking the torsion angles on a Ramachandran plot to see whether or not they fall within allowed or favored regions and checking for consistency with knowledge of motifs or binding sites. If this is not successful, try again with a different template.

This chapter describes methods of comparative modeling. We start with a short overview of the methods and considerations that apply to developing them:

(1) **Divide your target sequence into domains.** One way to do this is to use the information available in *Pfam*, a database of protein families and domains. Then, for the following steps, you treat each domain as a separate object.

(2) **Choose what level of detail you want.** Do you want full atomic detail? Or, do you want a coarser-grained representation? Often the

1 Perform sequence alignment against sequences of PDB structures to find a good target structure

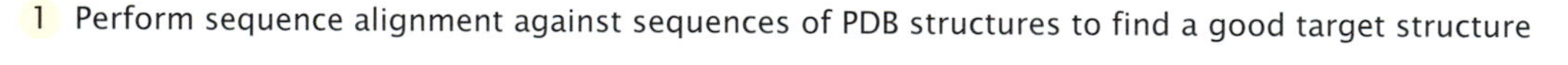

2 Get template structure

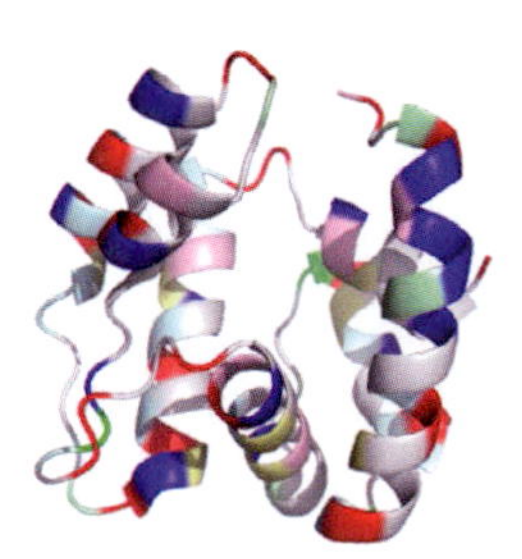

3 Mount target sequence on template structure and refine structure (colors are residue types)

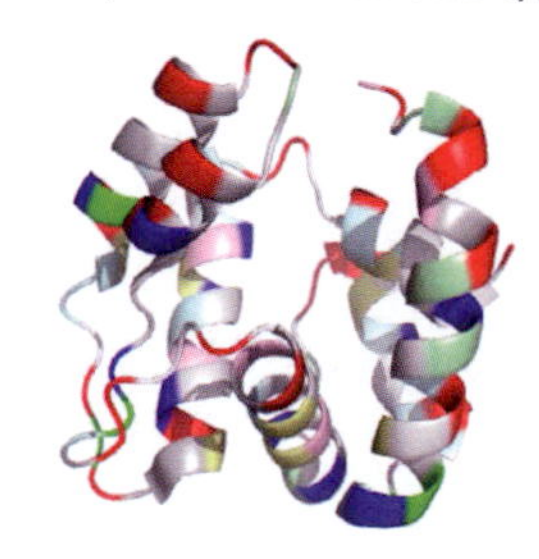

4 Generate sample of conformations and evaluate the energies to identify the best case for low energy and high score

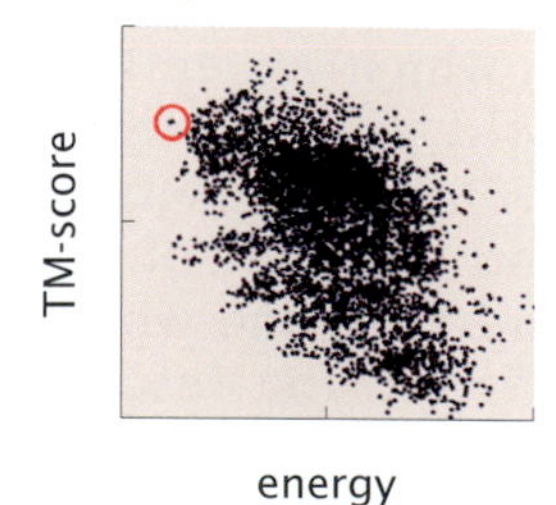

5 Further refine model

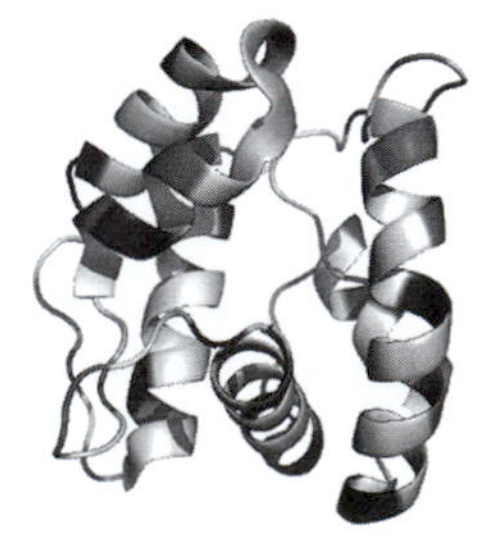

6 Does model satisfy required criteria? If not, then repeat.

Figure 11.1 The procedure for homology modeling. Identify a template protein. Superimpose the target sequence onto the template structure. Add in the atoms of the side chains and loops. Perform conformational sampling while evaluating the energy or score reflecting the "nativeness" of each conformation. Select the best final conformations in conjunction with further refining, clustering, or accounting for additional information.

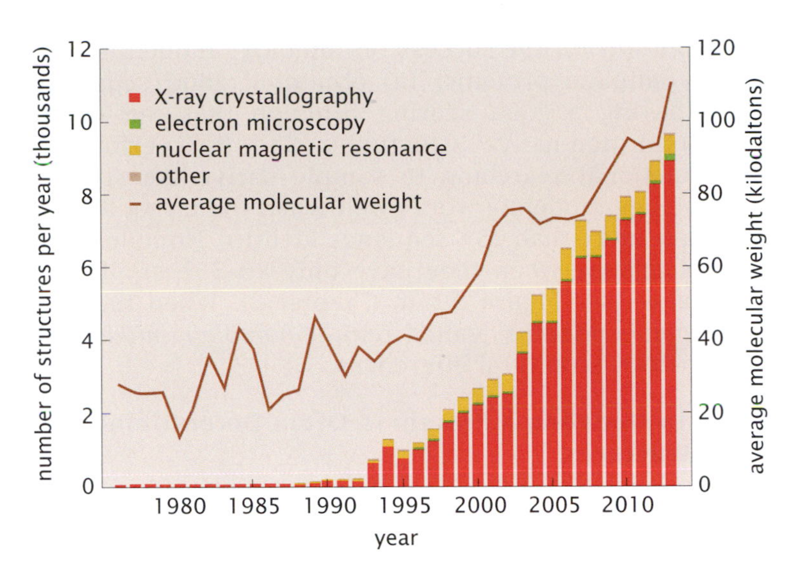

Figure 11.2 The number of protein structures in the Protein Data Bank (PDB) has grown steadily, to more than 100,000. The bar colors indicate the experimental methods used to obtain the structure. The average sizes of resolved protein structures are also increasing. (Adapted from *Nature*, 509:264–265, 2014. With permission from Macmillan Publishers Ltd.)

nature of the problem will make this clear. If you are looking at relatively small differences such as a mutation or an enzyme mechanism, you probably need atomic details. If you just want to know the fold class, you would need less detail.

(3) **Identify a homologous template protein.** Find possible template proteins in the PDB for each domain of your target protein. Currently, there are more than 100,000 structures in the PDB (Figure 11.2). To find template proteins for your target, search the PDB using the BLASTP module of the BLAST suite of tools. You just enter your query sequence in FASTA format, and BLASTP delivers structures that contain similar sequences, as well as their sequence alignment against your query sequence. Or, you can identify template proteins using *threading* or *fold recognition* (discussed later on).

(4) **Choose any additional information that you want to guide your modeling.** You may want to incorporate additional experimental knowledge. You might know its enzyme function, and which residues should be neighbors in the active site, for instance. Or you may have insights about trusting some parts of your homology model more than others. When your knowledge is in the form of internal distances, you can enforce those as restraints by adding spring-like energies.

(5) **Apply a *scoring function* to distinguish conformations that are native-like from those that are less native-like.** You can use database-derived *statistical potentials*, which are fast to compute, but approximate. Or, you can compute free energies using an all-atom physical potential.

(6) **Choose a sampling strategy.** To predict a native structure, you need to sample conformations broadly enough to find the native structure. In designing a protocol, you pick a sampling method (such as molecular dynamics or Monte Carlo; see Chapter 10), the degrees of freedom you want to explore, and how extensively you want to sample.

(7) **Choose how to *cluster* your results.** Computer simulations often generate many conformations that are quite similar. Usually, you just want to keep a few that are most representative of the most populated clusters.

(8) **Further refine your structures.** You can often improve fine-grained detail by performing energy minimization or molecular dynamics using atomistic physical potentials.

Critically important for the success of homology modeling are (i) finding relevant template proteins, (ii) obtaining good sequence alignments, (iii) having a good scoring function that can discriminate native-like from non-native structures, and (iv) performing sufficient conformational searching to sample native-like structures. In general, homology modeling works well when the target and template sequences have more than 35% sequence identity. Homology modeling becomes less reliable when they have only 20–35% similarity, called the *twilight zone* (see Figure 8.3 in Chapter 8). When they have less than 25% sequence identity (called *remote-homology modeling*), it can be difficult to draw structural inferences.

A First Step in Modeling a Protein Is Often Determining Its Secondary Structures

Structure-prediction methods often first determine what secondary structures a protein has and where they are located in the sequence. How you identify secondary structures depends, to some degree, on how you define them. Secondary structures are sometimes defined by their bond-torsion angles, or sometimes based on their hydrogen-bonding patterns. Along with the atomic coordinates of a protein, information about its secondary structures can be found in the PDB or in a more consistent way by using other software or servers, as illustrated in Boxes 11.1 and 11.2.

Box 11.1 Defining the Secondary Structures in a Protein of Known Structure

Suppose you already know a protein's structure, and you simply want to define the locations of its secondary structures. You can find these assignments by using DSSP [1], which is software and a database that provides secondary-structure assignments using an electrostatic model of hydrogen bonding (Figure 11.3). Secondary structures are described by a one-dimensional string of eight possible states: (H) α-helix, (E) β-strand, (G) 3_{10}-helix, (I) π-helix, (B) bridge (single-residue β-strand), (T) β-turn (hydrogen-bonded), (S) bend (local curvature $>70°$), and (C) coil for all others. Some residues are unassigned by DSSP. A useful alternative is STRIDE, which generally produces fewer short pieces of secondary structure.

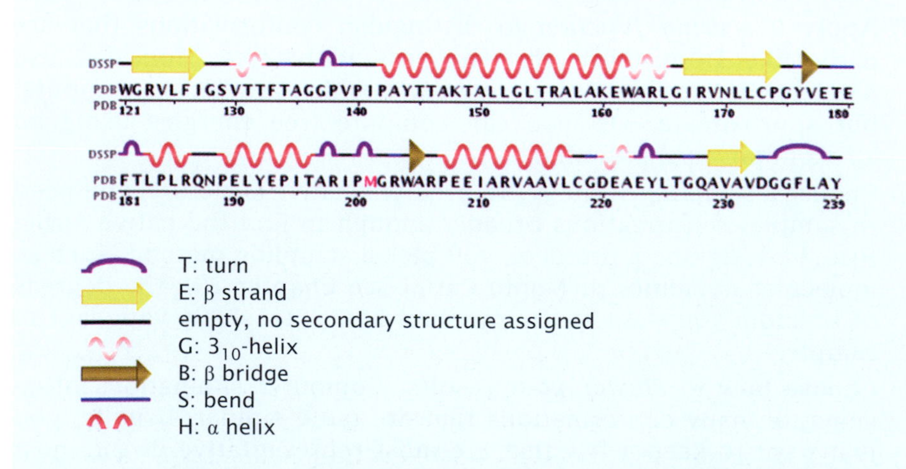

Figure 11.3 DSSP output shows the locations of secondary structures along the sequence.

Box 11.2 Predicting Secondary Structures When You Don't Know a Protein's Structure

If you don't know a protein's structure, you can estimate its secondary structure by submitting your amino acid sequence to PSIPRED [2] (PSI-blast-based secondary-structure PREDiction) or to other secondary-structure-prediction websites. PSIPRED is a protein-structure server that, among other things, accepts an amino acid sequence and returns a best estimate of secondary structure assignments. PSIPRED uses a neural-net machine-learning algorithm. Figure 11.4 shows an example of output from the PSIPRED server [3]. Predicting secondary structures from sequence alone can be done with accuracies up to 80%, if you use multiple-sequence information.

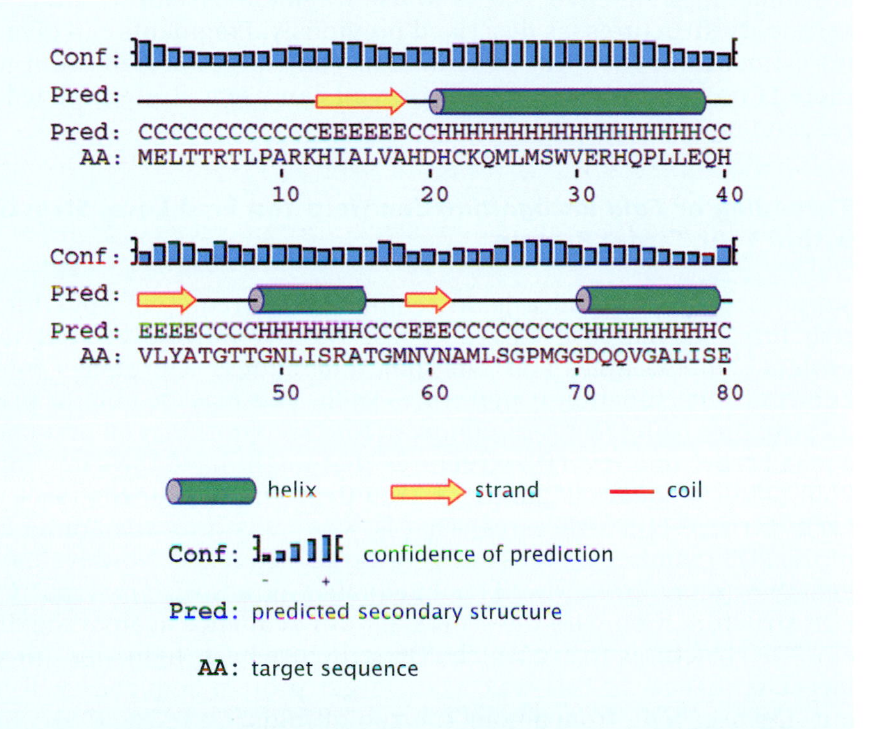

Figure 11.4 PSIPRED output gives the predicted locations of helix, strand, and coil regions for any protein sequence. It also gives estimated confidence levels. It is shown here for a segment of methylglyoxal synthase. The notation for the secondary structure is given in Figure 11.3. (From LJ McGuffin, K Bryson, andDT Jones. *Bioinformatics*, 16:404–405, 2000. By permission of Oxford University Press.)

You Can Assemble Protein Structures from *Fragments* Instead of Secondary Structures

Instead of computing a protein structure by assembling its secondary structures, you can assemble its *fragments*. A fragment is a peptide, typically about nine residues long, whose structure is defined by its string of (ϕ, ψ) backbone coordinates, rather than by labels such as helix, sheet, or turn. A fragment can be more irregular and diverse than a secondary structure. Here's how you can use fragments in homology modeling. First, break your target protein into 9-mer sequences. Use the *i*th target fragment as a query sequence to find similar sequences in the PDB. Once you have found one or more template sequences, take the backbone and side-chain coordinates of that *i*th *template* *fragment* as the structure of your *i*th *target fragment*. Now, you have a collection of target fragments, $i = 1, 2, 3, \ldots$, each having particular backbone and side-chain coordinates. Assemble your target fragments into a first-guess native structure. However, any first attempt at randomly squeezing rigid irregularly shaped fragments together is not likely to lead to a native-like structure. They won't pack well. Fragments will have steric clashes and unfavorable energetics. So, then run a conformational sampling procedure, tweaking the degrees of freedom extensively and scoring them, in order to find more native-like structures. Fragment assembly, first developed by D Baker and colleagues, is now widely used in protein-structure-prediction algorithms [4].

How can you compute a target structure when you can't find a good template? When sequence identity is in the *twilight zone*, two strategies are sometimes effective. One is to use fragment assembly, rather than secondary structures, as described previously. Fragments can give useful estimates of the local structures in your target protein even when there is no homologous protein. Or, you can use a strategy called *fold recognition* or *threading*.

Threading or *Fold Recognition* Can Help You Find Local Structures within Your Target Protein

Suppose you can't find a good template for predicting your target's structure. But suppose you can guess the overall fold of your target protein. For example, you can sometimes guess a protein's fold by knowing what function it performs. Then, you may be able to predict its structure using *fold recognition*. In fold recognition, you first choose a small "zoo" of a few PDB structures that might have the same fold as your target sequence (Figure 11.5). You then *thread* your sequence onto each of those PDB structures. That is, you substitute the amino acids of the PDB template by those of your target sequence. Now evaluate an energy or score for how well your sequence agrees with each particular PDB structure. Compare the scores of your sequence against the different PDB structures to see which PDB structure best "lights up" for your target sequence. In this way, your target protein sequence will "pick out" the best fold, from among the zoo of folds.

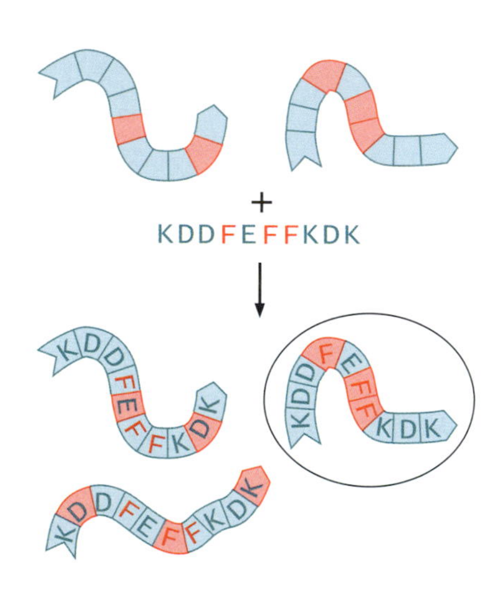

Figure 11.5 The threading procedure. Superimpose your target sequence onto each member of a "zoo" of different native structures. Score them to find the best match. (Top) The zoo of two possible folds. (Middle) The target sequence for which you want to find the folded structure. (Bottom) That sequence is superimposed onto each given fold. Here, the scoring function is just hydrophobic (*orange*) or polar (*cyan*). The fold at the bottom right is the best match for the sequence, because the three hydrophobic residues in the sequence match the hydrophobic positions in the fold (*red* letters on *red* squares and *blue* letters on *blue* squares).

There are some limitations to threading, however. First, your PDB zoo must have the correct fold in it. Your zoo must contain the "animal" you are searching for. Second, you may need to perform the threading simulations by shifting the starting residue on the template structure, one residue at a time, to establish the correct register between the sequence and structure. You must make decisions about how to score insertions/deletions between the query sequence and template fold. Third, as with all methods, you need a good scoring function.

Sequence Co-evolution Analysis Can Help You Predict Protein Structures

Here is another way to predict a protein structure. Suppose you don't know the structure of any template protein for your target protein. But suppose you do know many sequences that resemble your target's sequence, maybe from evolutionarily related organisms. You can draw inferences from those sequences alone, without knowing any structure. You first align the template sequences to construct a *multiple sequence alignment* (MSA). Now look at the columns in your MSA and look for

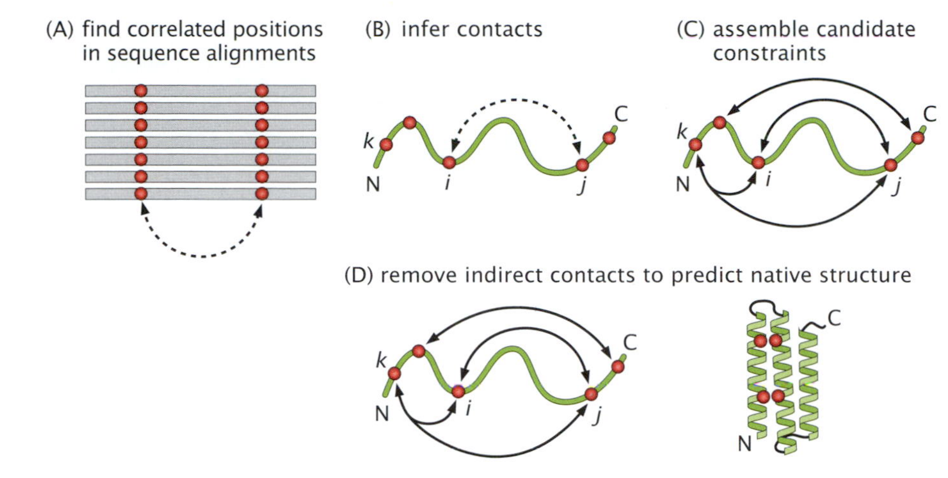

Figure 11.6 EVfold predicts native contacts by using information about correlated mutations from similar sequences. (A) First, align many sequences that are similar to the target's sequence. EVfold then finds a list of pairs of sequence positions (i, j) where the positions i and j are correlated with each other. (B) Strong evolutionary correlations indicate that i and j are likely to be adjacent to each other in the native structure. (C) Such inferred contacts can be used as constraints for predicting the protein's native structure. (D) Observed correlations may originate from contacts that are either direct (such as i with j) or indirect (such as k with j, due to their relationships with i). EVfold eliminates indirect couplings. (Adapted from DS Marks, TA Hopf, and C Sander. *Nat Biotechnol*, 30:1072–1080, 2012. With permission from Macmillan Publishers Ltd.)

correlations among them. For example, compare columns i and j. When sequence changes happen at position i, do sequence changes also happen at position j? Such residues i and j are said to *covary*, or undergo *correlated mutations* (see Figure 8.19 in Chapter 8). A common reason for evolutionary covariation, also called *co-evolution,* is that residues i and j are in contact with each other in the native structure of the protein. If amino acids i and j are adjacent in a native structure, then mutations that increase the size of i are likely to correlate with mutations that decrease the size of j. Finding such correlations can help you infer a protein's native structure.

One such strategy is *direct-coupling analysis* [5], which is implemented in EVfold by Marks et al. [6] (Figure 11.6). Figure 11.7 illustrates the EVfold prediction of the tertiary contacts for Ras, compared with those predicted by *mutual information*, which contains less global information (see Equation 8.5 in Chapter 8). It shows that direct-coupling analysis predicts the tertiary contacts well, which are then informative about the protein's native structure.

On the one hand, EVfold is fast, and can be used in situations where you don't have any template structures available, as is often the case for membrane proteins, for example. On the other hand, the success of EVFold depends on having a sufficient number of homologous sequences for your target protein.

STATISTICAL POTENTIALS ARE "ENERGY-LIKE" SCORING FUNCTIONS FOR SELECTING NATIVE-LIKE PROTEIN STRUCTURES

To predict a protein's structure, you need a good scoring function. You need a way to distinguish native-like from non-native conformations. How can you measure the "nativeness" of a conformation? First, if you knew exactly nature's true physical potential function, the principles of thermodynamics say that you could find protein native structures by seeking the conformation that is at the global minimum of free energy. But searching atomically detailed physical potentials is often too computationally expensive to find the global minimum. Another approach is to use *statistical potentials*, which provide a much faster—but approximate—way to sort conformations by nativeness. Statistical potentials tell you how closely a given conformation resembles average structural properties of the native proteins in the PDB.

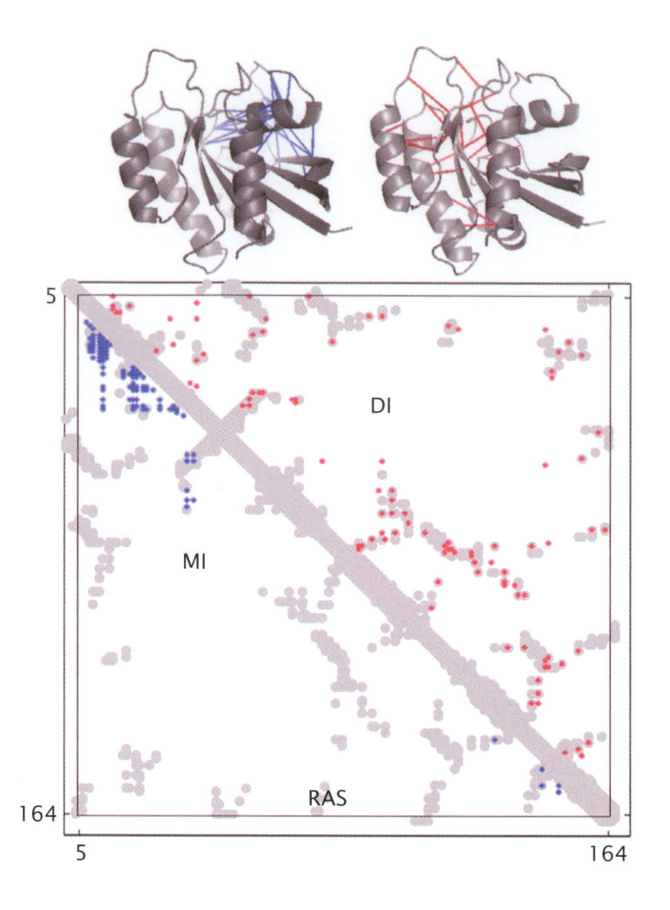

Figure 11.7 Comparing the residue–residue contacts determined by direct-coupling analysis (DI) and mutual information (MI). Applied to RAS (ribbon diagram). (*Gray*) The native structure contact map. (*Blue* dots) Contacts inferred from MI. (*Red* dots) Contacts inferred from DI. The same colors indicate the contacts on the structures. DI provides a more complete coverage of interresidue contacts than MI. (Adapted from DS Marks, LJ Colwell, R Sheridan, et al. *PLoS One*, 6:e28766, 2011.)

The PDB contains information about the conformational preferences of known native proteins. You can capture those observed native-like preferences—for example, for torsion-angle preferences or for preferences of types of amino acid contact pairings—by converting their statistical frequencies to energy-like functions called *statistical potentials* or *knowledge-based potentials*, first developed by Tanaka and Scheraga [7] and Miyazawa and Jernigan [8, 9]. To create a statistical potential, you first determine the probabilities of different component structures across the protein structures in the PDB (see below). Then you take the negative of the logarithm to convert these probabilities into free-energy-like quantities, the statistical potentials. You can use those potentials to evaluate different structural models.

For example, suppose you are interested in the distribution of ϕ angles for a certain type of amino acid across the PDB. You first define bins, $i = 1, 2, 3, \ldots, s$. Bin 1 might represent the you observe from 0° to 10°; call this ϕ_1. Bin 2 might represent angles 11–20°; call this ϕ_2; and so on. Now, suppose you search the PDB and find m_1 instances in which the system populates angular state $i = 1$, m_2 instances of state $i = 2$, etc. And suppose you make $M = \sum_i m_i$ total observations. Then, the probability that the system has angle $\phi = \phi_i$ is $p_i = m_i/M$.

You can convert any such probability p_i to an energy-like quantity $\Delta U(i)$ by taking the logarithm:

$$\frac{\Delta U(i)}{RT} = -\ln\left(\frac{p_i}{p_{i0}}\right), \tag{11.1}$$

where RT is the gas constant multiplied by temperature (to put U into units of energy) and p_{i0} is the probability of a reference state that is chosen to represent a random distribution. Such quantities are called *knowledge-based potentials* or *statistical potentials.*[1] Statistical potentials provide an approximate way to express the known structural preferences that you can observe in the PDB.

Why should we convert the conformational propensities into energy-like quantities by taking logarithms? Why couldn't we just work with the PDB frequency distributions themselves? There are mathematical advantages to working with quantities that can be added and subtracted from each other (energy-like quantities), rather than with quantities that are multiplicative (probabilities). Like money, energy-like quantities have the important properties of being additive and providing a common currency. For proteins, we have different types of forces, including the *local interactions* that dictate helices, turns, and dihedral angles; and we have the *nonlocal interactions* that dictate contacts, such as hydrophobic interactions, salt bridges, and hydrogen bonding among residues that are distant in the sequence but close in space. Statistical potentials give you a common unit for estimating how nature makes trade-offs between them.

Contact Potentials Express the Noncovalent Pairing Preferences of Amino Acids in Native Structures

Statistical potentials are also useful for describing the frequencies of different types of contacts in proteins (that is, nonbonded pairs of residues in close proximity) such as the numbers of glycine–lysine or tryptophan–alanine pairings found in the PDB. Such *contact potentials* are useful in scoring native-like protein structures, for modeling mutations, and for estimating binding propensities of one protein to another.

Here is how you make a statistical potential for noncovalent contacts. First, consider the following imaginary process in solution. You have an amino acid of type A in a dilute solution with two other types of amino acids, B and C. You want to know the probability $p(AC)$ that A and C form a pairwise contact, relative to the probability $p(AB)$ that A and B form a pairwise contact. You can express the difference in contact pairing free energies as the ratio of these relative populations (frequencies):

$$\Delta\Delta U = \Delta U_{AC} - \Delta U_{AB} = -RT \ln\left[\frac{p(AC)}{p(AB)}\right]. \qquad (11.2)$$

Now, the same idea applies inside a protein. You measure the noncovalent pairing frequencies of the amino acids that you observe across a set of protein native structures. First, construct a 20×20 matrix, where each row represents one of the 20 amino acids, and each column represents the other member of the pair. Start with a blank matrix. Now, select a representative dataset of proteins from the PDB, called

[1]Note that statistical potentials are not true physical energies or free energies or potentials of mean force, as might be inferred from using the Boltzmann equation 11.1. The canonical ensemble of Boltzmann physics describes how state probabilities change with temperature. But not all probability distributions have to do with temperature. Statistical potentials are based on probability distributions over a database; they are not taken from a thermal ensemble. For database potentials, you replace RT by a parameter that you adjust in order to balance these energy-like terms relative to other energies or energy-like terms.

the learning (or training) dataset.[2] The learning dataset should be large enough (typically, thousands of proteins) to provide statistically significant information on the occurrence of each type of amino acid pair, but should not contain redundant structures (for example, members of the same family) to avoid biases. Then, you go through the learning dataset, one protein at a time, one amino acid at a time, and count how many contacts you find of each type: (Ala, Gly), (Leu, Phe), etc. You now fill up the 210 elements of this matrix with these contact counts of each type. (You only have unique entries for the main diagonal and the upper half of the matrix, so $N(N+1)/2$ distinct pairs, where $N = 20$, because (Leu, Phe) has the same frequency as (Phe, Leu), for example.) If you normalize all the elements of the matrix by dividing by the total number of contacts you counted in the PDB, you will get the probabilities $p(AB)$ and $p(AC)$ that you need for Equation 11.2.

By counting the numbers of contacts for the 210 different types of amino acid pairs, you can establish contact potentials for the pairings of different types of amino acids. Contact potentials

$$\Delta U_{ij} = U_{ij} - U_{ij}^0 = -RT \ln\left(\frac{p_{ij}}{p_{ij}^0}\right) \tag{11.3}$$

are a generalization of Equation 11.1, where the indices i and j refer to the 20 different types of amino acids. Given a database of native protein contacts, Equation 11.3 provides a matrix of contact potentials ΔU_{ij}, relative to a reference state U_{ij}^0 (see Appendix 11A).

How should you choose the reference state? What pairing frequency p_{ij0} (corresponding to the energy U_{ij0}) would you expect, say for $i =$ tryptophan and $j =$ alanine, in some random collapsed polymer? After all, our goal with contact potentials is usually to distinguish that the lysozyme sequence folds into the lysozyme structure, rather than into some other arbitrary collapsed state. The reference state is chosen depending on the problem. If the goal is to compare folded structures, it should describe how often we expect pairing in some "random" compact protein structure. In this case, it needs to account for solvent interactions (Box 11.3). An example of a contact potential is given in Appendix 11A; others are given in [11].

Box 11.3 How Do You Extract Pairing Frequencies from a Database?

Consider sequence position i in a protein. To keep the math simple, consider just two types of amino acids, A and B, not 20. Figure 11.8 shows a residue of type A at the position marked X. The total number of neighbors m_i for an amino acid at position i is the sum over the numbers of the different possible types (AA, AB, and AW), so

$$m_i = m_{AAi} + m_{ABi} + m_{AWi}, \tag{11.4}$$

which sums to $m_i = 6$ in the figure. These are all within a first coordination shell, indicated by the dashed circle. Similarly, if a B residue

[2] Some datasets are *nonredundant*. This means that if there are identical or nearly identical copies of a protein, those copies are removed. Some proteins are more heavily studied than others, giving rise to many small variants of that protein in the PDB. However, it is better to use all structures with an appropriate weighting scheme [9, 10].

were at position i, its total number of neighbors would be $m_i = m_{BBi} + m_{ABi} + m_{BWi}$. The total number of contacts in this protein is

$$M = \frac{1}{2} \sum_i m_i, \tag{11.5}$$

where the sum is over all sequence positions i. The factor of $\frac{1}{2}$ corrects for the fact that this summation counts every contact twice.

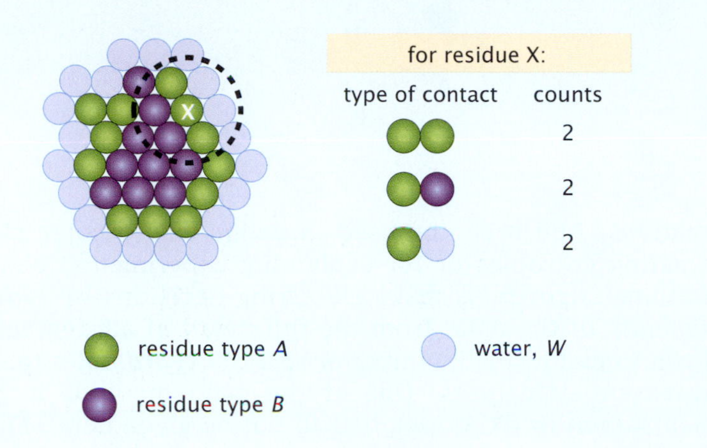

Figure 11.8 Counting contacts for statistical potentials. For two types of residues, *A* (*green*) and *B* (*purple*), plus water *W* (*light blue*), there are three types of contacts around the residue marked with a *white* X.

For counting m_i, we make a simplification. Instead of assuming the count of nearest neighbors around a residue depends on its position i, we assume that it only depends on what type, say A, of amino acid it is. (Each amino acid has a specific coordination number, which allows you to evaluate the number of contacts with water, each "water" representing a residue-sized cluster of waters.) This is called the *quasi-chemical* reference state. Developed by Miyazawa and Jernigan [8, 9], it provides the approximation that is needed for applying Equation 11.4 to compute statistical potentials.

OTHER COMPUTATIONAL TOOLS CAN ALSO HELP YOU PREDICT NATIVE STRUCTURES

In the following sections, we describe a few methods that are helpful for predicting protein structures.

Clustering Algorithms Can Separate Similar Conformations from Different Ones

Often, you want to *cluster* similar conformations together. You want a way to label conformations i and j as being either "essentially the same" or "essentially different" from each other. Clustering is useful for various purposes. It helps you choose a conformation that is representative, or average, of a whole class. It can pick out a representative conformation for visualization. Larger clusters sometimes indicate structures having greater conformational entropy and higher populations. Clustering can help discriminate native-like from non-native

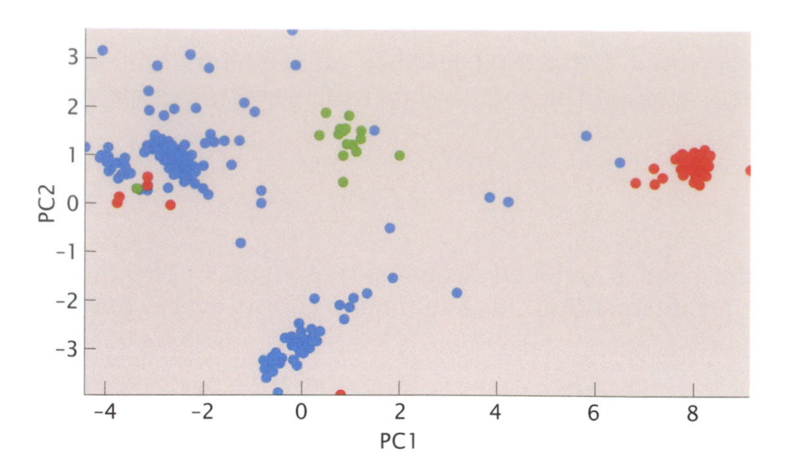

Figure 11.9 Protein structures cluster into conformational classes. The structures of many myoglobin proteins were clustered and compared by principal-component analysis (PCA). Of these proteins, 291 were from whale (*blue*), 52 from horse (*red*), and 17 from pig (*green*), including many mutants. Structures were superimposed and the deviations from the mean positions were subjected to PCA. The results are plotted along the two lowest principal components, PC1 and PC2. This shows how clustering can discriminate among different classes of protein structures. (From AA Rashin, MJ Domagalski, MT Zimmermann, et al. *Acta Crystallogr D Biol Crystallogr*, 70(2):481–491, 2014. Reproduced with permission of the International Union of Crystallography.)

conformations. And it can be used to define macroscopic states, say for estimating entropies or for explaining experimental observables. Computational algorithms make clustering decisions by *reducing the dimensionality* of the data, from the full detail of all degrees of freedom, down to just a small number of *reduced coordinates* that capture just the essential features. One of the main methods is *principal-components analysis* (PCA) (details will not be given here). Figure 11.9 shows how PCA can cluster similar protein structures together. It shows how the two lowest principal components can distinguish the structures of myoglobin from whale versus from horse or pig. This type of structure-based clustering, which is computationally efficient, can be performed by ProDy [12].

Databases of *Decoys* Can Help You Develop Energy-Like Scoring Functions

Suppose you are developing a scoring function. It may have adjustable parameters that you can tune to better discriminate between native and nonnative conformations. To tune your scoring function, you can use *decoys*. Decoys are non-native conformations that are sufficiently native-like that they can confuse an inaccurate energy function. You can improve empirical scoring functions for structure prediction by using both databases of decoys, such as those from recent CASP competitions [13], and databases of known structures.

To Compute Protein Structures, You Need Accurate Conformations of Side Chains and Loops

Sometimes, you will know a protein's backbone structure, and you will want to predict its side-chain rotamers. One computationally efficient algorithm for predicting side-chain rotamers is SCWRL [14].

Proteins often contain *loops*. Some loops have particular structures, and some are conformationally disordered stretches of chain. (Loops that are disordered are sometimes missing from PDB structures.) Loop conformations can be relevant to biological mechanisms or binding, and can play a role in capturing substrates or optimizing their binding. In a typical situation, you will know the fixed endpoints where the loop attaches to the rest of a more rigid protein, and you will want to

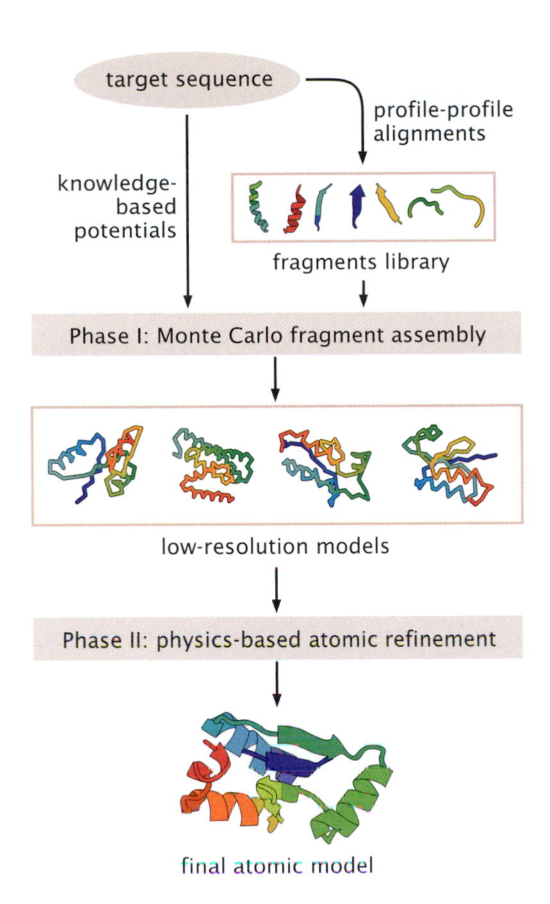

Figure 11.10 The protocol of the Rosetta protein-structure-prediction method, illustrating some now common features among structure-prediction methods. *Profile–profile alignments* refers to the use of MSAs to establish a good alignment of the target with the template. (Adapted from J Lee, S Wu, and Y Zhang. In DJ Rigden, editor, *From Protein Structure to Function with Bioinformatics*, pp 3–25. Springer Netherlands, Dordrecht, 2009; see also CA Rohl, CEM Strauss, KMS Misura, and D Baker. *Methods Enzymol* 383:66–93, 2004.)

explore the possible loop conformations that satisfy the endpoint constraints. There are various algorithms for predicting them. One that is particularly efficient and was inspired by algorithms for robotics is called kinematic loop closure [15].

Full Protocols Are Available for Predicting Structures

There are several accessible complete protocols for predicting protein structures, including *Modeller* [16], *Rosetta* [17], *I-Tasser* [18], and others. Figure 11.10 shows the flowchart for Rosetta.

CASP: A COMMUNITY-WIDE EVENT EVALUATES STRUCTURE-PREDICTION METHODS

How can you assess the quality of a structure-prediction method? In 1994, J Moult started a community-wide event called CASP (Critical Assessment of Structure Prediction) for systematic comparison and evaluation of computational methods for protein structure prediction [13]. CASP is a blind competition in which the amino acid sequences of many different target proteins are made publicly available, but structures are not. Participants make predictions of the target structures. After the competition, the structures are revealed, the performances of the participants are evaluated, and community-wide evaluations are published.

Prediction methods have improved over time. The increased success can be attributed to the growth in the number of structures in the PDB, improved MSA methods (which have enabled the detection of remote evolutionary relationships), and the use of fragment-assembly methods. In general, predictions are most accurate for proteins that are relatively small, are water-soluble (not membrane proteins), are single-domain, do not bind prosthetic groups, and have good templates in the PDB. At the present time, only a small fraction of protein structures can be predicted accurately.

ATOMISTIC PHYSICAL SIMULATIONS CAN PREDICT THE STRUCTURES OF SOME SMALL PROTEINS

For several reasons, it is important to develop atomistic physical simulation methods, of the type described in Chapter 10, for predicting protein structures. First, purely physical simulations would reduce our reliance on homology models and knowledge bases of native structures. In some cases, such as membrane proteins or other types of foldable polymers, databases are currently too small for homology modeling or statistical potentials. Second, atomistic modeling does not require sequence alignments. Third, often we want to know biological mechanisms and machine actions, including conformational changes, protein motions, binding modes, or states of partial disorder, and not just native structures. Physical potentials give the free energies for understanding the driving forces for conformational change.

Until recently, atomistic simulations have been too expensive computationally for predicting native structures from sequences. But, there is good reason for optimism. First, molecular dynamics (MD) simulations have recently been shown capable of folding some small proteins [19, 20] (Figure 11.11). Second, accurate atomistic simulations are reaching significantly longer simulation timescales [21]. And third, new methods can speed up MD structure-prediction simulations by harnessing generic physical insights and fuzzy knowledge, such as "make a hydrophobic core" or "make good secondary structures" [22].

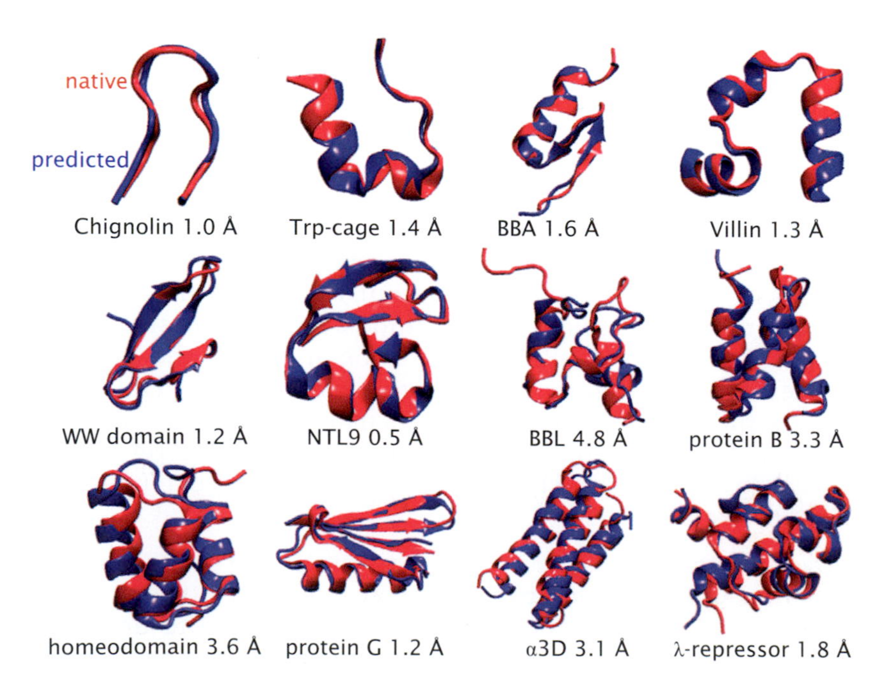

Figure 11.11 MD simulaions can fold some small proteins to near their native states. Using a high-performance custom-built computer called Anton [23], Shaw and coworkers observed reversible folding and unfolding. (*Red*) Experimental structures. (*Blue*) Computed structures. The proteins are listed, along with the RMSD (based on α-carbons) between the predicted and experimental structures. (Adapted from K Lindorff-Larsen, S Piana, RO Dror, and DE Shaw. *Science*, 334:517–520, 2011. With permission from AAAS.)

METHODS ARE AVAILABLE FOR PREDICTING THE STRUCTURES OF PROTEIN COMPLEXES, MULTIMERS, AND ASSEMBLIES

Proteins are commonly functional in multimeric states, complexes, or assemblies. For example, some 30–50% of proteins self-assemble to form symmetrical multimers. A first step in understanding the actions of multiprotein associations is to determine their structures. There are methods for computationally estimating the structures of protein–protein complexes. The first step in *protein–protein docking* is typically to try rigid-body rotations and translations of one protein relative to the other, while evaluating the paired structures using energies or scoring functions similar to those used for single-protein-structure prediction. Such rigid-body modeling can succeed in simple cases, such as for BPTI (bovine pancreatic trypsin inhibitor) binding to trypsin. But often two protein binding partners don't just dock rigidly. As they bind, they make many adjustments, such as in their side-chain conformations. It remains a challenge in most cases to correctly compute these adjustments.

One useful tool is a library of protein–protein interface structures taken from those observed in the PDB. PRISM (Protein Interactions by Structural Matching) is a server (and database) where protein–protein target interfaces can be compared with a database of template interfaces. Also useful are MD simulations that can be combined with flexible fitting algorithms to help refine low-resolution supramolecular structures/assemblies. The approach is to fit high-resolution X-ray structures resolved for different subunits/domains of the assembly onto the electron density maps resolved for the supramolecular systems, and allow for conformational fluctuations (sampled by MD) to optimize the match. A recent example is the refinement of low-resolution structure for the HIV-1 capsid [24].

Inspired by CASP©, there is a community-wide blind prediction experiment for protein–protein docked structures. Initiated in 2001 by I Vakser and S Vajda, CAPRI (Critical Assessment of Prediction of Interactions) is an event that evaluates methods for predicting the *pose* (that is, the relative position and orientation) of two proteins that bind to each other. Some prominent algorithms for protein–protein docking include ClusPro [25], Zdock [26], and HADDOCK© [27].

Complexes larger than protein–protein pairs, such as protein quaternary structures or multimeric structures, can be attempted using servers and databases. Some successes have been found using Swiss-Model or protein–protein docking algorithms such as CombDock, HADDOCK, and DockStar.

SUMMARY

Computational modeling is increasingly successful at predicting protein structures from their amino acid sequences. A common strategy is comparative modeling, where you use other known protein structures (templates) to help infer the structure of your target. You find a good template, align your target and template residues, and extract information from the known template structure; then you sample and evaluate many different conformations using knowledge-based potentials or other scoring functions. When you don't have a good template,

Table 11.A.1 Solvent-mediated contact potentials U_{ij}^{0}, in units of RT [9]

	A	C	D	E	F	G	H	I	K	L	M	N	P	Q	R	S	T	V	W	Y
A	−2.72	−3.57	−1.70	−1.51	−4.81	−2.31	−2.41	−4.58	−1.31	−4.91	−3.94	−1.84	−2.03	−1.89	−1.83	−2.01	−2.32	−4.04	−3.82	−3.36
C	−3.57	−5.44	−2.41	−2.27	−5.80	−3.16	−3.60	−5.50	−1.95	−5.83	−4.99	−2.59	−3.07	−2.85	−2.57	−2.86	−3.11	−4.96	−4.95	−4.16
D	−1.70	−2.41	−1.21	−1.02	−3.48	−1.59	−2.32	−3.17	−1.68	−3.40	−2.57	−1.68	−1.33	−1.46	−2.29	−1.63	−1.80	−2.48	−2.84	−2.76
E	−1.51	−2.27	−1.02	−0.91	−3.56	−1.22	−2.15	−3.27	−1.80	−3.59	−2.89	−1.51	−1.26	−1.42	−2.27	−1.48	−1.74	−2.67	−2.99	−2.79
F	−4.81	−5.80	−3.48	−3.56	−7.26	−4.13	−4.77	−6.84	−3.36	−7.28	−6.56	−3.75	−4.25	−4.10	−3.98	−4.02	−4.28	−6.29	−6.16	−5.66
G	−2.31	−3.16	−1.59	−1.22	−4.13	−2.24	−2.15	−3.78	−1.15	−4.16	−3.39	−1.74	−1.87	−1.66	−1.72	−1.82	−2.08	−3.38	−3.42	−3.01
H	−2.41	−3.60	−2.32	−2.15	−4.77	−2.15	−3.05	−4.14	−1.35	−4.54	−3.98	−2.08	−2.25	−1.98	−2.16	−2.11	−2.42	−3.58	−3.98	−3.52
I	−4.58	−5.50	−3.17	−3.27	−6.84	−3.78	−4.14	−6.54	−3.01	−7.04	−6.02	−3.24	−3.76	−3.67	−3.63	−3.52	−4.03	−6.05	−5.78	−5.25
K	−1.31	−1.95	−1.68	−1.80	−3.36	−1.15	−1.35	−3.01	−0.12	−3.37	−2.48	−1.21	−0.97	−1.29	−0.59	−1.05	−1.31	−2.49	−2.69	−2.60
L	−4.91	−5.83	−3.40	−3.59	−7.28	−4.16	−4.54	−7.04	−3.37	−7.37	−6.41	−3.74	−4.20	−4.04	−4.03	−3.92	−4.34	−6.48	−6.14	−5.67
M	−3.94	−4.99	−2.57	−2.89	−6.56	−3.39	−3.98	−6.02	−2.48	−6.41	−5.46	−2.95	−3.45	−3.30	−3.12	−3.03	−3.51	−5.32	−5.55	−4.91
N	−1.84	−2.59	−1.68	−1.51	−3.75	−1.74	−2.08	−3.24	−1.21	−3.74	−2.95	−1.68	−1.53	−1.71	−1.64	−1.58	−1.88	−2.83	−3.07	−2.76
P	−2.03	−3.07	−1.33	−1.26	−4.25	−1.87	−2.25	−3.76	−0.97	−4.20	−3.45	−1.53	−1.75	−1.73	−1.70	−1.57	−1.90	−3.32	−3.73	−3.19
Q	−1.89	−2.85	−1.46	−1.42	−4.10	−1.66	−1.98	−3.67	−1.29	−4.04	−3.30	−1.71	−1.73	−1.54	−1.80	−1.49	−1.90	−3.07	−3.11	−2.97
R	−1.83	−2.57	−2.29	−2.27	−3.98	−1.72	−2.16	−3.63	−0.59	−4.03	−3.12	−1.64	−1.70	−1.80	−1.55	−1.62	−1.90	−3.07	−3.41	−3.16
S	−2.01	−2.86	−1.63	−1.48	−4.02	−1.82	−2.11	−3.52	−1.05	−3.92	−3.03	−1.58	−1.57	−1.49	−1.62	−1.67	−1.96	−3.05	−2.99	−2.78
T	−2.32	−3.11	−1.80	−1.74	−4.28	−2.08	−2.42	−4.03	−1.31	−4.34	−3.51	−1.88	−1.90	−1.90	−1.90	−1.96	−2.12	−3.46	−3.22	−3.01
V	−4.04	−4.96	−2.48	−2.67	−6.29	−3.38	−3.58	−6.05	−2.49	−6.48	−5.32	−2.83	−3.32	−3.07	−3.07	−3.05	−3.46	−5.52	−5.18	−4.62
W	−3.82	−4.95	−2.84	−2.99	−6.16	−3.42	−3.98	−5.78	−2.69	−6.14	−5.55	−3.07	−3.73	−3.11	−3.41	−2.99	−3.22	−5.18	−5.06	−4.66
Y	−3.36	−4.16	−2.76	−2.79	−5.66	−3.01	−3.52	−5.25	−2.60	−5.67	−4.91	−2.76	−3.19	−2.97	−3.16	−2.78	−3.01	−4.62	−4.66	−4.17

you can use methods such as threading co-evolution or *de novo* methods, such as molecular simulations. An important development for validating prediction methods has been the community-wide blind-test evaluation events CASP and CAPRI. Protein structures can now be predicted well for small simple water-soluble single-domain proteins, but predicting more complex structures remains a research frontier.

APPENDIX 11A: THE MIYAZAWA–JERNIGAN CONTACT-POTENTIAL MATRIX

Table 11.A.1 is an example of a contact-potential matrix for the different types of amino acids. This one was derived from a set of 1168 diverse proteins by counting the number of contacts between every pair of residues having a distance between the centers of the side chains of 6.8 Å or less. More negative values indicate stronger preferences for residues of types *i* and *j* to be paired together in protein structures. It shows that hydrophobic attractions are strongest and polar interactions are weaker. The largest value in the matrix is for pairs of Cys, and this is because they are frequently paired up by disulfide covalent bonding. The matrix is symmetric. These 210 pairwise parameters can be approximated by a reduced set of 20 single-body parameters plus 4 two-body parameters [28] (see also [29]).

REFERENCES

[1] W Kabsch and C Sander. Dictionary of protein secondary structure: Pattern recognition of hydrogen-bonded and geometrical features. *Biopolymers*, 22:2577–2637, 1983.

[2] DT Jones. Protein secondary structure prediction based on position-specific scoring matrices. *J Mol Biol*, 292:195–202, 1999.

[3] LJ McGuffin, K Bryson, and DT Jones. The PSIPRED protein structure prediction server. *Bioinformatics*, 16:404–405, 2000.

[4] KT Simons, C Kooperberg, E Huang, and D Baker. Assembly of protein tertiary structures from fragments with similar local sequences using simulated annealing and Bayesian scoring functions. *J Mol Biol*, 268:209–225, 1997.

[5] F Morcos, A Pagnani, B Lunt, et al. Direct-coupling analysis of residue coevolution captures native contacts across many protein families. *Proc Natl Acad Sci USA*, 108:E1293–E301, 2011.

[6] DS Marks, TA Hopf, and C Sander. Protein structure prediction from sequence variation. *Nat Biotechnol*, 30:1072–1080, 2012.

[7] S Tanaka and HA Scheraga. Medium and long-range interaction parameters between amino acids for predicting three-dimensional structures of proteins. *Macromolecules*, 9:945–950, 1976.

[8] S Miyazawa and RL Jernigan. Estimation of effective inter-residue contact energies from protein crystal structures: Quasi-chemical approximation. *Macromolecules*, 18:534–552, 1985.

[9] S Miyazawa and RL Jernigan. Residue–residue potentials with a favorable contact pair term and an unfavorable high packing density term, for simulation and threading. *J Mol Biol*, 256:623–644, 1996.

[10] C Yanover, N Vanetik, M Levitt, et al. Redundancy-weighting for better inference of protein structural feautres. *Bioinformatics* 30:2295–2301, 2014.

[11] RL Jernigan and I Bahar. Structure-derived potentials and protein simulations. *Curr Opin Struct Biol*, 6:195–209, 1996.

[12] A Bakan, A Dutta, W Mao, et al. Evol and ProDy for bridging protein sequence evolution and structural dynamics. *Bioinformatics*, 30:2681–2683, 2014.

[13] J Moult. A decade of CASP: Progress, bottlenecks and prognosis in protein structure prediction. *Curr Opin Struct Biol*, 15:285–289, 2005.

[14] GG Krivov, MV Shapovalov, and RL Dunbrack Jr. Improved prediction of protein side-chain conformations with SCWRL4. *Proteins*, 77:778–795, 2009.

[15] DJ Mandell, EA Coutsias, and T Kortemme. Sub-Ångstrom accuracy in protein loop reconstruction by robotics-inspired conformational sampling. *Nat Methods*, 6:551–552, 2009.

[16] D Russel, K Lasker, B Webb, et al. Putting the pieces together: Integrative modeling platform software for structure determination of macromolecular assemblies. *PLoS Biol*, 10:e1001244, 2012.

[17] CA Rohl, CEM Strauss, KMS Misura, and D Baker. Protein structure prediction using Rosetta. *Methods Enzymol*, 383:66–93, 2004.

[18] J Yang, R Yan, A Roy, et al. I-TASSER: A unified platform for automated protein structure and function prediction. *Nat Methods*, 12:7–8, 2015.

[19] K Lindorff-Larsen, S Piana, RO Dror, and DE Shaw. How fast-folding proteins fold. *Science*, 334:517–520, 2011.

[20] H Nguyen, J Maier, H Huang, et al. Folding simulations for proteins with diverse topologies are accessible in days with a physics-based force field and implicit solvent. *J Am Chem Soc*, 136:13959–13962, 2014.

[21] A Perez, JA Morrone, C Simmerling, and KA Dill. Advances in free-energy-based simulations of protein folding and ligand binding. *Curr Opin Struct Biol*, 36:25–31, 2016.

[22] A Perez, JL MacCallum, and KA Dill. Accelerating molecular simulations of proteins using Bayesian inference on weak information. *Proc Natl Acad Sci USA*, 112:11846–11851, 2015.

[23] DE Shaw, MM Deneroff, RO Dror, et al. Anton, a special-purpose machine for molecular dynamics simulation. *Commun ACM*, 51:91–97, 2008.

[24] G Zhao, JR Perilla, EL Yufenyuy, et al. Mature HIV-1 capsid structure by cryo-electron microscopy and all-atom molecular dynamics. *Nature*, 497:643–646, 2013.

[25] SR Comeau, DW Gatchell, S Vajda, and CJ Camacho. ClusPro: An automated docking and discrimination method for the prediction of protein complexes. *Bioinformatics*, 20:45–50, 2004.

[26] BG Pierce, K Wiehe, H Hwang, et al. ZDOCK Server: Interactive docking prediction of protein–protein complexes and symmetric multimers. *Bioinformatics*, 30:1771–1773, 2014.

[27] SJ de Vries, M van Dijk, and AMJJ Bonvin. The HADDOCK web server for data-driven biomolecular docking. *Nat Protoc*, 5:883–897, 2010.

[28] O Keskin, I Bahar, AY Badretdinov, et al. Empirical solvent-mediated potentials hold for both intra-molecular and inter-molecular inter-residue interactions. *Protein Sci*, 7:2578–2586, 1998.

[29] H Li, C Tang, and NS Wingreen. Nature of driving force for protein folding: A result from analyzing the statistical potential. *Phys Rev Lett*, 79:765–768, 1997.

SUGGESTED READING

Dill K and MacCallum J, The protein-folding problem, 50 years on. *Science*, 338:1042, 2012.

Marks DS, Hopf TA, and Sander C, Protein structure prediction from sequence variation. *Nat Biotechnol*, 30:1072–1080, 2012.

Ofran Y and Margalit H, Proteins of the same fold and unrelated sequences have similar amino acid composition. *Proteins*, 64:275–279, 2006.

Zvelebil M and Baum JO, *Understanding Bioinformatics.* Garland Science, New York, 2008.

CHAPTER 12

Biological Actions Arise from Protein Motions

NATIVE PROTEINS HAVE CORRELATED MOTIONS

The shapes of native proteins fluctuate, due to natural thermal motions. Even though these movements occur in the presence of random solvent forces, the protein's motions are not random. Some motions are *collective*: the atoms move in concert with each other, not independently. Whole domains can move as a unit, for example. In *allosteric responses*, a stimulus at one location in the protein triggers an action elsewhere in the protein.

Concerted motions often underlie protein *mechanisms of action*. Different proteins encode different biological mechanisms because different protein shapes have different intrinsic motions. For example, the ribosome has a shape that leads to a natural ratcheting motion that is essential to catalyzing the addition of each amino acid during protein synthesis (**Box 12.1**). Collective motions depend more on a protein's overall shape than on its atomic details, so they can often be captured by coarse-grained models. In this chapter, we describe motions, mechanisms, and *elastic network models* (ENM).

Our knowledge of protein motions comes from various sources, including different structures resolved for the same protein under different conditions, for example at different stages of a functional cycle (see Figure 12.1 and **Figure 12.2**), in substrate-bound or -unbound forms, in

Box 12.1 Intrinsic Motions of the Ribosome Play a Role in Protein Synthesis

A ribosome is a molecular machine that pulls an mRNA molecule through itself as it adds amino acids to a growing protein. This action results from rotating and ratcheting of the large subunit relative to the small subunit during the translocation of the tRNAs along with the mRNA [1], illustrated in **Figure 12.1**. These motions can be computed from the ribosome's structure by an elastic network model [2, 3].

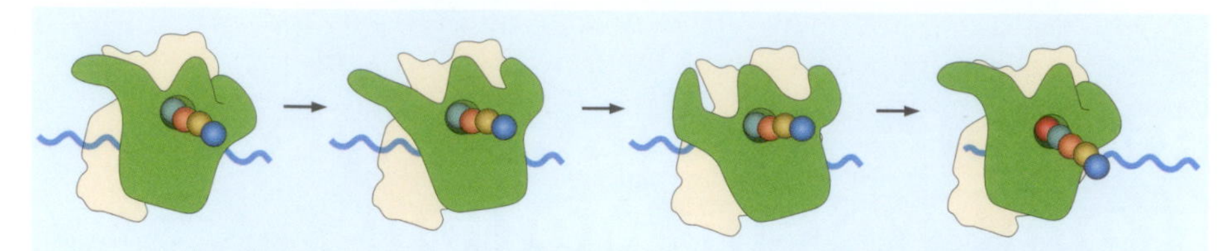

Figure 12.1 The ribosome undergoes ratchet like motions, translocating along the mRNA as it synthesizes proteins. In one ratchet step, starting from the left the 50S subunit (*green*) rotates versus the 30S subunit (*cream*) counterclockwise, before returning to the starting state on the right. During this cycle, the *red* amino acid has been added. This cycle repeats for each added amino acid (*colored* spheres). (For details see [3].)

the presence of inhibitors or other small molecules, in different multimeric states (as monomers, dimers, or oligomers), or even in different mutant forms, which present an ensemble of accessible conformers. Other sources of data indicating conformational variabilities are the experimental *B-factors* in X-ray crystal structures (a measure of the variances around the native atomic coordinates); NMR solution structures and order parameters (a measure of the fluctuations in residue positions, often measured in solution); electron microscopy, electron density maps from X-ray crystallography, and, to a lesser extent, Raman spectroscopy and inelastic neutron scattering. In addition, two types of computer modeling contribute to understanding protein dynamics and mechanisms: all-atom models, as described in Chapters 9 and 10, and coarse-grained (CG) ones such as the elastic network models that we describe here.

ELASTIC NETWORK MODELS USE BEADS AND SPRINGS TO DESCRIBE PROTEIN MOTIONS

At equilibrium, a protein fluctuates around its native structure because of Brownian motions. The motions are coordinated, in *collective modes of motion*, over larger parts of the protein. These modes can be described by ENMs, which originated in the statistical mechanical theory of macromolecular networks and rubber elasticity introduced by Flory, Erman, and others [4, 5]. ENMs treat molecular substructures as bead-like masses connected by springs (see Figure 12.2). ENMs were first extended to protein structural dynamics by B Erman, I Bahar, and

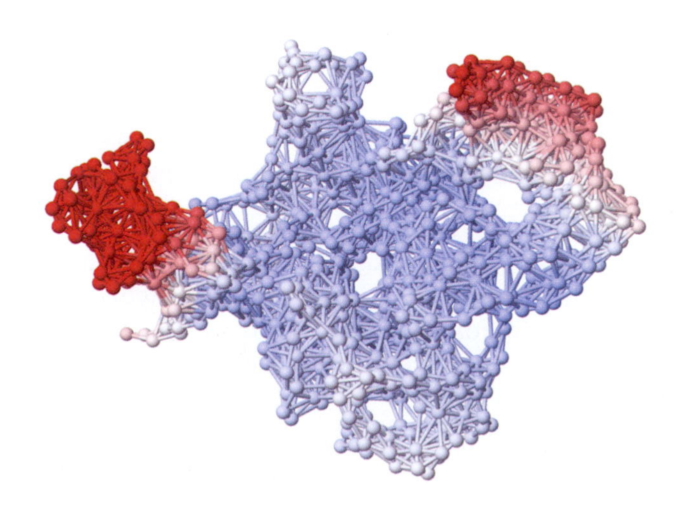

Figure 12.2 Representation of HIV-1 reverse transcriptase in an ENM. The structure is represented by a network of beads/nodes connected by springs (shown here as sticks). Each bead represents an amino acid, and springs connect amino acid pairs (bonded or nonbonded) within a cutoff distance of interaction. The structure is color-coded from *red* (most mobile) to *blue* (most rigid), based on cooperative motions predicted by the ENM.

coworkers [6, 7], inspired by a normal mode analysis with uniform harmonic potentials performed by Tirion [8]. The first ENM used for describing protein dynamics is the Gaussian network model (GNM), described here.

Consider a protein having N amino acids. In the GNM [6, 7], you represent each amino acid i $(1 \leq i \leq N)$ as a single bead, centered on its α-carbon. You can use the PDB coordinates $\mathbf{r}_i$ of the amino acids to define the $3N$-dimensional *configuration vector* of the protein,

$$\mathbf{r} = [\mathbf{r}_1,\ \mathbf{r}_2,\ \mathbf{r}_3, \ldots,\ \mathbf{r}_N]. \tag{12.1}$$

The configuration vector has $3N$ dimensions because each bead vector $\mathbf{r}_i$ has x-, y-, and z-components. Each vector $\mathbf{r}_i$ can deviate by $\Delta\mathbf{r}_i(t)$ from its equilibrium position $\mathbf{r}_i^0$:

$$\Delta\mathbf{r}_i = \mathbf{r}_i - \mathbf{r}_i^0. \tag{12.2}$$

The instantaneous changes in the positions of all residues at a given time are

$$\Delta\mathbf{r} = [\Delta\mathbf{r}_1,\ \Delta\mathbf{r}_2,\ \Delta\mathbf{r}_3, \ldots,\ \Delta\mathbf{r}_N]. \tag{12.3}$$

Now consider the separation between any pair of beads i and j. You can express their *equilibrium separation* as $\mathbf{r}_{ij}^0 = \mathbf{r}_j^0 - \mathbf{r}_i^0$ and the *instantaneous separation* as $\mathbf{r}_{ij} = \mathbf{r}_{ij}^0 + \Delta\mathbf{r}_{ij}$, where we define $\Delta\mathbf{r}_{ij} = \Delta\mathbf{r}_j - \Delta\mathbf{r}_i$ as the instantaneous fluctuation in the position of residue i (Figure 12.3). By instantaneous, we mean a specific microscopic configuration at time t.

In the GNM, each spring has a potential energy given by

$$U_{ij} = \gamma_{ij}(\Delta\mathbf{r}_j - \Delta\mathbf{r}_i) \cdot (\Delta\mathbf{r}_j - \Delta\mathbf{r}_i) = \gamma_{ij}(\Delta\mathbf{r}_{ij})^2, \tag{12.4}$$

where γ_{ij} is the force constant for the spring connecting residues i and j, and the dot indicates the dot product (scalar product) of vectors. Note that U_{ij} depends on the change $\Delta\mathbf{r}_{ij}$ in the inter-residue distance vector $\mathbf{r}_{ij}$; that is, if $\mathbf{r}_{ij}$ differs from $\mathbf{r}_{ij}^0$, then $U_{ij} > 0$ even if the magnitude $|\mathbf{r}_{ij}|$ is maintained. We implicitly assume here that the initial structure is a state of minimum energy. Now, to compute the total elastic energy U_{GNM} of the whole protein, you sum over all the springs and divide by two (to correct for double counting the springs), to get

$$U_{\mathrm{GNM}} = \frac{\gamma_{ij}}{2} \sum_i \sum_j (\Delta\mathbf{r}_{ij})^2. \tag{12.5}$$

In the GNM, the protein is treated as a microscopically uniform elastic material at equilibrium, meaning that we take all the springs to have equal spring constants $\gamma_{ij} = \gamma$. And usually there is no distinction between the springs connecting bonded groups and those that connect nonbonded groups nearby in space (Figure 12.4). Two amino acids that are far apart in the protein interact only weakly or indirectly, so we neglect those interactions, and we connect each residue to its neighbors within a first coordination shell. From X-ray crystal structures, it is found that first neighbors are usually located within a radius of 7 Å [9]. So, a cutoff distance of $r_c = 7$ Å is usually used in the GNM representation of proteins.

To describe all the neighbor pairings, we use the *Kirchhoff matrix* $\mathbf{\Gamma}$, also called the *adjacency matrix*, of the native structure of the protein

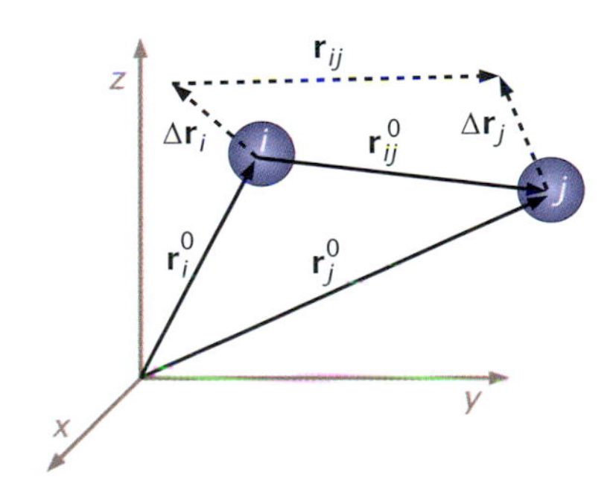

Figure 12.3 Defining the position vectors and their fluctuations. Beads i and j have initial positions $\mathbf{r}_i^0$ and $\mathbf{r}_j^0$. The subsequent displacements are $\Delta\mathbf{r}_i$ and $\Delta\mathbf{r}_j$, respectively, leading to the new separation $\mathbf{r}_{ij}$.

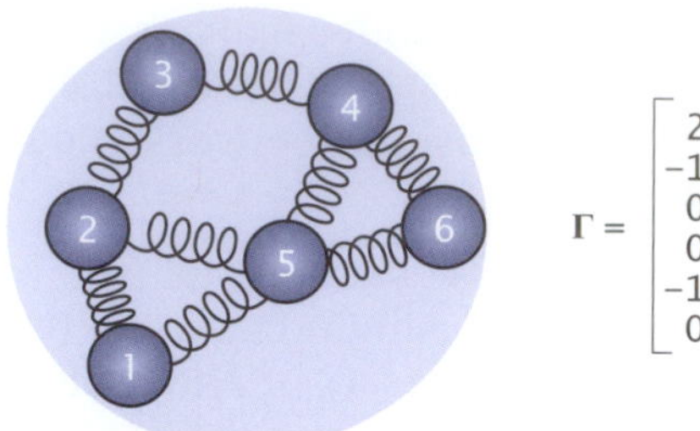

$$\Gamma = \begin{bmatrix} 2 & -1 & 0 & 0 & -1 & 0 \\ -1 & 3 & -1 & 0 & -1 & 0 \\ 0 & -1 & 2 & -1 & 0 & 0 \\ 0 & 0 & -1 & 3 & -1 & -1 \\ -1 & -1 & 0 & -1 & 4 & -1 \\ 0 & 0 & 0 & -1 & -1 & 2 \end{bmatrix}$$

Figure 12.4 Example of an adjacency matrix, Γ. A six-bead structure in the Gaussian network model. Beads that are within a cutoff distance of r_C are connected by springs. A bead typically represents an amino acid. The corresponding adjacency matrix represents the topology of inter-residue contacts. The rows and columns (1 to 6) correspond to the node numbers of the beads. The off-diagonal matrix element (i,j) is set to -1 when beads i and j are connected by a spring, and set to 0 otherwise. Each diagonal element is set equal to the negative sum of all off-diagonal terms on the same row (or column, since the matrix is symmetric). The numbers on the diagonal equal the coordination number z_i of each residue (also called the *degree of node i*).

(see Figure 12.4). The Kirchhoff matrix is symmetric (if i is a neighbor of j, then j must also be a neighbor of i). This matrix is the central structural quantity for computing the equilibrium motions of the network. The internal energy of the network at time t can be expressed in terms of Γ as (see Appendix 12C)

$$U_{\text{GNM}}(t) = \frac{\gamma}{2}\Delta\mathbf{r}(t)^T\mathbf{\Gamma}\Delta\mathbf{r}(t). \tag{12.6}$$

Some Motions of Different Parts of a Structure Are Correlated with Each Other

Suppose you want to know if movements of amino acid i are correlated with those of amino acid j. Such motions are important in *allostery* for example, where an action at site A in a protein causes an action at a distant site B. Sometimes the binding of a ligand triggers a conformational change elsewhere in the protein. To learn how the fluctuations in i and j are related to each other, you can compute the *cross-correlations* between them. Here is how you do that using the GNM. First, take the dot product of the fluctuation $\Delta\mathbf{r}_i$ with $\Delta\mathbf{r}_j$, to get the degree of alignment of those motions. Then, compute the Boltzmann average of that dot product over all the possible configurational changes:

$$\langle \Delta\mathbf{r}_i \cdot \Delta\mathbf{r}_j \rangle = \frac{1}{Q}\int (\Delta\mathbf{r}_i \cdot \Delta\mathbf{r}_j)e^{-U/RT}\, d(\Delta\mathbf{r}). \tag{12.7}$$

The integral in Equation 12.7, called the *generalized Gaussian integral* reduces to the expression

$$\langle \Delta\mathbf{r}_i \cdot \Delta\mathbf{r}_j \rangle = \frac{3RT}{\gamma}[\mathbf{\Gamma}^{-1}]_{ij}. \tag{12.8}$$

Here's how you use Equation 12.8. First, write down the Kirchhoff matrix $\mathbf{\Gamma}$ for your native protein from its interatomic contacts, as indicated in Figure 12.4. Next, take the *pseudoinverse* of that matrix.[1] The (i, j) element of that inverse matrix is $[\mathbf{\Gamma}^{-1}]_{ij}$. Multiply each element by $3RT/\gamma$ to get the cross-correlation between residues i and j. Incidentally, notice that by setting $i = j$, you can also determine the mean-square fluctuation (MSF) of residue i:

$$\langle (\Delta r_i)^2 \rangle = \langle \Delta\mathbf{r}_i \cdot \Delta\mathbf{r}_i \rangle = \frac{3RT}{\gamma}[\mathbf{\Gamma}^{-1}]_{ii}. \tag{12.9}$$

[1] By definition, $\mathbf{\Gamma}$ is not invertible, because the rows and columns each sum to zero. Instead, calculate the pseudoinverse. This means obtaining the eigenvalue decomposition and reconstructing the matrix after removing the contribution of the zero eigenvalue.

Here is an important implication of Equation 12.9. Γ^{-1} is directly related to the density of contacts that an amino acid has with its near neighbors in the protein (see Appendix 12B). So, Equation 12.9 shows that the MSFs of residue i depend on how sterically restricted it is by its surrounding residues. Residues on protein surfaces are less restricted, on average, than residues in protein cores. Note that *time* does not appear in these expressions. Equation 12.9 simply describes the average fluctuations of the protein structure at equilibrium. Such fluctuations are observable as structural variances in X-ray and NMR experiments.

PROTEIN MOTIONS CAN BE OBSERVED IN EXPERIMENTS AND PREDICTED BY THE GNM

Information about protein motions can be inferred from various types of experiments. For one, *Debye–Waller factors*, also called *B-factors*, are a measure of the fluctuations in atomic coordinates that are observed in the X-ray crystallographic determination of protein structures:

$$B_i = \frac{8\pi^2}{3}\langle(\Delta r_i)^2\rangle. \tag{12.10}$$

B-factors are proportional to the MSFs of atoms. B-factors of structurally resolved atoms are reported in the PDB. Comparing Equation 12.10 to Equation 12.9 shows that you can use the GNM to compute B-factors, given only the native coordinates. The GNM predicts relatively well the fluctuations of proteins (Figure 12.5) and those of RNAs. B-factors are often better predicted by the GNM than by atomistic molecular dynamics (MD) simulations because GNM is analytical and not subject to sampling inaccuracies. Furthermore, the GNM is much simpler and faster to compute, and is easy to interpret.

Now, to determine the *absolute magnitudes* of the motions, you need to know the value of γ, which uniformly rescales the size of all residue motions. In typical applications, γ is taken as a fit parameter, chosen

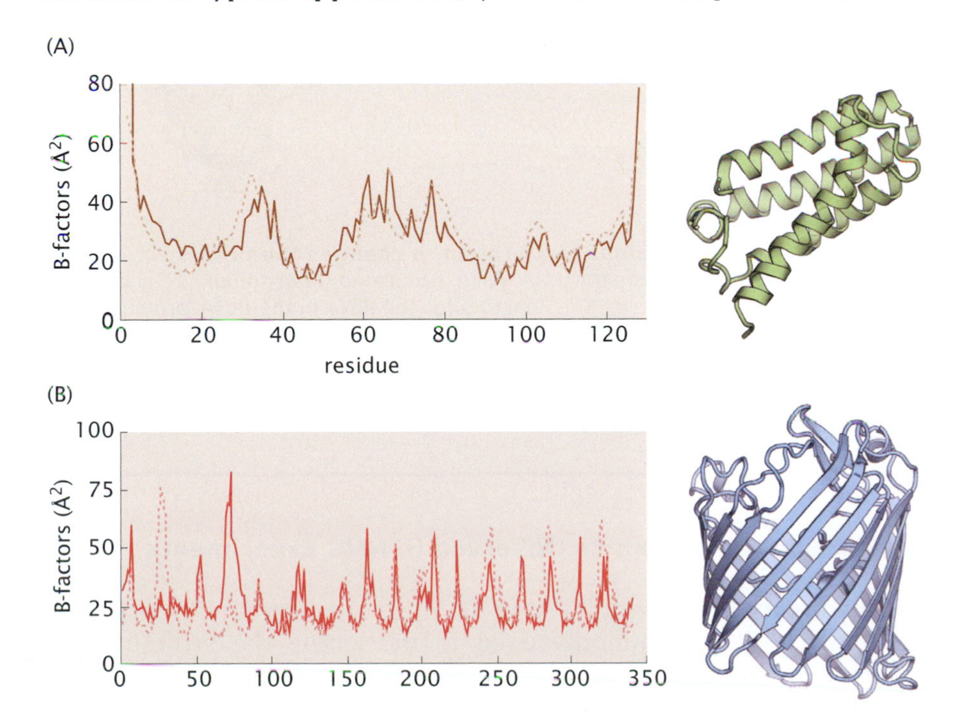

Figure 12.5 The GNM computes the equilibrium fluctuations of a protein from its X-ray structure. (Left) Experimental X-ray B-factors (*dotted* curve) for (A) ferricytochrome c' and (B) matrix porin, compared with the calculated B-factors from the GNM using the native structure and Equation 12.10 (*solid* curve). In general, the motions are smallest in the middle of secondary structure elements, and largest in the connecting loops. (Right) The corresponding protein structures.

so that that the average fluctuations of a protein's residues are approximately equal to the average B-factors. Across proteins, it is found that the average spring constant is $RT/\gamma = 0.87 \pm 0.46\,\text{Å}^2$, and $r_c = 7.3\,\text{Å}$ [10]. This means that a deformation of 1 Å in inter-residue distance increases the energy by $\sim$0.70 kcal mol^{-1}, although the range of the spring constants in proteins is quite broad.

There is a caveat. In a protein crystal, there are points where one protein contacts an adjacent protein. Not surprisingly, at those crystal-contact points, the motions are more restricted than the GNM would predict for the isolated protein [10] (Figure 12.6). By including the crystal packing interactions (for example, GNM calculations for a protein that includes its neighbors from the crystal lattice), Phillips and coworkers showed that the average correlation with experimental B-factors improved from 0.594 to 0.661 [10]. The motions inside a protein can also be changed when the protein binds to a substrate (Box 12.2).

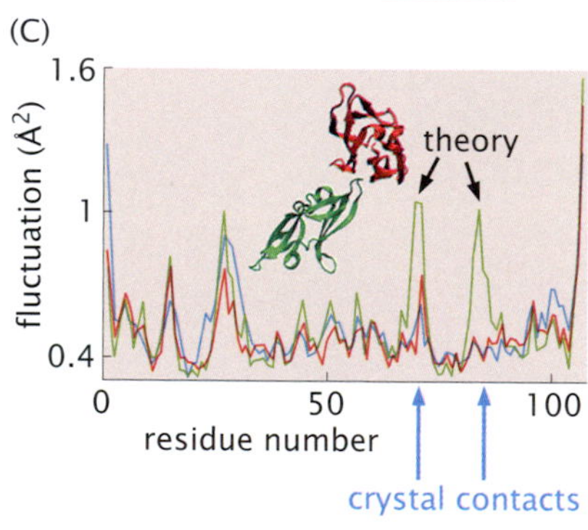

Figure 12.6 Proteins in crystals have dampened motions at sites of their intermolecular contacts. (A) A protein (*green*) surrounded by its neighbors (*red*) in a crystal. (B) Two crystal contacts, where a red molecule contacts the green one. (C, *green*) Protein fluctuations computed using the GNM for the isolated protein. (C, *blue*) Protein fluctuations computed with the GNM by taking into account the crystal contacts. (C, *red*) Fluctuations measured by experiments on the crystal. The two contact points have reduced fluctuations compared with the isolated protein. (Adapted from L Liu, LMI Koharudin, AM Gronenborn, and I Bahar. *Proteins*, 77:927–939, 2009.)

Box 12.2 A Protein's Motions Can Be Altered by Binding to a Substrate

When a glutamine-carrying tRNA (tRNA-Gln) binds to glutamine synthetase, the motions of the tRNA are altered (Figure 12.7). Such binding-induced changes in motions can often be predicted by an ENM. GNM results show that the motions of a given biomolecule (protein or DNA/RNA) derive from the shape of the whole structure of the complex.

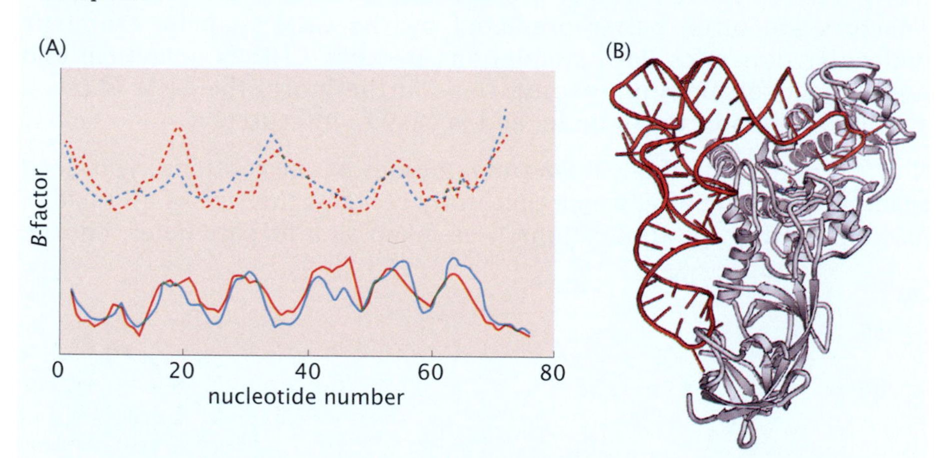

Figure 12.7 When tRNA binds to a protein, it changes the tRNA's fluctuations. (A) The B-factors for tRNA alone: calculated (*blue dotted curve*) and experimental (*red dotted* curve). The B-factors for the tRNA in the tRNA–protein complex: calculated (*blue solid* curve) and experimental (*red solid curve*) (B) The complex of tRNA with glutamine synthetase. The main changes upon binding are well described by the GNM. (Adapted from I Bahar and RL Jernigan. *J Mol Biol*, 281:871–884, 1998. With permission from Elsevier.)

Protein Fluctuations Can Be Observed in NMR Experiments

NMR spectroscopy is another way to measure protein fluctuations. In general, NMR is commonly used to determine the structures of biomolecules in solution. It has the advantages that you don't need to crystallize the protein and you don't have artifacts due to the protein–protein contacts in crystals. NMR is used to study relatively small proteins (typically 30–40 kDa), and provides ensembles of their structures.

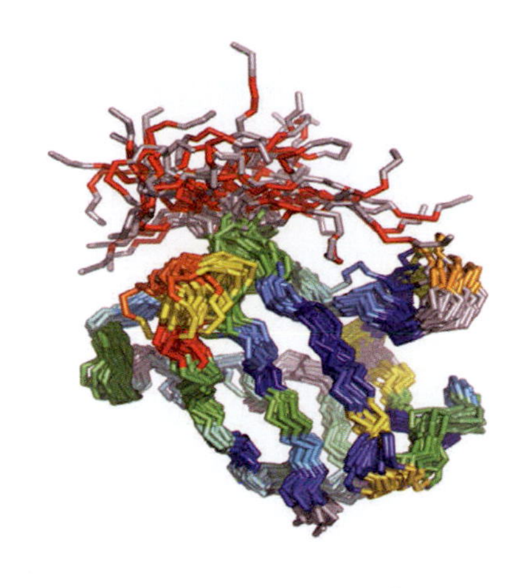

Figure 12.8 Native proteins are ensembles of structures, as seen here by NMR residual dipolar coupling experiments for ubiquitin. (From OF Lange, NA Lakomek, C Farès, et al. *Science*, 320:1471–1475, 2008. With permission from AAAS.)

Now, compare how different experimental methods see structures and their variations. Is a protein structure that is determined by NMR the same as the structure determined by X-ray crystallography? And are the fluctuations determined by the two methods the same? It is generally observed that where both methods are used on the same protein, the differences in the average structure are usually small. It is not surprising that crystal structures (from X-rays) resemble solution structures (from NMR), because protein crystals typically are 40–60% water by volume, and proteins in crystals are often even enzymatically active. The largest differences between X-ray and NMR structures are often at solvent-exposed regions, such as loops. But, this also shows how robust protein structures are, and often they can maintain the same structures in different environments.

But while the *average* structures are similar, the *fluctuations* seen by NMR (Figure 12.8) can be different from those seen in X-ray crystallography. The GNM contributes interesting insight here. Comparing (a) NMR structures of 64 proteins in solution with (b) B-factors of those same proteins in the crystalline state, and with (c) GNM calculations that explicitly leave out crystallographic packing contacts between protein molecules in crystal structures, shows that variations across the NMR structures are better predicted by the GNM (correlation coefficient of 0.76) than by experimental X-ray B-factors (correlation coefficient of 0.49) [11]. The GNM is a good predictor of the native motions of proteins in solution.

Protein Breathing Motions Can Be Observed in Hydrogen Exchange (HX) Experiments

Hydrogen–deuterium (H/D) exchange (HX) is another way to study structural changes in proteins, for example "breathing" motions that occur naturally during equilibrium fluctuations around the native structure. The use of HX began with Kai Linderstrom-Lang, and it was applied to protein folding intermediates by Udgaonkar and Baldwin [12] and Roder et al. [13], who monitored HX by NMR. One type of experiment measures *HX protection factors*. Each amino acid in a protein contains an amide group. An amide group has a nitrogen atom that is covalently bonded to two hydrogen atoms. Imagine putting an amide group by itself into a water solution in which H_2O is replaced by deuterated water, D_2O. The deuterium atoms will pop on and off of the solvent water molecules, and they will exchange with the protons on the amide groups, causing the amide group on the protein to become deuterated. You can determine the extent of protein deuteration either by mass spectrometry, since the masses of H and D are different, or by NMR, since protons are detectable by proton NMR and deuterium is not. Why would you want to measure this? It can tell you about when different atoms of the protein become buried in the folding process, or which protein atoms are protected from solvent by binding a ligand, for example.

Here is the kinetic scheme for HX exchange reaction for an amino acid in a protein:

$$\mathrm{p_{closed}} \underset{k_{\mathrm{cl}}}{\overset{k_{\mathrm{op}}}{\rightleftharpoons}} \mathrm{p_{open}} \overset{k_{\mathrm{ex}}}{\rightarrow} \text{exchange},$$

where p_{closed} and p_{open} represent the populations of protected and exposed conformers, in folded and exposed states of the protein. We are interested in the overall observed reaction rate k_{obs}. There are two limiting cases for the relationship between the rates.

1. **The *EX1* mechanism.** Here, exchange occurs under conditions in which, once the protein opens, exchange is so fast that the proton always exchanges with the solvent: $k_{cl} \ll k_{ex}$. In this limit, the rate of exchange you observe equals the opening rate: $k_{obs} = k_{op}$.
2. **The *EX2* mechanism.** In this case, exchange occurs under the opposite conditions, in which the protein re-closes in a much shorter time than it takes for the proton exchange to take place: $k_{cl} \gg k_{ex}$. In this case, you have

$$k_{obs} = \left(\frac{k_{op}}{k_{cl}}\right) k_{ex} = K_{op} k_{ex}. \tag{12.11}$$

In separate model-compound experiments, you can measure k_{ex}, which is a measure of the intrinsic speed of exchange, that is, how fast an amide group by itself in solution can exchange with the solvent. By taking the ratio of two measurements, you get the *HX protection factor* $P = 1/K_{op}$:

$$P = \frac{1}{K_{op}} = \frac{k_{ex}}{k_{obs}}. \tag{12.12}$$

Amino acids that are deeply buried in the protein's core will exchange slowly (k_{obs} is small) so their protection factors are large. On the other hand, amino acids that are on the protein's surface will exchange protons more rapidly with the solvent (k_{obs} is large) so their protection factors are small. In the native state, P for buried residues can be as high as 10^8, indicating that the protection of these amides is higher by eight orders of magnitude compared with that in the unfolded state. $P \approx 1$ for unfolded or solvent-exposed residues. $P \approx 10^3$ for residues in a molten-globule state. Interestingly, P measures more than just the depth of burial of an amino acid in the protein's core. Various specific motions of the protein cause even groups that have identical degrees of burial to "see" the solvent to different degrees. The GNM captures both aspects well.

The H/D exchange behaviors of proteins in the native state—or under mildly denaturing conditions (*EX2* conditions)—have been interpreted with the GNM [14]. By definition, in the GNM, all residues undergo Gaussian fluctuations, and the departure from mean positions entails an entropic cost. The entropy increase accompanying the fluctuation $\Delta\mathbf{r}_i$ of amino acid i is therefore found from the Gaussian distribution

$$W(\Delta\mathbf{r}_i) = \exp\left[\frac{-3(\Delta r_i)^2}{2\langle(\Delta r_i)^2\rangle}\right] \tag{12.13}$$

to be

$$\Delta S_i = R \ln W(\Delta\mathbf{r}_i) = -\frac{\gamma(\Delta r_i)^2}{2T[\mathbf{\Gamma}^{-1}]_{ii}}. \tag{12.14}$$

In the last equality, we have replaced $\langle(\Delta r_i)^2\rangle$ by $(3RT/\gamma)[\Gamma^{-1}]$, using Equation 12.9. Equation 12.14 implies that the ith residue is subject to a free-energy increase of

$$\Delta G_i = -T\Delta S_i = \frac{\gamma}{2}\frac{(\Delta r_i)^2}{[\Gamma^{-1}]_{ii}} \tag{12.15}$$

upon distortion of its coordinates by an amount Δr_i.

Equation 12.15 shows that for a given displacement $\Delta r_i = \Delta r$ of residue i, the free energy of exchange ΔG_i becomes inversely proportional to $[\Gamma^{-1}]_{ii}$, or to $\langle(\Delta r_i)^2\rangle$. Physically, this just means that residues having smaller fluctuations in the folded state have a stronger resistance to deformation. Figure 12.9 compares the experimentally measured HX protection factors for BPTI and staphylococcal nuclease with those predicted by the GNM using Equation 12.15, which gives $\Delta G_i \sim [\Delta\Gamma^{-1}]_{ii}$, with the proportionality constant scaled to match the absolute value of the data.

Overall, we can draw some general conclusions from ENMs. The higher the local packing density, the more highly constrained is the residue. Not surprisingly, motions are usually smaller among residues buried in protein cores and larger among residues on protein surfaces. Similarly, helices or strands in proteins, particularly their central portions, often have smaller MSFs, whereas loops often have larger motions (see Figure 12.5). This reflects their local packing densities. Helices and β-strands are usually closely packed. Loops are usually on protein surfaces and are often floppy. And beyond packing effects, as described later on, the GNM (or ENMs in general) can predict cooperative actions in the protein. These are motions of the entire molecule, also called

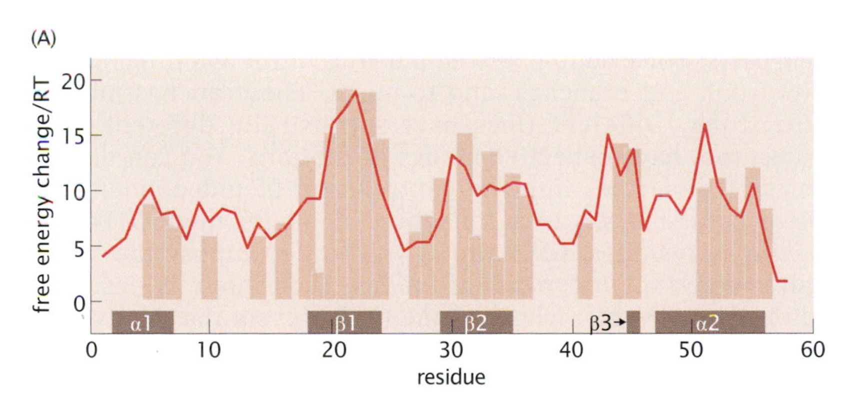

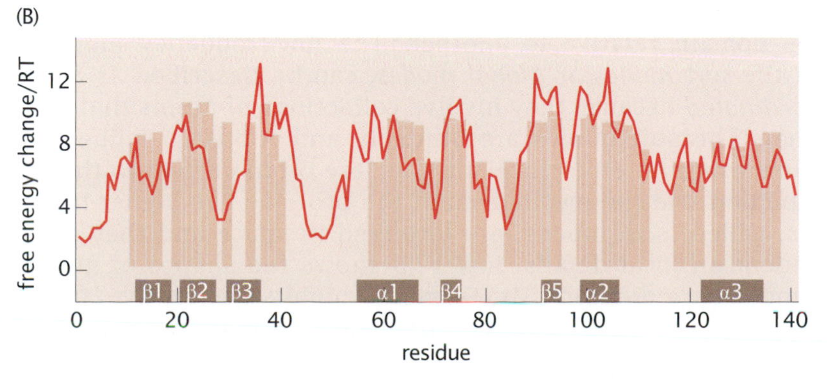

Figure 12.9 H/D exchange happens slowly inside secondary structures and faster at turn regions. The x-axis shows the residue position in (A) BPTI and (B) staphylococcal nuclease. (*Brown bars*) Experimental protection factors (higher bars indicate greater protection). (*Red lines*) GNM predictions using Equation 12.15. (Adapted from I Bahar, A Wallqvist, DG Covell, and R Jernigan. *Biochemistry*, 1998;37:1067–1075.)

global motions. They are uniquely defined by the three-dimensional structure, and they are robust to changes in small atomic details.

PROTEIN MOTIONS ARE RELEVANT TO MECHANISMS OF ACTION

Biological actions can result from concerted protein motions. One example is the ribosomal action shown in Figure 12.1. Motor proteins undergo large rotational or linear motions to haul cargo. Chaperones bind to client proteins that are folding, then undergo large changes to release those clients. Signaling proteins undergo allosteric motions to propagate signals over distances. Enzymes often undergo different important motions on different timescales. On fast timescales, they distort into transition-state structures by small motions that enable their catalytic actions. On slower timescales, enzymes often undergo larger collective motions of their domains to bind or release their substrates or products. These larger motions are inferred from experiments showing that catalytic rates depend on the solvent viscosity, indicating that the relevant motions involve the protein's sloshing through the solvent. An unprecedented level of detail in watching proteins in action can be obtained using time-resolved X-ray crystallography [15]. For example, upon the unbinding of CO from myoglobin, it is found that much larger conformational changes are observed in the protein during the unbinding process than would be inferred from just looking at the two endstates alone.

Protein Motions Can Be Decomposed into a Spectrum of Normal Modes

To make sense of motions, you decompose the structural dynamics into a collection of modes (see Appendix 12C) A mode of motion describes one particular collective way that a part of the protein moves relative to other parts. For example, a tree blowing in the wind has particular ways its trunk and branches tend to move. The branches move more than the trunk. Different trees have intrinsically different motions. Molecules, too, have collective modes of motions. You can dissect the dynamics of molecules into the contribution of independent modes. A protein's full motion is a collection of all its modes. For a protein having N residues, GNM allows you to decompose the structural dynamics into $N - 1$ independent modes. Each mode is described by an N-dimensional vector whose elements represent the relative extents of motion of each of the residues along a given mode axis. Different modes have different *frequencies*. The lowest-frequency modes are typically the largest-scale motions of the protein—for example, of one domain relative to another. The low-frequency modes, also called the *soft modes* or *global modes*, can be described as *concerted* or *coordinated* because they involve collections of atoms that all move together. The soft modes are uniquely and robustly defined by the protein's overall architecture and they are often critical to the biological function and the mechanism of the protein. The higher-frequency modes, called *local* modes, are localized in space and they are more dependent on local details. Box 12.3 shows the shapes of global and local modes are different: the former is highly distributive, it couples all residues; the latter shows sharp peaks corresponding to isolated sites.

Box 12.3 Hinge Bending Happens at Points of Minimal Global Motion. Substrate Recognition Happens at Points of Maximal Motion

Mode shapes can identify where motions are big, and where they are small. Figure 12.10A shows the GNM global-mode profile for the dimeric HIV-1 protease, that is, the normalized distribution of square displacements of residues in the global mode. The motion is distributed over several residues. The active site (residues 25–27) does not move much in the global mode. This is typical of enzymatic active sites, because the catalytic residues must have quite precise relative atomic positions and orientations in order to achieve chemical reactivity. In contrast, the recognition loop has high mobility, which is typical of substrate recognition loops. In this way, mode profiles can inform you about mechanisms. The curve in Figure 12.10C is a local-mode profile. This shows the small, fast, localized motions. They are often indicative of conserved residues in the core.

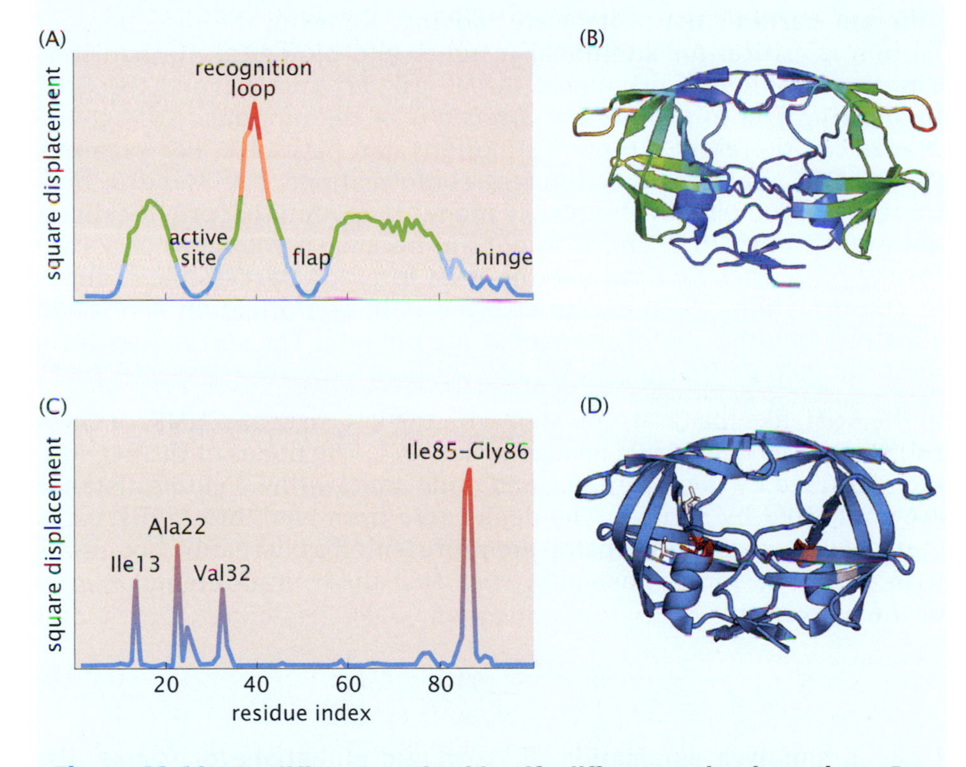

Figure 12.10 The different modes identify different molecular actions. For HIV-1 protease. For the slowest mode, (A) shows displacements versus sequence, and (B) shows them on the structure. For the fastest six modes, (C) shows displacements versus sequence, and (D) shows them on the structure. This shows that (1) the points of minimal mobility in the global modes are at hinge-bending centers; (2) the points of maximal mobility in the global modes are at substrate recognition sites; and (3) the maximal motions in the fast modes are those of buried residues that contribute to stability and to the chemical reaction. (A and C, adapted from I Bahar, AR Atilgan, MC Demirel, and B Erman. *Phys Rev Lett*, 80:2733–2736, 1998. With permission from APS.)

The GNM is a simplification. It treats proteins and amino acids as beads connected by springs, leaving out atomic details and energetics and solvation. So, why should it work at all? It works because it takes account of the full connectivity of the structure, bonded and nonbonded, through a representation that lends itself to a unique analytical solution defined by the entire network connectivity. Any network of beads and springs (or any solid-like material) will have correlated motions on a larger spatial scale, and a slower timescale, than the individual localized vibrations themselves have. For proteins, those larger-scale motions are often the most important mechanistic ones. These intrinsic motions are defined by the overall architecture. They do not depend much on the atomistic details of packing or hydrogen bonding, or even the identities of the amino acids. Proteins having similar shapes often have similar global motions and mechanisms, regardless of their sequence differences.

Directions of Motion Can Be Found Using the Anisotropic Network Model

On the one hand, the GNM tells you which parts of the structure are correlated, and which residues are moving in the different modes of motion. On the other hand, it doesn't tell you what *directions* the different parts of the protein are moving. Knowing the directions of motion is critical for additional insights into biological mechanisms. The anisotropic network model (ANM) [16, 17] goes beyond the GNM in allowing you to explore the *directions* of the motions. ANM global modes can represent a protein's motions and pathways. For example, if a protein populates two different conformations, say A and B, then there are prominent low-frequency modes of motion that provide direct pathways from A to B. This is important because if you know only conformation A, then you can use the ANM to predict accessible changes in conformations. Suppose an enzyme is in conformation A prior to substrate binding, and B after substrate binding. The soft modes from ANM can help find B from A [18].

In the ANM, like the GNM, you start with the known (X-ray, NMR, or cryo-EM) structure to construct an elastic network. The nodes of the network are identified by the α-carbons, and node pairs within a cutoff distance are connected by springs. The departures from equilibrium distances (between connected node pairs) are represented as harmonic potentials with a uniform force constant γ, such that the overall internal energy of the network is

$$U_{\text{ANM}} = \frac{1}{2}\sum_{ij}\gamma(r_{ij} - r_{ij}^{0})^{2}. \tag{12.16}$$

U_{ANM} is simply a summation of harmonic potentials (of vector distances) for all connected residues; it depends only on the changes in interresidue distances. In contrast, U_{GNM} is a function of scalar distances (that is, vector components), so it depends on changes in both distance and orientation (see Equations 12.4 and 12.5).

The second step is to perform a normal mode analysis (NMA) of this network. As described in Appendix 12D, the main ingredient in NMA is the Hessian matrix $\mathbf{H}$; it is the counterpart to the Kirchhoff matrix $\boldsymbol{\Gamma}$ in the GNM. For a system of N sites, evaluation of $\mathbf{H}$ in classical NMA means the calculation using the full atomic force field (after energy minimization) and taking second derivatives with respect to all atomic coordinates. This task is time-consuming and is performed numerically. The advantage of the ANM (versus

NMA) is that you don't need to energy-minimize the structure (you just take the PDB structure as is). And you don't need to numerically evaluate the second derivative of the potential, because the expression for U_{ANM} is simple enough to yield an analytical solution for $\mathbf{H}_{\text{ANM}}$. The ANM has become widely used because it is simple and fast to compute, and because it is now so broadly validated against experiments. The second derivative of U_{ANM} (with respect to x_i and y_j, for example) is

$$\frac{\partial^2 U_{\text{ANM}}}{\partial x_i \, \partial y_j} = -\frac{\gamma(x_j - x_i)(y_j - y_i)}{r_{ij}^2}, \tag{12.17}$$

where we have used the notation $x_{ij} = x_j - x_i$ for the x-component of the instantaneous distance vector $\mathbf{r}_{ij}$ between i and j (and similarly for y_{ij} and z_{ij}). The Hessian matrix is

$$\mathbf{H}_{ij} = \frac{\gamma}{r_{ij}^2} \begin{bmatrix} x_{ij}^2 & x_{ij}y_{ij} & x_{ij}z_{ij} \\ x_{ij}y_{ij} & y_{ij}^2 & y_{ij}z_{ij} \\ x_{ij}z_{ij} & y_{ij}z_{ij} & z_{ij}^2 \end{bmatrix}. \tag{12.18}$$

$\mathbf{H}_{\text{ANM}}$ can be written as a matrix composed of $N \times N$ submatrices $\mathbf{H}_{ij}$, each of which is size 3×3. The diagonal submatrices of $\mathbf{H}_{\text{ANM}}$ satisfy the identity

$$\mathbf{H}_{ii} = -\sum_{j, j \neq i} \mathbf{H}_{ij}. \tag{12.19}$$

Now you can calculate the spectrum of collective modes of motions predicted by the ANM as described in Appendix 12D. You obtain a set of $3N - 6$ normal modes, each expressed by a $3N$-dimensional vector, $\mathbf{u}_k$, the elements of which describe the displacement vector of each of the N beads in mode k. The global modes usually have the highest degree of collectivity (see Appendix 12F), that is, they are the most cooperative modes. These modes often provide insights into the functional movements the structure is predisposed to undertake if perturbed. Box 12.4 shows how the ANM captures the variations among the NMR models determined for a protein. Box 12.5 shows how ANM modes describe the mechanism of action of HIV-1 reverse transcriptase, and Box 12.6 shows that the gating of ions is achieved by a global mode. You can evaluate and visualize such motions and compare them to experimental data using the ProDy interface [19].

Box 12.4 A Protein's Computed Global Modes of Motion Correlate with Its Equilibrium Distribution of NMR Structures

Here we describe a relationship between, on the one hand, equilibrium motions that you can calculate, and, on the other, variations that you can observe across different equilibrium structures. The modes of motion that you can calculate from ANM can often predict the principal structural variations observed in equilibrium ensembles [20], such as those measured by NMR. Figure 12.11 shows the case of calmodulin. The green arrows in (A) show the directions of residue movements predicted by the softest (global) mode computed by the ANM from the known native structure. The purple arrows show the results of computing the largest principal component PC1, from a principal component analysis (PCA), across an ensemble of NMR structures of a protein. This illustrates how the modes of motion predicted by elastic network models are descriptive of ensembles of equilibrium structures.

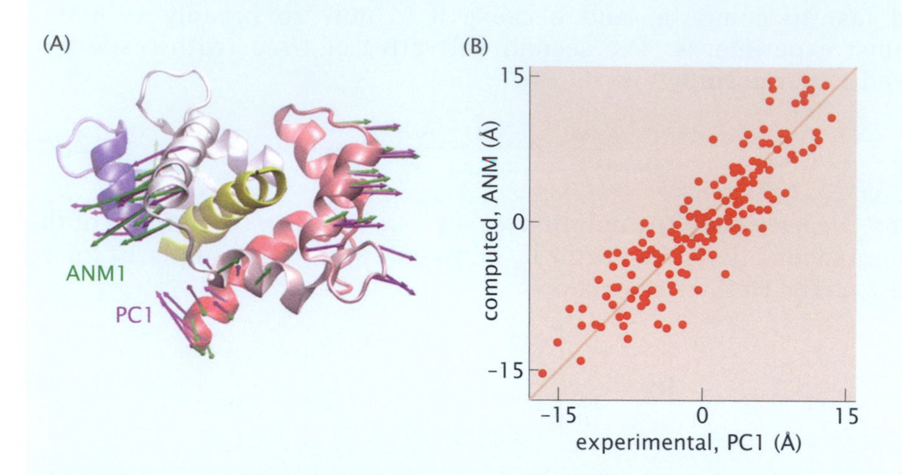

Figure 12.11 Global ANM modes of calmodulin correspond to the main structural variations observed in NMR experiments. (A) NMR structure of calmodulin (CaM; *light blue*) complexed with myosin light chain kinase (MLCK; *brown*). Here, 160 NMR model structures (*red dots*) were subjected to PCA to obtain the dominant structural variations along principal component 1 (PC1), shown by the *purple* arrows. The ANM analysis of one representative model gives the softest ANM mode (ANM1), shown by *green* arrows. (B) The computed ANM dynamics correlates with the NMR structural variations. (Adapted from A Bakan and I Bahar. *Proc Natl Acad Sci USA*, 106:14349–14354, 2009.)

Box 12.5 Reverse Transcriptase Walks Along RNA Following ANM Dominant Modes

With ENMs, you start from a static native structure—which is often readily available from the PDB—and compute motions of the protein that may be relevant to its biological mechanism of action. Typically, mechanistic motions arise from the softest modes. For example, HIV-1 reverse transcriptase (RT) "walks" along either an RNA or a DNA strand, synthesizing and laying down a second strand that can form base pairs to the first strand. Its mechanism of action can be inferred from ANM-predicted motions. Figure 12.12 compares the variations of about 200 RT structures resolved by X-ray crystallography in different forms (unbound, inhibitor-bound, or DNA/RNA-bound), with the conformations explored in the softest dynamical modes calculated by the ANM.

RT consists of two subunits: p66 and p51. The p66 subunit has two domains: a DNA polymerase and an RNase H domain. The DNA polymerase domain, which copies the RNA onto DNA, has four subdomains called *fingers, thumb, palm,* and *connection* (see Figure 12.12A). One of the softest modes is a motion of the thumb subdomain back and forth relative to the fingers, similar to a hand closing around a rope. The rope is the nucleic acid strand, and the "clamp" motion of the thumb and fingers is enabled by a hinge region at their interface with the palm subdomain (see step 2 in

Figure 12.12C). A second hinge, which is perpendicular to hinge 1, pulls the nucleic acid strands ahead one base in a processive motion when it opens (with the polymerase hinge open)—see Step 2 in Figure 12.12C. This motion translocates the polymerase along the RNA or DNA.

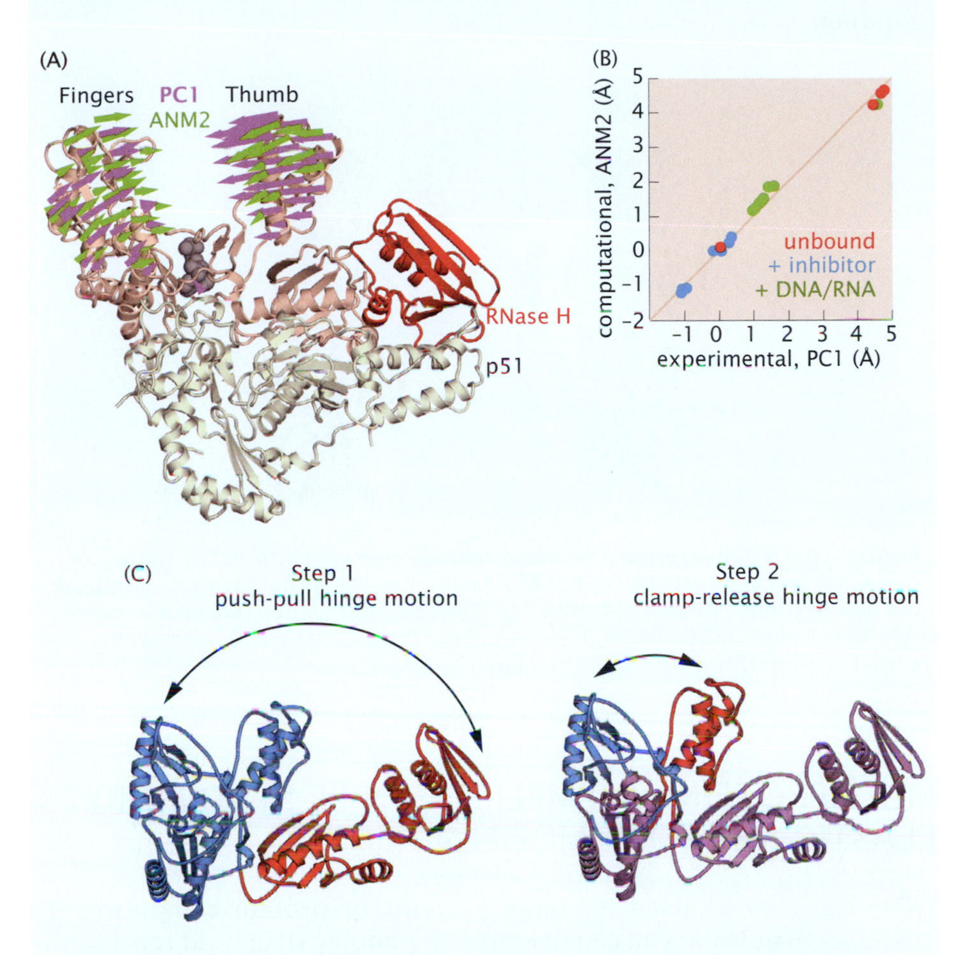

Figure 12.12 Functional motions of HIV-1 reverse transcriptase (RT). (A) (*Green* arrows) The ANM-predicted global motions of ANM mode 2 of the native structure. (*Purple* arrows) Principal component 1 (PC1) from the structural variations. (Spheres) A bound inhibitor near the hinge site. (B) The distribution of structures is predicted well by ANM soft mode 2. Different colors refer to different bound states. (C) The two softest modes predicted by the ANM, for the p66 domain clamp-release hinge motion that enables the tight binding required for the polymerase reaction. (A and B, adapted from L Meireles, M Gur, A Bakan, and I Bahar. *Protein Sci*, 20:1645–1658, 2011.)

Box 12.6 Soft Modes of Motion Couple Ligand Binding to Gating in a Membrane Protein

Nicotinic acetylcholine receptors (nAChR) are ligand-gated ion channels. They are implicated in neurological disorders, including Alzheimer's disease, attention deficit hyperactivity disorder (ADHD), and tobacco addiction [21]. Their ligand is the neurotransmitter acetylcholine (ACh). The binding of two ACh molecules to the extracellular domain triggers a cooperative structural change in the transmembrane domain, opening a central pore for conducting cations. ANM analysis shows that pore opening is favored by the softest mode of motion, a *quaternary twist*, shown in Figure 12.13 [22]. The allosteric coupling between the extracellular and transmembrane domains of AChR is regulates ion permeation upon ACh binding.

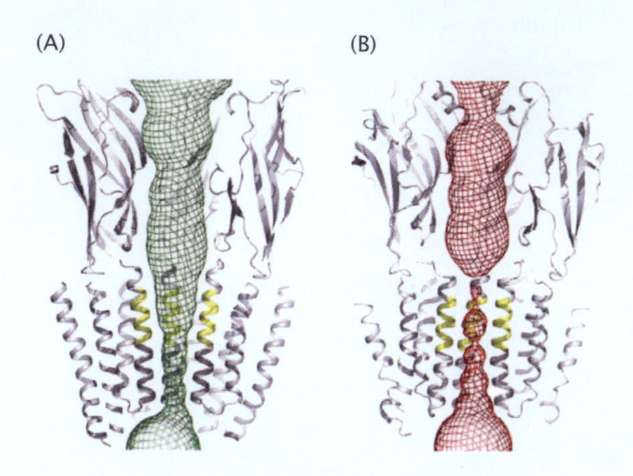

Figure 12.13 A quaternary twisting motion corresponds to the gating of the AChR ion channel. (A) Open conformation that allows ion permeation through the gate near the TM2 helices (*yellow*). (B) Closed conformation. (Adapted from A Taly, PJ Corringer, D Guedin, et al. *Nat Rev Drug Discov*, 8:733–750, 2009. With permission from Macmillan Publishers Ltd.)

MULTIPROTEIN ASSEMBLIES CAN BE STUDIED BY ELASTIC NETWORK MODELS

ENMs can also be used for large proteins or protein complexes. To model large objects, you can use a more granular structural representation of the molecules. In our modeling so far, we have taken each bead to represent one amino acid. Now, each bead will represent multiple amino acids. Does this altered granularity change the motions predicted by ENMs? Figure 12.14 shows that for influenza hemagglutinin A trimer [23], using one bead to represent 10 sequential residues yields nearly the same global mode as the one bead/one amino acid model. This is because global modes are relatively insensitive to fine-grained detail.

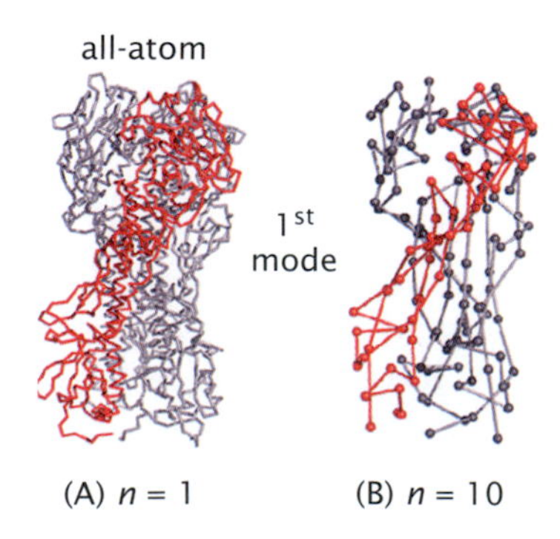

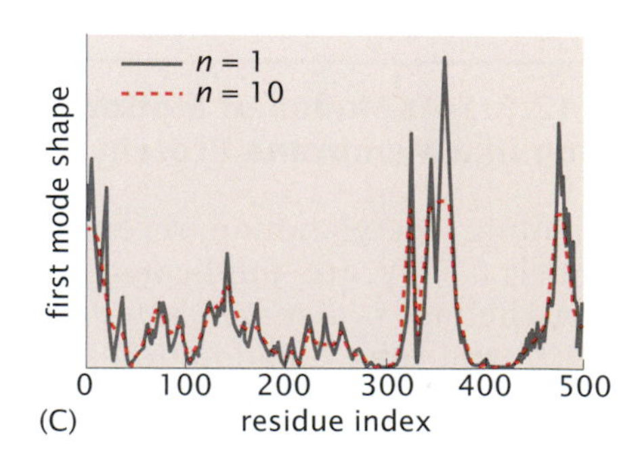

Figure 12.14 The global motions of hemagglutinin A are the same for different levels of coarse-graining. (*Red*) A trimeric subunit. (A) One bead represents one amino acid. (B) One bead represents 10 amino acids. (C) The fluctuations in the first ANM mode is essentially the same. (Adapted from P Doruker, RL Jernigan, and I Bahar. *J Comput Chem*, 23:119–127, 2002.)

Box 12.7 shows another large complex [24]: the capsid of the HK97 virus, which is composed of 420 icosahedrally arranged subunits.

Box 12.7 A Spherical Virus Breathes Along Radial Directions

Figure 12.15 shows the lowest ANM normal mode of the motions of a virus capsid. This large-scale motion, which is an expansion–contraction mode, arises from the spherical symmetry of the virus shell, rather than from more microscopic structural details.

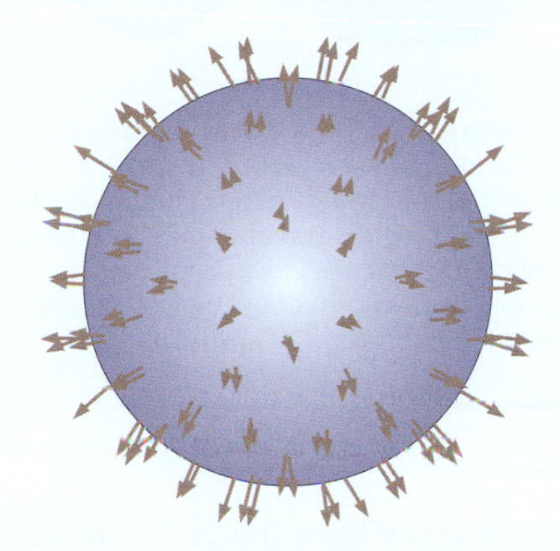

Figure 12.15 The first normal mode of the hepatitis B virus. The same first mode would be seen in other hollow spherical objects. (Adapted from F Tama and CL Brooks. *J Mol Biol*, 345:299–314, 2005. With permission from Elsevier.)

The *nuclear pore* is a complex of 456 protein molecules. It regulates the transport of molecules between the cytoplasm and the nucleus. ANM calculations show how its main motions are governed by its overall large-scale donut shape, rather than by small-scale details (Box 12.8).

Box 12.8 The Nuclear Pore Complex Is Toroidal. Its Motions Reflect This Symmetry

Figure 12.16 shows the calculated lowest dynamical normal modes of the nuclear pore complex computed with ANM, color-coded from blue to red with increasing mobility. It predicts that these motions arise from the main geometric features—the pore's toroidal overall shape and local cylindrical geometry, rather than from more microscopic structural properties. For the motions shown in Figure 12.16A and B, the cylinder can undergo these stretching–compression deformations centered about any two points on the opposite sides of the toroid. For example, for the bend shown in Figure 12.16A, the rotation axis could be oriented in any direction in the plane of the top digram. This figure also illustrates how ENMs are applicable to large structures.

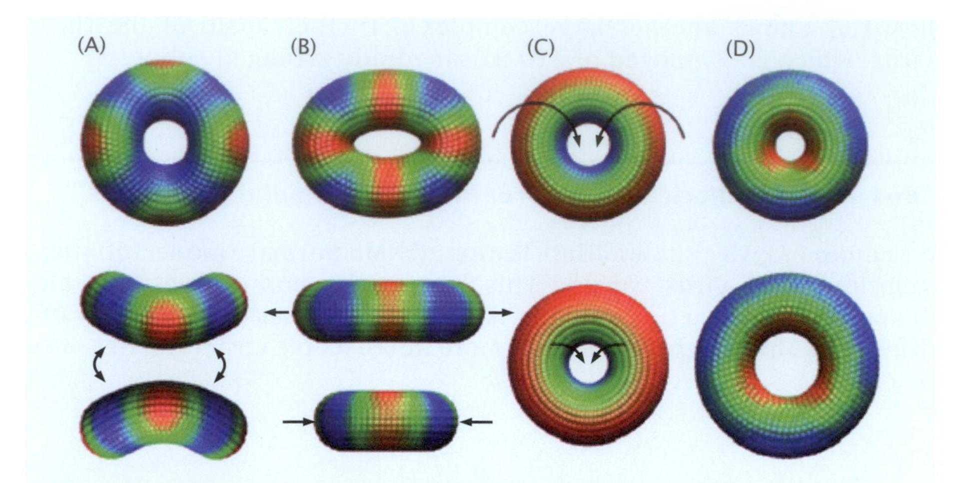

Figure 12.16 Slowest motions of the nuclear pore complex. (A) The first mode is a bending of the toroid. (B) The second mode is a symmetric stretching of the toroid. (C) The third mode is a rolling mode. The points on the top surface move along the toroid surface toward the central channel, and the points on the opposite side move away from the central channel toward the outside perimeter. (D) Mode 6 is a symmetric compression/expansion about the central vertical axis. (Adapted from TR Lezon, A Sali, and I Bahar. *PLoS Comput Biol*, 5:e1000496, 2009.)

Some large protein complexes have irregular shapes. An example is the bacterial chaperonin complex GroEL/GroES, which is a multi-subunit machine that helps misfolded or unfolded client proteins to fold up. Box 12.9 shows how a chaperone's dominant modes of motion are dictated by its overall shape.

Box 12.9 The Motions of the GroEL/GroES Protein Arise from Its Irregularly Shaped Interior Space

GroEL is a cylindrically shaped 14-protein structure that has an empty interior, large enough for client proteins to enter, so that they can then fold into their native structures. The co-chaperonin GroES closes over the cavity after binding the client protein. The GroEL/ES structure is hollow, so it has complex and diverse motions that may facilitate the uptake, folding, unfolding and refolding, and expulsion of client proteins (Figure 12.17).

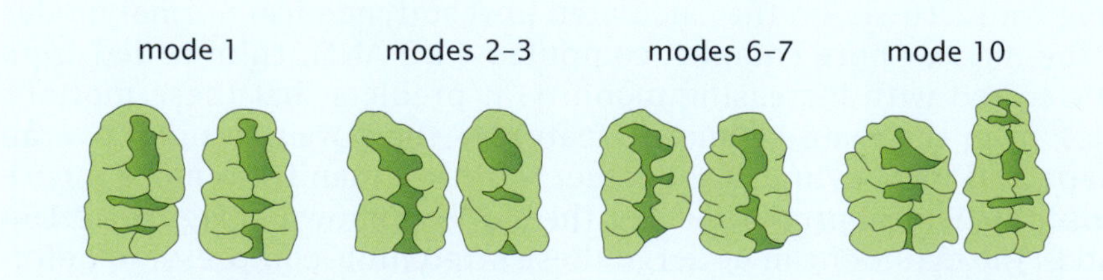

Figure 12.17 Some of the slowest normal modes of motion of GroEL/GroES, calculated using ANM. For each mode, the two extremes of structure are shown. (Adapted from O Keskin, I Bahar, D Flatow, et al. *Biochemistry*, 41:491–501, 2002.)

SUMMARY

Native proteins at equilibrium undergo thermal motions. Those motions are collective, not random. Encoded in a protein's native structure are particular motions that can facilitate its function. They can be modeled by supposing that the protein molecule is a miniature elastic network made up of amino acid beads with springs connecting the residue pairs within a first interaction distance. The network topology is the main determinant of the spectrum of collective motions. The soft modes, which have the lowest frequencies, are uniquely and robustly defined by the overall three-dimensional structure and often underlie mechanisms of biological function. Elastic network models are scalable to larger proteins, assemblies, and complexes. As a way to compute motions in proteins, the Gaussian network model has some advantages over molecular dynamics. It has conceptual and mathematical simplicity; it gives insights that are inherent to analytical modeling; and it is applicable to large systems, without large computational time or memory requirements.

APPENDIX 12A: HERE'S HOW TO EXPRESS THE ELASTIC FREE ENERGY IN TERMS OF THE ADJACENCY MATRIX

Consider a network of three beads (i, j, k) connected by two springs. The potential is a sum of two terms:

$$U = \frac{\gamma}{2}\left[(\Delta r_{ij})^2 + (\Delta r_{jk})^2\right] = \frac{\gamma}{2}\left[(\Delta \mathbf{r}_j - \Delta \mathbf{r}_i)^2 + (\Delta \mathbf{r}_k - \Delta \mathbf{r}_j)^2\right]$$

$$= \frac{\gamma}{2}\left[(\Delta \mathbf{r}_j - \Delta \mathbf{r}_i)\cdot(\Delta \mathbf{r}_j - \Delta \mathbf{r}_i) + (\Delta \mathbf{r}_k - \Delta \mathbf{r}_j)\cdot(\Delta \mathbf{r}_k - \Delta \mathbf{r}_j)\right]. \quad (12.A.1)$$

This *Rouse model* is commonly used to treat polymer dynamics (Figure 12.A.1). U can be expressed as the sum of three terms, $U = U_x + U_y + U_z$, with U_x given by

$$U_x = \frac{\gamma}{2}\left[(\Delta x_j - \Delta x_i)^2 + (\Delta x_k - \Delta x_j)^2\right]. \quad (12.A.2)$$

Alternatively, we can adopt the compact notation

$$U_x = \frac{\gamma}{2}\Delta \mathbf{x}^T \mathbf{\Gamma} \Delta \mathbf{x}, \quad (12.A.3)$$

where $\Delta \mathbf{x}^T = [\Delta x_1 \ \Delta x_2 \ \Delta x_3]$ and

$$\mathbf{\Gamma} = \begin{bmatrix} 1 & -1 & 0 \\ -1 & 2 & -1 \\ 0 & -1 & 1 \end{bmatrix}. \quad (12.A.4)$$

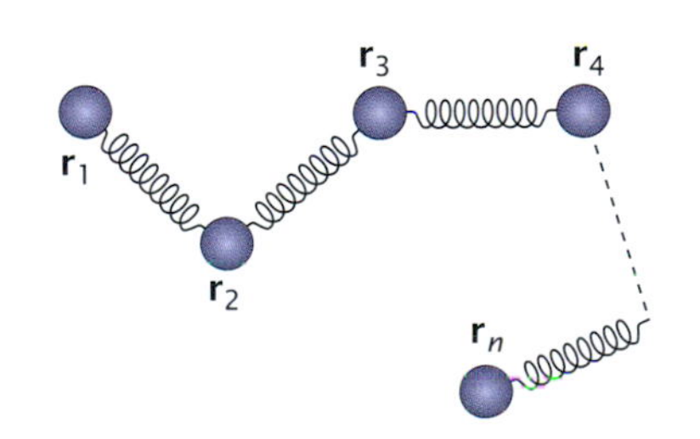

Figure 12.A.1 Bead-and-spring (Rouse) model of a polymer. Springs represent interactions between interacting units and are subject to Gaussian fluctuations. The vectors $\mathbf{r}_1, \mathbf{r}_2, \ldots, \mathbf{r}_N$ give the instantaneous positions of the N beads with respect to a Cartesian coordinate system.

You can verify that Equation 12.A.3 reduces to Equation 12.A.2 by substituting Equation 12.A.4. Similar forms hold for U_y and U_z. A general way to express U is

$$U = \Delta \mathbf{r}^T (\mathbf{\Gamma} \otimes \mathbf{E}) \Delta \mathbf{r}, \quad (12.A.5)$$

where $\Delta \mathbf{r}^T = [\Delta x_1 \ \ldots \ \Delta x_2 \ \ldots \ \Delta z_{3N}]$ is the $3N$-dimensional row vector of fluctuations, $\mathbf{E}$ is the 3×3 identity matrix, and $\otimes$ means that you take the *direct product*. The direct product is simply a matrix in which each term is the product of the two corresponding terms in the two matrices, $\mathbf{\Gamma}$ and $\mathbf{E}$, in this case. So here, the direct product is

$$\mathbf{\Gamma} \otimes \mathbf{E} = \begin{bmatrix} 1 & & & -1 & & & 0 & & \\ & 1 & & & -1 & & & 0 & \\ & & 1 & & & -1 & & & 0 \\ -1 & & & 2 & & & -1 & & \\ & -1 & & & 2 & & & -1 & \\ & & -1 & & & 2 & & & -1 \\ 0 & & & -1 & & & 1 & & \\ & 0 & & & -1 & & & 1 & \\ & & 0 & & & -1 & & & 1 \end{bmatrix}. \tag{12.A.6}$$

Note that for proteins, which have additional interactions because of their compact folded states, you could use instead the adjacency matrix $\mathbf{\Gamma}$ defined in Figure 12.4. Next, given this adjacency matrix $\mathbf{\Gamma}$, we compute the energy cost $U_{\text{GNM}}(t)$ for any particular set of fluctuations $\Delta\mathbf{r}(t)$, as follows:

$$U_{\text{GNM}}(t) = \frac{\gamma}{2}[\Delta\mathbf{r}(t)^T(\mathbf{\Gamma} \otimes \mathbf{E})\Delta\mathbf{r}(t)]. \tag{12.A.7}$$

Equation 12.5 in the main text is a compact notation for U_{GNM} provided that (i) $\Delta\mathbf{r}(t)$ is viewed as a N-dimensional vector representing the changes ($\Delta\mathbf{r}_i$) in the position vectors of the N beads, and (ii) the products between those elements are simply the scalar products of the vectors.

Now that you have an energy function, you can compute Boltzmann probabilities $p(\Delta\mathbf{r}_{ij})$ for the fluctuations in internode distances. Once you have probabilities, you can compute averages and correlation functions. To do this, first express the partition function as

$$Q = \int \exp(-U_{\text{GNM}}/RT)\, d(\Delta\mathbf{r}), \tag{12.A.8}$$

where $\int d(\Delta\mathbf{r})$ is a compact notation that means that you perform integrations over the change in each coordinate, that is, $\int d(\Delta\mathbf{r}) = \int \cdots \int d(\Delta\mathbf{r}_1)d(\Delta\mathbf{r}_2)\cdots d(\Delta\mathbf{r}_N)$. Inserting Equation 12.A.7 into Equation 12.A.8 leads to the free energy [25]:

$$\Delta G = -RT \ln Q = -\frac{3RT}{2} \ln\left[\left(\frac{2RT\pi}{\gamma}\right)^{N-1} \det(\mathbf{\Gamma}^{-1})\right], \tag{12.A.9}$$

where $\det(\mathbf{\Gamma}^{-1})$ means the determinant of the inverse of $\mathbf{\Gamma}$.

APPENDIX 12B: HOW IS $\mathbf{\Gamma}$ RELATED TO LOCAL PACKING DENSITIES?

Here, we show the relationship between the sizes of fluctuations and the number of near neighbors of a residue. Let's look at the proportionality $\langle\Delta\mathbf{r}_i \cdot \Delta\mathbf{r}_i\rangle \sim [\mathbf{\Gamma}^{-1}]_{ii}$ between the MSFs of residues and the diagonal elements of the inverse connectivity matrix in Equation 12.9. Recall that the ith diagonal element of $\mathbf{\Gamma}$ is equal to the *coordination number* z_i of residue i, that is, its number of nearest neighbors. This suggests that, to a first-order approximation, the diagonal elements of Γ^{-1} are inversely proportional to z_i; that is, the larger fluctuation of residue i is simply a manifestation of its lower coordination number. To see this, express Γ as a sum of two matrices Γ_1 and Γ_2, composed separately of the diagonal and off-diagonal elements, respectively, of Γ. Thus, the inverse of Γ can be written as

$$\begin{aligned}\Gamma^{-1} &= [\Gamma_1 + \Gamma_2]^{-1} = [\Gamma_1(\mathbf{E} + \Gamma_1^{-1}\Gamma_2)]^{-1} = (\mathbf{E} + \Gamma_1^{-1}\Gamma_2)^{-1}\Gamma_1^{-1} \\ &= (\mathbf{E} - \Gamma_1^{-1}\Gamma_2 + \ldots)\Gamma_1^{-1}, \end{aligned} \tag{12.B.1}$$

where $\mathbf{E}$ is the identity matrix of order N. The second line of Equation 12.B.1 is valid if the invariants of the product $(\Gamma_1^{-1}\Gamma_2)$ are small compared to those of $\mathbf{E}$, which is a valid approximation for a protein's Kirchhoff matrix. Thus, the diagonal matrix Γ_1^{-1} can be seen as the first term of a series expansion for Γ^{-1}. So, $\langle(\Delta r_i)^2\rangle$ scales with $[\Gamma_1^{-1}]_{ii} = 1/z_i$ to a first approximation. The matrix element at position i along the diagonal of Γ_1 just tells you how many neighboring beads are connected to bead i.

APPENDIX 12C: HOW DO YOU DETERMINE THE GNM MODES?

Here is how you use the GNM to compute the modes of motion accessible to a given protein. In the GNM, the normal modes are extracted by performing an *eigenvalue decomposition* of the Kirchhoff matrix [7]. Express the matrix Γ in the form

$$\Gamma = \mathbf{U}\Lambda\mathbf{U}^T, \tag{12.C.1}$$

where $\mathbf{U}$ is the matrix of the eigenvectors $\mathbf{u}_i$ of Γ. And Λ is the diagonal matrix of its eigenvalues λ_i, that is,

$$\mathbf{U} = [\mathbf{u}_0,\ \mathbf{u}_1,\ \mathbf{u}_2,\ \ldots,\ \mathbf{u}_{N-1}] = \begin{bmatrix} u_{0,1} & \cdots & u_{N-1,1} \\ \vdots & & \vdots \\ u_{0,N} & \cdots & u_{N-1,N} \end{bmatrix}, \tag{12.C.2}$$

where

$$\Lambda = \begin{bmatrix} \lambda_0 & 0 & 0 & \ldots & 0 \\ 0 & \lambda_1 & 0 & \ldots & 0 \\ 0 & 0 & \lambda_2 & \ldots & 0 \\ \vdots & \vdots & \vdots & & \vdots \\ 0 & 0 & 0 & \ldots & \lambda_{N-1} \end{bmatrix}. \tag{12.C.3}$$

The matrix $\mathbf{U}$ is called a *unitary* matrix. For unitary matrices, the transpose equals the inverse, that is, $\mathbf{U}^T = \mathbf{U}^{-1}$. The eigenvalues are proportional to the *frequencies* of the individual modes, and the eigenvectors of $\mathbf{U}$ define the *shapes* of the modes. The eigenvector $\mathbf{u}_k$ has N elements $u_{k,1}, u_{k,2}, \ldots, u_{k,N}$; they describe the (normalized) displacements of the N residues along the kth mode axis. The first eigenvalue λ_0 equals zero; it corresponds to the rigid-body translation of the protein, and the corresponding eigenvector has all elements equal to a constant, indicative of the uniform displacement of all residues in this mode. All other eigenvalues are organized in ascending order: $\lambda_0 \leq \lambda_1 \leq \lambda_2 \cdots \leq \lambda_{N-1}$ (equality holds exclusively for degenerate modes). The slowest (global) mode therefore has eigenvalue λ_1 and eigenvector $\mathbf{u}_1$, and is usually called *mode 1*.

The MSF of residues and their cross-correlations can be expressed as a sum of nonzero modes $1 \leq k \leq N-1$, by substituting these expressions into Equation 12.8:

$$\langle \Delta \mathbf{r}_i \cdot \Delta \mathbf{r}_i \rangle = \frac{3RT}{\gamma}[\mathbf{U}\mathbf{\Lambda}^{-1}\mathbf{U}^T]_{ii} = \frac{3RT}{\gamma}\sum_k [\lambda_k^{-1}\mathbf{u}_k\mathbf{u}_k^T]_{ii}. \tag{12.C.4}$$

So, the contributions of the kth mode to the MSF $\langle(\Delta r_i)^2\rangle$ of residue i and to the cross-correlation $\langle \Delta \mathbf{r}_i \cdot \Delta \mathbf{r}_j \rangle$ are

$$[(\Delta r_i)^2]_k = \frac{3RT}{\gamma}\lambda_k^{-1}[\mathbf{u}_k\mathbf{u}_k^T]_{ii} = \frac{3RT}{\gamma}\lambda_k^{-1}u_{ki}^2 \tag{12.C.5}$$

and

$$[\Delta \mathbf{r}_i \cdot \Delta \mathbf{r}_j]_k = \frac{3RT}{\gamma}\lambda_k^{-1}[\mathbf{u}_k\mathbf{u}_k^T]_{ij} = \frac{3RT}{\gamma}\lambda_k^{-1}u_{ki}u_{kj}, \tag{12.C.6}$$

respectively. In Equations 12.C.5 and 12.C.6, $[\mathbf{u}_k\mathbf{u}_k^T]_{ij}$ designates the ijth element of the matrix $\mathbf{u}_k\mathbf{u}_k^T$. The diagonal elements of this matrix yield the kth *mode shape.*

The lower-frequency modes (smaller λ_k) make larger contributions to observed correlations, or observed MSFs. So, these are *dominant* modes. Note that the mode profiles in Figure 12.10 are obtained by plotting the diagonal elements u_{ki}^2 against the residue index i, for $k = 1$ (upper panel), and for k summed up over the range $N-1 > k \geq N-6$ (that is, the fastest five modes).

APPENDIX 12D: NORMAL MODE ANALYSIS

Normal Mode Analysis Describes the Motions Near Equilibrium

Here, we describe the method called *normal mode analysis* (NMA). NMA was first applied to proteins in the 1980s by Go [26, 27], Karplus and Brooks [28], and Levitt et al. [29]. NMA allows you to extract the collective modes of motion accessible to a given protein near its equilibrium state.

NMA transforms your protein's fluctuations in Cartesian coordinates into collective motions called *normal modes* along a set of *normal coordinates.* NMA rank-orders the modes: the first modes describe the largest motions (the global modes), which collectively involve large portions of the molecule, while the last modes describe the most localized motions. Each normal coordinate is a linear combination of the Cartesian coordinates of all atoms of the molecule, and each normal mode represents a collective fluctuation of the molecule along the normal coordinate. Hence normal modes are called *collective modes.* NMA is quite general; it can be applied to any system of N particles near the equilibrium state (Figure 12.D.1).

The basis for NMA is that systems near their equilibrium states can be treated as being stable in approximately square-law energy wells. Any potential function U, including those that arise in all-atom forcefields,

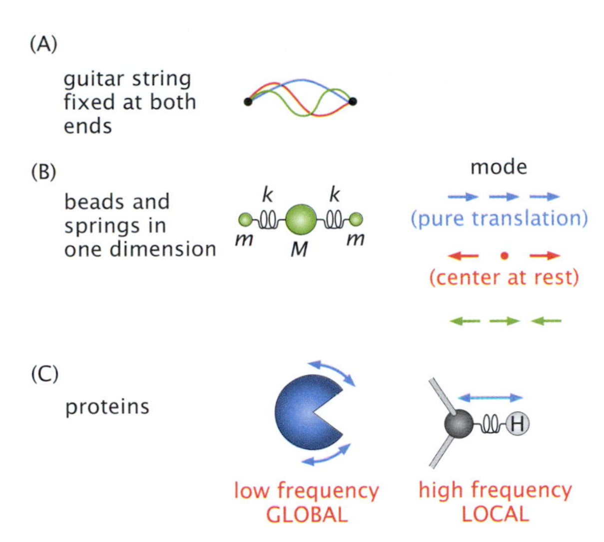

Figure 12.D.1 Examples of normal modes. (A) For strings, the three slowest modes are shown. (B) For a bead-and-spring model of three beads, two nonzero modes are shown. (C) For a protein, the left figure indicates that the slow modes are global, representing collective motions of large parts of the molecule. The right figure indicates that the fast modes represent more localized motions. (Adapted from PJ Steinbach. Introduction to macromolecular simulation. From online tutorial. http://cmm.cit.nih.gov/intro_simulation/node26.html. Courtesy of Peter J. Steinbach, National Institute of Health.)

can be expanded as a Taylor series near a given equilibrium state $\mathbf{r}^0$ as

$$U = U^0 + \sum_{q_i} \left.\frac{\partial U}{\partial q_i}\right|_{\mathbf{r}^0} (q_i - q_i^0) + \frac{1}{2} \sum_{q_i} \sum_{q_u} \left.\frac{\partial^2 U}{\partial q_i \partial q_j}\right|_{\mathbf{r}^0} (q_i - q_i^0)(q_j - q_j^0) + \ldots, \tag{12.D.1}$$

where $q_1, q_2, \ldots, q_{3N}$ is the set of variables corresponding to the $3N$ dimensional coordinates: $x_1, y_1, z_1, x_2, \ldots, z_N$ that define a given conformation, and the summations are performed over all coordinates. Let's take as our reference state $U^0 = 0$, the equilibrium state. By definition, at equilibrium, the first-derivative term in Equation 12.D.1 is zero. Hence, the first nonzero term in this Taylor expansion is the square-law contribution,

$$U = \frac{1}{2} \sum_{q_i} \sum_{q_u} \left.\frac{\partial^2 U}{\partial q_i \partial q_j}\right|_{\mathbf{r}^0} (q_i - q_i^0)(q_j - q_j^0). \tag{12.D.2}$$

So, near equilibrium, the dominant contributions to the energy are the quadratic terms. Thus, you can think of the protein at equilibrium as if the atoms or residues were connected by Hooke's-law springs. The second derivatives of the potential with respect to the spatial coordinates define the $3N \times 3N$ *Hessian matrix* $\mathbf{H}$. The Hessian matrix describes the "spring constants" for the coupled motions between i and j. You can write Equation 12.D.2 in compact (matrix) form as

$$U = \frac{1}{2} \Delta\mathbf{q}^T \mathbf{H} \Delta\mathbf{q} = \frac{1}{2} \sum_i \sum_j H_{ij}(q_i - q_i^0)(q_j - q_j^0), \tag{12.D.3}$$

where $\Delta\mathbf{q}$ is the $3N$-dimensional vector of atomic fluctuations away from equilibrium positions:

$$\Delta\mathbf{q} = \begin{bmatrix} q_1 - q_1^0 \\ q_2 - q_2^0 \\ \vdots \\ q_{3N} - q_{3N}^0 \end{bmatrix}. \tag{12.D.4}$$

$\Delta\mathbf{q}^T$ is its transpose (that is, the row vector), and the elements of $\mathbf{H}$ are defined as (compare with Equation 12.D.2)

$$H_{ij} = \left.\frac{\partial^2 U}{\partial q_i \partial q_j}\right|_{\mathbf{r}^0}. \tag{12.D.5}$$

So, $\mathbf{H}$ is a measure of stiffness (effective force constant) with respect to changes in structure. It is a measure of the resistance to deformation along various degrees of freedom.

There Is an Inverse Relationship between Covariance and Stiffness

The correlation between the movements of individual atoms is expressed by the covariance matrix $\mathbf{C}_{(3N\times 3N)} = \langle \Delta\mathbf{q}\,\Delta\mathbf{q}^T\rangle$. The term in angle brackets represents a statistical mechanical average, which may be evaluated as

$$\begin{aligned}\mathbf{C}_{(3N\times 3N)} &= \langle \Delta\mathbf{q}\,\Delta\mathbf{q}^T\rangle \\ &= \frac{\int_{-\infty}^{\infty} \Delta\mathbf{q}\,\Delta\mathbf{q}^T \exp(-\Delta\mathbf{q}^T\mathbf{H}\Delta\mathbf{q}/2RT)\,d\Delta\mathbf{q}}{\int_{-\infty}^{\infty} \exp(-\Delta\mathbf{q}^T\mathbf{H}\Delta\mathbf{q}/2RT)\,d\Delta\mathbf{q}}.\end{aligned} \tag{12.D.6}$$

Using the Gaussian integrals $\int_{-\infty}^{\infty} e^{-ax^2}\,dx = \sqrt{\pi/a}$ and $\int_{-\infty}^{\infty} x^2 e^{-ax^2}\,dx = \frac{1}{2}\sqrt{\pi}\,a^{-3/2}$ in the denominator and numerator, respectively, $\mathbf{C}_{(3Nx3N)}$ reduces to the simple relationship

$$\mathbf{C}_{(3N\times 3N)} = RT\mathbf{H}^{-1}. \tag{12.D.7}$$

Equation 12.D.7 means that the fluctuations of two sites tend to be larger when they are subject to weaker constraints. In the simplest case of a structure of two nodes connected by an elastic spring of force constant γ (where $U = \gamma(\Delta x)^2/2$), this equality reduces to $\langle(\Delta x)^2\rangle = RT/\gamma$.

You Can Find the Collective Modes Using NMA

NMA requires the evaluation of the Hessian $\mathbf{H}$. This is the only quantity you need to perform NMA. The entire mode spectrum is defined by the eigenvalue decomposition of $\mathbf{H}$. Its eigenvalue decomposition yields both a complete orthonormal set of basis vectors (the eigenvectors that describe each type of collective motion) and the corresponding eigenvalues that define their frequencies.

In classical NMA, it is not always simple to evaluate H_{ij}. You have a system of $3N$ degrees of freedom, subject to a complex force field, and, prior to the evaluation of the second derivative of U with respect to all internal coordinates (see Equation 12.D.5), you need to make sure that U is an energy minimum. In practice, you evaluate $\mathbf{H}_{ij}$ numerically. First, you do an energy minimization and check that the change in U is zero with respect to changes in any degree of freedom. Then you evaluate the change in the energy U with respect to successive incremental changes in the two coordinates $\mathbf{q}_i$ and $\mathbf{q}_j$. Overall, this can be computationally expensive.

The advantage of the ANM, relative to NMA, is that you can write **H** in a closed form. There is no need to do an energy minimization, or to evaluate derivatives. The mathematical simplicity (and ensuing efficiency) has led to the broad use of ENMs, after the realization that global modes are insensitive to detailed interactions at atomic level. Thus, the mathematical simplicity would not cause any loss in the accuracy of predicted global modes. This is because the global modes are defined by the overall architecture, and can thus be accurately obtained by models like ENMs that take rigorous account of the overall inter-residue contact topology.

Suppose you have **H**. Next, we proceed to its eigenvalue decomposition. You can use any eigenvalue decomposition software (freely available online). Eigenvalue decomposition of **H** means rewriting it as a product of three matrices:

$$\mathbf{H} = \mathbf{U}\Lambda\mathbf{U}^{-1}, \tag{12.D.8}$$

where **U** is a unitary matrix (not to be confused with the energy U) composed of the $3N$ orthonormal eigenvectors of **H**, as

$$\mathbf{U} = [\mathbf{u}_1, \quad \mathbf{u}_2, \quad \ldots, \quad \mathbf{u}_{3N}]_{3N\times 3N}. \tag{12.D.9}$$

$\mathbf{U}^{-1} = \mathbf{U}^T$ by definition, and $\boldsymbol{\Lambda}$ is the diagonal matrix of eigenvalues λ_1, $\lambda_2, \ldots, \lambda_{3N}$:

$$\boldsymbol{\Lambda} = \begin{bmatrix} \lambda_1 & 0 & 0 & \ldots & 0 \\ 0 & \lambda_2 & 0 & \ldots & 0 \\ 0 & 0 & \lambda_3 & \ldots & 0 \\ \vdots & \vdots & \vdots & & \vdots \\ 0 & 0 & 0 & \ldots & \lambda_{3N} \end{bmatrix}, \tag{12.D.10}$$

where the lowest (nonzero) mode is conveniently assigned index 1. Note that **H** has 6 zero eigenvalues (associated with rigid-body translations and rotations), and therefore the nonzero mode index varies in the range $1 \leq k \leq 3N-6$, with the upper limit corresponding to the highest-frequency modes. The kth eigenvector

$$\mathbf{u}_k = \begin{bmatrix} u_{k,1} \\ u_{k,2} \\ u_{k,3} \\ \vdots \\ u_{k,3N} \end{bmatrix} = \begin{bmatrix} \Delta x_1 \\ \Delta y_1 \\ \Delta z_1 \\ \vdots \\ \Delta z_N \end{bmatrix}_k \tag{12.D.11}$$

describes the x-, y-, and z-displacements of the N residues (or atoms) along mode k. By substituting Equation 12.D.11 into Equation 12.D.8, **H** may be written in terms of the eigenvalues and eigenvectors as

$$\mathbf{H} = \sum_{k=1}^{3N-6} \lambda_k \mathbf{u}_k \mathbf{u}_k^T. \tag{12.D.12}$$

The eigenvector λ_k serves as the force constant for displacement along the mode (this is why low-frequency modes are called “soft modes”). Using Equations 12.D.7 and 12.D.8, the covariance matrix **C** may be expressed in terms of the eigenvalues and eigenvectors of **H** as

$$\mathbf{C}_{3N\times 3N} = RT\mathbf{H}^{-1} = RT[\mathbf{U}\boldsymbol{\Lambda}\mathbf{U}^{-1}]^{-1} = RT\mathbf{U}\boldsymbol{\Lambda}^{-1}\mathbf{U}^{-1}. \tag{12.D.13}$$

The last equality follows from the relation $\mathbf{ABC}^{-1} = \mathbf{C}^{-1}\mathbf{B}^{-1}\mathbf{A}^{-1}$ for inverting the product of three matrices **A**, **B**, and **C**. Because **H** has zero eigenvalues, it is not invertible. Instead, we evaluate the pseudoinverse, leading to

$$\mathbf{C}_{3N\times 3N} = \sum_{k=1}^{3N-6} \lambda_k^{-1}\mathbf{u}_k\mathbf{u}_k^T. \qquad (12.D.14)$$

The inverse eigenvalue $1/\lambda_k$ serves as the weight of the square displacement along mode k. Given that $\mathbf{u}_k$ is a unit directional vector (that is, it has magnitude 1), the overall size of the motion in mode k is determined by $1/\lambda_k$, and the lowest-frequency modes make the largest contribution to the overall motion.

APPENDIX 12E: MEAN-SQUARE FLUCTUATIONS IN INTERNAL DISTANCES DEPEND ON THE NETWORK CONNECTIVITY

You can obtain the MSFs in the distance $\mathbf{r}_{ij}$ between any pair of residues i and j, if you know the MSFs $\langle\Delta\mathbf{r}_i^2\rangle$ and $\langle\Delta\mathbf{r}_j^2\rangle$ and their cross-correlations using

$$\begin{aligned}\langle\Delta\mathbf{r}_{ij}^2\rangle &= \langle(\Delta\mathbf{r}_j - \Delta\mathbf{r}_i)\cdot(\Delta\mathbf{r}_j - \Delta\mathbf{r}_i)\rangle \\ &= \langle\Delta\mathbf{r}_i^2\rangle + \langle\Delta\mathbf{r}_j^2\rangle - 2\langle(\Delta\mathbf{r}_i\cdot\Delta\mathbf{r}_j)\rangle \\ &= [\mathbf{\Gamma}^{-1}]_{ii} + [\mathbf{\Gamma}^{-1}]_{jj} - 2[\mathbf{\Gamma}^{-1}]_{ij}. \end{aligned} \qquad (12.E.1)$$

Therefore, knowledge of the connectivity matrix $\mathbf{\Gamma}$ is sufficient to compute the equilibrium distance fluctuations.

Note that there is a difference between *mobility* and *flexibility*. Mobility refers to rigid-body translational and rotational motions, for example of domains, which maintain fixed internal inter-residue contacts. Such domains can move in space, but they are not "flexible" internally. In contrast, flexibility refers to internal deformations, such as at hinge regions, of parts of the chain that have fixed centers of mass in space.

APPENDIX 12F: HOW CAN YOU COMPARE ONE MOTION WITH ANOTHER?

Suppose you want to know the extent to which one computed motion resembles another. For example, you want to know whether a given mode of motion, $\mathbf{u}_k$, accessible to a structure A correlates with the conformational change observed in experiments. Look at the *overlap*, also called the *correlation cosine* [30],

$$I_k = \frac{\Delta\mathbf{q}_{AB}\cdot\mathbf{u}_k}{|\Delta\mathbf{q}_{AB}|}, \qquad (12.F.1)$$

between the eigenvector $\mathbf{u}_k$ and the experimentally known direction of deformation $\Delta\mathbf{q}_{AB} = \mathbf{q}^{(B)} - \mathbf{q}^{(A)}$. The overlap describes the structural change A to B between two substates of a given protein, determined after optimal structural alignment of the two substates. The potential contribution of subsets of modes to such a transition may be deduced from the *cumulative overlap* $\left(\sum_k I_k^2\right)^{1/2}$, starting from the slowest modes. Note that this summation equals 1 if it is performed over all $3N-6$ modes/eigenvectors, which form a complete orthonormal basis set for the $3N-6$ dimensional space of conformational changes. In many cases, a few soft modes are sufficient to reach a value close to 1, indicating that the structural change from A to B is favored by the intrinsic dynamics of A.

Another quantity of interest is the *degree of collectivity* κ_k for mode k [31]:

$$\kappa_k = N^{-1} \exp\left[-\sum_{i=1}^{N} \alpha(\Delta r_i)^2\Big|_k \log(\alpha\Delta r_i)^2\Big|_k \right], \quad (12.F.2)$$

where α is the normalization constant in $\sum_i \alpha(\Delta r_i)^2\Big|_k = 1$. The higher the degree of collectivity, the more broadly the mode is distributed over a larger number of residues.

REFERENCES

[1] J Frank and RK Agrawal. A ratchet-like inter-subunit reorganization of the ribosome during translocation. *Nature*, 406:318–322, 2000.

[2] F Tama, M Valle, J Frank, and CL Brooks 3rd. Dynamic reorganization of the functionally active ribosome explored by normal mode analysis and cryo-electron microscopy. *Proc Natl Acad Sci USA*, 100:9319–9323, 2003.

[3] Y Wang, AJ Rader, I Bahar, and RL Jernigan. Global ribosome motions revealed with elastic network model. *J Struct Biol*, 147:302–314, 2004.

[4] PJ Flory, M Gordon, and NG MacCrum. Statistical thermodynamics of random networks. *Proc R Soc Lond A*, 351:351–380, 1976.

[5] A Kloczkowski, JE Mark, and B Erman. Chain dimensions and fluctuations in random elastomeric networks. 1. Phantom Gaussian networks in the undeformed state. *Macromolecules*, 22:1423–1432, 1989.

[6] I Bahar, AR Atilgan, and B Erman. Direct evaluation of thermal fluctuations in protein using a single parameter harmonic potential. *Fold Des*, 2:173–181, 1997.

[7] T Haliloglu, I Bahar, and B Erman. Gaussian dynamics of folded proteins. *Phys Rev Lett*, 79:3090–3093, 1997.

[8] M Tirion. Large amplitude elastic motions in proteins from a single-parameter, atomic analysis. *Phys Rev Lett*, 77:1905–1908, 1996.

[9] I Bahar and RL Jernigan. Inter-residue potentials in globular proteins and the dominance of highly specific hydrophilic interactions at close separation. *J Mol Biol*, 266:195–214, 1997.

[10] S Kundu, JS Melton, DC Sorensen, and GN Phillips Jr. Dynamics of proteins in crystals: Comparison of experiment with simple models. *Biophys J*, 83:723–732, 2002.

[11] LW Yang, E Eyal, C Chennubhotla, et al. Insights into equilibrium dynamics of proteins from comparison of NMR and X-ray data with computational predictions. *Structure*, 15:741–749, 2007.

[12] JB Udgaonkar and RL Baldwin. NMR evidence for an early framework intermediate on the folding pathway of ribonuclease A. *Nature*, 335:694–699, 1988.

[13] H Roder, GA Elöve, and SW Englander. Structural characterization of folding intermediates in cytochrome *c* by H-exchange labelling and proton NMR. *Nature*, 335:700–704, 1988.

[14] I Bahar, A Wallqvist, DG Covell, and RL Jernigan. Correlation between native-state hydrogen exchange and cooperative residue fluctuations from a simple model. *Biochemistry*, 37:1067–1075, 1998.

[15] F Schotte, M Lim, TA Jackson, et al. Watching a protein as it functions with 150-ps time-resolved X-ray crystallography. *Science*, 300:1944–1947, 2003.

[16] AR Atilgan, SR Durell, RL Jernigan, et al. Anisotropy of fluctuation dynamics of proteins with an elastic network model. *Biophys J*, 80:505–515, 2001.

[17] F Tama and YH Sanejouand. Conformational change of proteins arising from normal mode calculations. *Protein Eng*, 14:1–6, 2001.

[18] D Tobi and I Bahar. Structural changes involved in protein binding correlate with intrinsic motions of proteins in the unbound state. *Proc Natl Acad Sci USA*, 102:18908–18913, 2005.

[19] A Bakan, LM Meireles, and I Bahar. ProDy: Protein dynamics inferred from theory and experiments. *Bioinformatics*, 27:1575–1577, 2011.

[20] A Bakan and I Bahar. The intrinsic dynamics of enzymes plays a dominant role in determining the structural changes induced upon inhibitor binding. *Proc Natl Acad Sci USA*, 106:14349–14354, 2009.

[21] A Taly, PJ Corringer, D Guedin, et al. Nicotinic receptors: allosteric transitions and therapeutic targets in the nervous system. *Nat Rev Drug Discov*, 8:733–750, 2009.

[22] A Taly, M Delarue, T Grutter, et al. Normal mode analysis suggests a quaternary twist model for the nicotinic receptor gating mechanism. *Biophys J*, 88:3954–3965, 2005.

[23] P Doruker, RL Jernigan, and I Bahar. Dynamics of large proteins through hierarchical levels of coarse-grained structures. *J Comput Chem*, 23:119–27, 2002.

[24] F Tama and CL Brooks. Diversity and identity of mechanical properties of icosahedral viral capsids studied with elastic network normal mode analysis. *J Mol Biol*, 345:299–314, 2005.

[25] I Bahar, AR Atilgan, MC Demirel, and B Erman. Vibrational dynamics of folded proteins: Significance of slow and fast motions in relation to function and stability. *Phys Rev Lett*, 80:2733–2736, 1998.

[26] N Go, T Noguti, and T Nishikawa. Dynamics of a small globular protein in terms of low-frequency vibrational modes. *Proc Natl Acad Sci USA*, 80:3696–3700, 1983.

[27] T Noguti and N Go. Collective variable description of small-amplitude conformational fluctuations in a globular protein. *Nature*, 296:776–778, 1982.

[28] BR Brooks and M Karplus. Harmonic dynamics of proteins: Normal modes and fluctuations in bovine pancreatic trypsin inhibitor. *Proc Natl Acad Sci USA*, 80:6571–6575, 1983.

[29] M Levitt, C Sander, and PS Stern. Protein normal-mode dynamics: Trypsin inhibitor, crambin, ribonuclease and lysozyme. *J Mol Biol*, 181:423–447, 1985.

[30] OA Marques and YH Sanejouand. Hinge-bending motion in citrate synthase arising from normal mode calculations. *Proteins*, 23:557–560, 1995.

[31] R Brüschweiler. Collective protein dynamics and nuclear spin relaxation. *J Chem Phys*, 102:3396–3403, 1995.

SUGGESTED READING

Bahar I, Lezon TR, Yang LW, and Eyal E, Global dynamics of proteins: Bridging between structure and function. *Annu Rev Biophys*, 39:23–42, 2010.

Cui Q and Bahar I, editors, *Normal Mode Analysis: Theory and Applications to Biological and Chemical Systems*, Chapman & Hall/CRC Press, Boca Raton, FL, 2006.

CHAPTER 13

Molecular Modeling for Drug Discovery

DRUGS OFTEN ACT BY BINDING TO PROTEINS

Proteins play a central role in how drugs work. Drugs usually inhibit or activate proteins, replace them, or block their interactions. For example, drugs against HIV, such as saquinavir or tipranavir, are designed to inhibit a protease that the virus makes to reproduce itself inside cells. Some brain diseases—including Huntington's, epilepsy, and ALS (amyotropic lateral sclerosis)—result from deregulation of neurosignaling by glutamate, a neurotransmitter, in synapses in the central nervous system. So, drugs such as memantine have been designed to regulate the activities of NMDA, an ionotropic glutamate receptor. And diabetics take insulin because their own pancreas does not provide enough of this protein to regulate their blood sugar levels.

More than 2000 drugs of three main types are approved and sold in the US: (1) *Pharmaceuticals* are small molecules, made by chemical syntheses. You usually take them as pills. These constitute about 90% of today's approved drugs. (2) *Biologics* (also called *biopharmaceuticals*) are proteins, made by genetic engineering methods. You inject them. Some biologics are *monoclonal antibody* proteins (mAbs) that bind to specific molecular targets. And some biologics, like insulin or growth hormone, act to replace a missing functionality in your body. (3) Increasingly, mid-sized molecules, such as peptides or macrocycles, are designed to disrupt protein–protein interactions by sandwiching between two proteins that would otherwise bind to each other.

First, we consider small-molecule drug discovery. Figure 13.1 shows four examples of important small-molecule drugs:

Penicillin was one of the first antibiotics. Discovered by Alexander Fleming in 1928, it prevents the growth and division of bacterial cells. It binds to a bacterial enzyme called DD transpeptidase, preventing bacteria from forming the peptidoglycan cross-links in their cell walls. It has been said that most of the people in the world today would not be alive without penicillin, because their parents or grandparents would have died from infections.

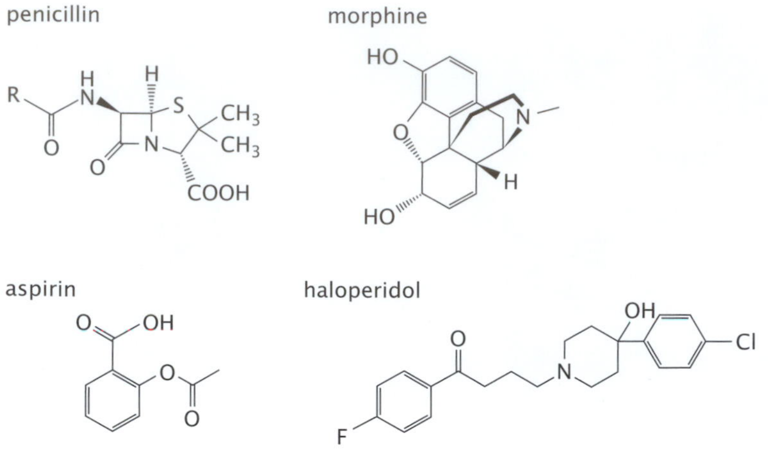

Figure 13.1 A few important drugs. On penicillin, R represents different possible chemical groups.

Morphine is a drug for severe pain. Its main action is considered to be binding to receptors on presynaptic neurons in the central nervous system, inhibiting the release of neurotransmitter molecules.

Aspirin relieves peripheral pain, such as in muscles, in headaches or in arthritis, and fights inflammation. It acts by irreversibly inactivating the cyclooxygenase (COX) enzyme by acetylating a serine in the enzyme's active site, suppressing the enzymatic production of lipids such as prostaglandins and thromboxanes.

Haloperidol is an antipsychotic. It was among the first drugs to control schizophrenia, allowing the de-institutionalization of people who otherwise would have been hospitalized. It binds to, and antagonizes, the dopamine receptor D2.

PHARMACEUTICAL DISCOVERY IS A MULTISTAGE PIPELINE PROCESS

Drugs are developed through a series of steps, resembling a pipeline (Figure 13.2). The front end of the pipeline, called *discovery*, has the following components:

Target ID and validation: ID refers to identifying a target. The first step is to find a *target*, usually a protein in the organism where you want the drug to act. And *validation* means confirming that the target plays a role in the disease. You do this, for example, by using knowledge of the basic biology of the disease; by knocking out genes and observing the impact; and by genome-wide association studies (GWAS), which compare the common genetic variations among individuals to find those variants associated with the disease. It also means confirming that the protein is not too similar to any human protein.

A *compound library* is a collection of potential drug molecules. Libraries can be obtained from commercial sources, or publicly available sites, or may just be collections of small molecules that a pharmaceutical company has previously developed for other purposes.

Assay development: Once you have a target, you need a fast way to learn how and whether your potential drug will affect the function of your target. For example, you can *screen* a ligand library by assaying their effect in a cell line that overexpresses the target protein.

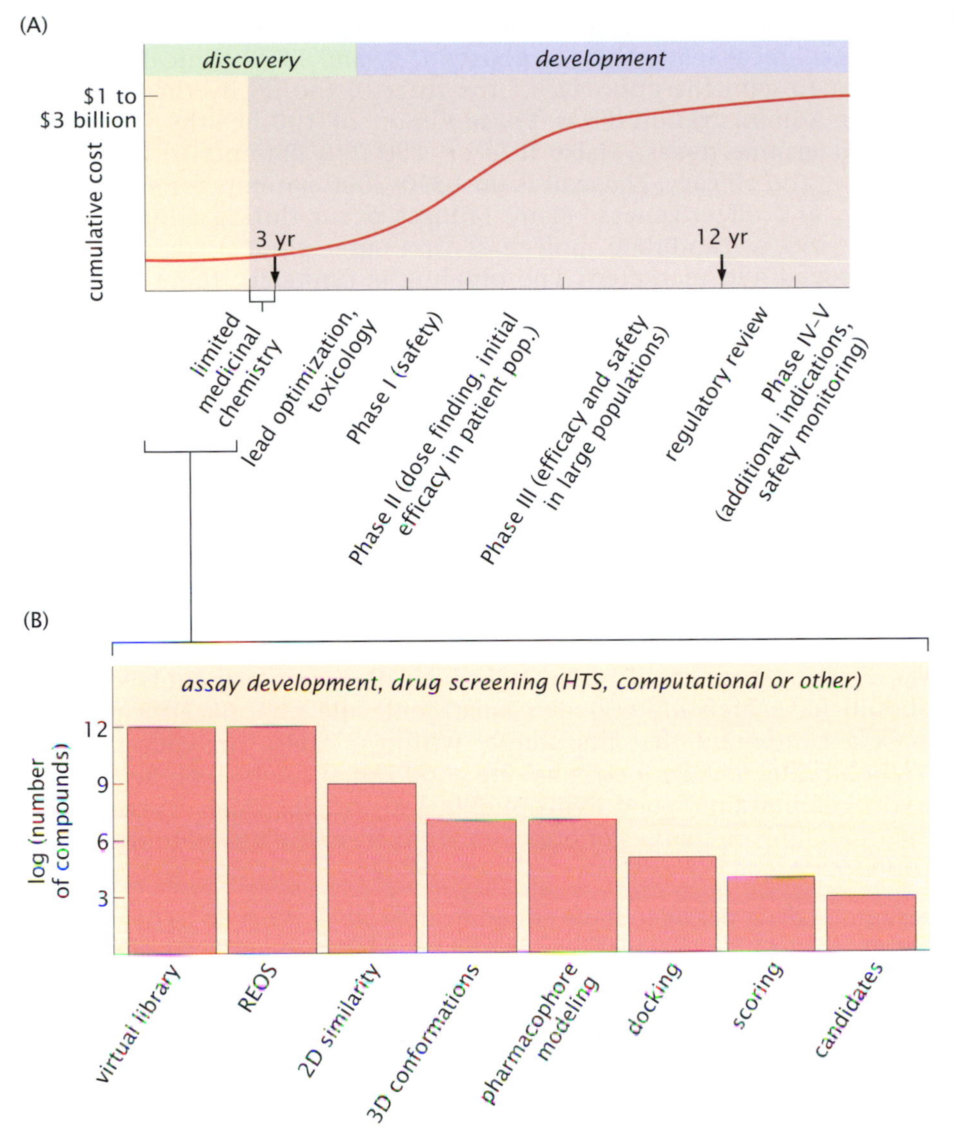

Figure 13.2 The slow expensive process of drug discovery and development. (A) The process can take many years. The costs are particularly high during the late stages, that is, for the clinical trials. (B) Computational drug discovery contributes mostly in early discovery. It begins with large libraries of candidates and weeds out unlikely candidates. Very few molecules make it all the way through the pipeline to become successful drugs on the market.

With these components in hand, drug discovery consists of the following steps:

1. *Hit identification:* You search libraries of small molecules, seeking *hits*, that is, molecules that affect the target. Hits can often be found using *high-throughput screening* (HTS), the testing of whole libraries at a time using automated methods in the lab. Or you can synthesize a library of compounds using *combinatorial chemistry*, a fast and systematic way to generate a large number of molecules in a single reaction process. Or you can find hits by *virtual screening*, that is, using computers to search databases of small molecules to find ones you think might be effective.
2. *Lead optimization:* Once you have hits, you systematically make chemical modifications to them to improve their properties. Typically, you would make thousands of compounds (single or in libraries), optimizing them for potency, for selectivity against close homologs, for specificity against pharmacologically important targets, for cell permeability and activity, and, critically, for efficacy and safety in animal models.

3. The back end of the pipeline is called *drug development*. It consists of preclinical tests, then of phases I, II, and III of clinical trials, in order to gain the approval of the drug in the US by the Food and Drug Administration (FDA). Phase I is on 20–100 healthy volunteers to determine doses. Phase II is on 100–300 patients to determine safety and efficacy. Phase III is on 1000–3000 patients for safety, efficacy, and effectiveness. Many failures occur during clinical trials: only 15% of candidate molecules that enter clinical trials are ever approved and marketed. The pipeline is funneled: there are many candidate molecules at the beginning, but very few ever make it to market as drugs.

DESIGNING A DRUG REQUIRES OPTIMIZING MULTIPLE PROPERTIES

What Properties of a Molecule Are Drug-Like?

To become a drug, a candidate molecule must satisfy various criteria. First, a drug should bind tightly to its target protein (that is, it should have high *affinity*). For small-molecule pharmaceuticals, you seek a compound that fits snugly within a cavity on your protein. Tight binding usually means having good *shape complementarity,* good hydrogen bonding, good hydrophobic and van der Waals interactions, and matching of charges of opposite signs from the ligand to the target protein.

Second, a drug needs to have adequate *solubility.* It needs to dissolve in water, so that it can flow through the bloodstream and enter cells. So, the molecule must have some charge or polarity. In addition, somewhat paradoxically, a typical drug must also dissolve sufficiently in nonpolar environments so that it can cross cell membranes. So, a drug should be somewhat hydrophobic and not too highly charged.

Third, a drug must meet the *ADMET criteria* (Adsorption, Distribution, Metabolism, Excretion, Toxicology). It must be taken up into the bloodstream. It needs to reach its intended target site in the body. It should not chemically degrade into breakdown products before it reaches its target. It should not be excreted from the body too quickly. And it should not be too toxic. That is, a drug should be *selective*: it should bind much more strongly to the target protein, and more weakly to other proteins and surfaces, since unintended binding is the basis for unintended drug side effects. In addition, a drug must be *formulated,* that is, it must be developed in a surrounding liquid or solid medium that protects it from degradation, so that it has a sufficiently long shelf life.

There are simple "rules of thumb" that can help you determine if a chemical entity is likely to satisfy these *drug-like* properties. For example, *Lipinski's Rule of Fives* says that absorption or permeation is likely to be poor when the ligand has more than five hydrogen bond donors; or the molecular mass of the ligand is greater than 500; or the calculated $\log P$ is greater than five (where P is the octanol/water partition coefficient, which is the ratio of the equilibrium concentration of the compound in octanol to that in water); or the sum of the nitrogen and oxygen atoms in the compound is greater than 10 [1]. Alternative rules of thumb are Jorgensen's Rule of Three [2, 3], or REOS (Rapid

Elimination Of Swill) (Table 13.1). These specifications are only broad guidelines; some drugs do not follow the rules.

Table 13.1 REOS rules for "drug-likeness." This table shows the range of properties of most of the drugs on the market. These ranges are rules of thumb for discovering new drugs.

Property	Min	Max
Molecular weight	200	500
log *P*	−5	5
Hydrogen-bond donors	0	5
Hydrogen-bond acceptors	0	10
Formal charge	−2	2
Rotatable bonds	0	8

(From WP Walters and M Namchuk *Nat Rev Drug Disc*, 2:259–266, 2003.)

Applying such rules can also reduce the problem of *nuisance binders*. Nuisance molecules are ligands that do bind to the target protein, but will never become drugs. Nuisance molecules have steep dose–response curves, show little effect on biological activity, and are highly sensitive to assay conditions. They bind nonspecifically to other proteins and bind to each other, forming colloidal aggregates that can be seen in microscopes. A goal in drug discovery is to filter out these nuisance molecules early. The previous rules of thumb can often help.

What Are the Properties of a "Druggable" Protein?

Not all proteins are good targets for drug design. Good target proteins are called *druggable*. While druggability is not always easy to predict, a protein is often druggable if it has a concave hydrophobic site where a small molecule can fit tightly, like a hand into a glove. Figure 13.3 shows a ligand binding to a concave hydrophobic site where there is favorable matching of shape and hydrophobic interactions between the ligand and the protein. Assessing target druggability is important for estimating whether a *drug-discovery campaign* will yield a potent drug-like molecule within a reasonable time. Only about 10% of the proteins in the human genome are thought to be druggable targets, and only half of those are relevant to disease [4].

However, sometimes there are ways to find drugs even for seemingly undruggable targets. One approach is to find molecules that bind *covalently* to the target. And sometimes a binding site is exposed only when there are appropriate motions in the protein. In those cases, druggability can be determined using molecular dynamics simulations in the presence of probe molecules representative of drug-like fragments [5]. In some cases, protein surfaces that are too flat or too flexible can be undruggable. But, in those cases, as described later in this chapter, large flat flexible molecules called *macrocycles* can sometimes bind well.

Figure 13.4 shows the distribution of protein targets for which there are approved drugs on the market.

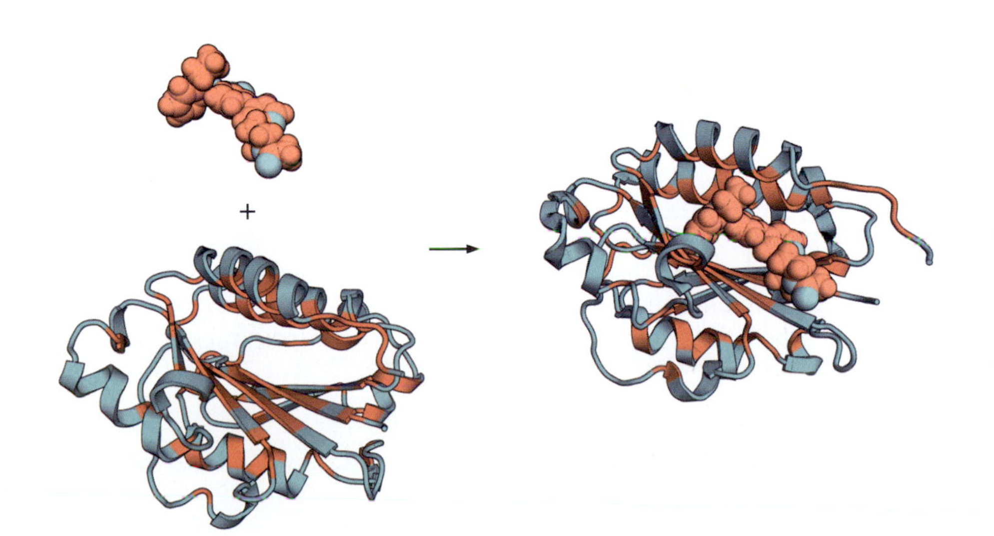

Figure 13.3 A drug binds to a druggable site on a protein. A hydrophobic ligand (intercellular adhesion molecule 1, *orange*) binds to a site (*mostly orange*) on a protein (leukocyte glycoprotein, *green* and *orange*) that is concave and hydrophobic, giving a good match of shapes and energetics. (Adapted from MP Crump, TA Ceska, L Spyracopoulos, et al. *Biochemistry*, 43:2394–2404, 2004.)

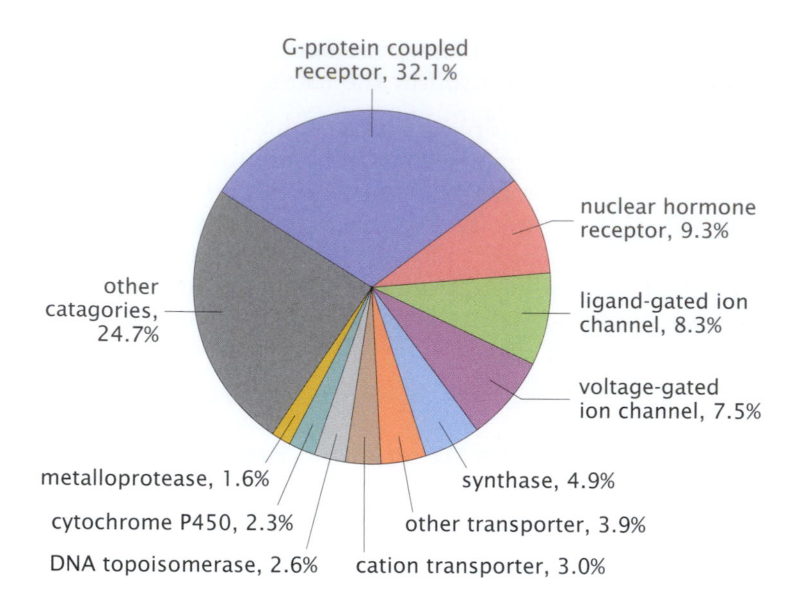

Figure 13.4 The distribution of functions that are performed by the protein targets of approved drugs. (From I Bahar, TR Lezon, A Bakan, and IH Shrivastava. *Chem Rev*, 110:1463–1497, 2010.)

In the following sections, we describe computational methods for discovering drugs. There are two basic strategies in *computer-aided drug discovery* (CADD): *ligand-based* or *target-based*, depending on what data are available (Figure 13.5).

In *ligand-based* approaches (also called *small-molecule similarity methods*), you start by knowing the activities of some ligands or drugs against the disease. You don't necessarily know the structure of the target protein. By comparing the structures and properties of different ligands with each other, using statistical correlations and pattern recognition, you can often guess what other ligands might also be worth exploring. Ligand-based approaches are widely used—both when you don't know a target protein structure and even when you do.

In *target-based* approaches, you start with the atomic structure of a target protein that is relevant to the disease of interest, or of a structural homolog to it. Then you seek small molecules that bind to the target. You do this using computational methods that work with the atomic

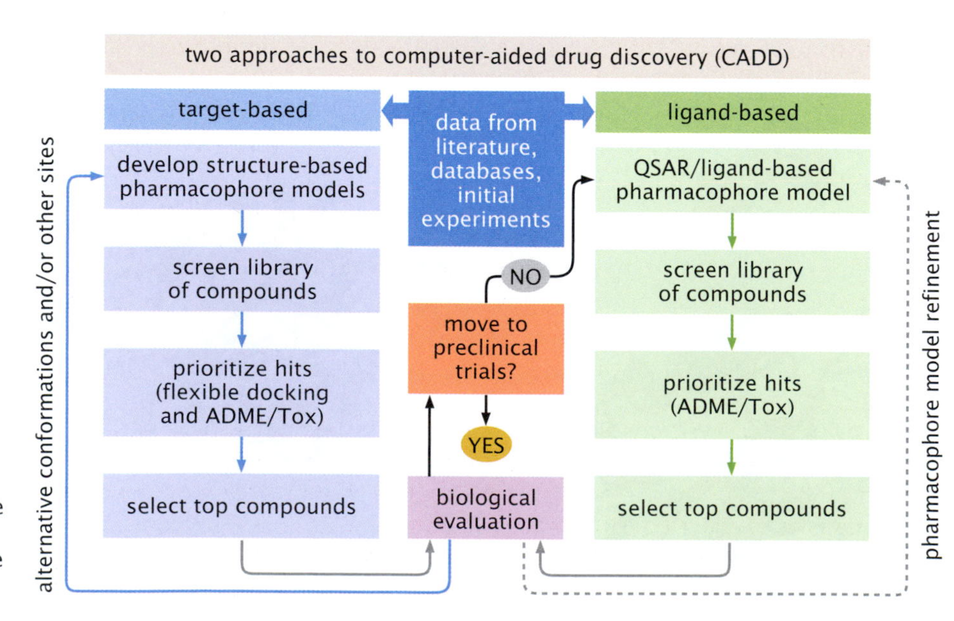

Figure 13.5 Two approaches to computer-aided drug discovery: target-based and ligand-based. Ligand-based methods use knowledge of existing ligands to find new ones. Target-based discovery uses knowledge of the structure of the target protein. Drug discovery campaigns typically use both approaches.

coordinates of the ligand and the protein. Target-based approaches are also called *rational drug discovery* or *structure-based design*. Drug discoverers often use both ligand- and target-based methods by iteratively finding candidate molecules, testing them in binding and activity assays, keeping the good ones, and then improving them in various ways.

LIGAND-BASED DISCOVERY USES KNOWN LIGANDS TO DESIGN NEW ONES

Much of drug discovery is done without knowing a target protein structure. For example, relatively few structures are yet known for membrane proteins, yet many important nervous-system drugs have been developed to act on membrane proteins. Then, ligand-based discovery is a good starting point.

A principle of ligand-based design is that *ligand molecules having similar structures are likely to have similar biological actions*. So, to discover new drugs, you modify active ligands that you already know. If you already know a set of small molecules that affect an enzyme of interest, that can be a good starting point for finding a drug that can activate or inhibit the enzyme. A *targeted library* is a set of chemicals based upon a fixed molecular structure, called a *scaffold*, on which systematic variations are made.

Methods for ligand-based drug design are described in the following sections; they include *quantitative structure-activity relationship* modeling (QSAR), similarity searching, substructure matching, and ligand-based pharmacophore modeling. It is critical that you have a good compound library. It should have a broad diversity of chemicals, provide as much coverage as possible of the space of chemical moieties that could be important, and be as representative as possible of *drug-like* molecules.

If You Know the Biological Activities of Some Ligands, You Can Estimate the Activities of Others Using QSAR Methods

Often, the first step in finding ligands that might affect a biological activity is to make a QSAR analysis of ligands that are already known. In QSAR, you have a collection of small molecules that cause different levels of some type of biological or chemical activity. For example, you might measure the midpoint concentration c at which each different compound causes either some toxicity to a cell, or some enzyme activity, or some reduction in the respiration rate of a cell, or some level of antimalarial activity. In QSAR, you assume that c is some empirical function of the physicochemical properties of the ligands, such as the following:

$$\log(1/c) = a\log P - b(\log P)^2 + dx + \cdots + e, \tag{13.1}$$

where a, b, d, and e are parameters used to fit the activity c as a function of these terms. Here, P is the octanol–water partition coefficient and x may be some quantity representing the size or shape or charge or other measurable property of the compound. You look for correlations by statistical methods such as linear regression or machine learning. QSAR gives you a way to predict activities of other ligands you have not yet tested. QSAR is a broad collection of methods aimed at correlating the physical properties of ligands with their biological or chemical activities.

For exploring databases of drugs, it is common to use either one-dimensional (1D), two-dimensional (2D), or three-dimensional (3D) representations of chemicals (Figure 13.6). A *molecular descriptor* is the physical information conveyed about a molecule by its representation in 1D, 2D, or 3D. The 1D representation tells you the types of atoms and molecular weight. The 2D representation tells you the numbers and types of bonds and connectivities and the numbers of hydrogen bond donors and acceptors, and helps you estimate the octanol–water partition coefficient. The 3D structure helps you estimate the conformational energies, the isomeric states of the bonds, and solvent accessibilities of the atom types.

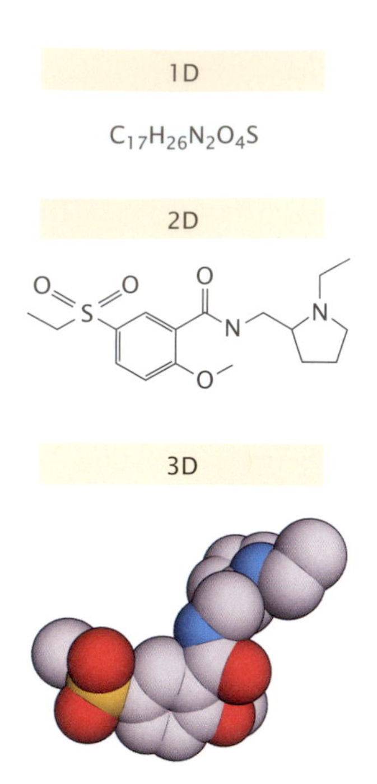

Figure 13.6 The 1D, 2D, and 3D representations of a small molecule. 1D representations describe the chemical composition. 2D representations give the types and connectivities among the atoms. 3D representations are the spatial coordinates of the atoms. (Adapted from J Bajorath. *Nat Rev Drug Discov*, 1:882–894, 2002. With permission from Macmillan Publishers Ltd.)

Similarity Searching Seeks Compounds Similar to the Good Ones You Know Already

If you know a ligand that has good biological activity, how do you find similar ligands? First, what does it mean to be "similar"? One definition of similarity is a count of the numbers and types of chemical groups in common on the two different molecules. You can do *similarity searching* by using a *2D fingerprint*. You identify the individual chemical substituents of the molecule using the 2D description above. For example, Figure 13.7 shows a molecule having a carbonyl oxygen, a carboxyl group, and other chemical moieties. Once you have a list of such descriptors, you can assemble them into a 1D binary string, in which a 0 means that descriptor is not present in your molecule, or a 1 means that descriptor is present. Then, you can establish similarities between molecules by making a fast computational comparison of two binary strings. To compute the chemical similarity of one molecule to another, you can use the *Tanimoto coefficient*, which is the ratio

$$T(a, b) = \frac{N_c}{(N_a + N_b - N_c)}, \tag{13.2}$$

where N_a and N_b are the numbers of bits (nonzero entries) in the strings representing the molecules a and b that are being compared, and N_c is the number of bits the two molecules have in common. Compounds having Tanimoto coefficients greater than 0.85 tend to have similar biological activity.

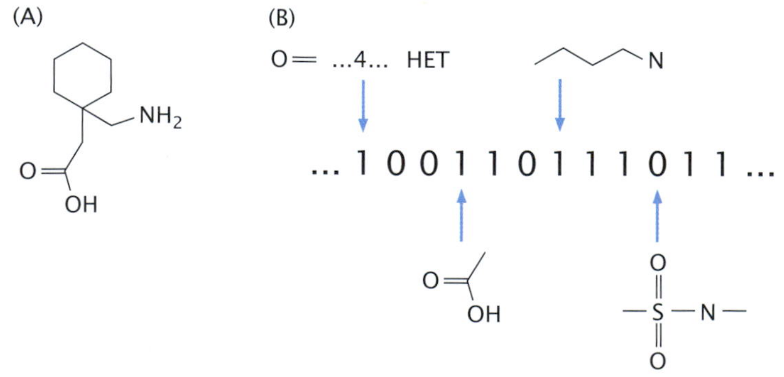

(A) NH_2 O OH

(B) O= ...4... HET N

... 1 0 0 1 1 0 1 1 1 0 1 1 ...

O OH O S N O

Figure 13.7 The 2D fingerprint of a molecule. (A) Molecular structure. (B) A string of 0's and 1's represent the presence or absence of molecular descriptors contained in the molecule. For example "O= ... 4 ... HET" is the code for an atom that has at least four neighbors and includes both a double-bonded oxygen and a heterocycle. (Adapted from the Indiana Cheminformatics Education Portal. https://icep.wikispaces.com/Characterizing+2D+structures+with+descriptors+and+fingerprints.)

A Pharmacophore Is a 3D Arrangement of Properties of Some Atoms

Another approach to similarity searching is *pharmacophore modeling* (Figure 13.8). A pharmacophore is some particular 3D spatial arrangement of different chemical substituents that correlates with measurable biological activity. A pharmacophore is described by the relative 3D positioning of the functional groups (for example, hydrogen-bond donors, acceptors, or hydrophobic or aromatic groups).

Once you have a pharmacophore model that captures the essence of a chemical structure, you can screen databases to find additional molecules that have a similar spatial arrangement of functional groups. Often, you don't need precise interatomic specifications in your pharmacophore; rather, you can specify ranges of values (for example, spherical regions) for the spatial positions of specific groups of atoms. For one thing, small molecules naturally possess some inherent flexibilities. Also, when your learning set of ligands is diverse, you may need to construct more than one pharmacophore model.

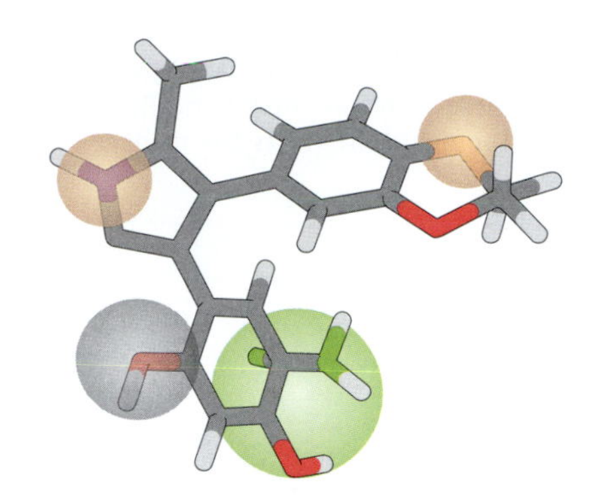

Figure 13.8 Example of a pharmacophore, a 3D arrangement of particular chemical groups that are relevant to a particular biological activity. (Adapted from DR Koes, and CJ Camacho. *J Chem Inf Model*, 51:1307–1314, 2011. With permission from American Chemical Society.)

Some Drugs Are Developed by Linking Fragments Together

One way to design a drug, called *fragment-based* discovery, is to link together smaller component pieces. Suppose you know two small molecules that each bind weakly to different sites nearby to each other in the target protein. Then, by covalently linking the two small molecules together with a *tether* (that is, a molecular linker that properly positions them and orients them in the protein), you can often make a tighter-binding larger ligand (Figure 13.9). Successful fragment-based discoveries have been targeted towards PDK1, a kinase, and Hsp90, a chaperone. Now, we describe target-based approaches, where you know the structure of a target protein.

TARGET-BASED DISCOVERY DESIGNS DRUGS BY USING THE STRUCTURE OF A TARGET PROTEIN

In *target-based* or *structure-based* drug design, you start with a known 3D atomic structure of your target protein, or of a protein that you expect to have a highly similar structure. You perform a computer search to identify possible binding sites on the protein. Your algorithm tries to fit a proposed ligand in different orientations (called *poses*) into each possible binding site on the target, to find a good "hand-in-glove"

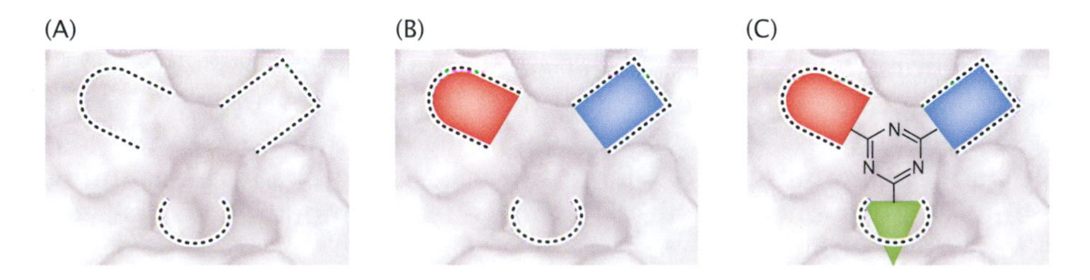

Figure 13.9 Fragment-based drug discovery. (A) A binding site has three pockets. (B) Individual fragments, one *red* and one *blue* are found that bind to two different pockets. (C) To create a tighter binder, the two fragments are linked together covalently (the *black* ring in the center), allowing, in this case, for an additional fragment (*green*) that can bind to a third site. (From TL Blundell and S Patel. *Curr Opin Pharmacol*, 4:490–496, 2004. With permission from Elsevier.)

(A) trypsin binding site

(B) benzamidine docked onto trypsin

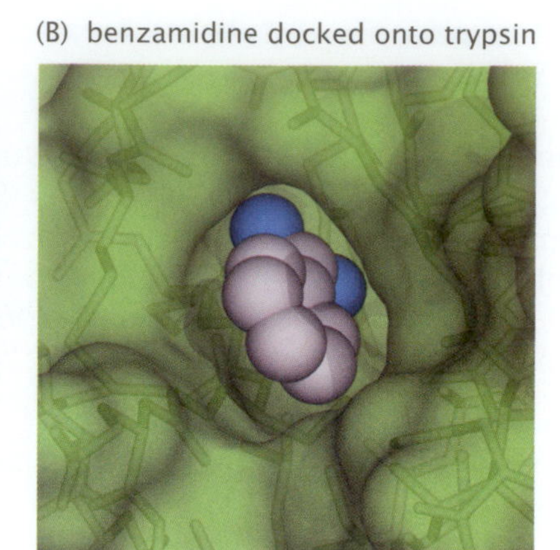

Figure 13.10 Illustrating structure-based drug design. (A) Find a cavity in a protein. (B) Find a small molecule that fits the cavity.

binding arrangement (Figure 13.10). Then, you repeat this procedure for many different ligands, until you find ones that you expect will bind well. These methods use energies, or energy-like scoring functions, to select the best ligands in their best poses.

In general, there are two main approaches to structure-based design, described in the following subsections:

1. *Docking* is computationally fast. Medicinal chemists need to sort through large lists of possible ligands very quickly, typically making decisions about ligands in hours to days, not weeks to months. But docking is crude because it approximates the true physical energies with simplified energetics and solvation, and performs limited conformational searching, usually by treating the protein as rigid.
2. *Atomistic simulations* are intended to be more accurate, but they are computationally slow. They treat the physical energies and solvation more accurately (by the methods of Chapters 9 and 10), and they aim to sample the ligand and protein degrees of freedom extensively.

Docking Is a Fast Way to Find Ligands that Bind to a Given Protein Structure

The first docking algorithm, DOCK, was developed by ID Kuntz and colleagues in 1982 [6]. It sought tight-binding ligands by a fast search for shape complementarity between a rigid ligand and a rigid binding site. It then evaluated the different poses with a scoring function that matches ligand with protein nonpolar regions, hydrogen-bond donors and acceptors, and charges. Other popular molecular docking packages include Glide [7], AutoDock [8], ICM [9], and GOLD [10]. Docking methods are used (1) for high-throughput lead finding, to identify possible active compounds from large libraries, and (2) for determining the binding pose, once you already know a lead compound, so that you can improve its binding.

How can you evaluate the quality of your docking method? Box 13.1 shows how to make *enrichment plots*, which represent the quality of a prediction model, based on the ratio of good to bad predictions from that model.

Box 13.1 You Can Evaluate Virtual Screening Methods Using Enrichment Plots

Here is how to make a *receiver operating characteristic* (ROC) curve, also called an *enrichment plot* (Figure 13.11). First, you determine the following numbers. Your total experimental dataset has P positives (active compounds) and N negatives (inactive compounds). Now, compare your model predictions with experimental data by counting the following:

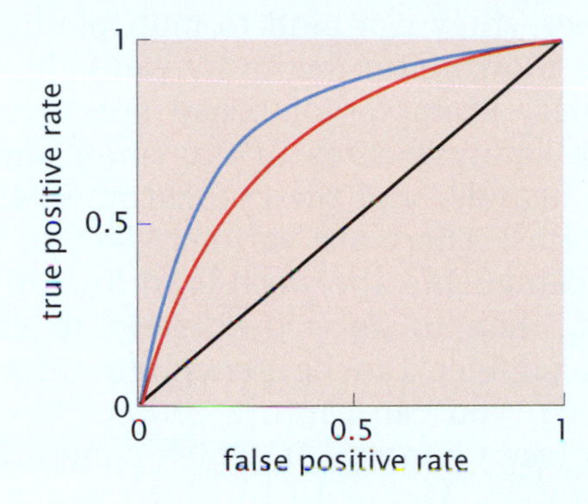

Figure 13.11 An enrichment plot indicates how well a model makes predictions. The best model is the curve that comes closest to the upper left corner, indicating a high ratio of rates of true positives to false negatives. Here the *blue* model is better than the *red* model, and both are better than random (*black line*).

P_T: true positives (correctly predicted to be a hit);
N_T: true negatives (correctly predicted to be a miss);
P_F: false positives (predicted as a hit, but is a miss);
N_F: false negatives (predicted as a miss, but is a hit).

A ROC plot shows the rate of finding true positives, P_T/P, on the y-axis (as you work through a series of ligands), as a function of the rate of finding false positives, P_F/N, on the x-axis, where $P = P_T + N_F$ and $N = N_T + P_F$. If the model gives results that fall onto the diagonal line, $y = x$, it indicates that the model is no better than random coin flips for picking hits versus misses. But if your method gives results like those in Figure 13.11, with curves above the diagonal line, it means the model is better than random at selecting hits from misses. The closer the line comes to passing through the upper left corner, the better is the model for enrichment.

On the one hand, docking is an important tool. It is fast enough to search large databases to prioritize a small number of ligands that are worthy of further experimental testing. It can commonly identify correct poses, and it can do so rapidly, screening million-compound libraries in a week. Docking often finds novel ligands and contributes insights into binding mechanisms.

On the other hand, there is a need for methods—even those that are computationally much slower—that can predict binding poses and affinities more accurately. There are limitations to the simplest docking assumption that good binders can be found by first finding a steric fit, and then by optimizing the energy. And the scoring functions are crude: they often neglect many degrees of freedom of the ligand and the protein, they neglect the changes in entropies, and they use simplified models of solvation. So, docking methods often do not predict correct binding affinities.

Including Protein Flexibility Can Improve the Modeling of Ligand Binding

Docking methods can often be improved by treating protein flexibilities. For one thing, you can find more native-like poses by allowing for protein flexibility. For another thing, some proteins have broad substrate specificities—they can bind to multiple ligands. Sometimes, simulating proteins allowing for flexibility can help you identify these substrate specificities. Examples of broad specificity include active-site loops near the catalytic sites of enzymes, antibodies that can recognize multiple ligands, and the cytochrome P450 family of electron transport proteins. There are various ways to introduce protein flexibilities into docking. One approach is to find an ensemble of protein conformations, then to dock the ligand to each conformation separately. Such ensembles can be generated from sets of different crystal structures. Or, you can capture protein flexibility using elastic network models (see Chapter 12), which compute the large modes of motion of the protein. Figure 13.12 illustrates a case where modeling the flexibility of a loop helped find the correct ligand binding to its site.

Molecular Dynamics Simulations Can Calculate Binding Free Energies

In principle, the most accurate way to compute the affinities of ligands for proteins is to apply an all-atom physical model, with a good model of the solvent, and to fully sample both the ligand and protein conformations, using detailed-balance-preserving methods such as molecular dynamics or Monte Carlo simulations (see Chapter 10). Historically, a first step in that direction was taken in the early 1980s, soon after the emergence of DOCK. That step combined molecular mechanics minimizations, with Poisson–Boltzmann or generalized Born and surface-area-based solvation modeling (called MM/PBSA or MM/GBSA).

More recently, molecular dynamics has been used instead of molecular mechanics to seek accurate binding affinities. These are *alchemical*

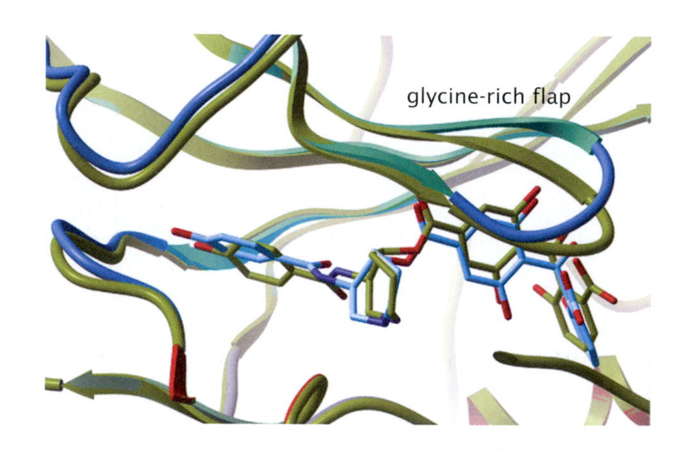

Figure 13.12 Drug binding is associated with a loop movement in the protein. Computations show that the drug balanol (the stick figure in *blue and red*) binds to a protein kinase (*olive green and light gray*) when the loop changes conformation from its unbound structure (*olive-colored*) to its bound structure (*green and blue*). (Adapted from CN Cavasotto and RA Abagyan. *J Mol Biol*, 337:209–225, 2004. With permission from Elsevier.)

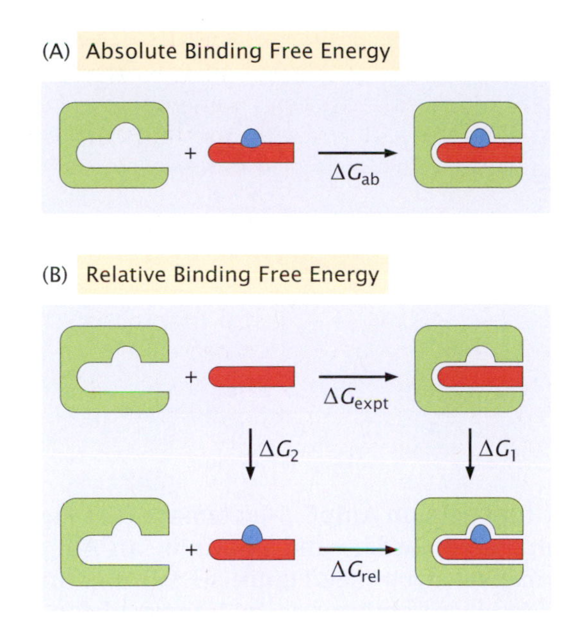

Figure 13.13 Absolute binding free energy (ABFE) and relative binding free energy (RBFE). (A) ABFE is the free-energy difference ΔG_{ab} of the bound, relative to unbound, state of the ligand (*red with blue bump*) to the protein (*green*). (B) In RBFE, you begin with a known experimental binding free energy ΔG_{expt} of the *red* compound. You perform simulations to obtain ΔG_1, the free energy of adding the *blue* chemical structure to the red compound in water and ΔG_2, the free energy of adding the *blue* part to the *red* compound in the protein. From the thermodynamic cycle, you compute $\Delta G_{rel} = \Delta G_{expt} + \Delta G_1 - \Delta G_2$, the free energy of binding the *red* + *blue* molecule to the protein. It works best when the ligand is similar to the reference compound.

free-energy methods. There are two main strategies: (1) *relative binding free energies* (RBFE) and (2) *absolute binding free energies* (ABFE) (Figure 13.13). To compute RBFEs, you begin with experimental knowledge of the binding free energy of some reference ligand, then you compute the *differences* in free energies $\Delta\Delta G$ that would result from converting that reference ligand into your ligand of interest. Computing ABFEs is more difficult. Here, the term "absolute" does not mean that you are computing some absolute free energy G. It just means that the calculated binding free energy is not based on experimental knowledge of the binding affinity of a reference compound.

In the method of *relative binding free energies*, the ligand of interest should be very similar to the reference ligand. The differences between the ligands should be sufficiently small that the sampling requirements will be minimal for morphing one into the other. You can then compute the relative binding free energy of interest by using the thermodynamic cycle in Figure 13.13. Because RBFEs calculate only *differences*, they are faster to compute and subject to smaller errors than ABFEs. Figure 13.14 gives examples in which calculated binding affinities correlate with experimental measurements. Accurate calculations for arbitrary targets and ligands remains a research frontier.

Structure-Based Methods Have Helped to Discover New Drugs

Structure-based design has contributed to the development of several important drugs. Among the first were inhibitors of HIV proteases in the 1990s, leading to anti-AIDS therapeutic agents that include amprenavir and nelfinavir. Other drugs developed using structure-based design include captopril for regulating blood pressure, based on knowledge of the active site of carboxypeptidase; dorzolamide for glaucoma, based on the structure of carbonic anhydrase; the anti-influenza drugs zanamivir and oseltamivir based on the neuraminidase structure; cimetidine, a prototypical H2-receptor antagonist; raltitrexed, a thymidylate synthase inhibitor; selective COX-2 inhibitors known as nonsteroidal anti-inflammatory drugs (NSAIDs); selective serotonin reuptake inhibitors (SSRIs) and other antidepressants; benzamidine, a

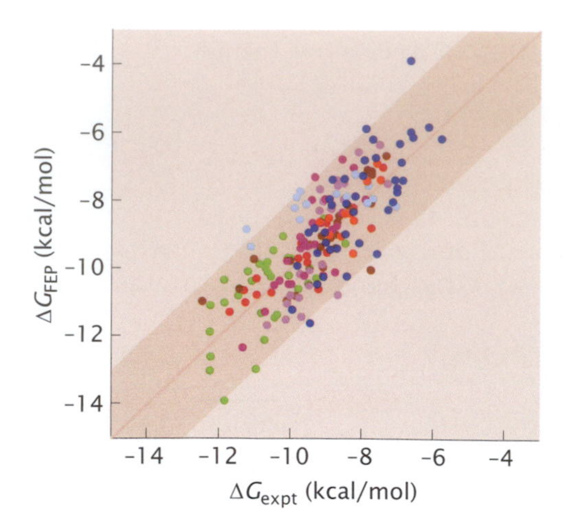

Figure 13.14 Binding affinities from molecular simulations versus experiments. RBFEs on eight different target proteins, tested with 11–42 ligands each. (Adapted from L Wang, Y Wu, Y Deng, et al. *J Am Chem Soc*, 137:2695–2703, 2015.)

trypsin inhibitor that acts on AmpC β-lactamase (see Figure 13.10); valproic acid for bipolar disorder; and imatinib, an Abl tyrosine kinase inhibitor, an anticancer drug (see Figure 13.17). Imatinib was the first selective tyrosine kinase inhibitor to be approved for the tratment of cancer.

A MAJOR CLASS OF DRUGS IS THE BIOLOGICS

Small molecules are not the only therapeutic drugs. A growing class of therapeutics comprises the *biologics* or *biopharmaceuticals*. These therapeutics are themselves proteins or peptides. Until recently, companies were known as pharmaceutical or biotechnology companies, depending on whether they produced small-molecule drugs or protein drugs. But now, typical drug companies develop both. Table 13.2 shows that seven of the top eight best-selling drugs in 2013 were proteins. More than 130 biologics are on the market, and the market for protein drugs is growing faster than the market for small-molecule drugs.

A major advance that enabled the rapid growth in biotechnology has been recombinant DNA technology. The first biologic was human insulin, a small protein that is used to treat diabetes. In the early days, insulin was produced by purification of the protein from animal

Table 13.2 Seven of the eight best-selling drugs in 2013 were proteins

Trade name	Type	Therapy for what disease
Humira®[a]	Monoclonal antibody	Arthritis, colitis, psoriasis
Enbrel®[b]	Recombinant protein	Arthritis, psoriasis
Advair®[c]	Corticosteroid (small molecule)	Asthma, chronic bronchitis, emphysema
Remicade®[d]	Monoclonal antibody	Rheumatoid arthritis, immune diseases
Rituxan®[e]	Monoclonal antibody	Non-Hodgkin's lymphoma, chronic lymphocytic leukemia, rheumatoid arthritis
Lantus®[f]	Recombinant insulin analog	Diabetes
Avastin®[g]	Monoclonal antibody	Angiogenesis inhibitor
Herceptin®[g]	Monoclonal antibody	Breast cancers over expressing the receptor HER2

Adapted from http://www.forbes.com/sites/peterubel/2014/10/16/the-best-selling-biologic-drugs/
[a]Abbvie Inc.; [b]Amgen Inc.; [c]GlaxoSmithKline plc; [d]Janssen Biotech, Inc.; [e]Biogen and Genentech USA, Inc.; [f]Sanofi-Aventis U.S. LLC; [g]Genentech USA.

sources, such as cows and pigs, which often caused unwanted side effects. Now, however, the gene for insulin simply is inserted by recombinant DNA technology into bacteria, which are grown up on a large scale, rapidly and cheaply. Human insulin was the first *recombinant-DNA* biologic, approved by the US FDA in 1982. Now, most biologics are produced by recombinant DNA and grown up in simple convenient organisms. The main commercial biologics are monoclonal antibodies (mAbs) (Figure 13.15), but other types of proteins and peptides are also therapeutics.

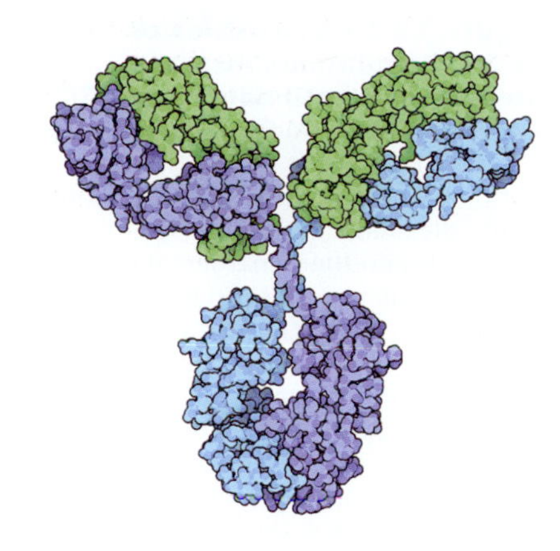

Figure 13.15 Monoclonal antibodies are a major class of therapeutic proteins. (From DS Goodsell. pdb101.rcsb.org/motm/136.)

Biologics have some advantages over small-molecule pharmaceuticals. Biologics can interact more specifically, leading to better targeting and fewer side effects. For example, interleukin 2 (IL-2) is a small protein that stimulates the immune system, useful for treating AIDS and cancer. Its efficacy comes from its precise differential binding to different protein complexes. Monoclonal antibodies are well tolerated by the body, and don't generate immune reactions, because the body doesn't see them as being foreign. Biologics can perform more complex tasks than small molecules can. And protein biologics can replace or augment deficient proteins. Interestingly, biologics can often be approved faster by regulatory agencies and can have stronger patent protection, leading to economic advantages for the companies that develop them. In addition to insulin and growth hormone, other biologic drugs have been developed for multiple sclerosis, for hepatitis C, for immunodeficiency diseases, and as coagulation factors for people with hemophilia. Biologic vaccines protect against infections, and some are used as medical diagnostics.

There are some challenges in developing biologic drugs. First, biologics tend to be expensive. Second, while pharmaceuticals can be formulated as pills that you swallow, biologics must be injected into your body so they are not degraded by the digestive system. Third, proteins are finicky and problematic in various ways. They can undergo covalent degradation such as deamidation, resulting in unwanted heterogeneity. And proteins can denature, aggregate, precipitate, or change properties when they adsorb to surfaces. These instabilities can depend on temperature, pressure, pH, and other processing and storage conditions.

Protein theory and modeling can help meet these challenges. For example, it is a challenge to formulate a solution of a biologic drug to have the proper protein concentration. On the one hand, you want a solution to contain a high concentration of protein, to increase its efficacy. On the other hand, highly concentrated protein solutions have high viscosity, making them difficult to manufacture and difficult for patients to inject. Figure 13.16 shows that the viscosities of antibody solutions increase as a steep function of antibody concentration, and they depend sensitively on the amino acid sequence.

CHALLENGES AND RECENT DEVELOPMENTS IN DRUG DISCOVERY

Drug Resistance Can Be Caused by a Protein Mutation Near a Drug-Binding Site

Sometimes, a drug loses its effectiveness over time. Such drugs no longer mitigate the disease that they once targeted effectively. This

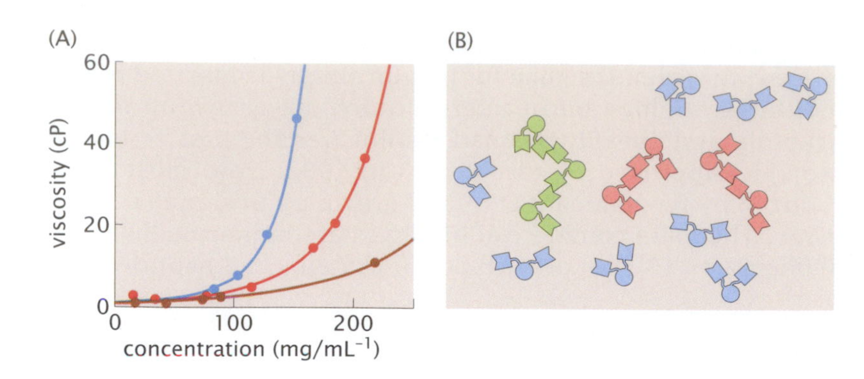

Figure 13.16 Viscosities of antibody solutions increase steeply with antibody concentration. (A) Solution viscosity as a function of the concentration of antibody (immunoglobin G). Different antibodies have different viscosities, depending on their charged and hydrophobic residues. (B) Proposed mechanism in which the high viscosities come from chaining together antibody molecules. Monomeric proteins are shown in *blue*, dimers in *red*, and a trimer in *green*. (A, from VK Sharma, TW Patapoff, B Kabakoff, et al. *Proc Natl Acad Sci USA*, 111:18601–18606, 2014; B, adapted from JD Schmit, F He, S Mishra, et al. *J Phys Chem B*, 118:5044–5049, 2014.)

happens because organisms (or tumors) can become *drug-resistant*. At first, the drug acts effectively on a target protein in a targeted cell. But, because the cell grows and duplicates, the cell and its protein can evolve. The evolved protein has a modified structure that is no longer affected by the original drug. A cartoon of a mechanism is shown in Figure 7.15 in Chapter 7.

For example, cancer cells became resistant to imatinib, which is a small-molecule drug that binds to the ATP-binding site of Bcr–Abl tyrosine kinase. Imatinib loses its potency when threonine 315 near the binding site of the cancer-cell kinase mutates to isoleucine (**Figure 13.17**). The isoleucine clashes sterically, preventing imatinib from inhibiting the protein. Adding a second drug along with imatinib can overcome that drug resistance. An inhibitor called GNF-2 overcomes the drug resistance by binding to a distant allosteric site [11]. The term *polypharmacology* refers to situations where multiple drugs are applied in concert. Polypharmacology can target multiple sites on a single protein, or multiple protein targets in a biochemical pathway.

Quantitative Systems Pharmacology Goes beyond "One Target–One Mechanism"

Drug discovery of the future is likely to go beyond the "one-target/one-drug/one-mechanism" paradigm. It will tackle whole biochemical pathways at a time. After all, it is pathways—not individual proteins—that are the elemental units of functionality in a cell. *Quantitative systems pharmacology* refers to the use of modeling and computation to guide

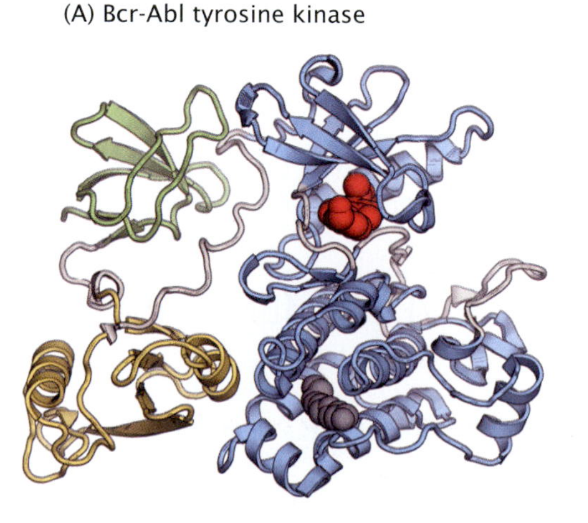

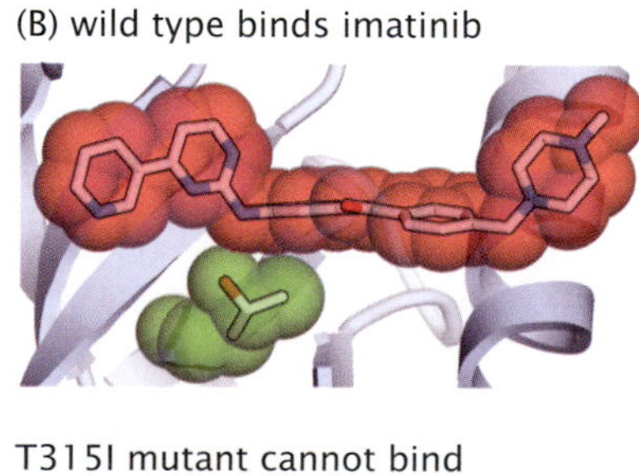

Figure 13.17 Polypharmacological targeting of Bcr–Abl tyrosine kinase. (A) Bcr–Abl tyrosine kinase. Imatinib (*red*) targets the ATP-binding site. A second drug, GNF-2 (*gray*), binds to the myristoylation site of the Bcr–Abl tyrosine kinase. Combined, imatinib plus GNF-2 lock the kinase into an inactive conformation. (B) When the threonine is mutated to isoleucine (*yellow*), it interferes with binding to imatinib, reducing the latter's inhibitory action. (A, adapted from L Skora, J Mestan, D Fabbro, et al. *Proc Natl Acad Sci USA* 110:E4437–4445, 2013; B, adapted from ME Gorre, M Mohammed, K Ellwood, et al. *Science* 293:876–880, 2001. With permission from AAAS.)

an understanding of how drugs affect networks at the cell level as well as at the tissue and organism levels. Future drug discovery could entail different sites on different proteins in different pathways in different tissues, and possibly delivered at different times.

Peptides and Macrocycles Can Interfere with Protein–Protein Interactions

Today's drugs are mainly pharmaceuticals (small, <500 Da) or biologicals (large, 150,000 Da). Pharmaceuticals have the advantage that they are cheap and taken orally, but the disadvantages that many targets are undruggable, and that they may have unwanted side effects. Biologics have the advantage of high specificity, effectiveness, and minimal off-target side effects, but their disadvantages are their high cost, the fact that they need to be injected, and the challenges in formulations, stability, and membrane permeation. A growing third class of drug comprises peptides and macrocycles (mid-sized, 500–5000 Da), which can often be cheap *and* taken orally. The power of macrocycles is that they can sandwich between flat protein–protein interfaces, due to their large size and flexibility, circumventing the druggability problem, while also providing high specificity.

Protein–protein interfaces are important because the basic functional units of cells are pathways, and pathways are made up of protein–protein interactions. Traditionally, targetting protein–protein interfaces has been challenging because those interfaces tend to be large and flat and flexible, not small, rigid, and tightly concave. Protein–protein contacts tend to bury 1000–3000 Å, whereas druggable sites are only around 300–1000 Å. For example, while kinases have been broadly used as targets, the flat surfaces near the active sites of phosphatases have been a challenge for designing high-affinity ligands. Figure 13.18 shows some contact regions in protein–protein pairs and their estimated contributions to the binding stability. The few critical residues responsible for most of the binding affinity are called *hotspots* (shown in red in Figure 13.18).

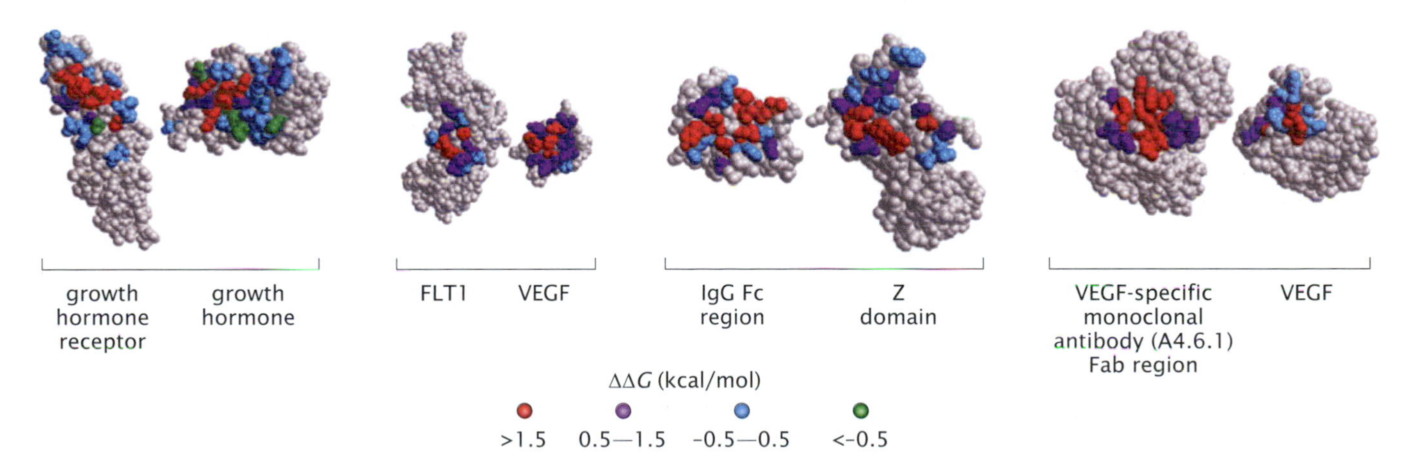

Figure 13.18 Four examples of protein–protein-interface hotspots. The coloring indicates how much each residue contributes to the binding free energy. (*Red*) Amino acid sites at which mutations have a large effect on the protein–protein binding affinities. The *purple* sites have less effect and the *blue* sites even less effect. (*Green*) Sites where mutations have little effect on the protein-binding affinities. VEGF is vascular endothelial growth factor. FLT1 is VEGF receptor 1. IgG is immunoglobulin G. The Z domain is the IgG-binding domain of protein A. (From JA Wells and CL McClendon. *Nature*, 450:1001–1009, 2007. Reprinted by permission from Macmillan Publishers Ltd.)

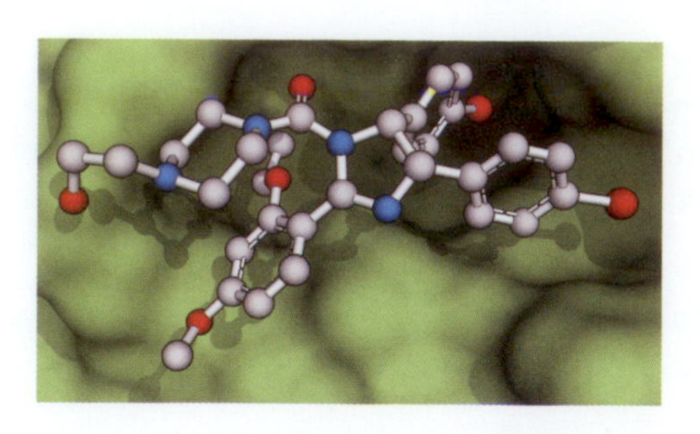

Figure 13.19 Binding of a protein–protein-interaction inhibitor at the p53–Mdm2 interface. Mdm2 (*green*) inhibits the tumor suppressor protein p53 (not shown), stimulates its degradation, blocks its transcriptional activity, and promotes its nuclear export. Nutlin (the stick structure) is a drug that binds to the interface, preventing the inhibition of p53 by Mdm2. (Adapted from LT Vassilev, BT Vu, B Graves, et al. *Science*, 303:844–848, 2004. With permission from AAAS.)

Recently, protein–protein interfaces have begun to yield to drug development efforts [12]. Some successes result from finding a hydrophobic cavity, such as at the interface between the tumor suppressor protein p53 and its natural inhibitory substrate Mdm2 (Figure 13.19). In other cases, protein–protein interfaces can be disrupted by peptides or macrocyclic molecules. Another useful way to disrupt protein–substrate (or domain–domain) interactions is through allosteric inhibition of target proteins. An allosteric inhibitor alters the structural dynamics of the target protein, blocking the protein's ability to undergo its natural conformational switching, or altering the protein's binding elsewhere in the protein.

Sometimes Biological Activity Correlates Better with Ligand–Protein Off-Rates than with Binding Affinities

Much of drug discovery is based on correlations between a protein's biological actions and its equilibrium binding affinities to ligands. However, in a few cases, biological actions correlate better with ligand *off-rates* than with affinities. The binding affinity $K = k_{\text{on}}/k_{\text{off}}$ is the ratio of the *on-rate* constant to the *off-rate* constant. A given binding affinity could be achieved in different ways: by a fast on-rate and a fast off-rate, or by a slow on-rate and slow off-rate, for example. Why should the off-rate matter? In essence, the longer the ligand sticks to the protein, the more biological action that results from it. Examples include long-acting β_2-adrenergic receptor agonists used in asthma therapy (for example, salmeterol and indacaterol). Metaphorically, you pay for a hotel room based on how long you stay in it (that is, your "off-rate"), not based on a ratio of rates. Even so, biological activities do appear to correlate quite generally with binding affinities.

You Can Sometimes *Repurpose* an Old Drug for a New Medical Indication

Some drugs hit more than one target protein. A drug that hits off-target proteins can cause adverse side effects and problems with drug safety. But, in other cases, *polypharmacology* (a drug hitting multiple proteins or multiple sites on the same protein) is advantageous. For example, some psychiatric medications have increased efficacy because they hit multiple target proteins. A powerful computational strategy [13] leverages predictive polypharmacology to find ways for *repurposing* old drugs. "Repurposing" refers to finding new targets for known drugs. One example is imatinib, which was originally developed as a kinase inhibitor (see previous discussion), and was later found to be an effective leukemia drug. The idea is to search for similarities between

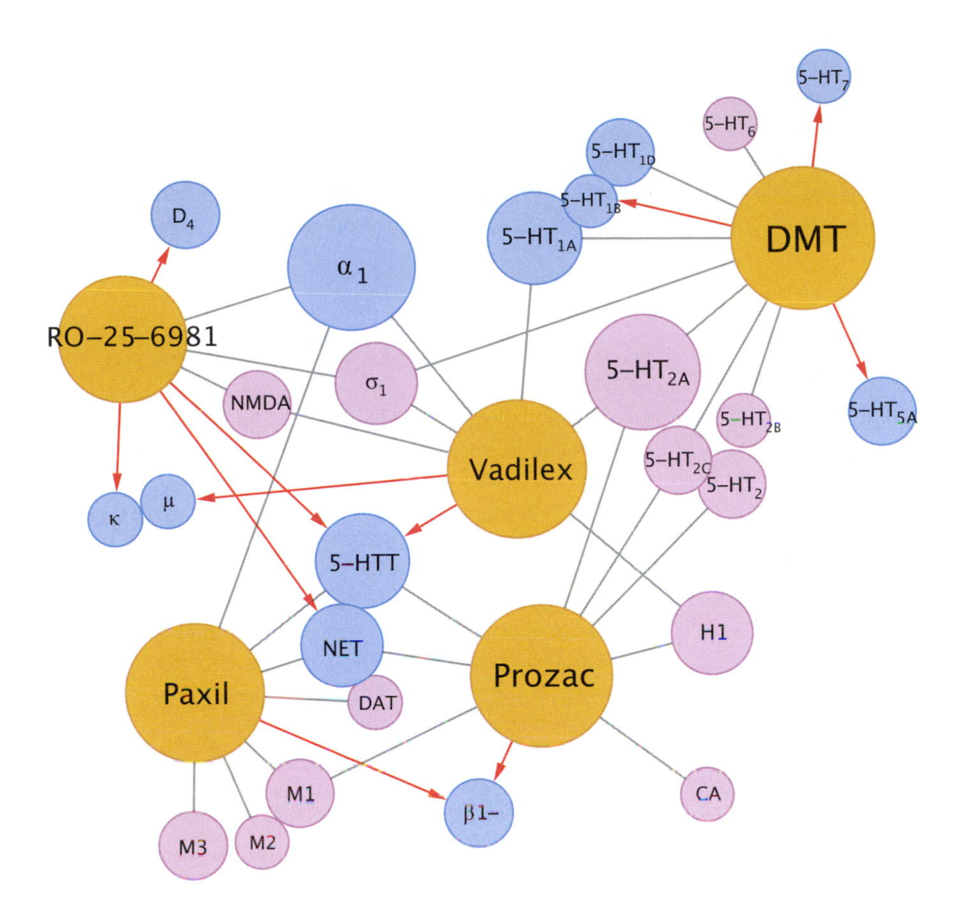

Figure 13.20 Finding ways to repurpose old drugs by their chemical resemblance to the known ligands of proteins. This graph shows some of the 3665 known drugs (*yellow*) and hundreds of target proteins (*purple*). *Gray lines* indicate which drugs act on which target proteins. The *blue circles* and *red arrows* indicate new off-target proteins to which old drugs might bind, based on similarities among ligand structures. (Adapted from MJ Keiser, V Setola, JJ Irwin, et al. *Nature*, 462:175–181, 2009. With permission from Macmillan Publishers Ltd.)

drugs, or between targets, or between pharmacological actions, to identify new links between drugs and targets. For example, suppose two drugs d_1 and d_2 interact with the same set of target proteins T_1 and T_2. Suppose you now learn that target T_3 interacts with one of the drugs, say d_1. Then, it is likely that protein target T_3 will also interact with drug d_2. You can draw these inferences [14] from large databases of known drug–target interactions, such as DrugBank [15] and STITCH [16]). A method called the similarity ensemble approach (SEA) [17] looks at ligand similarities to identify repurposable drugs (Figure 13.20).

SUMMARY

Computational modeling of ligands and proteins is increasingly important in drug discovery. In early-stage discovery, you need fast ways to search large databases of ligands to seek chemicals with drug-like properties that hit your target protein. You can use ligand-based methods to seek new candidate drugs based on ligands you already know. Or you can use target-based discovery if you know the structure of the protein you want the drug to inhibit or activate. In target-based discovery, docking methods can help you find binding sites and poses quickly, while atomistic simulations can help in refining structures of computing affinities. An important class of drugs comprises biologics, which are often monoclonal antibodies. Another growing class of drugs comprises peptides and macrocycles, which are larger than typical small-molecule compounds, and which can target protein–protein interfaces. Key developments are happening in polypharmacology, in leveraging allosteric interactions, in repurposing of drugs, and in quantitative systems pharmacology. There is an increasing reliance upon computational tools throughout drug discovery and development.

REFERENCES

[1] CA Lipinski, BW Dominy, and PJ Feeney. Experimental and computational approaches to estimate solubility and permeability in drug discovery and development settings. *Adv Drug Deliv Rev*, 46:3–26, 2001.

[2] WL Jorgensen. The many roles of computation in drug discovery. *Science*, 303:1813–1818, 2004.

[3] N Huang and MP Jacobson. Physics-based methods for studying protein–ligand interactions. *Curr Opin Drug Discov Devel*, 10:325–335, 2007.

[4] AL Hopkins and CR Groom. The druggable genome. *Nat Rev Drug Discov*, 1:727–730, 2002.

[5] A Bakan, N Nevins, AS Lakdawala, and I Bahar. Druggability assessment of allosteric proteins by dynamics simulations in the presence of probe molecules. *J Chem Theory Comput*, 8:2435–2447, 2012.

[6] ID Kuntz, JM Blaney, SJ Oatley, et al. A geometric approach to macromolecule–ligand interactions. *J Mol Biol*, 161:269–288, 1982.

[7] RA Friesner, RB Murphy, MP Repasky, et al. Extra precision Glide: Docking and scoring incorporating a model of hydrophobic enclosure for protein–ligand complexes. *J Med Chem*, 49:6177– 6196, 2006.

[8] GM Morris, R Huey, W Lindstrom, et al. Autodock4 and AutoDockTools4: Automated docking with selective receptor flexiblity. *J Comput Chem*, 16:2785–2791, 2009.

[9] MAC Neves, M Totrov, and R Abagyan. Docking and scoring with ICM: The benchmarking results and strategies for improvement. *J Comput Aided Mol Des*, 26:675–686, 2012.

[10] ML Verdonk, JC Cole, MJ Hartshorn, et al. Improved protein–ligand docking using GOLD. *Proteins*, 52:609–623, 2003.

[11] J Zhang, FJ Adrián, W Jahnke, et al. Targeting Bcr–Abl by combining allosteric with ATP-binding-site inhibitors. *Nature*, 463:501–506, 2010.

[12] JA Wells and CL McClendon. Reaching for high-hanging fruit in drug discovery at protein–protein interfaces. *Nature*, 450:1001–1009, 2007.

[13] MJ Keiser, V Setola, JJ Irwin, et al. Predicting new molecular targets for known drugs. *Nature*, 462:175–181, 2009.

[14] MC Cobanoglu, C Liu, F Hu, et al. Predicting drug–target interactions using probabilistic matrix factorization. *J Chem Inf Model*, 53:3399–3409, 2013.

[15] C Knox, V Law, T Jewison, et al. DrugBank 3.0: A comprehensive resource for “omics” research on drugs. *Nucleic Acids Res*, 39:D1035–D1041, 2011.

[16] M Kuhn, D Szklarczyk, S Pletscher-Frankild, et al. STITCH 4: Integration of protein–chemical interactions with user data. *Nucleic Acids Res*, 42:D401–D407, 2014.

[17] MJ Keiser, BL Roth, BN Armbruster, et al. Relating protein pharmacology by ligand chemistry. *Nat Biotechnol*, 2:197–206, 2007.

SUGGESTED READING

Arkin MR, Tang Y, and Wells JA, Small-molecule inhibitors of protein–protein interactions: Progressing toward the reality. *Chem Biol*, 21:1102–1014, 2014.

Jorgensen WL, The many roles of computation in drug discovery. *Science*, 303:1813–1818, 2004.

Sousa SF, Fernandes PA, and Ramos MJ, Protein–ligand docking: Current status and future challenges. *Proteins*, 65:15–26, 2006.

INDEX

D

E

I

J

K

L

M

N

O

P

Q

R

S

T

U

V

W

X

Z